Standards and Applications of Mobile Learning

Standards and Applications of Mobile Learning

Edited by **Henrik Olsen**

Willford Press

New York

Published by Willford Press,
118-35 Queens Blvd., Suite 400,
Forest Hills, NY 11375, USA
www.willfordpress.com

Standards and Applications of Mobile Learning
Edited by Henrik Olsen

International Standard Book Number: 978-1-68285-305-4 (Hardback)

Printed in China.

Contents

Preface

Mobile learning is an innovative mode of learning and education that uses mobile devices and mobile technology for delivering and sharing information and databases with the purpose of imparting education and training. Some of the chapters included in this book talk about the significant concepts related to mobile learning such as future trends in m-technologies, m-learning applications, social impact of current and next-generation mobile technologies, mobile assisted language learning, etc. The extensive use of examples and case studies make this book a great source of reference for students and professionals alike.

This book unites the global concepts and researches in an organized manner for a comprehensive understanding of the subject. It is a ripe text for all researchers, students, scientists or anyone else who is interested in acquiring a better knowledge of this dynamic field.

I extend my sincere thanks to the contributors for such eloquent research chapters. Finally, I thank my family for being a source of support and help.

Editor

1

Towards Development of Web-Based Assessment System Based on Semantic Web Technology

Hosam F. El-Sofany[1], and Samir A. El-Seoud[2]
[1]Cairo Higher Institute for Engineering, Computer Science, and Management, Cairo, Egypt
[2]Technische Universität Darmstadt, Darmstadt, Germany

Abstract—**The assessment process in an educational system is an important and primordial part of its success to assure the correct way of knowledge transmission and to ensure that students are working correctly and succeed to acquire the needed knowledge. In this study, we aim to include Semantic Web technologies in the E-learning process, as new components. We use Semantic Web (SW) to: 1) support the evaluation of open questions in e-learning courses, 2) support the creation of questions and exams automatically, 3) support the evaluation of exams created by the system. These components should allow for measuring academic performance, providing feedback mechanisms, and improving participative and collaborative ideas. Our goal is to use Semantic Web and Wireless technologies to design and implement the assessment system that allows the students, to take: web-based tutorials, quizzes, free exercises, and exams, to download: course reviews, previous exams and their model answers, to access the system through the Mobile and take quick quizzes and exercises. The system facilitates generation of automatic, balanced, and different exam sheets that contain different types of questions covering the entire curriculum, and display gradually from easiness to difficulty. The system provides the teachers and administrators with several services such as: store different types of questions, generate exams with specific criteria, and upload course assignments, exams, and reviews.**

Index Terms—**E-learning, M-learning, Semantic Web, Educational Systems, Evaluation Systems, Interactive Applications.**

I. Introduction

Traditional assessment modes in crowded classrooms can be a serious load on the instructors [13, 14]. Considering roles of the instructors in digital age, use of internet and even wireless technologies in the assessment can be very useful starting point for the instructors in order to be successful in integrating new technologies to the courses. Moreover, this integration process will support their professional career development.

Rouet et al. in [15] conducted a study to investigate the impact of web-based and paper based delivery of quizzes on students' performance and perceived satisfaction. They found that students who studied with printed documents and undertook paper based quiz were superior to the students who studied with web documents and took web-based quiz. Although the students expressed preference on the paper materials and paper based quiz they appreciated the availability of the course web site. As authors suggested while integrating web technologies into assessment phase of the instruction, ease of use and legibility of the tests should be taken into considerations. Besides accessibility of the online content and test of the courses are important factor to support students anytime, anywhere access to the instruction.

In recent years, most universities recognized the educational excellence of the wireless campus. This is a great opportunity for the students to access information more easily at any area of the campus using their laptops. In addition, with the promotion of Palm, Pocket PC and Mobile Phones as learning media, the number of practices which deals with integrating such devices into instruction has increased in the universities. Thus, the students can reach the course scores, discussion forums, information systems and tests by using mobile phones and PDA devices whenever and wherever they want.

Seppälä and Alamäki in [16] conducted a study on comparing the effectiveness of face-to-face, internet and mobile based instruction. As a result of their study, they suggested that innovative internet and mobile solutions can be useful for academic teaching because of providing possibilities for open teaching.

Use of WAP or SMS based tests through PDA, PALM, mobile phone or computer in higher education has been promoted due to the fact that they support the learning process of the students, offer exercise media and provide the opportunity to test the expected learning level achieved [17-19]. However, as stated by Homan and Wood [20], the cost of such services for the students should be decreased so that these new learning media can be widely used. Homan and Wood analyzed the achievement level and views of the students on the paper based and mobile based tests in internet based wireless and traditional classrooms. They found that there was no significant difference between students' achievement level and the students had positive views about the wireless test conducted on PDA, however the students stated that they wanted to use the infrastructural opportunities of the university, instead of their own devices.

The results of the most researches in the literature have focused on the students' achievement level in web or paper based tests in the university level education. This study aims to determine and measure the effects of Web-based and Mobile-based assessments on the achievement levels, performance and perceptions of the students in internet-assisted courses in the Qatari schools.

The aim of this study is not to provide all the features contained in Course Management Systems such as Moodle or Blackboard, but to providing the basic support for carrying out the assessment processes.

The main objectives of this research project are:

1) Design and implement the web-based assessment system, using Semantic Web (SW), and Mobile technologies. The project allows the students:

1-1 To take: web-based tutorials, quizzes, free exercises, and exams (monthly, midterms and finals).

1-2 To download: course reviews, previous exams and their model answers.

1-3 To access the system through wireless devices (such as, Mobile phone, PDA- Personal Digital Assistants, and Tablet PC) and take quick quizzes and exercises.

2) Develop new algorithms to improve the assessment process in the schools:

To improve the assessment process in the schools, by developing new algorithms and techniques to generate balanced; exam sheets (monthly, midterms and finals), quizzes, and exercises, that containing different types of questions, covering the entire curriculum, and displaying gradually from easiness to difficulty. The exam sheets produced by the system take into account the different levels of the students from excellent, good, to fair, and avoids any mistakes of language and non-clear terminologies.

3) Provide the teachers and the administrators with several management services. In addition of the traditional admin services such as: login, change password, add, update, delete the basic data for the system (levels, subjects in each level, school, classes, teachers and students data), the project provided the following services to the teachers and administrators:

3-1 Store different types of questions: *multiple choices, matching, filling in the blanks, true or false, parallel lists, etc.*

3-2 Generate exam sheets (monthly, midterms and finals) with specific criteria: *the number of questions, total score, chapters limit, lessons limit, and time limit.*

3-3 Upload *course assignments, previous exams, and their model answers.*

3-4 Make sure that students have submitted their *assignments, quizzes, and homework.*

4) Determine and measure the effects of Web-based and Mobile-based assessments on the achievement levels, performance and perceptions of the students in internet-assisted courses in the schools:

To assist the educational system for transferring from traditional learning to computer-based and mobile-based learning, by allowing students and instructors to participate in remote learning communities using personal computers and mobiles at home or at school.

II. Analysis and Design Methods

We build first the general architecture of the system, set the graphical user interface and design the structure of the database that stores the required data and information. In order to achieve our goals we defined the following tasks:

A. System Analysis

The analysis phase is the main phase in which the assessment system requirements are identified in more details. The goal of this phase in the system development is to represent the system goals into defined functions and operations of the intended applications. The assessment system requirements are documented in a complementary set of artifacts (flowcharts, use case diagrams, system sequence diagrams, ...). Each artifact provides a different perspective of the system under design and contains distinct requirements. The combination of these perspectives establishes the tasks that are to be accomplished by the system.

- **Information Gathering Techniques:** By assisting of some specialist employees from the SEC (Supreme Education Council), all the system materials are checked to match the SEC curriculum. The following are some information gathering techniques that used in the assessment system analysis.
  - *Interviews*: The most common technique for gathering requirements is to sit down with the clients (teachers, students, parents, and school administration) and ask them what they need. Our discussion would help us plan out ahead of time, based on the type of requirements that we are looking for.
  - *Prototyping*: We use this technique for gathering requirements. In this approach, we gather preliminary data and requirements that we use to build the initial version of the system.
  - *UML* (Universal Modeling Language) use-cases is used also in the analysis of the assessment system as a model tool, in which the functional requirements are extracted and described along with a scenario of the flow of events.

The system requirements are classified in two types: *functional* and *non-functional* requirements.

- **Functional requirements:** The assessment system allows different types of users to interact with the system according to given privileges. There are three main users of the system: *administrator*, *teacher*, and *student*, as shown in Figure 1. The system provides the following applications (services) to each user:

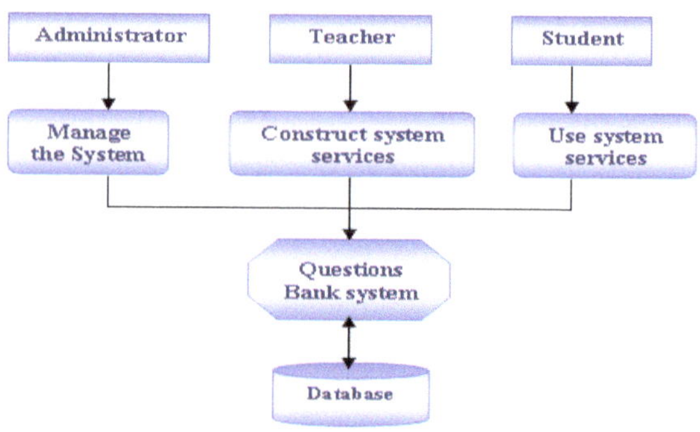

Figure 1. The three main users of the system

1. *The administrator application*: This application provides the school administrator with simple tools to manage the information of the assessment system. The following are some features provided by the administrator application:
   - Activate/inactivate the system.
   - Import and export the system' database
   - Send automatic email immediately containing the new password whenever the user changes his/her password.
   - Administer the basic data and information of the system: levels, subjects (or courses) teachers, and students.
2. *The teacher application*: This application provides the teachers with various tools to construct the system services. The following are some features provided by the teacher application:
   - Change account password
   - Manage (add, update, delete and display) all the system services (tutorials, reviews, exam questions, quizzes, previous exams, and model answers).
   - Print his/her exam sheets (monthly, midterms and finals) with specific criteria: the number of questions, total score, chapters limit, lessons limit, and time limit.
   - As mentioned before, the system helps the teachers to print automatic, balanced, and different exam sheets that contain different types of questions covering the entire curriculum, and display gradually from easiness to difficulty.
3. *The student application*: This application provides the students with interactive tools to use the services of the system. The following are some features provided by the student application:
   - Change account password
   - Download assignments, revision documents, previous exams and their model answers, and other files that are uploaded by the teachers.
   - Take web-based tutorials, quizzes, free exercises, exams (monthly, midterms and finals).
   - Access the assessment system through wireless devices (such as, Mobile phone, PDA and Tablet PC) and take quick quizzes and exercises.

- **Non-functional requirements:** In this part, we have verified and proved the system' aspects and interfaces that are not directly related to the functional behavior, such as: usability, reliability, supportability, performance, and security.
- **Hardware and software resource:** We have used some software during our first prototype of the system development such as: Windows XP professional, Windows Server O/S, Microsoft office Excel, Microsoft Word, SQL Server database system, MS Visual Studio.Net, and Photo Shop.

  To extend our research work, by using Semantic Web and Mobile technologies, we used other new languages and tools including: XML- eXtensible Markup Language, XML Schemas, RDF- Resource Definition Framework, RDF Schemas, OWL- Web Ontology Language, and Java 2ME.

B. System Design

Any successful web-based system should have an easy to use graphical user interface (GUI) that facilitates its use and flexibility. Thus the proposed system should possess an attractive GUI that is suitable to the potential users. The interface should be tested by its users and updated whenever necessary.

- **System architecture:** The assessment system interface is entirely web-based, and doesn't necessitate any technical skills from the potential users. The system is divided into three applications, called *administrator*, *teacher*, and *student* application. As shown in Figure 2, the system is appropriate for on-campus teaching, with access to the Internet, or off-campus teaching using the local network.
- **Database design:** Database is an important component in our system. We have used Microsoft SQL Server database as a database management system (DBMS). The DBMS stores the subjects in each school level, the tutorials, the reviews, the previous examinations, the exams questions, the quizzes, the student grades, and the basic information of the administrators, teachers, and students. We have chosen the entity-relationship model to represent the database design. In [6] we introduced the basic entity-relationship diagram, that contains the main entities (tables) used for creating the system database.

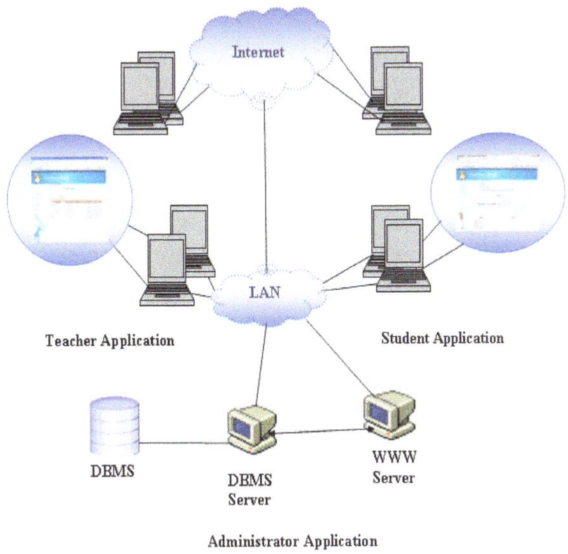

Figure 2. System architecture

- **Using Semantic Web as a new methodology for E-learning research :**

  The Semantic Web appears as a promising technology for implementing E-learning. The Semantic Web constitutes an environment in which human and machine agents will communicate on a semantic basis [8]. One of its primary characteristics, viz. shared understanding, is based on ontologies as its key backbone. Ontologies enable the organization of learning materials around small pieces of semantically annotated learning objects [9]. Items can be easily organized into customized learning courses and delivered on demand to the user, according to her/his profile and business needs.

Semantic Web architecture: The term "Semantic Web" encompasses efforts to build a new WWW architecture that enhances content with formal semantics. That means, content is made suitable for machine consumption, as opposed to content that is only intended for human consumption. This will enable automated agents to reason about Web content, and produce an intelligent response to unforeseen situations.

"*Expressing meaning*" is the main task of the Semantic Web. In order to achieve that objective several layers of representational structures are needed. They are presented in the Figure 3, among which the following layers are the basic ones [8]:

- The XML layer, which represents the structure of data,
- The RDF layer, which represents the meaning of data,
- The Ontology layer, which represents the formal common agreement about meaning of data,
- The Logic layer, which enables intelligent reasoning with meaningful data.

The important property of the Semantic Web architecture (i.e., common-shared-meaning and machine-processable metadata), enabled by a set of suitable agents, establishes a powerful approach to satisfy the E-Learning requirements. The process is based on semantic querying and navigation through learning materials, enabled by the ontological background. The Semantic Web can be exploited as a very suitable platform for implementing an E-learning system, because it provides all means for E-learning: ontology development, ontology-based annotation of learning materials, their composition in learning courses and active delivery of the learning materials through E-learning portals. Table 1, shows the suggested advantages to the possibility of using the Semantic Web for realizing the proposed E-learning requirements.

- **The proposed Semantic Web framework for assessment system:**

Our basic Semantic Web framework illustrated in Figure 4, provides the student with different types of assessment services. In this research study we focus on the implementation of these services that allows the students to take: web-based tutorials, quizzes, free exercises, exams (monthly, midterm, and final), to download: course reviews, previous exams and their model answers, to access the system through the Mobile and take quick quizzes and exercises. On the other hand, the system allows the teacher to store different types of questions, to check the contents of the site and update it regularly, to generate exams with specific criteria, to make sure that students have submitted their assignments, and to make students up to date by posting different announcements.

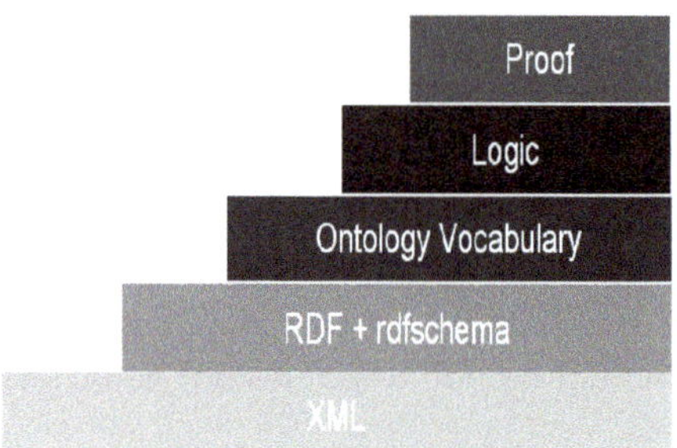

Figure 3. Layers of the Semantic Web

TABLE I.
ADVANTAGES OF USING SEMANTIC WEB AS A TECHNOLOGY FOR E-LEARNING

E-learning Semantic Web		Requirements
Knowledge items (learning materials) are distributed on the web, but they are linked to commonly agreed ontologies. This enables construction of a user-specific course, by semantic querying for topics of interest.	Pull – student determines agenda	Delivery
Software agents on the Semantic Web may use a commonly agreed service language, which enables co-ordination between agents and proactive delivery of learning materials in the context of actual problems. The vision is that each user has his own personalized agent that communicates with other agents.	Reactionary – responds to problem at hand	Responsiveness
User can describe the situation at hand (goal of learning, previous knowledge,...) and perform semantic querying for the suitable learning material. The user profile is also accounted for. Access to knowledge can be expanded by semantically defined navigation.	Non-linear – allows direct access to knowledge in whatever sequence makes sense to the situation at hand	Access
The Semantic Web offers the potential to become an integration platform for all business processes in an organization, including learning activities.	Symmetric – learning occurs as an integrated activity	Symmetry
Active delivery of information (based on personalized agents) creates a dynamic learning environment that is integrated in the business processes.	Continuous – learning runs in parallel to business tasks and never stops	Modality
The Semantic Web will be as decentralized as possible. This enables an effective co-operative content management.	Distributed – Content comes from the interaction of the participants and the educators	Authority
A user (using its personalized agent) searches for learning material customized for her/his needs. The ontology is the link between user needs and characteristics of the learning material.	Personalized – content is determined by the individual user's needs and aims to satisfy the needs of every user	Personalization
The Semantic Web enables the use of distributed knowledge provided in various forms, enabled by semantically annotation of content. Distributed nature of the Semantic Web enables continuous improvement of learning materials.	Dynamic – content changes constantly through user input, experiences, new practices, business rules and heuristics	Adaptively

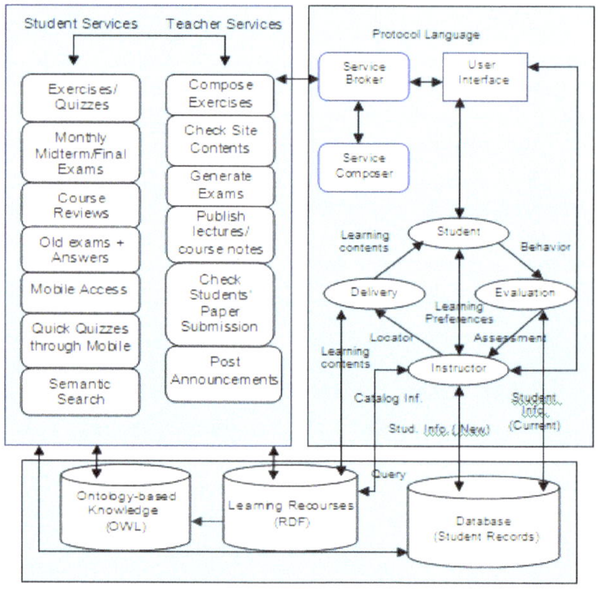

Figure 4. Basic Semantic Web framework for assessment system

- **Using Mobile technology as a new stage for E-learning research:**

M-learning means the provision of education and training materials and courses on wireless devices: portable computers, PDA's, mobile telephones and Tablet PC. Our learning system has functions to enter and to organize the learning objects in the database and to make them available for students and teachers using the learning management systems (LMS) portals in which the learning process and its activities have been structured [4,5]. Most of the e-learning student applications have been made available for users of mobile computers, the laptop and the PDA. The learning object management systems are now compatible for all kind of user devices, the conventional desktop PC as well as the mobile devices laptop, PDA and mobile telephone. *As mentioned before, one of the main objectives of this study is to use mobile technology to provide schools with M-learning system that help in student assessment process.*

- **The advantages:** The advantages of mobile learning refer mostly to the mobile user: a great flexibility, an improved learning schedule is possible, increased productivity during dead moments and just in time learning.
- **Difficulties and limitations:** Nevertheless they have very small screens, limited memory capacity and the large diversity of mobile devices obstruct a good learning experience. In the educational field we notice that the learning materials must answer to specific conditions. We will have to get around the technical restrictions so we can create a good learning experience. The usage of video, audio, clear interfaces and divided courses must contribute to this. Furthermore we will have to adapt the content to the needs of the mobile user. Because the user has a fragmentized time schedule, we will have to be sure that the learning object (quiz, exercise, ...) are not to long. Dividing the knowledge in smaller modules offers a solution.

C. *Exam Generation Algorithms*

As mentioned before, one of the main objectives of the proposed study is to improve the assessment process in the schools, by developing new algorithms and techniques to generate balanced; exam sheets, quizzes, and exercises, that containing different types of questions, covering the entire curriculum, and displaying gradually from easiness to difficulty. One of these innovative and intelligent algorithms is the "ManageQuestion" algorithm that used for managing the *difficulty* of the exam (quiz and exercise) questions. The algorithm uses the *difficulty attribute* set by the user as input parameter. The teachers set the *difficulty levels* of their questions at creation phase using a numbering system. The value assigned to a specific question identifies how difficult (or easy) it would be for a student to reply correctly to the question. Our algorithm can generate automatically different exams with different levels of difficulties. Obviously, teachers are not always as accurate as would be desirable in estimating the difficulty of a question (as they don't know the exact level of the students using the system) our algorithm can update the difficulty and the easiness of the questions based on the responses given by the students. The algorithm has the ability to detect the difficulty/easiness of a question based on students reply and adjust the difficulty value accordingly. As a result, some of the feedbacks of the students in a given course are used to adjust the level of the questions. The following are the some functions of our algorithm:

- Get_Difficulty(): This method determines the difficulty of a specific question.
- Calc_Margn(): This method calculates the lower and the upper bounds of error percentage in the students' responses.
- Calc_Error(): This method calculates the percentage of wrong responses to a question, as a ratio between errors and a constant value named Max_Num_Access. The involved variables are reset after each difficulty evaluation.
- Mark_Ques(): This method activates a Boolean field in the register of a question if the difficulty level reaches the thresholds.

Algorithm: ManageQuestion(difficulty):

Access random question from the "questions" database table:

```
Check the Max_Num_Access of the question
  Get_Difficulty();
  Calc_Margn();
  Calc_Error();
  Recalculate_Difficulty();
  if (Error < LowerBound) then
       Difficulty--;
  else if (Error > UpperBound) then
  Difficulty++;
  if Difficulty is out-of-bounds then
  Mark_Ques();
```

Four difficulty degrees have been considered rangy from level one to level four. If a question has a value higher than four, it could be too difficult or it could be ambiguous. Similarly, if a question has a value of 0, the question could be too easy, or too obvious, and nearly all the students can answer it correctly. The questions that are out of bounds are marked, so they cannot be used for any exam (or quiz), and the teachers must update them. Table 2 shows the upper and lower bounds for determining the difficulty of the questions [1, 2].

TABLE II.
QUESTIONS LEVELS

Difficulty	Lower bound (%)	Upper Bound (%)
1	10	30
2	30	50
3	50	70
4	70	90

III. SYSTEM IMPLEMENTATION

Database implementation: The current database is designed to have the following main tables: Level, Subject, ReviewFile, Quiz, ExamFiles, Questions, StudentScore, Student, Teacher, and Administrator. During the development of the system, data for the testing of different functionalities was needed to be imported into the database tables. The tables are filled by importing the data from an MS Excel sheet that contains the basic data from the school's database or during the run time of the system.

Ontology implementation: As mentioned in our proposed Semantic Web model, Figure 4, the ontological knowledge is added to the learning resources as a resource for contextual learning, and it searched by means of queries. The student's performance is measured by the evaluation component, and the result is stored in the student records database. The data in the database can be used by the instructor component to locate a new content. Searching learning resources and sequencing a course can be done using a knowledge base of learning resources and a delivery component. To implement the knowledge base, first of all, the leaning resources have to be described by means of *metadata*. The metadata consists of the contextual knowledge of the learning resources, i.e., ontology in our model. It contains the general representation of the structural knowledge on specific domains, such as Science, Mathematics, English, and so on. The ontology can be used for adaptive learning to retrieve the context of a course and to structure the contents. Also the metadata actually consists of the framing description of each learning object of a subject, i.e., the modularized content, which is linked to the concept of the ontology. For instructors to be able to sequence courses and create exercises adaptively, the suitability of different approaches has to be analyzed based on the relationships between the resources and their descriptions.

System requirements implementation: To prove the functionality of the assessment system, we have implemented several web-based applications as components of the system. The assessment system is divided into three sub-systems: *the administrator*, *the teacher*, and *the student* sub-system. Figure 5, shows a snapshot of our system interface that developed as a grant project [6].

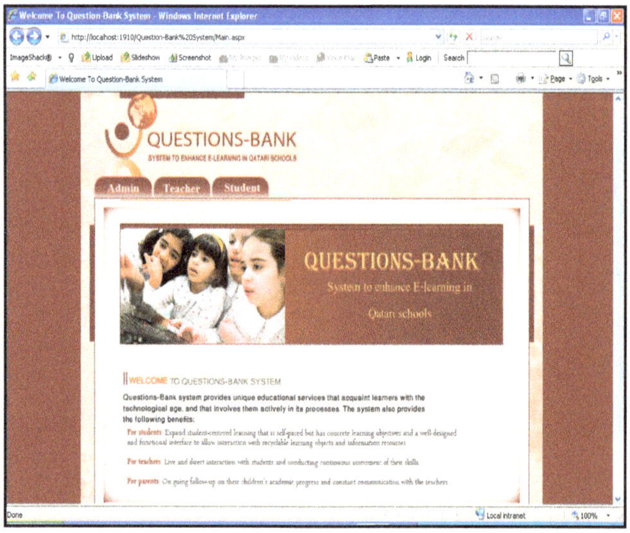

Figure 5. The assessment sub-systems [6]

i. **The administrator sub-system:** This sub-system provides the school administrator with web-based services to manage the information of the assessment system. the following services are available in the administrator sub-system:
   - **Login:** username and password is the most common authentication method.
   - **System management:** provides the school administrator with web-based services to add (update, or delete) the basic data needed for the other two sub-systems, such as: levels, subjects in each level, school, classes, teachers and students data
   - **Change password:** The system provides the user to change his/her account password.

ii. **The teacher sub-system:** This sub-system provides the teacher with web-based tools to manage the information of the assessment system. In addition to the login, and change password, the following services are available in the teacher sub-system:
   - **Tutorials administration:** Teacher uses this service to upload the MS Word, PP presentation, video, etc. files that contain the tutorials for each subject. Also teacher can delete specific tutorial (s) from the database and can display the stored tutorials.
   - **Reviews administration:** Teacher uses this service to upload (add) the MS Word, PP presentation, video, etc. files that contain the reviews for each subject. Also teacher can delete specific review(s) from the database and can display the stored reviews. The teacher can store the following data for each review: level name, subject name, review title, review date, review file, and the review answer file (if exist).
   - **Previous exams administration:** Teacher uses this services to upload (add) the previous exams and their model answers for each subject. Also teacher can delete specific exam(s) from the database and can display the stored exams. The teacher can store the following data for each previous exam: level name, subject name, exam type (midterm, final), exam title, exam date, exam file, and the exam answer file (if exist).
   - **Exam administration:** Teachers can use this on-line tool to add, update, and delete questions and answers for their exams (quizzes, and exercises). The questions are inserted into the database dynamically, so new questions can be added during the configuration of exam or quiz. The teacher can store the following data for each exam/quiz: level name, subject name, chapter number, lesson number, quiz title, quiz type (multiple choices, true or false, fill in the blank, …), question text, question image file, choice1, choice2, choice3, choice4, correct answer, question weight (difficulty), question score, and question explanation.
   - **Print exam sheets:** Teachers can use this on-line tool to print balanced exam sheets (monthly, midterms and finals), quizzes, and exercises, that containing different types of questions, covering the entire curriculum, and displaying gradually from easiness to difficulty. The exam sheets produced by the system take into account the different levels of the students from excellent, good, to fair, and avoids any mistakes of language and non-clear terminologies.
   - **Mobile quizzes and exercises:** Teachers can use this on-line tool to add, update, and delete questions and answers for their Mobile quizzes and exercises.
   - **Print students' scores:** Teacher uses this tool to print the grade reports for the students who took the web-based and mobile-based quizzes and exams created by the system.

iii. **The student sub-system:** This sub-system allows the student to: 1) download tutorials, reviews, previous exams and their model answers, 2) to take: web-bases exams, quizzes, and free exercises, 3) to take mobile-based quizzes and exercises. In addition to the login, and change password, the following services are available in the student sub-system:
   - **Download reviews:** Student can use this service to download the reviews files for his/her subjects. The student select the subject name, and the teacher name (as input) from the available list of values, the

system search for the required query, and displays all the stored reviews from the database.

- **Download tutorials:** Student can use this service to download the tutorials files for his/her subjects.
- **Download previous exams:** Student can use this service to download previous exams and their model answers for his/her subjects. The student select the subject name, and the teacher name (as input) from the available list of values, the system search for the required query, and displays all the stored exams from the database.
- **On-line quiz:** After a student login into the system from the student login page, he/she can select the quiz, where the questions will be selected randomly by our algorithm from the database. The quiz will be over either whenever the student finishes it and select the "Finish" button, or whenever a pre-allocated time will be over. The system will calculate the number of correct answers and display the result. Finally, the student score will be saved in the corresponding database for possible tracking. When the student finishes the quiz, a record will be created in the StudentScore table, and stored in the database. Teachers can consult this table to track the activities of their students.
- **On-line free exercises:** Student can use this service to take free web-based exercises for his/her subjects by setting his/her interested criteria: the number of questions in the exercises, chapter number, lessons limit, font size, and time allowed.
- **Wireless quizzes and exercises:** Student can use this service to access the system (not just anytime, but anywhere) through wireless devices (such as, Mobile phone, PDA- Personal Digital Assistants, and Tablet PC) and take quick quizzes and exercises.

IV. System Testing and Evaluation

Materials and methods: we have used 2*3 factorial design methods in this study to test and to evaluate the assessment system on school site and to get the feedback of students, teachers and parents who are the potential users of the system. Dependent variable of the study is the scores obtain from the students who take the web-based quizzes, exams, tutorials, and free exercises by the system in different months. The students are divided into two groups, including *experimental group* and *control group.*

Participants: in the initial test of this study we selected 30 students from preparatory-three level (class 9-A) to form experimental group and 30 students (class 9-B) to form control group. Experimental group took paper based exam, web-based exam and mobile based exam respectively for 3 weeks. Control group took only paper based exam for 3 weeks. Each exam consisted of 10 true-and-false questions and is scored by 10 points.

Survey: The perceptions of the students in the experimental group who undertook paper based exam, web-based exam and mobile based exam are gathered with the survey. The survey is validated and revised by experts. The survey consisted of 13 questions, 11 of which are 5-point Likert type questions (1 = Strongly Disagree, 2 = Disagree, 3 = Undecided, 4 = Agree and 5 = Strongly Agree) and 2 of which are open ended. These open ended questions are designed to ask students to explain which exam they liked most and which exam they liked least.

Procedure: The initial study conducted on the students from 2 classrooms for 3 weeks. All exams are including true-and-false questions. In the first week, paper based exam is applied to both groups. The scores of paper based exam are announced to the students one week later. In the second week, web-based exam is given to the experimental group, while paper based exam is given to the control group. 10 questions are broadcasted together on one screen in web-based exam. The feedback is given on the next screen to inform students about their answers for each question, the correct answer of the questions and the number of correct and incorrect answers and their scores. In the third week of the study, control group is informed on the scores they received from paper based exam which they took in previous week and they were given paper based exam in the third week. The students in the experimental group are given mobile based exam which are formed of 10 true-and-false questions. The questions in the mobile based exam are displayed sequentially. Before the mobile based exam ended, they have the opportunity to go back to the previous questions and change their answers. Feedback page which are displayed at the end of mobile based exam informed the students about number of correct and incorrect answers and their scores. The students enter the web-based exam and mobile based exam with their student ID numbers and passwords and they have right to use these systems only once. Each exam has 10 true-and-false questions, the numbers of correct answers are calculated as the *achievement score* and there is no time limitation.

Results: To measure the achievement and student' perceptions in paper based, web-based and mobile based exams, we have applied two-factor ANOVA (Analysis of Variance) for repeated measurements to find out whether there is significant difference between the scores of the students. Table 3 summarizes the ANOVA results, and Table 4 summarizes the means and standard deviations of the answers that computed

TABLE III. The two way ANOVA for repeated measures of students' scores in paper, web, and mobile based test.

Source of variation	Sum of Square	df	Mean Square	F	p
Between Groups	142.290	37			
Group (Experimental/control)	.490	1	.50	.125	.726
Error	140.980	35	3.94		
Within Group	167.989	77.20			
Paper/web/mobile based Exam	41.195	1.85	22.24	12.38	0
Group * Paper/web/mobile based Exam	7.890	1.85	4.23	2.39	.103
Error	119.700	66.7	1.78		
Total	311.213	113.4			

TABLE IV. STUDENTS' PERCEPTIONS ON PAPER, WEB, AND MOBILE BASED EXAMS.

Criteria	$\overline{X}$	S
1. I prefer the paper based Exam.	1.85	0.99
2. I liked the web based Exam.	4.65	0.59
3. I liked mobile based Exam.	3.30	1.42
4. Use of different media, such as internet and WAP increased my attention to the course.	4.45	0.60
5. If I am to undertake Exam in other courses, I would prefer web based Exam.	4.30	0.80
6. If I am to undertake Exam in other courses, I would prefer mobile based Exam	3.15	1.31
7. It was easy to use mobile based Exam.	3.70	1.30
8. It was easy to use web based Exam.	4.70	0.57
9. Paper based Exam was the best in offering feedback on my answers to the questions and my scores.	1.90	1.12
10. Web based Exam was the best in offering feedback on my answers to the questions and my scores.	4.25	1.12
11. Mobile based Exam was the best in offering feedback on my answers to the questions and my scores.	3.50	1.19

A. *Anticipated Results and Evaluation Criteria*

- This research study used a mixed methods approaches which emphasized on qualitative data collection [10]. The project data is collected from the following specific resources:
  i. Primary data will collect from the first prototype of "Questions-Bank" project that implemented in three Qatari schools, and presented in the IMCL'09 and ICL2009 international conferences. This includes the articles by Hosam F. El-Sofany, et al [6, 7], and the article by Samir A. El-Seoud et al [21].
  ii. Qualitative data is collected from Alwakra School (as a performance site) through the project documents, system testing and implementation, scheduled interviews with involved project participants, and analysis of open (free-response) questions from two questionnaires. Moreover, we used external material such as books and research articles covering relevant topics for this study.
  iii. Secondary data collected from the E-learning literature. A lot of research literature on E-learning in conjunction with school education is available; and relevant elements in relation to E-learning and customization for corporations are drawn out in this study. The search strategy included electronic databases and hand searches of some published books on E-learning. We used databases like ACM Digital Library, IEEE Xplore, ISI Web of Science and CiteSeerX. In addition, we used Google Scholar that provided a simple way to broadly search for scholarly literature across many disciplines and sources [12].
  iv. Supplementary data is taken from the PSUT, as a collaborative academic institution in this study

The Quantitative Research Approach: Two electronic questionnaires are given to the students who follow the two classes (9-A and 9-B). These questionnaires are conducted primarily to get an indication of how the students perceived the overall quality in the courses. A mix between free-response questions, dichotomous questions, multiple-choice questions, checklists and rating questions were used. The instrument designs for these questionnaires are based on Cooper & Schindler (chapters 12-13) [11]. The questionnaires are distributed to the students as part of the mandatory exercise program. All 30 students in the first group answered the first questionnaire and the other 30 students in the second group answered the second questionnaire. At the end of the semester the two groups are taken the final exam in three subjects (Mathematic, Science, and English language) using the project.

B. *Evaluation Criteria*

We have evaluated the system according to the following main criteria:

- **Usability:** Is the system providing a coherent and flexible user interface that can be used easily with all users? Has the system an obvious component which allows the user to navigate easily? Are the system applications easy to use by students, teachers, and parents? Is the system adaptive to user needs? Is the tracking component working properly and produce the expected results
- **Supportability:** Can the system be easily modified or extended easily? Can teachers and administrators add materials easily?
- **Contents and Design:** Are the design and contents of the system satisfactory to student, teachers and parents? What are their feedbacks? Does the system cover well the topics? Are they useful and meet their goals? Quality of texts, images, sounds, and clips? We compare the progress of students.
- **Performance:** Are the performance requirement concerned with quantifiable attributes of the system, such as response time, throughput availability and accuracy?
- **Flexibility:** Is the system flexible, since it will be used regularly with a new set of initial data and can accommodate changes in database such as modifying the contents?
- **Security:** Is the system preventing unauthorized users to access the system?

REFERENCES

[1] Hosam F. El-Sofany, Ahmad Hasna, Jihad Jaam, Fayed Ghaleb, and Samir A. El-Seoud, "A Web-Based E-Learning System Experiment". *International Journal of Computing & Information Sciences (IJCIS)*, Vol. 4, No. 1, P22-29, April 2006.

[2] Fayed F. M. Ghaleb, Sameh S. Daoud, Ahmad M. Hasna, Jihad M. Jaam, and Hosam F. El-Sofany, "A Web-Based E-Learning System Using Semantic Web Framework". *Journal of Computer Science* 2 (8): 619-626, 2006, ISSN 1549-3636. Science Publications U.S.A, 2006.

[3] Fayed Ghaleb, Sameh Daoud, Ahmad Hasna, Jihad Jaam, Samir A. El-Seoud, and Hosam El-Sofany, "E-Learning Model Based On Semantic Web Technology". *International Journal of Computing & Information Sciences (IJCIS)*, Vol. 4, No. 2, P63-71, August 2006.

[4] Hosam F. El-Sofany, and Samir A. El-Seoud, "Towards the Development of an M-Learning System: A New Stage to Enhance Higher Education", *International Journal of Interactive Mobile Technologies* (iJIM), Vol. 3, Issue 3 , Pages 1-6, doi:10.3991/ijim.v3i3.719, July 2009.

[5] Samir Abou El-Seoud, Ashraf M.A. Ahmad and Hosam Farouk El-Sofany, "Mobile Learning Platform Connected to Moodle using J2ME", International *Journal of Interactive Mobile Technologies* (iJIM), Volume 3, Issue 2, P: 46-54, April 2009.

[6] Hosam F. El-Sofany, Samir A. El-Seoud, F.F.M. Ghaleb, Shaima Ibrahim, and Noor Al-Jaidah, "Questions-Bank System to Enhance E-Learning in School Education", *International Journal of Emerging Technologies in Learning* – iJET, Volume 4, Issue 3, Pages 8-19, , September 2009.

[7] Hosam Farouk El-Sofany, Noor Al-Jaidah, Shaima Ibrahim, and Salha Al-kubaisi, "Web-based Questions-Bank System to Improve E-Learning Education in Qatari School", *Journal of Computer Science* 5(2): P: 79-108, 2009, ISSN 1549-3636, Science Publications U.S.A, March 2009.

[8] Berners-Lee, T. what the Semantic Web can represent.

[9] Nejdl, W. Learning Repositories – Technologies and Context. In Proc. of ED-MEDIA 2001 World Conference on Educational Multimedia, Hypermedia & Tele-communications, 2001.

[10] Creswell, J. W.. *Research Design: Qualitative, Quantitative, and Mixed Methods Approaches* (second edition). Thousand Oaks, CA: Sage Publications, 2003.

[11] Cooper, D. R., & Schindler, P. S. *Business Research Methods* (8th. edition). Boston, MA: Irwin McGraw-Hill, 2008.

[12] Google Scholar.

[13] Mercier F, David B, Chalon R, Berthet, JP. Interactivity in a large class using wireless devices. Proceedings of Mlearn, Bracciano (Rome), Italy pp. 129-132, 2004.

[14] Kim CS. Work in progress – Mobile computer based classroom assessment. Proceedings of 35th ASEE/IEEE Frontiers in Education Conference, October 19 – 22, Indianapolis.

[15] Rouet MM, Ney M, Charles S, Boidin GL. Students' performance and satisfaction with Web vs. paper-based practice quizzes and lecture notes. Comput. Educ. 53: 375-384, 2009.

[16] Seppälä P, Alamäki H. Mobile learning in teacher training. J. Comp. Assist. Learn. 19(3): 330-335, 2003.

[17] Dawabi P, Wessner M, Neuhold E. Using mobile devices for the classroom of the future. Proceedings of Mlearn 2003 Conference on Learning with Mobile Devices, London pp. 14-15, 2003.

[18] Scornavacca E, Marshall S. TXT-2-LRN: Improving students' learning experience in the classroom through interactive SMS. Proceedings of the 40th Hawaii International Conference on System Sciences, Hawaii, 2007.

[19] Wentling L, Park J, Peiper C. Learning gains associated with annotation and communication software designed for large undergraduate classes. J. Comp. Assist. Learn. 23(1): 36-46, 2007.

[20] Homan S, Wood K. Taming the mega-lecture: Wireless quizzing. Syllabus Magazine.

[21] Samir Abou El-Seoud, Hosam Farouk El-Sofany, and Yahia Al-Halabi, "Wireless Questions-Bank System to Enhance M-Learning in School Education". *International Journal of Interactive Mobile Technologies* (iJIM), Vol. 4, Issue 1, Pages 4-8, , January 2010.

AUTHORS

Hosam F. El-Sofany is with the Cairo Higher Institute for Engineering, Computer Science, and Management, Cairo, Egypt.

Samir A. El-Seoud is with Technische Universität Darmstadt, Darmstadt, Germany

2

Using Concept Mapping for Needs Analysis for a Social Support System in Learning Network

Danish Nadeem, Slavi Stoyanov and Rob Koper
Center for Learning Sciences and Technologies (CELSTEC), Heerlen, The Netherlands

Abstract—**This paper describes the needs analysis of tool for connecting people in a Learning Network using Group Concept Mapping method. We conducted an empirical study to identify the most relevant features of such a tool. There were 11 experts from 7 different countries with technical and non-technical background who participated in the study. Each of the participants were invited by email and given a brief description about the scenario for the need of a tool in an online Learning Network to find people for help (social support system). This tool can be developed on mobile devices to search for peer learners. The participants generated 153 unique ideas about the need for social support system in Learning Network. These ideas were sorted and rated, and further analysis with Concept Mapping produced ten relevant clusters of interest. These clusters (concepts) suggest the features which are relevant for a tool that would support learners in connecting with others in a Learning Network. The concept mapping methodology described in this paper has also been applied in other projects; one of them is about the "investigation of pedagogic areas in which mobile learning is an appropriate solution". We intend to publish the work in the future publication.**

Index Terms—**concept mapping, learning network, social support system, software needs analysis**

I. Introduction

In general, there are several contexts in which people look for other people, pursuing shared common interests or social conversation [7]. The tendency is also that quite often people tend to seek the tacit knowledge held by individuals instead of explicit knowledge stored in formal documents [2, 9]. Research has claimed that technology could catalyze collaboration and knowledge sharing by enhancing the traceability and visibility of information, which resides in people's mind, especially the knowledge which is not well-documented, like best practices and lessons learned [13]. Social support system in learning networks would be an added value for a learner to know about people and decide whom and how to interact with, pursuing particular learning goals [5]. It would give several benefits to a learner:

(1) A learner could connect and make social ties with other learners either for question in hand or for future contact in similar context. (2) It is possible that in connecting people to other people, one can share and exchange expertise about the information which is not documented like people's experiences, interests and also tacit knowledge. (3) A learner could share with others different perspectives on a topic, comparing and constructing meaning, which would lead to effective learning for her/him. (4) A learner would not only depend on experts but s/he can seek help from any suitable person available in learning networks, where people with different levels of expertise exist.

II. Methodology

In this empirical study we wanted to gather opinion of experts on the requirements of Social Support System (SoSuSy) in Learning Network. For this purpose we have chosen Concept Mapping [8] as appropriate method because it includes both qualitative techniques and multivariate analysis approaches [4]. The term is also known as Group Concept Mapping (GCM) in some of the related studies [6, 11]. Concept Mapping provides a structured approach to identify experts' opinion on a given domain. We used Concept Mapping to collect, structure, analyze and interpret data from the experts of the domain on a particular project. The method identifies the experts' opinion in the form of clusters on a given subject using advanced statistical techniques as referred above. Group Concept Mapping differs from other approaches in the following ways, as reported in [6]: (a) there is only one round of structuring the data as the participants work independently and anonymously to each other; (b) in data structuring, coding the information in different ways is not a problem as it is the advanced analysis included in the methodology that identifies common patterns in the data; (c) consensus is not forced, it emerges from the data; and (d) visualizations, as substantial part of the analysis, allow for grasping at once the emerging structures in the data and help their interpretation. These characteristics provide a good basis to choose this methodology. The Concept Mapping provides a visual map of useful and important concepts which can be used for the development of a project under hand. Traditionally, Concept Mapping has been used for getting direction in project developments that are related to business management, planning, evaluation and identifying health care policies [10, 8, 12]. In this study we report our first attempt to use Concept Mapping for the purpose of software systems needs analysis. We expect that the conceptual domains derived from the Concept Mapping process will be the guiding principles for the developers of the social support systems for desktop or mobile versions. The inputs taken from diverse people with varying expertise can be combined to understand the important features to implement while developing the system. Also, the ideas generated about the important concepts in the concept map will provide a cross-validation for reliable and efficient systems that meet the expert opinion.

Concept Mapping consists of: generation of statements by the experts (brainstorming); rating and sorting of statements; and interpretation of the concept map (data analysis). While all three stages may be done in person, in this study the first two stages were done by mail due to participants' busy schedules and different locations. All concept mapping analyses were accomplished using The Concept System© (Version 1.75, Concept Systems Incorporated, Ithaca, NY).

A. *Participants*

On the basis of research problem we selected a total of 11 different participants from 7 different countries across different continents. Each participant came from a variety of educational backgrounds, namely: computer sciences, business management, human resources, education sciences and journalism. The idea was to gather varied expert opinion on the topic of social support system, from a diverse group of people. An email invite was sent to each of the participants with the brief description of the requirements of social support systems in Learning Network. They were asked to think-aloud about the need for such a system in Learning Network that searches for people. Furthermore, a trigger statement was given to help them in generating the ideas: "*What defines a good Social Support System in Learning Network*". This trigger statement helped them to generate statements about the features or characteristics, that they thought are necessary to develop such a system in Learning Network.

B. *Procedure*

Idea generation requires the participants to individually generate ideas in respond to a focus (trigger) statement. We gave them a scenario description and the following focus statement: *"What makes a good social support system?"*

We asked the participants to consider a scenario: *"Imagine you are interested in learning Piano. You already have some skills in playing piano, like the rules for building major and minor piano scales from any note on the piano keyboard and you also learned how to combine these scales to form major and minor chords. Now you want to improvise, let's say you are interested in playing Arpeggios, a technique in which the notes of chord are played independently rather than together. As you are learning by yourself, it will be helpful if you knew who else is learning similar things in Piano. You can ask him to give you advice, share his experiences of learning Piano techniques etc, may be you already have a friend or a friend of your friends or somebody totally out of your personal/professional network who is interested in learning Piano. How will you find that person? How will you know who is suitable person for you to contact and share knowledge about Arpeggios technique. Let's say you are on Facebook, where you have your network of friends, you could ask there. What if the person you are looking for is not on Facebook, but say Orkut, LinkedIn or other social or professional networking sites What you need is a service to search people in different networks (professional/social) available over the web that recommends the people with their contact details, knowledge and interests so you can choose whom to contact for your purpose". This service (basically a software application for mobile phones or desktop computers) facilitates search for people associated with Arpeggios in the network. The service looks for others who have interests, knowledge and skills in playing Arpeggios and gives you a list of people suitable for you. It may happen that you already know one of the recommended people through your social networks), but never knew s/he could also play Arpeggios as well. The need for such a system is to find sufficiently qualified people who have interest to help others in learning. To accomplish this, the system needs to first "know-about" people's interests and competence, then rank them according to their levels of expertise and finally recommend them to the learners. To fulfill the requirements of finding people, a "social support system" has to be provided in a Learning Network. When you are asked to think-aloud of what you expect from such a system that searches people based on your problems or interests, what are the features or characteristics, you think are necessary to develop such an application.*

Then some hint statements were given to the participants and they were asked to think in similar lines to generate their own statements.

- assists learners in finding suitable people.
- allows knowledgeable people to answer learner's question on a given topic.
- supports socializing of learners with other learners.
- matches people according to their common learning interests.
- affects the productivity of individuals and groups in a learning scenario.

C. *Sorting and Rating of ideas*

The participants generated a total of 153 statements These generated ideas in the first phase, were sent back to the experts for first sorting and then rating according to separate instructions for the two idea structuring activities. With regard to sorting, we asked the participants to use the following guidelines provided with a sorting recording sheet: (a) group the statements for similarity in meaning; (a) place each statement in one group only; (c) place each statement somewhere (not one pile comprising all statements); (d) place a statement in own group if it seems unrelated to the other statements; do not make groups of statements called miscellaneous. We also suggested the participants to give a name of each of the group of statements.

For the next activity in this phase, we asked the experts to rate each of the statements on a 1-to-5 scale of importance and feasibility (for importance, 1 = Relatively Unimportant; 5 = Very Important; for feasibility, 1 = Least Feasible; 5 = Most Feasible). The meaning of importance is straightforward: how important is the each feature of social support system? Feasibility is about how likely those features are implemented? In the rating instruction, we advised the participants to use the full range of rating values, emphasizing on relative rather than absolute rating.

III. RESULTS

A. *Data Analysis*

In Figure. 1, there are 10 different clusters shown, which depicts how they were grouped together by hierarchical cluster analysis [1, 8]. The hierarchical cluster analyses, which are contiguous but not overlapping with each other [3] gives the number of clusters indicating the experts' opinion (see Figure. 1). To determine the right number of clusters we modified the heuristic called "20-to-5"to the heuristic "12-to-5", because the experts in this study produced 12 clusters in average. The classical "20-to-5" heuristics is based on the fact that most of the participants in Group Concept Mapping (GCM) projects make between 5 and 20 clusters. We began the analysis in this study with the 12-cluster solution checking at each step whether the solution from the merging of clusters makes sense until we arrived at the 10-cluster solution. Other criteria were bridging or anchoring values of the statement in a particular cluster. The bridging/anchoring value is between 0 and 1. A low bridging/anchoring value means that more people have grouped the statement together with others in its vicinity. Statements with low bridging / anchoring value better represent the meaning of a particular cluster's content than those with a higher value. Statements with a relatively high bridging value within a cluster play more the role of bridges to other clusters, while statements with relatively low bringing value play the role of anchors for the statements within a cluster.

The main point of interpretation of the cluster map is that all participants come to figure out well the interrelationships among the clustered statements. It is aimed that everyone in the group has a clear picture of the project through the concept map. The clusters shown are depicting the expert opinion on various features that a Social Support System should have with varied degree of importance (shown as multiple layers on each cluster).

1. **Technical features**: Focusing the state-of-art technologies for development like web 2.0 applications and Social applications.
2. **Showing search similarity**: The feature focuses on the search options where people would like to know the search results similar to what they are looking for.
3. **Visibility**: The feature focuses on the how social help system provides visibility of learners in the Learning Network.
4. **Business application**: The social help system as a tool for collaboration for further interests and commercial interests among people.
5. **Communication among learners**: The tool supports effective sharing of knowledge among learners facilitating communication via email, instant messengers or phones.
6. **Learning community and connection among varied people**: The tool supports forming learning communities and enhancing communication among the participants.
7. **Facilitating learning and engagement**: The tool provides support for bringing people to engage on a learning task and enhance learning by social engagement
8. **Interface design**: Easy to use features and user-interface to search for suitable people in a network of learners.
9. **Effects on society**: The overall effects on society as a useful feature for supporting learning by socialization.
10. **Output and Solution execution**: The feature focusing on additional support like finding not only people but also relevant learning resources, supporting external collaboration, other embedded support.

Apart from the traditional descriptive statistics, the GCM approach applies some specific types of analyses such as multidimensional scaling (MDS) and hierarchical cluster analysis (HCA). The input for the multidimensional scaling is the creation of a total square similarity matrix from the sorting data across the participants. The matrix is binary and symmetric, showing the number of people that group together in their sorts each pair of statements. The combined matrix sums individual matrices of the participants. An individual matrix consists of as many rows and columns as are the statements. A cell indicates whether two statements are grouped together or not. '1' in the cell indicates that the two statements are sorted together by the participant, '0'that they are not.

MDS transforms the total square similarity matrix into a map depicted as a coordinated matrix. From the coordinates, MDS can compute the distances between all pair of statements (points) and can show this as matrix of distances between points. The closer the statements to each other are, the more similar in meaning they are, which also mean that more people sorted these statements together. The extent to which each of the distances between the statements on the map deviate from the values of the total similarity matrix which is used as input to the map is measured with the "stress index" [4]. In principle, the lower the value of the stress index is, the better the overall fit between the map and the input matrix is. A meta-analytical study across a broad range of concept mapping projects indicated that around 95% of concept mapping projects would produce value of the stress index in the range between 0.205 and 0.365.

There is a weak relationship between the two values (see Figure 2.). It is also interesting to see how end-users and experts view the different features of the Social Support System. The clusters 'showing search similarity,' Facilitating learning and engagement' and 'Interface design' represent the largest margins in scores on the two scales. There are relatively small differences in scores of clusters such as, 'Technical features', Communication among learner, 'visibility' and 'output and solution execution'.

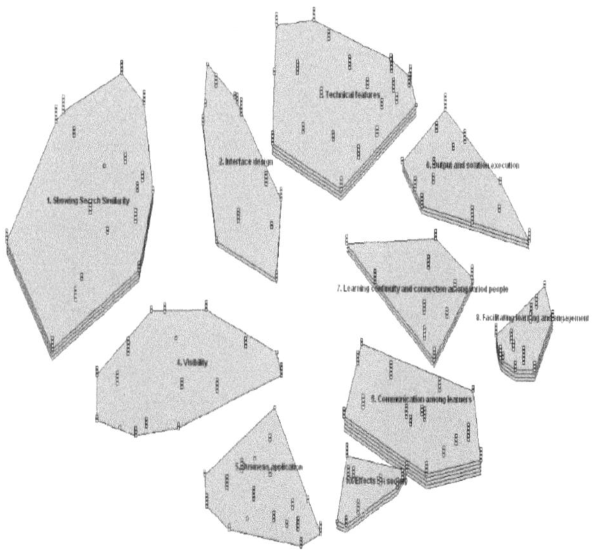

Figure 1. Cluster Map: eliciting the focus on 10 required features of a social support system

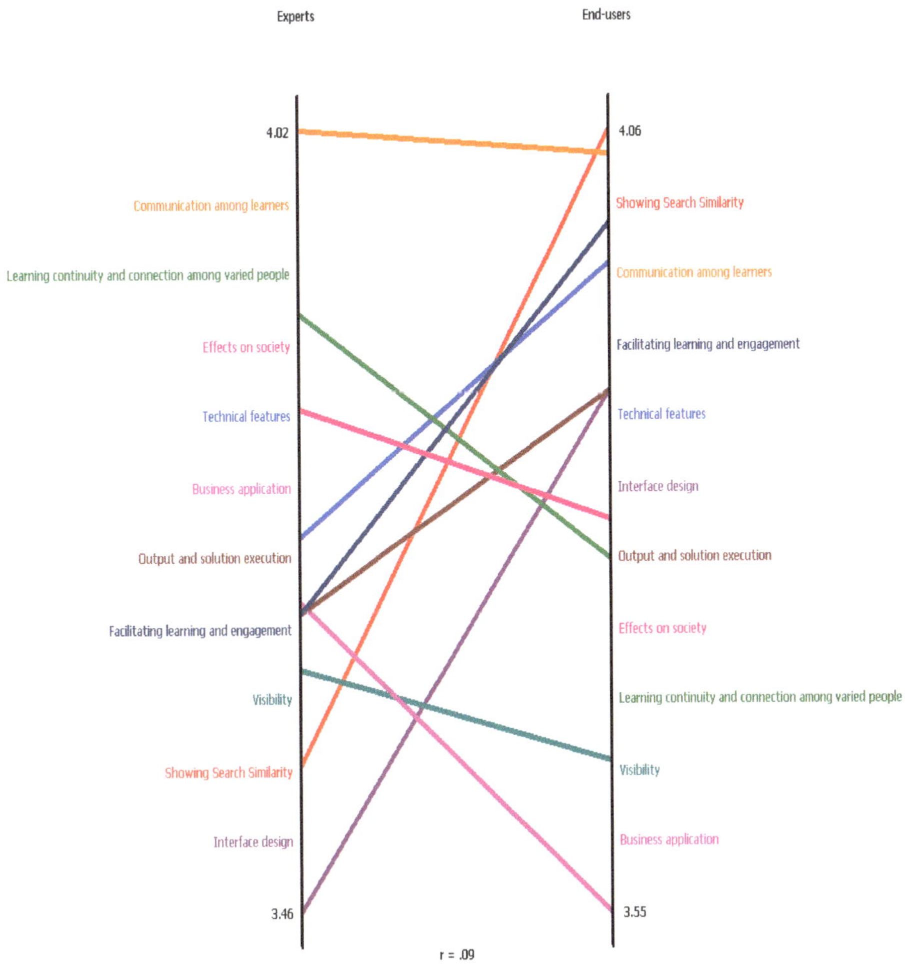

Figure 2. The ladder graph: Also termed as pattern match in Figure 2 provides a visual comparison of the clusters on importance and feasibility between the end-users and experts view

A specific analysis that compares the statements on importance and feasibility within a particular cluster is "go-zone" (Figure 3.). Go-zone identifies statements that score high on both importance and feasibility, thus suggesting where we should look first when planning the implementation of changes in education and training. Go-zone is a bi-variate graph that shows the average ratings for importance and feasibility on each statement within a specific cluster. For example, something that is considered to be very important could be deemed not to be very feasible. The graph is divided into quadrants based upon the mean rating values of importance and feasibility. The upper right quadrant represents issues that are above average on both variables ("go to" – very important and very feasible).

IV. Discussion and Conclusion

The results of this study suggested 10 different features that can be developed for Social Support system (figure 1.). The system itself can be developed for mobile devices or desktop learning environment. The cluster "Showing search similarity" is differently rated on the scale of *importance* and *feasibility* (figure 2.) by the experts and end-users, this also indicates how the different users, view the need of this feature in Social Support System.

Group Concept Mapping (GCM) in the SoSuSy project proved to be an efficient way and appealing to the participants approach for data collection, structuring and analysis. Idea generation, sorting and rating are activities that the participants are used to. Eleven participants generated 153 unique ideas (statements). Sorting and Rating were time consuming activities, but not difficult for the participants from conceptual point of view. In contrast to some other similar techniques (e.g affinity diagram or card sorting), GCM applies some rigorous statistical techniques for data analysis such as multidimensional scaling and hierarchical cluster analysis to identify emerging structures in the data. The visualizations provided by the software (Concept System, 2010) support the interpretation of the results. The clusters indicates the possible features for the social support system that can be developed and later validated according to their importance as we found in the ladder graph in figure 2. The main point of interpretation of the cluster map is that all participants come to figure out well the interrelationships among the clustered statements. It is aimed that everyone in the group has a clear picture of the project through the concept map. The clusters shown are depicting the expert opinion on various features that a Social Support System should have with varied degree of importance (shown as multiple layers on each cluster). The development these features as depicted in cluster map, will be the guideline for developing the Social Support System for a Learning Network. The clusters identified, will also provide a way to evaluate any such system which needs and early evaluation to meet the user needs.

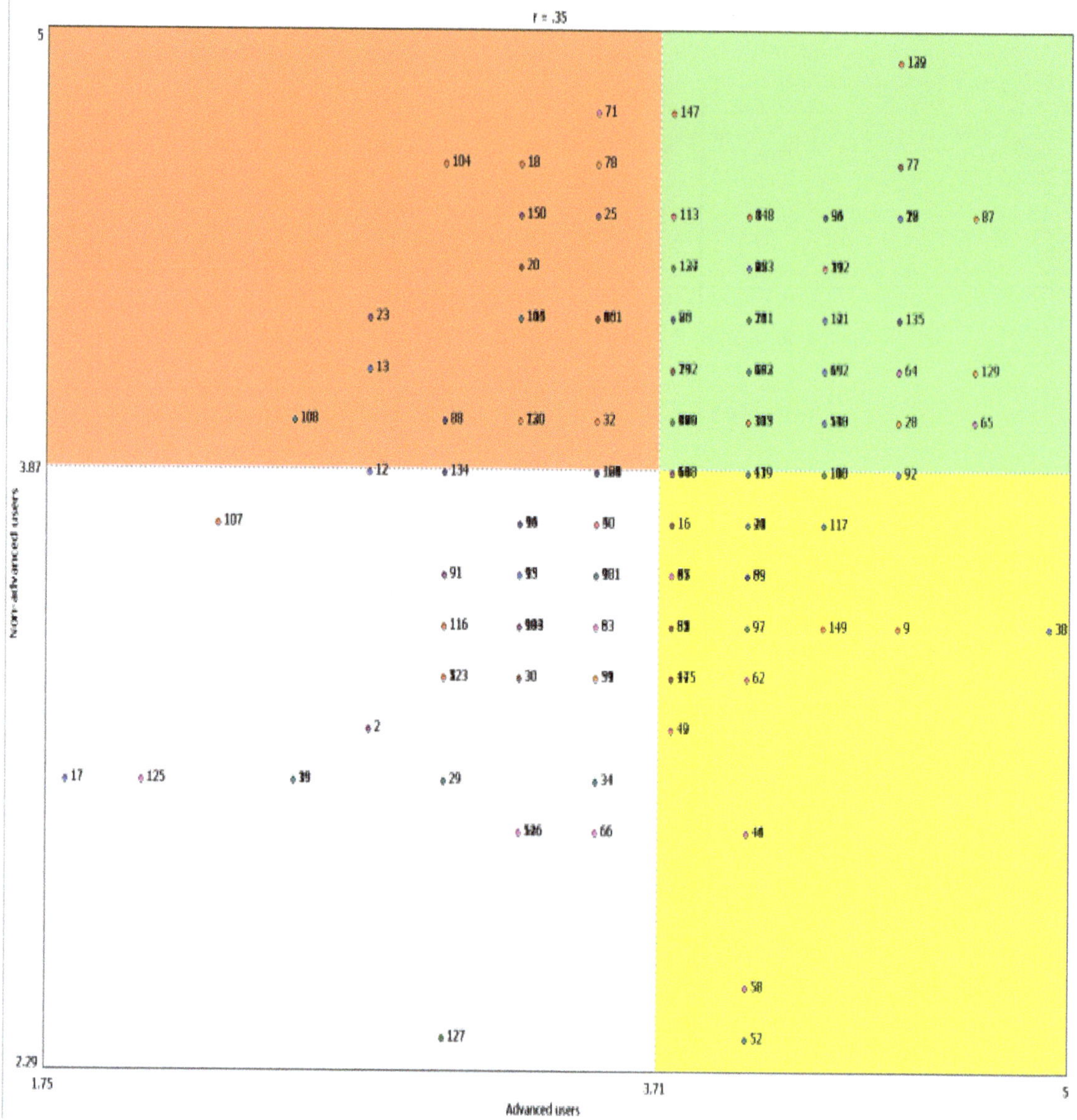

Figure 3. An example of a go-zone comparing the statements in the cluster Social Support system on importance and feasibility.

ACKNOWLEDGMENT

This paper has been (partly) sponsored by the TEN-Competence Integrated Project that is funded by the European Commission's 6th Framework Programme, priority IST/Technology Enhanced Learning. Contract 027087 (www.tencompetence.org).

REFERENCES

[1] Everitt, B. (1980). Cluster Analysis (2nd Edition). New York, NY: Halsted Press, A Division of John Wiley and Sons

[2] Hertzum M, Pejtersen AM. The Information-Seeking Practices of Engineers: Searching for Documents as well as for People. *Information Processing & Management.* 2000;36(5):761-778.

[3] Kane, M., & Trochim, W. (2007). Concept mapping for planning and evaluation. Thousand Oaks, CA: Sage Publications.

[4] Kruskal, J.B. and Wish, M. (1978). Multidimensional scaling. Beverly Hills, CA: Sage Publications.

[5] Nadeem D, Stoyanov S, Koper R. Social support system in learning network for lifelong learners: a conceptual framework. *International Journal of Continuing Engineering Education and Life-long Learning.* 2009;Vol. 19, N(4/5/6):337–351.

[6] Stoyanov S, Kirschner P. Expert concept mapping method for defining the characteristics of adaptive e-learning: Alfanet project case. ,. *Educational technology research and development.* 2004;52(2):41-56.

[7] Terveen L, McDonald DW. Social matching: A framework and research agenda. *ACM Trans. Computer-Human Interaction.* 2005;12(3):401-434.

[8] Trochim W. An introduction to concept mapping for planning and evaluation. *In W. Trochim (Ed.) A Special Issue of Evaluation and Program Planning.* 1989;12:1-16.

[9] Wilson T. Information-seeking Behaviour: Designing Information Systems to meet our Clients' Needs. ACURIL: Association of Caribbean University, Research and Institutional Libraries. XXV Conference, San Juan, Puerto Rico. 1995.

[10] Wolf . J, Boevink W. Mapping the quality of life of people with severe and enduring mental health problems. Ment Health Care. 1999;2:228–231.

[11] Wopereis, I.G.J.H., Kirschner, P.A., Paas, F., Stoyanov, S., & Hendriks, M. (2005). Failure and success factors of educational ICT projects: a group concept mapping approach. British Journal of Educational Technology, 36, 681-684.

[12] Watson GR. What is...concept mapping? Med Teach. 1989;11:265–269.

[13] Yimam, D. & Kobsa, A. (2000). DEMOIR: A Hybrid architecture for expertise modeling and recommender systems. In Proceedings of the 9th IEEE international Workshops on Enabling Technologies: infrastructure For Collaborative Enterprises, WETICE. pp. 67-74. IEEE Computer Society, Washington, DC.

AUTHORS

Danish Nadeem works at Centre for Learning Sciences and Technologies (CELSTEC), Open University of the Netherlands as researcher and technology developer. He obtained his M.Sc. in Cognitive Science from University of Osnabrueck, Germany. He has a solid background in computer science, artificial intelligence (AI) and computational linguistics. His research interests are in the fields of cognitive science, AI, human computer interaction (HCI) and semantic web/desktop technologies. (e-mail: danzinde@gmail.com).

Slavi Stoyanov is a Senior Researcher at the Centre for Learning Sciences and Technologies, Open University of the Netherlands. He holds a PhD on Instructional Technology from the University of Twente, The Netherlands. His major research interests include technology-enhanced learning, individual differences in learning, learning to solve ill-structured problems and cognitive mapping. (e-mail: slavi.stoyanov@ou.nl).

Rob Koper holds a PhD in Educational Technology from the OUNL. Since 1998 he has been a Full Professor in Educational Technology, specifically in e-learning technologies. He was responsible for the development of educational modelling language (the predecessor of IMS learning design). As the Director of RTD in learning technologies, he led the OUNL's contribution to the IMSLD specification activities. His research focuses on self-organised distributed learning networks for lifelong learning, including RTD into software agents, educational semantic web, interoperability specifications and standards.(e-mail:rob.koper@ou.nl)

A short version of this paper was presented in the IEEE EDUCON 2010 conference in Madrid and selected for submission to this journal. It received a co-award as a Best Student Paper of the conference.

Unified M-Learning Model through Interactive Education Satellite

A Proposal for an Arab Homeland Education Satellite

Ghassan F. Issa, Shakir M. Hussain and Hussein Al-Bahadili
Petra University, Amman, Jordan

Abstract—**in this paper, we propose a unified and interactive mobile learning (M-Learning) model to help with expanding and spreading education in the Arab Homeland countries. The model utilizes a new competitive spot beam satellite communication technology, which enables efficient channel allocation, where communication channels can be allocated to specific and precise areas. The proposed model is referred to as the interactive Arab education satellite (IAESat).The communication satellite can efficiently and effectively cover the entire Arab Homeland and reaches a wide area and mobile users that cannot be reached otherwise. The model implements existing interactivity components to enhance the learning process and meet international standards in education.**

Index Terms—**E-Learning, M-Learning, interactive learning, education satellite, spot beam communication.**

I. Introduction

Electronic learning (E-Learning) is about the transmission of learning content using communication and information technologies (CIT). Conventional learning involves identifying of information, conceptualizing, and making meaning to enhance user's knowledge base, understanding and skills, as well as finding the time and space for learning is left to the individual [1, 2]. The total E-Learning solution comprises the integration of three elements: content, technology and services [3]. This concept is also underpinned by the assumption that learners will be responsible for the cognitive tasks that will lead to learning.

In E-Learning, the information and communication systems, whether networked or not, serve as specific media to implement the learning process, and content is delivered via the Internet, Intranet/Extranet, audio or video tape, satellite TV, and CD-ROM. It can be self-paced or instructor-led and includes media in the form of text, image, animation, streaming video and audio. E-Learning definition abounds to [4]:

- The convergence of the Internet and learning, or Internet-enabled learning.
- The uses of network technologies to create, foster, deliver, and facilitate learning, anytime and anywhere.
- The delivery of individualized, comprehensive, dynamic learning content in real time, aiding the development of communities of knowledge, linking learners and practitioners with experts.
- A phenomenon delivering accountability, accessibility, and opportunity to allow people and organizations to keep up with the rapid changes that define the Internet world.
- A force that gives people and organizations the competitive edge to allow them to keep ahead of the rapidly changing global economy.

E-Learning is also known as distance learning (D-Learning) [5]. However, a recent form of learning, namely, mobile learning (M-Learning) was introduced due to the tremendous advancement in Internet technologies, and the exponential growth in the processing power, availability, and affordability of wireless mobile devices while becoming more affordable.[6, 7].

The Arab-Homeland consists of 22 countries occupying an area close to 10 millions Km^2 and housing more than 350 millions citizens. Figure (1) shows the geographical maps for the Arab-Homeland, and Table (1) lists the name, area, population, population density, the gross domestic product (GDP), and income per capita (IPC) for all Arab-Homeland countries [8]. When examining the figures in Table (1), differences in area, population, population density, GDP, and IPC, between Arab countries appear quite obviously.

Many of the Arab-Homeland countries have large areas with low population, meaning that the population is wide spread over large areas of land. The GDP of Saudi Arabia close to 600 billions compared to Comoros Island with only 0.7 billions. However, many countries have low GDP and therefore lack proper education. Overall, the annual GDP of the Arab countries is exceeding 2.5 Trillions. Table (1) also shows that the IPC is ranging from $90149 in small country like Qatar to $6347 large country like Egypt. With its diverse demography, the Arab Homeland constitutes an excellent example for launching mobile-based education to provide education to all people in such diverse area. However, providing mobile education requires high performance connectivity for remote locations in countries with large area, small population, and with low income. This would be infeasible through conventional technologies due to high initial and operating costs. Therefore, the most obvious solution is to use the newly evolving approach that is based on establishing connectivity through interactive satellite system. In this paper, in order to enhance, meet the needs of, and widen the education system in all Arab countries, we propose a unified and interactive M-Learning model that utilizes a satellite-based communication channels, which is referred to it as the interactive Arab-Homeland education satellite (IAE-

Sat). This model looks very promising as it can provide education anywhere anytime efficiently and cost-effectively.

This section provides an introduction to the general domain of this paper. The rest of the paper is organized as follows: Section 2 reviews previous experiences in using satellite-based education. In addition, in Section 2, some advanced satellite communications technologies are discussed. The unified M-Learning model and the main components of the proposed IAESat are discussed in Section 3. Section 4 describes the operation procedure and lists the advantages of IAESat. Finally, in Section 5 conclusions and recommendations for future work are pointed-out.

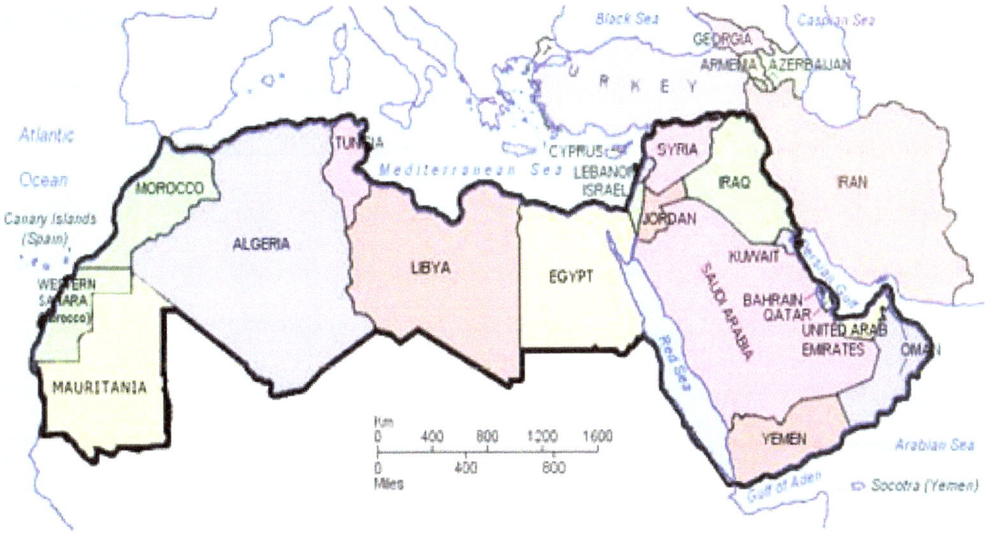

Figure 1. The geographical map of the Arab-Homeland.

TABLE I.
LIST OF ARAB-HOMELAND COUNTRIES AND THEIR AREAS, POPULATIONS, POPULATION DENSITIES, GDP, AND IPC.

Country	Area (Km^2)	Population	Population Density	GDP ($ Billion)	IPC
Bahrain	750	1 234 596	1646	28.275	27214 (2009)
Kuwait	17820	3 566 437	167.5	140.589	3984 (2010)
Oman	309550	2 845 000	9.2	74.431	25984 (2009)
Qatar	11437	1 696 565	123.2	102.147	90149 (2010)
Saudi Arabia	2149690	257 313 766	12	618.744	23701 (2010)
Arab Emirates	83600	4 675 593	97	182.876	40175 (2010)
Yemen	555000	23580 000	44.7	58.218	2457 (2009)
Gaza Strip	360	1604235	4118	0.77	3100 (2009)
Iraq	438317	31234000	71.5	111.5	3570 (2009)
Jordan	92300	6407085	56.4	35.3	5956 (2010)
Lebanon	10452	4224000	404	58.576	14988 (2010)
Syria	185180	22505000	118.3	105.238	5043 (2009)
Egypt	1002450	79089650	82.3	469.604	6347 (2010)
Algeria	2381740	35423000	14.6	255.189	7104 (20010)
Mauritania	1030700	3291000	3.2	6.326	2037 (2009)
Libya	1759541	6420000	3.6	96.138	14884 (2010)
Morocco	710850	32200000	71.6	193.15	48886 (2010)
Tunisia	163610	10432500	63	86.086	8254 (2009)
Sudan	2505810	43939598	16.9	54.681	2464 (2010)
Somalia	637657	9359000	14.6	5.731	600 (2009)
Djibouti	23200	818159	37.2	1049	2549 (2010)
Comoros Island	2235	691000	275	722	1159 (2009)
Total	13041549	852550184		4454.569	

II. Review of Previous Experiences

This section presents an overview of three satellite based educational programs, namely, the Italian satellite (ItalSat) [9], the Indian national satellite (InSat) system [10], and the Indian educational satellite (EduSat) [11]. An overview on recent advances in satellite communications technologies is provided here. Furthermore, since the proposed solution can be embedded with existing satellite projects such as the Arab satellite (ArabSat) and the multimedia exchange network over satellite (MENOS) hub [12].

A. *Satellite-based educational programs*

Three well-known satellite-based educational programs are briefly described below, one adopted by the Italian government, and the other two adopted by the Indian government.

1) *Italian Experience in ItalSat*

The concept of multiple spot beam communications was successfully demonstrated in 1991 with the launch of ItalSat [8, 9], which is a satellite-based education program developed by the Italian Research Council (IRC). With six spot beams operating at 30 GHz (uplink) and 20 GHz (downlink), the satellite interconnects TDMA transmissions between ground stations in all the major economic centers of Italy. It does this by demodulating uplink signals, routing them between up- and downlink beams, and combining and remodulating them for downlink transmission.

In ItalSat, laser beams can also be used to transmit signals between a satellite and the earth, but the rate of transmission is limited because of absorption and scattering by the atmosphere. Lasers operating in the blue-green wavelength, which penetrates water, have been used for communication between satellites and submarines.

2) *Indian Experience in INSat*

The Indian National Satellite System (INSat) is a series of multipurpose geo-stationary satellites launched by the Indian Space Research Organization (ISRO) to satisfy the telecommunications, broadcasting, meteorology, education and search and rescue needs of India [8, 10]. Commissioned in 1983, INSat is the largest domestic communication system in the Asia Pacific region. It is a joint venture of the Department of Space, Department of Telecommunications, India Meteorological Department, and India Radio and Doordarshan. The overall coordination and management of INSat system rests with the secretary-level INSat Coordination Committee.

InSAT satellites provide 199 transponders in various bands (C, S, Extended C and Ku) to serve the television and communication needs of India. Some of the satellites also have the Very High Resolution Radiometer (VHRR), CCD cameras for metrological imaging. The satellites also incorporate transponder(s) for receiving distress alert signals for search and rescue missions in the South Asian and Indian Ocean Region, as ISRO is a member of the Cospas-Sarsat program.

3) *Indian Experience in EduSat*

With the success of the InSat-based educational services, a need was felt to launch a satellite dedicated for educational service and ISRO conceived the EduSat project in October 2002 [8, 11]. EduSat is the first exclusive satellite for serving the educational sector. It is specially configured to meet the growing demand for an interactive satellite-based distance education system for the country through audio-visual medium, employing Direct-To-Home (DTH) quality broadcast. The 1950 kg EduSat is launched from Satish Dhawan Space Centre (SDSC) SHAR, Sriharikota, into a Geosynchronous Transfer Orbit (GTO) by ISRO's Geosynchronous Satellite Launch Vehicle (GSLV). From GTO, EduSat reaches the 36,000 Km high Geo-Stationary Orbit (GSO) by firing, in stages, its on board Liquid Apogee Motor (LAM). In GSO, the satellite is co-located with KALPANA-1 and INSat-3C satellites at 74° east longitude. EduSat carries five Ku-band transponders providing spot beams, one Ku-band transponder providing a national beam and six extended C-band transponders with national coverage beam. It joins the InSat system that already has more than 130 transponders in C-band, extended C-band and Ku-band providing a variety of telecommunication and television services.

B. *Advances in satellite communications technologies*

Through the last decades, there has been tremendous development in satellite communications technologies [13, 14]. Satellite communications started with using multiple access based mainly on time division multiple access (TDMA), moving to bandwidth reuse through the use of spot beams, then the extraordinary hopping technology, and finally the low orbit systems technology. A brief definition for each of these technologies is provided below and further details can be found in [13, 14].

1) *The Time Division Multiple Access (TDMA) Technology*

Communications satellite systems have entered a period of transition from point-to-point high-capacity trunk communications between large, costly ground terminals to multipoint-to-multipoint communications between small and low-cost stations. The development of multiple access methods has both hastened and facilitated this transition. Satellite communication technology was based on using the TDMA technique for user allocation on the communication channel. With TDMA, each ground station is assigned a time slot on the same channel for use in transmitting its communications; all other stations monitor these slots and select the communications directed to them. By amplifying a single carrier frequency in each satellite repeater, TDMA ensures the most efficient use of the satellite's onboard power supply.

2) *The Bandwidth Reuse through Spot Beam Technology*

A spot beam is a satellite signal that is specially concentrated in power normally sent by a high-gain antenna so that it will cover only a limited geographic area on Earth. Beam widths can be adjusted to cover areas as large as the entire United States or as small as a state like Maryland. In addition, satellite antennas have been designed to transmit several beams in different directions using the same reflector.

Spot beams are used so that only earth stations in a particular intended reception area can properly receive the satellite signal. Spot beams allow satellites to transmit different data signals using the same frequency. Because satellites have a limited number of frequencies to use, the ability to re-use a frequency for different geographical locations (without different data interfering with each other at the receiver) allows for more local channels to be carried, since the same frequency can be used in several regions.

3) Frequency Hopping Technology

A method for interconnecting many ground stations spread over great distances was demonstrated in 1993 with the launch of NASA's Advanced Communications Technology Satellite (ACTS). The satellite uses what is known as the hopping spot beam technique to combine the advantages of frequency reuse, spot beams, and TDMA. By concentrating the energy of the satellite's transmitted signal, ACTS can use ground stations that have smaller antennas and reduced power requirements.

4) Low Orbit Technology

The latest development in satellites is the use of networks of small satellites in low earth orbit (2000 Km or less) to provide global telephone communication. The Iridium system, for example, uses 66 satellites in low earth orbit, while other groups have or are developing similar systems. Special telephones that communicate with these satellites allow users to access the regular telephone network and place calls from anywhere on the globe. Anticipated customers of these systems include international business travelers and people living or working in remote areas.

C. The MENOS Hub

MENOS is a revolutionary networking concept used to exchange multimedia content over satellite [12]. It is intended primarily for professional broadcasters, allowing them to share video and audio material among several sites scattered across a large geographical area. It has been designed to provide these broadcasters not only with the fastest and most cost-effective technologies to perform the media exchange, but also with a complete range of tools to facilitate the related coordination tasks and improve people collaboration across the network.

In traditional satellite contribution systems, television and radio material is exchanged as real-time transmissions from one ground station to another. This requires the reservation of a satellite segment for fixed time duration, a manual line-up procedure, and expensive uplink equipment. At the receive site, the transferred material needs to be used on the fly or recorded. The coordination between the two stations, or between the stations and the network operating center, must typically be done via terrestrial or mobile telephony.

MENOS is fundamentally different with IP as the core-protocol, all exchanged material transmits through a central hub station, which also provides permanent two-way satellite IP connectivity among all remote stations. The multimedia content can be transmitted in real-time or be transferred as data files. It can also be retained in the central hub station for archiving and later access by other stations. The reservation of the bandwidth and the line-up procedure are automatic and the uplink stations are smaller and much less expensive than traditional systems. In general, the two-way IP connectivity is ideal for voice over IP (VoIP) coordination channels, e-mail exchange, Intranet and Internet access and other collaboration tools.

1) MENOS System Architecture

A MENOS system consists of a redundant central platform (hub) connected to a number of remote sites, each equipped with a satellite interactive terminal (SIT). The terminal is able to transmit or receive data to and from the hub. The data can be exchanged between the hub and the terminal or between two terminals via the hub. From the terminal to the hub, the data is transmitted either on a dedicated single channel per carrier (SCPC) carrier, or on a return channel shared dynamically in time and frequency with other terminals. Low rate data, such as Internet and Intranet exchanges, VoIP, radio exchange and low bit rate file transfers, are typically sent using the multiple frequency time division multiple access (MFTDMA) channels, while real time television transmission and fast file transfers are operated in SCPC.

All transmissions are first received by the central hub. If necessary, the transferred material can be automatically archived in the central hub for later use by the remote stations. Data is transferred from the hub to the stations on one of two MFTDMA multiplex carriers, i.e., multiple channel per carrier (MCPC). The first one regroups all the video transfers and the second one regroups the Internet data, the file transfers and the VoIP calls.

Different types of MENOS remote stations are available, depending on the type of applications performed at the remote site.

- Data SITs only provide data and VoIP connectivity and can be used for Internet, Intranet, VPNs, and interactive collaboration tools
- Radio SITs provide all the service of a Data SIT in addition to radio exchange services
- Television SITs provide all the service of a Data SIT in addition to television exchange services

MENOS terminals can also be integrated into mobile units, in the form of digital satellite news gathering (DSNG) trucks or flyaway kits.

III. The Proposed Interactive Arab Education Satellite (IAESat)

This section provides a description of the proposed interactive Arab education satellite (IAESat) for all Arab Homeland countries. The proposed solution can be accomplished as a new project or can be embedded with existing projects such as ArabSat and MENOS hub.

A. The IAESat model

The generic framework of the unified M-Learning model described in [15] is used in IAESat as this framework was developed a generic methodology for building a satellite TV based Interactive M-Learning (STV-IML) system. The framework consists of three major components; these are:

1. Centralized Broadcasting Center (CBC), which consists of several sub-components such as live and recorded broadcasting facility, earth station, a satellite channel connected to a well-known satellite, video servers, and video storage devices, web servers, short-message-session (SMS) servers, and so forth.
2. Client side (educational site), where several educational sites can be connected to the CBC and can share the resources for reception and broadcasting of educational material. The typical setting required by an educational site is minimal and consists of a satellite dish connected to a set top box with Digital Video Recorder (DVR). It may require Internet con-

nection. An educational site can operate its own media production and auditing units. A mobile broadcasting van can also be used for live and mobile broadcasting and live event coverage.

3. Communication and broadcasting channels: Four types of communication channels were proposed in the STV-IML framework for broadcasting, reception, and interaction, namely, interactive channel, download channel, satellite reception on client side, and satellite broadcasting.

In order to implement the unified M-Learning model, first it is required to establish a satellite broadcasting capabilities by constructing an earth station connected to a spot beam satellite. A broadcasting center and a production and auditing centers are to be established and may well be integrated with a centralized M-Learning center managed by educational sites. Viewers can interact with education programs using available interactivity components.

B. *Components of IAESat*

The proposed IAESat educational satellite consists of the following main components:

1. Communication satellite with spot beaming technology
2. Remote stations with VSAT terminals
3. Main hub station with archiving capabilities
4. Centralized broadcasting station
5. Centralized content development and class scheduling
6. Distributed and mobile broadcasting stations
7. Interactive and non-interactive class rooms

Figure (2) depicts the main components and configuration of the proposed system and in what follows a brief description is given for each component.

1) *Communication satellite with spot beam technology*

IAESat is based on using spot beam technologies because they enable IAESat to deliver more local channels to specific, precisely defined areas, which improves its ability to compete with cable broadcasters. The communication satellite uses beams that cover the entire Arab Homeland. To provide local broadcast coverage, channels intended for only one local area are scrambled so users elsewhere cannot view them. The new spot beam satellites are designed to project spot beams at selected areas, allowing the same radio frequencies to be reused in different areas, thus increasing the channel capacity. For example, by focusing over a very small area, like Amman, the frequency can be reused in several other places like Baghdad, Cario, Riyad, etc. Where these cities are used as examples, and will not actually share the same spot beam frequency. With highly powerful spot beam satellites, we can direct different spot beam projections to the Arab Homeland and deliver a lot more local programming to those specific areas. Each spot beam can have an approximate diameter of 500 Km, making them powerful satellites.

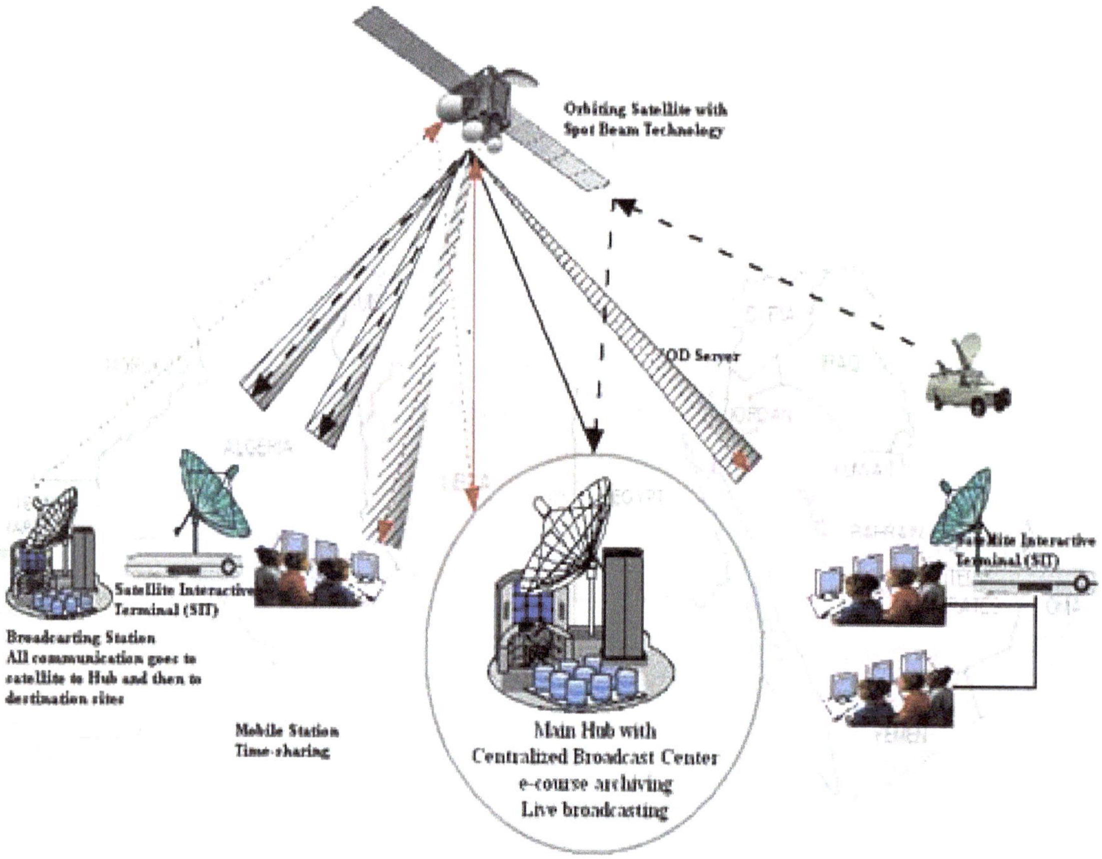

Figure 2. The main components of the IAESat.

2) Remote stations with VSAT and SIT terminals

A typical remote station constitutes an interactive receiving end (classroom) consisting of a variety of components such as a satellite dish, special satellite receiver, computers, router, multimedia equipment, E-Learning software. The two distinguish components that can be used as remote stations are: the Very Small Aperture Terminal (VSAT) and the Satellite Interactive Terminal (SIT) which are explained below:

- VSAT is a low cost business terminal with small antenna. It is a two way data terminal or one-way data link depending on the situation. The most common VSAT configuration is the Time Division Multiplexing (TDM)/TDMA star network. VSAT has a high bit rate outbound TDM carrier from the hub to the remote earth stations, and one or more low or medium bit rate TDMA inbound carriers. Remote user sites have several low bit rate Data Terminal Equipments (DTEs) operating at 1.2 to 9.6 Kbits/s. These are connected through the VSAT network to a centralized host processor.
- SIT is a satellite receiver with broadband Internet connection. The multiplexer of SIT exchanges video content with the multiplexer of the MENOS hub for:
  - Live TV contribution/distribution (reserved channel).
  - High speed streaming on reserved channel or best effort file transfer.
  - Store and forward: short time storage in the SIT or long term archiving in the network hub. These exchange sessions are synchronized and activated automatically by the hub's Multimedia Reservation Server (MRS). The TV SIT is connected to the MENOS hub via two satellite subsystems: a Multiple Frequency TDMA (MF-TDMA) broadband subsystem for data and voice communication, and a Reservation Access Multiple Access (RAMA) subsystem for video and fast file transfers.

3) Main hub station with archiving capabilities

Part of the VSAT network, is the hub, which is a centralized high performance earth station (with an antenna of up to 9 m in diameter). The proposed communication network is constructed using a STAR topology with the hub at the center of the star. The hub plays an important role in enhancing the performance and efficiency of the proposed M-Learning systems, as:

- The hub is considered as the heart of communication between the satellite and the remote stations.
- All stations are connected to the hub at all times.
- The hub provides remote stations with broadband internet services, and voice services.
- The hub works as a centralized operational center that provides collaboration and media exchange between connected remote sites. It also performs automatic configuration of transmission and multiplexing of transmitted signals for live broadcasting from one station to other stations.
- The hub can be used to broadcast programs to all the connected stations, or it can be used to direct specific programs to specific stations utilizing the spot beam technology in the satellite.
- All types of broadcast are archived at the hub facilities for later use.
- Any remote station that requires communication with a different station must go through the hub facilities first.

4) Centralized broadcasting station

Located at the center of the star network along with hub facilities, this station broadcasts the unified E-Learning material to the rest of the network. Broadcasts include live classes as well as pre-recorded E-Learning material. This center contains an archive of all produced material and exchanged media. Any member station can search through the central station contents, whether archived or live, and can join the sessions. The hub facility decodes the contents to make it available for the client station.

5) Centralized content development and scheduling

Since the objective of this project is to produce and distribute unified E-Learning materials to member educational institutions, it is very important to establish a centralized commission for the purpose of organizing content development using well-document quality procedures. The core responsibilities of the commission include:

- Solicitation of highly qualified lecturers in the region
- Subject coordination and syllabus creation
- Adjustment of syllabus based on feedback
- Maintain copyrights and quality of produced materials
- Production of actual live or pre-recorded lessons
- Arranging for online examinations
- Produce statistics and research studies, and organize conferences
- Develop and maintain live-lessons schedules

6) Distributed and mobile broadcasting stations

These stations are owned by individual countries, universities, or any interesting institutions. These stations can also help members of this education system with developing and producing their own specialized materials, and communicate with the main hub through, for example, the orbiting satellite, which then retransmit the signals to the specified destinations.

7) Interactivity and non-interactive classrooms

User interaction is an important component and factor that needs to be carefully considered in designing and building a M-Learning system, like IAESat. This is because it targets wide and different categories of users of different backgrounds. Fortunately, the enormous advancement in hardware and software technologies enable system developer to build a system that can meet all rage of users, services, and applications interactivity.

Interactivity components may include TV-based technologies; such as: traditional TV, interactive TV (ITV), and IP based TV (IPTV). To be truly interactive, the viewer must be able to alter the viewing experience or return information to the broadcaster. This "return path" or "back channel" can be established by telephone, mobile SMS (text messages), radio channels, digital subscriber line (ADSL), cable channels, etc [15]. Satellite viewers can return information to the broadcaster via their regular telephone or ADSL lines or other data communication technologies [15]. They will be charged for this service according to the national billing standard. Further discussion on user interaction technologies can be found in [15].

IV. Operation of IAESat

All operations of the IAESat go through the main hub which constitutes the heart of the star network. Signals and broadcast requests leaves broadcast stations directly to the orbiting satellite, which in turns send those signals to the hub to determine their destinations. The hub sends the signals back the satellite with proper destination information. The satellite using its spot beam capabilities, resend the signal to the proper destination.

The main features of the proposed project include:

- Use of advanced satellite communication technologies, namely, the spot beam technologies, which enables customization and personalization of satellite communication and reuse of satellite frequencies.
- Use of VSAT terminals is an easy and non-expensive solution to build remote classes.
- Use of star network with a hub station has several advantages such as overcoming the limitations with time sharing, transmission configuration.
- Establish a high commission on E-Learning insures proper management of production and distribution of contents using quality standards.

V. Conclusions

This paper presented a description of a unified M-Learning model through interactive education satellite to widen education in the Arab Homeland States, which is referred to as IAESat. The model utilizes the new spot beam satellite communication technology to ensure allocation and customization of local channels to specific and precisely defined areas. Using spot beam technology improves the satellite capability to compete with other existing wire and wireless communications technologies. The communication satellite can efficiently and cost-effectively cover the entire Arab Homeland and reaches a wide area and mobile users that cannot be reached otherwise. The model also implement advanced interactivity component to meet the standards of the learning process.

References

[1] Sarojni Choy. Benefits of E-Learning Benchmarks: Australian Case Studies. The Electronic Journal of E-Learning (EJEL), Vol. 5, Issue 1, pp. 11 – 20, 2007.

[2] Evelyn K. Kahiigi, Love Ekenberg, Henrik Hansson, F. F. Tusubira, and Mats Danielson. Exploring the E-Learning State of Art. Electronic Journal of E-Learning (EJEL), Vol. 6, Issue 2, pp. 149-160, 2008.

[3] P. Henry. E-Learning Technology: Content and Services. Journal of Education and Training, Vol. 43, No. 4, pp. 249-255, 2001.

[4] Marguerite Leanne Koole. "The Framework for the Rational Analysis of Mobile Education (FRAME) Model: An Evaluation of Mobile Devices for Distance Education". Master Thesis, Athabasca University, Alberta-Canada, 2006.

[5] D. S. Sethy. Learning at a Distance through Mobile Devices: A Global Choice of Learners in the Open and Distance Learning (ODL) System. Journal of Educational Multimedia and Hypermedia, Vol. 20, No. 1, pp. 83-95, 2011.

[6] Norbert Pachler (Editor). Mobile Learning: Towards a Research Agenda. WLE Centre, Institute of Education (IOE), London, 2007.

[7] Korneliya Yordanova. Mobile Learning and Integration of Advanced Technologies in Education. Proceedings of the International Conference on Computer Systems and Technologies (CompSysTech 07), pp. IV.23-1-6, 2007.

[8] http://en.wikipedia.org/wiki/Arab_League. Last visited on 16/03/2011.

[9] http://www.italsat.it. Last visited on 16/03/2011.

[10] http://india.gov.in/sectors/science/indian_national.php. Last visited on 16/03/2011.

[11] http://isro-news.blogspot.com/2010/08/edusat.html. Last visited on 16/03/2011.

[12] http://www.arabsat.com/pages/Default.aspx. Last visited on 16/03/2011.

[13] Gerard Maral andMichel Bousquet. Satellite Communications Systems: Systems, Techniques and Technology. John Wiley, 5th Edition 2009.

[14] Bruce R. Elbert. Introduction to Satellite Communication. Artech House, Inc., 3rd Edition 2008.

[15] Ghassan F. Issa, Shakir M. Hussain and Hussein Al-Bahadili, "A Framework for Building an Interactive Satellite TV Based M-Learning Environment", International Journal of Interactive Mobile Technologies (iJIM), Vol. 4, No. 3, pp. 19-24, 2010.

[16] M. S. Islam, J. B. Alam and S. M. L. Kabir. "Satellite Based Internet Education Delivery and E-Learning Objects Evaluation: SOI Asia and Codewitz Perspective. Proceedings of the International Conference on Teaching and Learning (iCTL 2007), INTI International University College, Malaysia, 15-16 November, 2007.

Authors

Ghassan F. Issa (gissa@uop.edu.jo). He received his B.E.T degree in Electronic Engineering from the University of Toledo, Ohio, in 1983, and B.S.EE in Computer Engineering from Tri-State University, Indiana in 1984. He received his M.S. and Ph.D. in Computer Science from Old Dominion University, Virginia, in 1987 and 1992 respectively. He was a faculty member and department chair of Computer Science at Pennsylvania College of technology (Penn State) from 1992 – 1995. He also served as faculty member and the dean of Computer Science at the Applied Science University in Amman, Jordan from 1995-2007. Currently he is an associate professor and the dean of Computer Science at Petra University in Amman, Jordan. His research interest covers block cipher, and authentication.

Shakir M. Hussain. He received his B.A. degree in statistics from University of Al-Mustansiriyah, Iraq, in 1976 and M.Sc. degrees in Computing and Information Science from Oklahoma State University, USA, in 1984. In 1997 he received the Ph.D. degree in Computer Science from University of Technology, Iraq. From 1997 to 2008 he was a faculty member at Applied Science University, Jordan. Currently, he is a faculty member of Information Technology at Petra University, Jordan. His research interest covers block cipher, key generation, authentication, and data compression. He is a member of ACM.

Hussein Al-Bahadili received his B.Sc degree in Engineering from University of Baghdad in 1986. He received his M.Sc and PhD degrees in Engineering from University of London in 1988 and 1991, respectively. His field of study was parallel computers. He is currently a faculty member of Information Technology at Petra University, Jordan. He is a visiting researcher at the Wireless Networks and Communications Centre (WNCC) at University of Brunel, UK. He is also a visiting researcher at the Centre of Osmosis Research and Applications (CORA), University of Surrey, UK. He has published many papers and book chapters in different fields of science and engineering in numerous leading scholarly and practitioner journals, and presented at leading world-level scholarly conferences. His research interests include parallel and distributed computing, wireless communications, computer networks, cryptography and network security, data compression, image processing, and artificial intelligence and expert systems.

4

Laptops vs. Desktops in a Google Groups Environment: A Study on Collaborative Learning

S.L. Abrantes[1], L.B. Gouveia[2]
[1] Polytechnic Institute of Viseu, Viseu, Portugal
[2] University Fernando Pessoa, Porto, Portugal

***Abstract*—Current literature on m-learning refers to the lack of studies on real use of m-learning applications and how they can compete with current desktop counterparts. The study consists of an experiment involving one hundred and twelve students of higher education and a set of learning activities that they have to accomplish. This study has the main objective to validate if the students that use laptops or desktops are in the flow experience and which of them are more in the flow experience, when using Google Groups. The used approach is based on the flow experience introduced by [1]. It was possible to conclude that students have experienced the flow state both by students using laptops or desktops, but having the laptop students a more positive effect in the flow experience.**

***Index Terms*—comparing mobile with desktop, flow experience, mobile devices**

I. Introduction

Computers are becoming increasingly indispensable in our current habits, and as a result, the market has increasingly invested in this reality. 10 years ago, a mobile device only served to send text messages and for calling. Today, because of the large advances in technology, these mobile devices let us do almost everything a computer does, besides telephone calls and sending text messages.

With the advancement of mobile technologies, the use of a computer is no longer restricted to a computer lab, a classroom or an office. We can transport them anywhere and anytime we want [2].

Technological applications and the way they are used has advanced in such a way that the manipulation of learning objects is no longer limited to a desktop, but extended to the use of mobile devices, to provide a greater range of application and obtain the benefits that mobile computing offers to the education sector. This results in the establishment of a new area of activity, related with the use of mobile technologies in learning, named m-learning.

II. Collaborative Environments

Collaborative learning can be seen as an act that results in a coordinated process of building and solving a particular problem [3].

Collaborative learning provides an environment that can animate and enrich the learning process. The participation of various people in a collaborative environment permits the creation of an educational system more realistic in a particular social context, thereby increasing the effectiveness of the system. This type of environment helps to sustain the interest of the student, providing a more natural habitat [4].

To learn in a collaborative environment, we need to follow these characteristics [5, 6]:

- Develop and share a common goal;
- Contribute to the understanding of the problem;
- Work, respond and understand the issues of other members;
- Responsibilities for all elements of the group;
- Dependency between group members so that everyone understands that the group's success depends on everyone.

Also [7] reports that through the collaborative environment, participants gain a deeper learning, a shared understanding, critical thinking and the retention of long-term learning as the main benefits for such learning activity.

The collaborative environments contain behaviors that improve learning. These environments contribute in a positive way for both situations where the participants are physically or through technology communicating with each others [8].

[9] defines collaborative learning as a situation where two or more people try to learn something in common and together. Each element of this definition can be interpreted in several ways:

- "Two or more persons" may be interpreted as a pair of people, a small group (3 to 5 persons), one class (20 to 30 people), community (a hundred or a thousand people) or a company (one hundred thousand people), and so on.
- "Learn something" can be interpreted as an accompaniment to a course, a determine lecture of a discipline, solving a problem, as many other ways.
- "Together" can be interpreted as different forms of interaction: face to face or through the new technologies of information and communication.

A group of people can never reach a perfect consensus of all of life, they need only to reach a reasonable consensus in order to continue the job they are doing [10].

The use of information activities has been considered crucial to the success of collaborative activities [11].

In nowadays, we see daily information activities, since we go to the Internet and we see appealing symbols about some new news, the publicity that we receive in our homes on promotion of a product, from receiving in our mobile phone SMS to inform us of new promotions, etc.. Due to a competitive society that we live in, it is crucial that there are such information activities, so that our society can survive and strive.

The same is true in education, that is, if there is a greater volume of information activities the greater is the students' attention. If students have information about what is happening in a particular subject the greater is the interest of students, as demonstrated by [11].

It is necessary that the group members are aware what each is doing, so that the collaboration between them can succeed [12].

In a collaborative environment, it is necessary to have social awareness of other members, this is, if they are reachable or not, if they are well prepared or not, if they can be disturbed. This social knowledge is essential because we can act according to their situation, for example, if an element is sick maybe we should save the discussion for another day [13].

Information services have been developed in collaborative environments, in order to monitor and notify members of the group if any work has been done during the group work [14, 15].

Since the notion of cooperation is inherent in collaborative learning, research can also be applied to collaborative learning environments. Both the cooperative and collaborative learning are built around the idea of socially constructed knowledge [16].

The two terms (cooperative learning and collaborative) are therefore often used synonymously, there is a considerable ambiguity [17].

Sometimes the collaborative and cooperative environments can be interpreted in the same way, but these two types of environments have different aspects.

[18] make a distinction between cooperative and collaborative learning. They indicate that cooperative learning is a protocol, which at the beginning the initial task is subdivided into subtasks, so that the various participants are able to develop them independently. Collaborative learning describes situations where two or more subjects are built synchronously and interactively in order to reach a common solution to a problem [18].

Cooperative learning generally leaves the authority structures unchanged. The end is defined in the beginning by an instructor, who also describes the means by which the objective will be achieved and evaluate the whole process [16]. As reported by [17] who defended that cooperative learning is based on the use of small groups, so that students can work together to maximize the learning of them self's and to others.

Collaborative learning is relatively cooperative, but it takes all participants a step forward: involving participants in a self-reflective process that often generates a series of questions, "meaning" and "power" and that forces them to confront issues that are implicit in any process of learning in the classroom, but are rarely explicitly defined and treated [16].

III. Technology Environments

Imagine that we are living in a time where cars could not go faster than 25 km / h; where letters were only written on paper; and where computers were only used for writing text. How could one live in this global economy and in this century? If you can not work with obsolete tools and services when it comes to survival, how can we continue to support an educational system that ignores the new researches on learning and continues to "educate" using outdated tools? [19].

For a long time, people thought that teaching consisted on the transmission of knowledge and on the training of the memory, and instil in students the values of society. People felt that learning was to acquire knowledge through a process of attention, memorization and reproduction of it, which is an individual and homogeneous task, that can be standardized [20].

"*The integration of the computer in education is now a reality impossible to ignore and that we must understand.*" [20]. That also brings new ways of working with data, information and knowledge and to relate with others. We can say that computers and networks have turn learning in a more open and freely activity.

The use of computers in teaching fits the constructive approach of learning (although other approaches are possible). Although the computer is not a technology designed for the education system, its characteristics of interaction and its capacity to deal with information, makes it a very useful and promising tool [20].

Teaching using computers is an added value to our society; it is certainly very appealing to students and more motivating, stimulating them to interact with different situations and depictions of real life, forcing the student to think creatively and independently about new subjects and materials.

Information technologies are rapidly changing the way we live. Computers, calculators and other technologies for processing information, help our brains to create knowledge from data and available information. Until recently they have being used in the majority of cases, for repetitive tasks (writing and printing), performing these in a more rapid and with minimum possible of errors. Since the evolution of computers, they are already capable of being used for tasks far more complex than in the past. These tasks have a direct implication on how to create a better learning environment [21]:

Access to unlimited information: computers allow users to access a vast amount of information;

Interactive teaching: computers may be chosen for interactive learning environments allowing students to learn at their own pace;

Multimedia: since the arrival of the CD-ROM, these have enabled us to integrate teaching with voice, video, text, graphics and music getting sights and sound from real world;

Simulation: The computers allow students to simulate different kind of experiences. This allows the exploitation of various kinds of experiences without students being limited to physical environments;

Virtual Reality: Allows users to create experiences in an environment in three dimensions and try even new approaches and perspectives for interaction;

Distance learning: The technology allows students to learn at any point in time, at any place, without having to be in a particular geographical place (at least at the same time);

New connections: Computer networks allow students to connect with each other in order to share a common knowledge among them.

When using technology for education we should use this in a fair and moderate way. We must not use it too much, because it may lead to cases where the users are so addicted to the technology that they cannot release it or its use superposes to the learning. Also if users are far removed from it, they never benefit from its advantages [22] (as the technology becoming a barrier).

We can list the main advantages that contribute to technology as an asset to a school environment as [23]:

The technology is attractive: You cannot think of using a particular technology if it is not attractive. The attractiveness of technology is achieved by the mode of operation, and its appearance;

The technology is available: A particular technology has no value to society if it is not used. The technology should not be locked behind a door where nobody can access it;

Technology is addictive: This is an ambiguous feature of technology. This feature reflects the learning effort that users do to use the technology and also place some restrictions for future change.

Educational technologies can be considered simply as a set of information technologies. However, important is what we do with these technologies, it is the way we are encouraged to use this set of technologies that becomes the ultimate challenge for the learning outcome (Buchan, 2008).

Technologies for education are a key part of a learning environment [22].

IV. Mobile Devices

Quin cited by [24] states that m-learning is the interaction of mobile computing (small applications, portable, and wireless communication devices) with e-learning (learning facilitated and supported through information and communication technologies).

M-learning is not a substitute of e-learning, but a subset of the e-learning environment. However, m-learning can improve some of the e-learning advantages. This new form of learning is a method with potential, as it enables students to learn anywhere, away from the traditional classrooms [25].

There is a widespread use of mobile devices in an m-learning environment, in our modern world: mobile phones, PDA's, MP3 players, portable gaming devices, Tablet PCs and laptops, which predominate in our everyday lives. From children to older people, they are increasingly linked with each other, communicating through communication technologies, something that didn't happen just few years ago.

There are a number of mobile devices that can be considered for an m-learning environment [24]:

A. IPod

The media player from Apple, allows users to download music, books, audio, podcasts, photos and video from the Internet. It also includes an address book and a calendar that syncs with Microsoft Outlook or Outlook Express. It can also serve as a storage device.

With the iPod, students can download podcasts of relevant educational materials, along with audio and video lectures. Although most models have a small screen, future versions will probably have bigger screens, so that users can read e-books on them. The iPod video (iPod Touch), for example, takes a step in this direction. And recently Apple has launched the iPad that has a bigger screen, offering the user a much more convenient way for reading (and interact with them!) electronic materials.

With the iPod, students can exchange files, review materials for a particular discipline, prepare them self's for exams, show their work to others and share the results of a project, with their colleagues.

Pros: With 87% of the market share, the iPod has proven its popularity among students. Apple's iTunes U (http://www.apple.com/education/itunes-u/), allows teachers to upload their lessons for students to download these materials, so they can study from them.

Cons: First, consider the cost. An iPod cannot be accessible to all students, and also because this device requires an application owned by Apple, the iTunes. We should also consider the screen, these are generally too small to be used by sophisticated applications or even to read large amounts of text (although this will probably be changed in the future versions, we can see this change already in the iPod touch and iPad) and also because these devices do not record sound (their major critique).

B. MP3 Players

This digital music player reads music and audio files. Some of these models have an integrated voice recorder.

Students can use these MP3 players to download and listen to podcasts and audio lessons. Students can also review the materials for a particular course, study for exams, stay informed about course contents, listen to audio books, and with some devices, record lectures.

Pros. MP3 players are compact and light. They have an excellent audio quality and they are upgradeable and expandable.

Cons. An MP3 player may be replaced with other devices that also play audio files.

C. PDA

The PDA combines the computing power and Internet access in one single system, with a calendar, notepad, address book, and also productivity tools. It is a device integrated with Bluetooth, Wi-Fi and a mini USB interface.

A PDA plays audio, video, Flash animations, allows editing of text documents, allows users to access their e-mail and also web contents; supports instant messaging and text messages, and can be used to store information.

These PDA's provide support for collaborative learning environments. Students can use them to present projects, write documents in Word, and take notes in a classroom.

Pros: The PDA's have a big screen (for a portable device) that makes reading easier. They also combine the various types of computing and communications tools in one single device. Data entry is possible through the on-screen keyboard, a pen, or through external peripherals.

Cons: The PDA's are big when compared to other mobile devices. They are not efficient for the introduction of long e-mails or text, without the use of an extra input peripherals device.

D. USB drive

The USB drive is a storage device that connects easily to multiple computers and other types of devices.

The USB drive is ideal for storing work files, audio and video. Students can share files for collaborative work. They can also copy files from this drive to school computers and vice versa, and send their work to the teachers.

Pros: The drive is small and portable and the USB interface is compatible with all newer computers. It works well for transporting files from home to school and vice versa. There are applications with the autonomy to run in a USB Drive.

Cons: A USB drive is a device with just one purpose only. Other devices can also serve for storing information.

E. E-Book Readers

E-book readers are used to download text-based materials. They can store hundreds of e-books, newspapers and magazines. The zoom and the search function are one of the fundamental characteristics of these types of devices. There are recently strong activity and visible technology enhancements in this kind of devices, which turn them a more viable alternative to paper.

Pros: The e-book reader has a large screen for reading, and also has a light to facilitate the reading in dark places. The digital marker allows users to mark their texts, and the search function enables users to easily find a particular text. An e-book reader can also store the entire contents of books from various courses.

Cons: An e-book reader is a device with only one purpose, with limited computing capabilities. These may require proprietary file formats and there are a limited number of e-books available today, although the market is rapidly evolving.

F. Smart Phone

A smart phone combines the capabilities of a PDA, USB drive, MP3 player in one single compact system.

Students can download audio, video lectures and podcasts to their Smart Phones. They can play audio, video, flash animations, view and edit text documents, access e-mail and Web contents, send instant messages, send SMS and use the phone to storage files.

Pros: Smart phones can also be used in collaboration environments. Users can also access global information. These devices can support collaborative learning.

Cons: The small screen makes Web browsing and reading difficult. The small keyboards or the virtual keyboards make writing text inefficient for long emails and texts. Finally, some smart phones cost as much as a normal PC with only a fraction of their capacity.

G. Ultra-Mobile PC (UMPC)

The UMPC have the entire main features of a tablet PC, but on a much smaller size device. They offer support for audio, video, games, Internet and other types of communications and networking applications. They have Bluetooth, Wi-Fi and also Ethernet controllers.

Students can download audio, video lectures and podcasts for their UMPC, surf the Web, send emails, send instant messages, send text messages and also log into sites of distance learning courses.

The UMPC allows users to participate in collaborative learning environments.

Pros: These ultra-small, ultra-portable PC's have a 7'' touch screen, which is great for Web browsing and viewing multimedia contents. The small size makes these devices great for travelling.

Cons: These units are expensive, costing more than a high-powered PC. Due to its small size, most UMPC do not have a full-size keyboard.

H. Laptop / Tablet PC

The most complete system of all the mobile devices. Laptop/Tablet PC came with Bluetooth, Wi-Fi and Ethernet. These devices offer additional features such as handwriting and voice recognition.

Students can download audio, video lessons, podcasts, browse the Web, send emails, send instant messages, send text messages and log into the course website at home or while they are on the road. These devices are great for collaborative learning.

Pros: The Laptop/Tablet PC are very good for students who need to take their work with them. They provide greater power and capacity of all other mobile devices.

Cons: The Laptop/Tablet PC are still relatively expensive, and its size makes it more difficult to transport when compared with other mobile devices.

V. Online Discussion Forums

In the growing context of the use of digital media to support the business of teaching and learning in higher education institutions, there has also been an increased use of online discussions. This enhance was due to the increasing use of the information technologies and communication, in the context of the courses in higher education, which are related with their school activities, applications of discussions and interactions with issues associated with objects of knowledge.

This new reality changes the processes of teaching and learning and allows students to interact with teachers and other colleagues on various issues, more openly, more often and more easily. One of the advantages of online discussions, is that it leaves recorded everything that was written to then be analyzed and discussed whenever possible [26], providing a memory of the work done, giving the opportunity, among others, to evaluate the interactions or even to analyze the results of the work done.

These online discussion forums are important factors for the virtual communities and can be considered an excellent collaborative tool, so that students can make the best environments possible for learning.

The virtual learning communities, where students can interact with the content, with technology, and more importantly, with each others, provides a powerful approach in distance learning environment [27].

For a collaborative learning to be successful, it is crucial that students feel part of a learning community, where their contributions can add knowledge to the community in which he is involved and where the spirit of community is promoted through social interactions [28].

The virtual learning communities have the potential to solve problems in a distance learning environment [29].

There are many online discussion forums, available on the Internet, free, that allow students to work collaboratively, to discuss various topics with each other. Google Groups (Figure 1) and Yahoo Groups (Figure 2) are just two examples of online discussion forums that can be found.

Google Groups (http://groups.google.pt/) is a service created by Google, which allows participants to discuss about a particular subject of interest among the participants of this group. Participants can either discuss using the electronic mail (e-mail) or by using a Web page provided by Google, requiring, for this last case, an e-mail account at gmail. Google Groups besides allowing discussions of various users online, this service also allows the creation of Web pages for the group, where the users can adjust the visual aspect of each page, or even insert images and change the background colors. It also allows file sharing among group members and also access to personal information of each group member.

Yahoo Groups (http://groups.yahoo.com) is a service similar to Google Groups. It is a discussion group developed by the company Yahoo. Yahoo groups as like Google groups offer a set of groups of interest, in which users can register themselves in. The Yahoo Groups allows participants to share photos with each other and share a calendar of events.

VI. The Flow Experience

An aspect related with the interaction of the users with collaborative environments has to see with the flow experience introduced by Csikszentmihalyi (1975). The experience of the flow means the sensation that people feel when they are completely involved in what they are doing, that is, people like the experience and want repeat it [30]. This means that for students to be involved with collaborative environments, it is necessary that they presence the flow state.

The theory of the flow allows us to measure the interaction of users with the computer systems, verifying if these are more or less playfulness [31].

The flow experience is used in this article to characterize the interaction between the human and the new technologies [31].

When one is in the presence of the flow experience, this will bring to the users, a sense of pleasure of what he is doing. This satisfaction will encourage the user to repeat the task again [32].

Csikszentmihalyi says that a person who is in the presence of the flow state has the following characteristics [1, 33]:

Clear goals and immediate feedback;

Equilibrium between the level of challenge and personal skill;

- Merging of action and awareness;
- Focused concentration;
- Sense of potential control;
- Loss of self-consciousness;
- Time distortion;
- Autotelic or self-rewarding experience.

For a person to be in the presence of the flow experience it is necessary a balance between the level of challenge and personal skill [30] (Figure 3).

The sensation of an excellent experience in the accomplishment of any daily task is our reason of living. If we do not feel this excellent experience with our everyday tasks, we will question our self, if it is worth living [30].

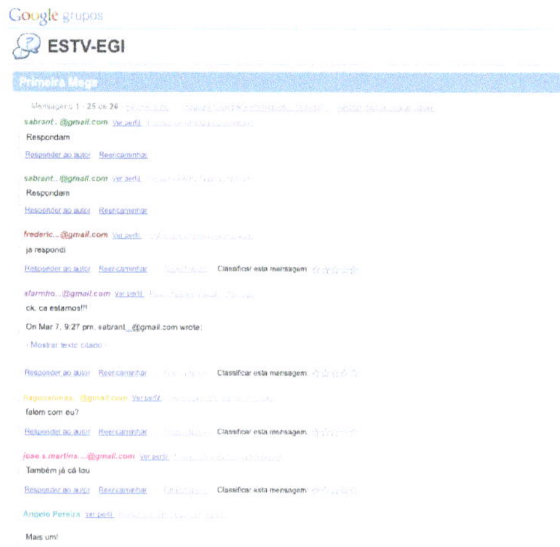

Figure 1. Google Groups

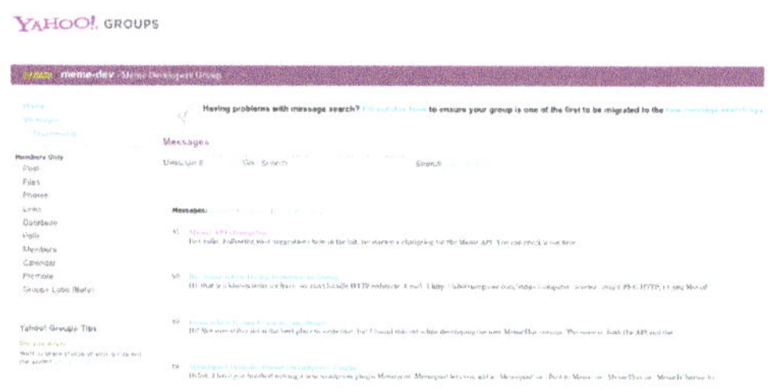

Figure 2. Yahoo Groups

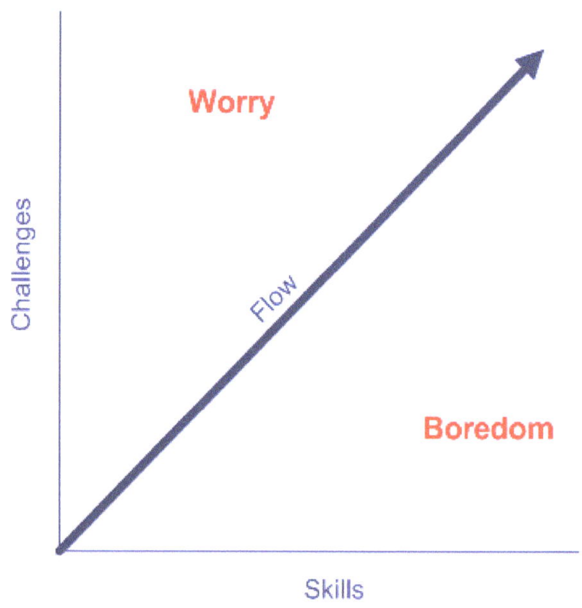

Figure 3. Flow Experience [30].

Previous researches have used the flow experience to measure playfulness, involvement, satisfaction and other states with the involvement in computational environments [31, 34-37].

Trevino and Webster (1992) define four dimensions for the flow experience:

- Control;
- Attention Focus;
- Curiosity;
- Intrinsic Interest.

There is one more dimension, sense of time, that is also important to measure the flow state [38].

A. Control

Individuals should experience, feelings in control, within computer interactions [1].

B. Attention Focus

Attention focus is another important element of flow. When individuals are in the flow state, their minds are narrowed to what they are doing, filtering out irrelevant thoughts and perceptions [32].

C. Curiosity

Curiosity is aroused when in the flow state. The curiosity sensation can be aroused through varied, new and admirable stimulations. For example, the new technologies will be able to cause this sensation of curiosity through colors and sounds [32].

D. Intrinsic Interest

When people feel they are in the flow state, these are involved for the amusement and pleasure [32].

E. Sense of time

When people feel they are in the flow state, there is a perceptual transformation of time, characterized by the sensation of time slowing down or speeding up [38].

People who interact with computers, with an entertainment spirit, transmit a much more positive experience, of those, who are in the computer for obligation [32].

VII. The Study

To evaluate the flow experience and to verify its occurrence in collaborative tools, an experience was carried through involving students from a university school. The main tool used was Google Groups, for this experience. This section presents the efforts carried through experience, the data obtained, as well as the statistical procedures applied.

Previously to this study, a test with five students was done, to analyze the effectiveness of the survey. From this previous study, we concluded that some questions were ambiguous for the population in the study.

After the accomplishment of the project given by the teacher, in witch they used Google Groups, the students answered the questions of the survey.

The survey was passed through the Internet with the help of "LimeSurvey" Web-based tool. The data collection was performed in the first week of November of 2009.

The Instruments used were Google Groups, Google Docs and Facebook and a survey consisting on some questions, in order to verify, in the end of the study, if the students were in the presence of the flow state. This survey will use the four dimensions: control, attention focus, curiosity and the intrinsic interest [32], as well as the dimension sense of time [38]. Beside these questions, this survey also contains other generic questions. All the related questions from this survey were built on a Likert scale of five points, since one (I totally disagree) up to five (I totally agree). Two questions for each dimension were elaborated.

A. Sample

This study intends to determine if the students inquired are in the flow state. The data has been collected through one hundred and twelve surveys of students. The surveys have been submitted to a rigorous test, having not excluded any individual; therefore, the sample consisted on one hundred and twelve valid surveys. The criteria of exclusion of inquiries were: students who had not discriminated their sex or age in the survey; students with incoherent answers throughout the survey (e.g answers that always presented values in the extremities of the scales, or incompatible); students who left 80% of the survey in blank. Once, one hundred and twelve valid inquiries were obtained, the sample is considered sufficiently satisfactory.

The statistical treatment of the data and the respective procedure [39, 40], that will be announced next, was carried through the software "S.P.S.S. - Statistical Package will be Social Science" (version 12.0 for Windows, http://www.spss.com/):

- Descriptive Statistics of the variables in the study;
- Evaluation of the index of internal consistency (Cronbach's alpha) for the dimensions of the flow experience;
- Correlation between the variables of the flow;
- Factor analysis in order to reduce the number of variables.

B. Analysis

This study was composed of 78.57% males and 84,82% had ages between sixteen and twenty four years. Most of the students have already used discussion forums in a fairly way.

The majority of the respondents used the laptop (72.32%) to access the tools for the project development, followed by the Desktop (27,68%).

We verified that Cronbach's alpha is always superior to 0.7, being able to conclude that the data is related to one same dimension, that is, the questions of the survey for the use of Google Groups, allowed us to determine if the individual finds himself in the presence of the flow experience, for students using a laptop or a Desktop.

To determine how the variables are correlated with each of the different devices used (laptop and Desktop), a correlation matrix was created for both types of the devices, where the correlation coefficient, R, is presented, that is a measure of the linear association between two variables. We can conclude from the correlation analysis that the correlation between the variables, for laptops, has a greater number of variables positively correlated than the desktop.

After the studies mentioned previously, we used the factor analysis in order to reduce the number of variables, both for laptops and desktops.

The extraction of the factors is given by considering the percentage of variance explained by the factors (Table 1).

To set the number of components to be retained, we choose, by default, those that have eigenvalues greater than one. If the total variance explained by the factors retained is less than 60%, then, at least, one more factor should always be selected. Thus, for this case study, two factors were retained in each type of device. For the mo-

TABLE I.
NUMBER OF FACTORS TO BE RETAINED (MOBILE DEVICE AND DESKTOPS)

Component	Mobile Devices Initial Eigenvalues		
	Total	*% of Variance*	*Cumulative %*
1	2,371	47,422	47,422
2	,881	17,625	65,047
3	,707	14,136	79,184
4	,631	12,613	91,797
5	,410	8,203	100,000

Component	Desktop Initial Eigenvalues		
	Total	*% of Variance*	*Cumulative %*
1	2,374	47,475	47,475
2	1,053	21,053	68,528
3	,704	14,077	82,604
4	,565	11,301	93,905
5	,305	6,095	100,000

bile device, it appears that the first factor explains 47.422% of the total variation and the second 17.625%, both explaining 65.047% of the total variation that exists in the five original variables. For the Desktop, the first factor explains 47.475% and the second 21.053%, explaining both, 68.528% of the total variation.

The matrix of components after rotation (Varimax method) aims to exaggerate the value of the coefficients that relates each variable to the factors retained, so that each variable can be associated with only one factor. The higher the value of the coefficient that relates one variable to a component, the greater is the relationship between them. We present below the matrix of components after rotation (Table 2) and the bold factor associated with each variable.

TABLE II.
THE MATRIX OF COMPONENTS AFTER ROTATION

	Mobile Device Component		Desktop Component	
	1	*2*	*1*	*2*
Concentration	,411	**,614**	**,751**	,001
Control	**,653**	,317	,011	**,955**
Curiosity	**,874**	,057	**,714**	,461
Intrinsic Interest	**,705**	,383	**,841**	,155
Sense of time	,033	**,877**	**,694**	,121

Having concluded the following for the case of the laptops:

- Factor group 1: (Intrinsic Interest, Control and Curiosity)
- Factor group 2: (Attention Focus and Sense of time)

And for the case of the desktops:

- Factor group 1: (Attention Focus, Sense of time, Intrinsic Interest and Curiosity)
- Factor group 2: (Control)

VIII. CONCLUSIONS

In order to evaluate the use of mobile devices and desktops and the potential of mobile devices in collaborative environments versus desktops, it was performed an experiment involving students of higher education. This study has the main objective to validate if the students that use laptops or desktops are in the flow experience and witch of them are more in the flow experience.

Most people all around the world use mobile devices. Due to the advance of the new technologies, and its size, users can carry them anywhere; can connect with a wide range of information to anywhere whenever they go.

Despite the widespread use of mobile devices today, there is a lack of reference to identify the advantages and disadvantages of the m-learning in collaborative environments, this is, we can not see the m-learning as an extension of e-learning but a rupture in the process of teaching and learning.

The analysis of data allows us to conclude that the majority of the students were males, had ages between sixteen and twenty-four years and that most of the students have already used discussion forums.

When going further to the analysis of the data, we verified that the variables described all the same characteristic (threw the determination of the Cronbhach's alpha), that is, the variables describe the flow experience.

We can conclude from the correlation analysis that the correlation between the variables, for laptops, has a greater number of variables positively correlated than the desktop.

From the factor analysis it was possible to isolate two factors that explain the majority of the total variation. Such factors had been Factor group 1: (Intrinsic Interest, Control and Curiosity), Factor group 2: (Attention Focus and Sense of time) for the laptops and Factor group 1: (Attention Focus, Sense of time, Intrinsic Interest and Curiosity) Factor group 2: (Control) for the desktops.

In order to determine the presence of the flow experience for each type of device, it was verified that, on average, the students were above value three (Likert scale of five points), that is, the majority of the students, in each of the different devices used (laptop and desktop), are in the presence of the flow experience, for the five variables mentioned for this study (attention focus, curiosity, control, intrinsic interest and sense of time). We can also see, that the average of the five variables associated with the flow experience, for students who used the laptops, were greater than those using the desktop to access the tools of the project development.

From this study we can conclude that the flow experience exists for people that use Google Groups, both for people that used the laptop or even the desktop, but having a more positively effect for users of the laptop. With this we can say that mobile users interact with Google Groups, with a more entertainment spirit and sense of involvement and satisfaction then the users that have used the desktop to access Google Groups. Considering that people use mobile device for m-learning and desktops for e-learning, we can conclude that people that use m-learning have a more positive effect on learning, when using Google Groups, than the people that use e-learning.

With these statements we can say that Google Groups is a good way for students to learn when using laptops and desktops, but having a more positive effect for the laptop users. We can also say that, m-learning when associated with the usage of Google Groups, it is a good tool for students to learn.

REFERENCES

[1] M. Csikszentmihalyi, *Beyond Boredom and anxiety*. San Francisco, CA, 1975.

[2] D. Arman, I. Kori, L. Felix, S. Keith, and B. Kellogg, "GeneyTM: designing a collaborative activity for the palmTM handheld computer," in *CHI 2001*, Seattle, 2001, pp. 388-395.

[3] J. Roschelle and S. D. Teasley, "The construction of shared knowledge in collaborative problem solving," in *Computer supported collaborative learning* Berlin, Germany: Springer: Springer-Verlag, 1995, pp. 67-97.

[4] V. S. Kumar, "Computer-supported collaborative learning: issues for research," 1996.

[5] G. Salomon, "What does the design of effective CSCL require and how do we study its effects?," *SIGCUE Outlook*, vol. 21, pp. 62-68, 1992.

[6] J. Landsberger, "Cooperative & Collaborative Learning," 2008.

[7] D. R. Garrison, T. Anderson, and W. Archer, "Critical thinking and computer conferencing: a model and tool to access cognitive presence," *American Journal of Distance Education*, vol. 15, pp. 7-23, 2001.

[8] D. D. Curtis and M. J. Lawson, "Exploring collaborative online learning," *Journal for Asynchronous Learning Networks*, vol. 5, pp. 21-34, 2001.

[9] P. Dillenbourg, Traum, C. Hoadley, and J. Rochelle, "The long road from a shared screen to a shared understanding," Stanford, 1999, pp. 12-15.

[10] H. H. Clark and S. Brennan, "Grounding in communication," *Perspectives on Socially Shared Cognition*, pp. 127-149, 1991.

[11] C. C. Liu, S. Y. Tao, and J. N. Nee, "Bridging the gap between students and computers: supporting activity awareness for network collaborative learning with GSM network," *Behaviour & Information Technology*, vol. 27, pp. 127 - 137, 2008.

[12] C. Gutwin, S. Greenberg, and M. Roseman, "Supporting awareness of others in groupware," in *Conference companion on Human factors in computing systems: common ground* Vancouver, British Columbia, Canada: ACM, 1996.

[13] K. Tollmar, O. Sandor, and A. Schmer, "Supporting social awareness @ work design and experience," in *Proceedings of the 1996 ACM conference on Computer supported cooperative work* Boston, Massachusetts, United States: ACM, 1996.

[14] C. Y. Jang, C. Steinfield, and C. Pfaff, "Supporting awareness among virtual teams in a web-based collaborative system: the teamSCOPE system," *SIGGROUP Bull.*, vol. 21, pp. 28-34, 2000.

[15] W. Prinz and T. Gross, "Ubiquitous Awareness of Cooperative Activities in a Theatre of Work," in *Fachtagung Arbeitsplatzcomputer: Pervasive Ubiquitous Computing*, A. Bode and W. Karl, Eds.: APC, 2001, pp. 135-144.

[16] L. M. B. Gouveia, "A Visualisation Design for Sharing Knowledge.," in *Computing Department* vol. Phd Lancaster: Lancaster University 2001.

[17] D. W. Johnson and R. T. Johnson, "Cooperation and the use of technology," in *Handbook of research for educational communications and technology, Simon & Schuster Macmilla* New York, 1996, pp. 1017-1044.

[18] P. Dillenbourg and D. Schneider, "Collaborative learning and the internet," 1995.

[19] G. Caine and R. Caine, "Natural Learning: The Basis for Raising and Sustaining High Standards of Real World Performance.," *Position Paper: Natural Learning Research Insitute.*, 2007.

[20] M. C. Duarte and J. L. C. Silva, "O Computador no Ensino/Aprendizagem das ciências : um nova forma de utilização," in *Revista Portuguesa de educação*. vol. 2, 1995, pp. 69-78.

[21] J. Cleveland, ""The Changing Nature of Learning." Background Information for the Community Learning Enterprise Design Workshop," in *On Purpose Associates*, 1996.

[22] J. Buchan, "Tools for survival in a changing educational technology environment," *ASCILITE 2008*, pp. 100-109, 2008.

[23] P. Boytchev, "Technology enhanced natural learning," in *Technology Enhanced Learning Workshop* Sofia, Bulgaria, 2005.

[24] J. R. Corbeil and M. E. M. Valdes-Corbei, "Are You Ready for Mobile Learning?." vol. 2: Educase, 2007, pp. 51-58.

[25] P. Mellow, "The Media Generation: Maximize learning by getting mobile," in *Proceedings for ASCILITE 2005 Conference: Balance, Fidelity, Mobility: maintaining the momentum?*, 2005.

[26] K. A. Meyer, "Evaluating online discussions: Four different frames of analysis," *Journal of Asynchronous Learning Networks*, vol. 8, pp. 101-114, 2004.

[27] H. McLellan, "The Internet as a virtual learning community," *Journal of Computing in Higher Education*, vol. 9, pp. 92-112, 1998.

[28] S. Benford, J. Bowers, L. E. Fahlén, C. Greenhalgh, and D. Snowdon, "User embodiment in collaborative virtual environments," in *Proceedings of the SIGCHI conference on Human factors in computing systems* Denver, Colorado, United States: ACM Press/Addison-Wesley Publishing Co., 1995.

[29] N. Augar, R. Raitman, and W. Zhou, "From e-Learning to Virtual Learning Community: Bridging the Gap," in *Advances in Web-Based Learning – ICWL 2004*, 2004, pp. 301-308.

[30] M. Csikszentmihalyi, "Towards a Psychology of Optimal Experience," in *Annual Review of Personality and Social Psychology*, 1982.

[31] L. K. Trevino and J. Webster, "Flow in computer-mediated communication," *Communication Research*, vol. 19, pp. 539-573, 2005/10/10/ 1992.

[32] J. Webster, L. K. Trevino, and L. Ryan, "The dimensionality and correlates of flow in human-computer interaction," *computer game research*, vol. 9, pp. 411-426, 1993.

[33] M. Csikszentmihalyi, *The psychology of optimal experience*: Harper Collins, 1990.

[34] H. Chen, R. T. Wigand, and M. Nilan, "Exploring Web users' optimal flow experiences," *Information Technology & People*, vol. 12, 2000.

[35] J. Ghani and S. Deshpande, "Task Characteristics and the Experience of Optimal Flow in Human-Computer Interaction," *The Journal of Psychology*, vol. 128, pp. 381-391, 1994.

[36] T. P. Novak and D. L. Hoffman, "Measuring the Flow Experience Among Web Users," Vanderbilt University, 1997.

[37] T. P. Novak, D. L. Hoffman, and Y. Yung, "Measuring the Customer Experience in Online Environments: A Structural Modeling Approach," *Marketing Science*, vol. 19, pp. 22-42, 2000.

[38] K. McKenna and S. Lee, "A Love Affair with MUDs: Flow and Social Interaction in Multi-UserDungeons," 2005.

[39] P. A. Pereira, *Complementos de Estatística*, 2002.

[40] M. Pestana and J. Gagueiro, *Análise de dados para Ciências Sociais - A complementaridade do SPSS*, 2005.

AUTHORS

S.L. Abrantes is with the Institute Polytechnic of Viseu, Viseu, Portugal (e-mail: steven@di.estv.ipv.pt).

L.B. Gouveia is with Faculty of Science and Technology, University Fernando Pessoa, Porto, Portugal (e-mail: lmbg@ufp.edu.pt).

5

Persuasive Mobile Device Sound Sensing in a Classroom Setting

Arttu Perttula[1], Jari Multisilta[2] and Pauliina Tuomi[1]
[1] Tampere University of Technology, Pori, Finland
[2] University of Helsinki, Helsinki, Finland

Abstract—**This paper presents an idea on how to utilize mobile phones to support learning in the classroom. The paper also tries to initiate discussion on whether we can create new kinds of learning applications using mobile devices and whether this could be the way we should proceed in developing 21st century learning applications. In this study, a mobile phone is programmed to function as a collective sound sensor. To achieve an appropriate learning atmosphere, the designed system attempts to maintain the noise level at a comfortable tolerance level in the classroom. The main aim of the mobile application is to change student behaviour through persuasive visualizations. The prototype application was piloted during spring 2012 with a total of 72 students and two teachers. The results, based on observations and interviews, are promising and several subjects for future work arose during the pilot study.**

Index Terms—**classroom, mobile, learning, persuasion**

I. Introduction

There has been a lot of discussion and research on mobile learning or m-learning, and learning theories for eLearning, web-based learning, and mobile learning [1, 2, 3, 4, 5, 6, 7]. The first research pilots on mobile learning were based on transferring electronic content to mobile phones. Soon, it was realized that mobile learning should not just copy the "old" practices of eLearning but the strengths of the mobile device should be harnessed in some other way for learning purposes. With the technology developing at a rapid pace, we soon had mobile phones with digital cameras. The mobile phone appeared as a tool for creating digital recordings of learning activities, for example on field trips [8].

The next innovations in mobile devices are sensors [9]. In general, a sensor is a small device that observes its environment and reports back to a remote base station [10]. Many mobile phones already have a GPS[1] feature that can be used to add location information for mobile applications. Some devices have motion sensors that monitor the acceleration and movements of the device in three dimensional space. It is also possible to use a heart rate monitor with a mobile phone for monitoring physical exercise. In addition, mobile phones can observe the ambient light or sound level from the surroundings of the device.

The data collected from the surroundings of the device can be used to determine something about the context where the device and the user are. The context can then be used in personalizing the mobile services we use with our mobile devices. There are already applications that can suggest a good restaurant or store nearby, based on our location. Could context and sensor data be used in creating new kinds of mobile learning applications? Sharples [11] argues that "the main barriers to developing these new modes of mobile learning are not technical but social". According to Sharples [11], we need more understanding of context and the role of new mobile technologies in the learning process.

This paper presents a novel idea of how to utilize mobile phones to support learning in the classroom. The paper also tries to initiate discussion on whether we can create new kinds of learning applications using mobile devices and whether this could be the way we should proceed in developing 21st century learning applications. The 21st century skills and learning applications can be defined in terms of knowledge, skills, attitudes, values and ethics, and they describe the skills that are needed in understanding information in the information society [12]. In this study, a mobile phone is programmed to function as a collective sound sensor. To achieve an appropriate learning atmosphere, the system presents persuasive visualizations to a target group of students. Thus it attempts to maintain noise levels at pleasurable tolerance level in the classroom.

According to Bistrup et al. [13], harmful effects of noise on children consist of both auditory and non-auditory effects. Auditory effects are reduced or impaired hearing or temporary threshold shifts or tinnitus. This is probably not a relevant problem in classroom settings but when it comes to the non-auditory effects of noise, which are mainly cognitive effects, there is something to improve. Noise can negatively affect children's learning and language development, can disturb children's motivation and concentration and can result in reduced memory and reduced ability to carry out more or less complex tasks. Noise may provoke a stress response in children that includes increased heart rate and increased hormone response, and noise may disrupt sleep and hinder the necessary recovery of the body and brain. As an indirect effect of noise, a raised voice may lead to hoarseness and vocal nodules [13]. In other words, noise in classrooms is a problem and furthermore a challenging research problem.

Bistrup et al. [13] have provided guidelines for achieving good classroom acoustics. In addition to these instructions, mobile technology could be used to prevent the noise problem. Instead of traditional mobile learning applications that are learning tools or information providers, this novel solution has a different approach to support

[1] The Global Positioning System provides location information based on satellites.

and enhance learning. The main aim of the collective sound sensing mobile application is to change student behaviour through persuasion. More detailed purposes of the application are: 1) to keep the soundscape in the classroom appropriate for learning, 2) teach students to work as a team, 3) teach them to take others into consideration, 4) teach proper behaviour and 5) provide something fun to create and enrich a positive learning experience. Furthermore, it can be used to teach sound pressure levels. The prototype application was piloted during the spring of 2012 with a total of 72 students and two teachers. The results based on observations and interviews are promising and several topics for future work arose during the pilot study.

II. Context

This study tries to answer the following research question: how can we reduce noise in the classroom and thus support learning? To attain an appropriate learning atmosphere, the designed system attempts to keep the noise level at pleasurable tolerance level in the classroom. The hypothesis is that through mobile persuasion, learning results will improve. To understand the context of the study better, mobile learning and persuasion are described in more detail below.

A. Mobile Learning

Lonsdale et al. [14] suggested that contextual information should be used in mobile learning to deliver the right content and services to the user. This could help to overcome the limitations of mobile devices related to user interface and bandwidth. A context, according to Lonsdale et al. [14], is understood as a combination of an awareness of current technical capabilities and limitations and the needs of learners in the learning situation.

The concept of context is also related to the concepts of adaptive, pervasive and ubiquitous learning. For example, Syvänen et al. [15] presented experiences of developing an adaptive and context-aware mobile learning system that they also describe as a pervasive learning environment. They define the pervasive learning environment as a single entity formed by the overlapping of "mental (e.g. needs, preferences, prior knowledge), physical (e.g. objects, other learners close by) and virtual (e.g. content accessible with mobile devices, artefacts) contexts."

Sharples et al. [6] explains that computer technology and learning are both ubiquitous. The concept of mobile learning is closely related to ubiquitous learning and context-aware ubiquitous learning. A context-aware ubiquitous learning system integrates authentic learning environments and digital (virtual) learning environments, which in turn will enable the learning system to interact more actively with the learners. This is possible because of current mobile devices and sensors, such as RFID[2]. [16] It can be said that learning in a classroom can be ubiquitous.

Yau and Joy [17] presented three different types of context-aware mobile learning applications: location-dependent, location-independent and situated learning. They also present a theoretical framework for mobile and context-aware adaptive learning. In their framework, they use contexts such as the learner's schedule, learning styles, knowledge level, concentration level and frequency of interruption.

B. Mobile Persuasion

Many people do not need external pressure or rewards to do things like reading books; the rewards are built-in, or intrinsic [18]. Malone and Lepper [19] have outlined different types of intrinsic motivators. Three of these motivators are group-level intrinsic motivators: competition, cooperation, and recognition. They involve interaction among people. According to Fogg [18], competition is perhaps the most powerful group-level intrinsic motivator. However, not everyone is competitive by nature, but in most situations and for most people, competition is energizing and motivating. Cooperation is another motivator, one that seems to be built into human nature. When people belong to a work group, most of them cooperate. Finally, people are intrinsically motivated by recognition. [18]

The system presented in this study utilizes these three motivators in particular. Classrooms can compete against each other to find out the best learning atmosphere. Also, students can challenge each other to obtain positive collective results. To maintain a suitable noise level, cooperation is essential. Students and teachers can discuss the study results together. Thus, by offering public recognition, the system can increase the likelihood that the group of students will adopt the target behaviour.

In addition to group-level intrinsic motivators, the system involves four other mobile persuasion principles: kairos, information quality, social facilitation and social comparison. Mobile devices are ideally suited to leverage the principle of kairos; offering suggestions at opportune moments to increase the potential to persuade [18]. The designed system in this study visualizes the current noise level in real-time with suggestive images. Visualizations also take information quality into consideration; computing technology that delivers current and relevant information has greater potential to create behaviour change. Furthermore, when it comes to social facilitation, people are more likely to display the target behaviour if they know they are being observed via computing technology or if they can discern via technology that others are displaying the behaviour along with them. According to the social comparison principle, people will have greater motivation to display the target behaviour if they are given information, via computing technology, about how their performance compares with the performance of others, especially others who are similar to them. With their ubiquity, mobile phones have already demonstrated their value as a persuasion tool in several cases, for example education and health. [18]

III. Related Work

Several studies have been made about mobile learning [1, 2, 3, 4, 5, 6, 7] and persuasion [18]. The intention of this study is to change student behaviour through persuasion based on the collective mobile sound sensor. Thus it is more relevant to present some related research concerning mobile sound sensing and noise for background information. There are also noise prevention studies, which utilize external sound sensors instead of mobile phones (e.g. [20]). Commonly, noise measurement or prevention studies are conducted in urban environments and are outside the learning context.

[2] Radio-frequency identification is a wireless short distance system that uses electromagnetic fields to transfer data.

Maisonneuve et al. [21] presented an approach called NoiseTube for the assessment of noise pollution involving the general public. The goal of this project was to turn GPS-equipped mobile phones into noise sensors that enable citizens to measure their personal exposure to noise in their everyday environment. Thus, each user can contribute by sharing their geo-localised measurements and further personal annotation to produce a collective noise map.

The MobGeoSens system [22] utilizes built-in microphones and other sensors attached to mobile phones to collect different kinds of pollutant levels in an urban environment. It also adds the dimension of spatial localization to the data collection process and provides the user with both textual and spatial cartographic displays. While collecting the data, individuals can interactively add annotations and photos which are automatically added and integrated in the visualization log.

Within the Nericell project [23], audio recordings from the built-in microphone of a smartphone constitute the input of a honk-detection algorithm, which in turn feeds an estimator of current traffic conditions. Sensors and radios are used to detect bumps and potholes, braking and honking, and to localize the phone in an energy-efficient manner. The authors investigated the influence of background noise and the sensitivity of the microphones on the performance of the honk-detector.

CenceMe [24] combines the inference of the presence of individuals using sensor-enabled mobile phones with the sharing of this information through social networking applications like Facebook and Twitter. The system also captures audio signals. The data is processed on the mobile phone to determine whether it contains voices or just background noise.

Misra et al. [25] present several examples of how mobile phone microphones can be used for music applications. Using simple feature extraction methods, parameters can be found that sensibly map to synthesis algorithms to allow expressive and interactive performance. For example, a blowing noise can be used as a wind instrument excitation source. Also, other types of interactions can be detected via microphones, such as striking.

The BikeNet project [26] presented how average noise levels can be used to influence daily decisions like the choice of cycling route. The prototype derives estimations of the actual noise levels in the immediate vicinity of the cyclist. The system provides quantitative guidance to cyclists about the healthiness of a given route in terms of pollution levels, allergen levels, noise levels, and roughness of the terrain. These measurements, together with data from cyclist performance measurements, are correlated to create a holistic picture of the cycling experience.

Furthermore, in addition to mobile phone sound sensing, for example Bistrup et al. [13] and Klatte et al. [27] have studied the effects of classroom acoustics on performance and well being in elementary school children. According to the authors, the acoustical conditions in classrooms often do not suit the specific needs of young listeners. They found that reverberation had a significant effect on speech perception and short-term memory of spoken items. The results highlight the importance of good acoustical conditions in classrooms.

Based on the background research, this paper presents a novel approach. Different solutions can be compared but currently there are no applications with this kind of implementation and purpose. The following chapter describes the prototype in detail.

IV. Prototype Implementation and Design

An initial approach (Figure 1.) was present decibel[3] readings on a television screen (Figure 2.). Basically, it presents the suitability of the current soundscape in a classroom. The solution utilizes one mobile phone for sound monitoring and one external display for visualization (Figure 2.). In this version, decibels were divided into three categories: quiet, tolerable and noisy. Numbers were colored green, yellow or red based on those categories. Threshold limits were adjustable. There were some difficulties with this approach: numbers were delayed couple of seconds, it might take too much attention, there are already commercial solutions like this and probably it is not persuasive enough.

The second version was designed to solve those grievances. Basically, only visualizations (Figure 3.) are different and there is a gamification[4] element; too noisy environment causes "a game over" visualization and sound level measuring ends up. Thus, this custom-made application can be considered also as a persuasive game, data visualization or interactive teaching material. Graphics (Figure 3.) were created to be suitable and entertaining for different age groups from 6 to 18 years. It was also topical with the seasonal Easter time decorations when the pilot study was carried out. Furthermore, any child or young person could draw it. Hence, it was easy to identify with and the design was attractive to the target group.

The Python for S60[5] application processes two sound files simultaneously: it records two seconds of the present soundscape to one WAV-format file (Figure 6.) and meanwhile it calculates decibel readings (Equation 1.) from another file recorded previously. Every two seconds it swaps these files. Thus, there is a two-second delay between the captured sounds and the visualized image (Figure 3.), which is still eligible for this purpose. The view of the mobile phone is transferred in real time from

Figure 1. An initial approach with colored decibel readings based on three categories: quiet, tolerable and noisy

[3] The decibel (dB) is a logarithmic unit that indicates the ratio of a physical quantity relative to a specified or implied reference level. [28]
[4] Gamification enhances non-game contexts by game design elements, thinking and mechanics. [29]
[5] The Python for S60 (PyS60) is Nokia's version of the Python programming language to its Symbian S60 software platform. [30]

the phone's s-video output to a data projector or a large television screen. Thus, everyone in the classroom is able to monitor it.

Without proper calibration, sensor devices produce data that may not be representative or can even be misleading [21]. In this case, for example the length of the s-video cable determines the place of the phone (Figure 4.). The best location would be probably in the middle of the classroom. The application is not able to be aware of the distance of the sound source. Thus it is also more proper to calculate estimate sound pressure values and visualize it in pictures than in concrete numbers. In addition, to ensure decent functioning while a lesson a charger cable should be connected to the phone also.

The prototype version is manually adjustable. Firstly, the main user, in this case a teacher, is able to adjust two suitable decibel-reading limits (Figure 5.) from the options menu. This may vary depending on the mobile phone utilized and the location of the phone in the classroom. If the sound pressure level exceeds the first limit, the second view is visualized (Figure 3.). If the sound pressure level also exceeds the second limit, the visualization ends up in the final phase (Figure 3.) and this is visualized until the sound monitoring is restarted. The initial prototype had only these two decibel reading limits (Figure 6.).

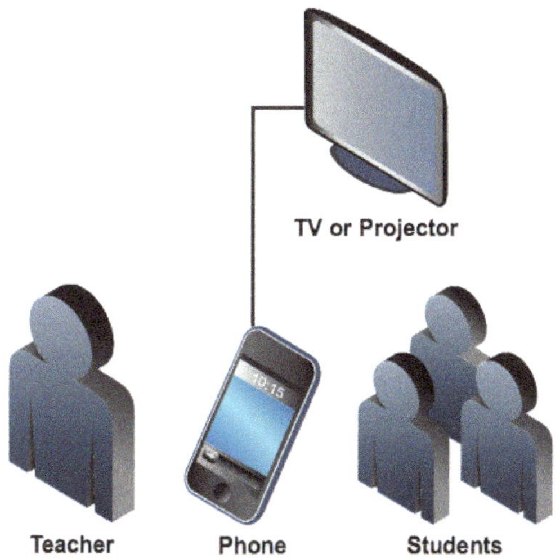

Figure 2. The system architecture of the collective sound sensor

Figure 3. Three views of the prototype mobile application (from left to right): 1) measured sound pressure level is tolerable, 2) the sound pressure level is not appropriate and 3) the sound pressure level has exceeded the tolerable zone

Because one unintentional loud sound could trigger "the game over" event in this case, a manually adjustable limit exceeding option was added. For example, if the second sound pressure level is ten times above the selected limit, only after that will the visualization show the final phase. Once manual calibration is done, a teacher only has to start recording before a lesson by selecting start from the options menu.

Noise level or loudness (Figures 6. and Equation 1.) is generally measured as the equivalent continuous sound level (Leq). Measured in decibels, Leq captures the sound pressure level of a constant noise source over the time interval T that has the same acoustic energy as the actual varying sound level pressure over the same interval. A human ear perceives the loudness differently depending on the frequency of the sound. Standard weighted scale frequency functions have been developed to reflect human perception. [21] The mobile application contains a signal processing algorithm which measures the loudness level of the microphone recording the environmental sound over two seconds at a chosen interval. A commonly used [21] A-weighting filter is applied to the recorded sound and the equivalent sound level Leq, measured in dB(A), is then computed (Equation 1.).

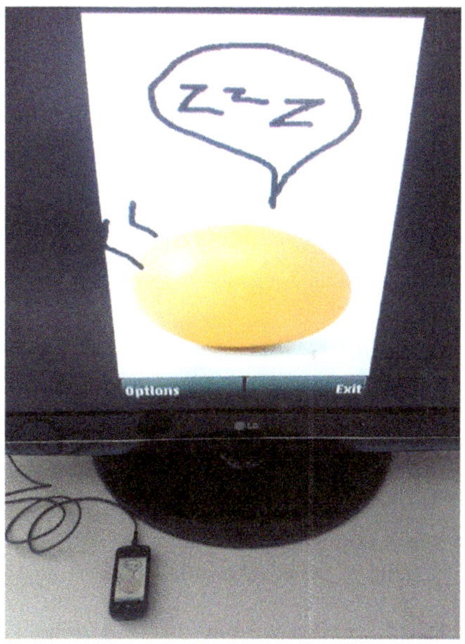

Figure 4. A demonstration setup with a phone and a television

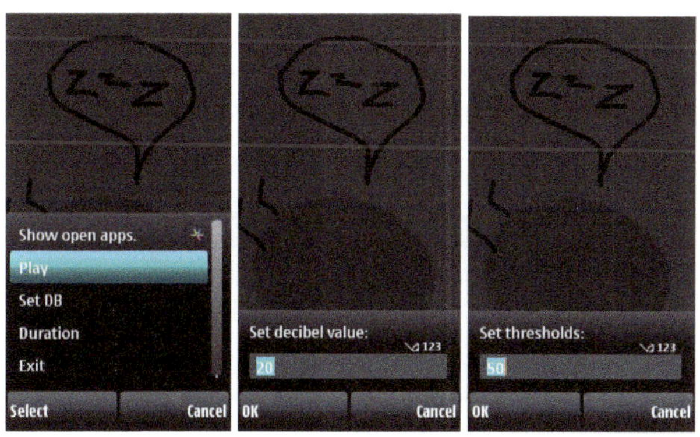

Figure 5. Menu as seen on the screen

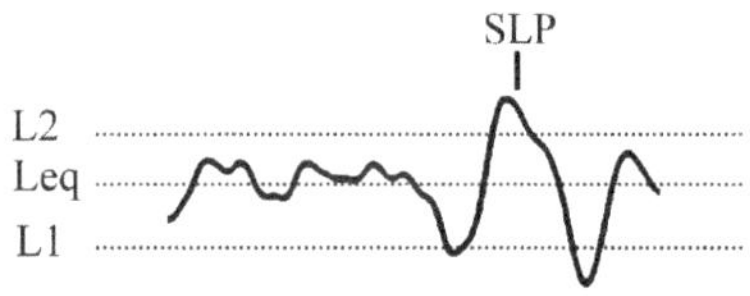

Figure 6. Recorded sound level pressure (SLP) and threshold limits L1 and L2. Presented wave and levels are examples

$$L_{eq} = 10log\frac{1}{T}\int_{0}^{T}\frac{p^2}{p_0^2}dt \quad . \tag{1}$$

The idea is related to the concept of opportunistic sensing, where the data collection stage is fully automated with no user involvement. The benefit of opportunistic sensing is that it lowers the burden placed on the user, allowing overall participation by a population of users to remain high even if the application is not that personally appealing. This is particularly useful for community sensing, where per user benefit may be hard to quantify and only accrue over a long time. [31]

V. Pilot Study

A five-day pilot test was conducted in the spring of 2012 (March 26-30) with the cooperation of Meri-Pori (Pori, Finland) upper level school and upper secondary school. Two teachers (mother tongue and physics) and four classes (two secondary schools and two high schools) of students (altogether n = 72, girls and boys, age 13-18) participated in the pilot. The system was installed in the classroom beforehand and the teachers were told to conduct as ordinary a lesson as possible. The teachers did not tell the students what was going on. A mobile phone was connected to a television or a video projector depending on the classroom equipment and the entire system was visible to the students.

A. *Research Methods*

In this case, the interest lies in the experience. In other words, how the teachers and students experienced the pilot situation. Does the system affect the behaviour of the students and hence the atmosphere of the class? This can be seen as a user experience study. Currently, the most common methods of assessing user experience are through subjective self- reporting and through objective reports [32].

In this study, objective reports were based on observations performed by teachers. This method was selected because it enabled the observation of participants during ordinary lessons. The teachers were interviewed after each lesson. In particular, the teachers tried to observe behaviour and social interaction. Self-reporting included informal conversations with teachers, although the teachers talked to the students and reported back on the students' thoughts. It is difficult to measure the effects on the learning process; hence only teachers provided study feedback.

B. *Results and Findings*

Based on the observations, high school students (age 16-18) were not excited about the application. All in all, it did not have an effect on them at all. They wondered about the system for a while but then no longer paid any attention to it. According to the teachers, noise is not a relevant problem among this age group hence the application was not meaningful. Also, visualization might be more appealing to younger students. Secondary school students were eager to know about the system. Although there was no animation and the graphics were simple, secondary school students (age 13-15) reported that the visualization was fun and entertaining. However, one teacher indicated that the application did not disrupt the students' learning process because of it did not require a lot of attention. Secondary school students quickly learned the meaning and function of the application.

Students tried to maintain an appropriate sound level in the classroom because they did not want to break the visualized egg. The application clearly had a persuasive effect on the students. They learned the differences between proper and inappropriate sound levels. In fact, they perceived the experiment as a game. If someone was too noisy the others instructed the disruptive student to be quieter. Team spirit improved significantly during the pilot study. They perceived the application as such a challenge that they discussed among themselves during breaks how they could keep the egg unbroken for as long as possible. During the pilot week, the students asked the teacher if they would have the experiment during the next lesson. And if the answer was no, they were disappointed. Other classes would have liked to try the system, too.

The teachers reported that it would be useful and interesting if it were possible to locate the noise more specifically in the classroom because it is not always obvious who the troublemaker is. Using three or more mobile phones to detect noise could fulfill this task, making it possible to locate the sound source approximately; similarly to the way an indoor positioning system utilizes three or more Wi-Fi hotspots. The teachers also had another kind of concern. They became anxious about whether their teaching performance could be recorded and monitored via this system. A class with the most suitable sound levels during lessons possibly indicates a skillful teacher.

Regarding the technology, the study revealed that the phone's (Nokia XpressMusic 5800) microphone or software is probably optimized for speech. For example, sounds from whistling and hand clapping had a random effect on decibel readings, whereas speech and shouting worked more precisely. In other words, the mobile phone utilized and the quality of its microphone might have an effect on the final experience. In any case, in this pilot study case it did not have significant impact on the results.

VI. Discussion and Ideas

Based on the study results, it is obvious that the target group should be determined more precisely. In its current state, the application could also be useful among kindergarten (age 2-5) and primary school (age 6-12) students. Students could draw and upload their own character images or animations to the application. In this way, the application could have an even greater impact on the target group.

Actually, various studies have shown that noise levels in day-care rooms and play areas can be considerable and that these may have negative effects [33, 34, 35, 36, 37, 38] In addition to kids, these effects concern teachers as well. Noise levels in day-care centres are occasionally high, possibly resulting in health repercussions: auditory fatigue and stress being the two main negative effects. [33,

34, 38] The stress of the day-care centre employees (in this case caused by the noise) can eventually lead to burnout and work fatigue. Work fatigue can be described as being a serious stress symptom disorder that appears trough physical and mental fatigue, lack of professional confidence and cynical attitude towards work [39, 40]. Overall, stress can be seen as one of the most significant factors behind work fatigue. [41, 42, 43]

Furthermore, high noise levels may lead to irritability, difficulty communicating, reduced intelligibility and even to a reduction in verbal acquisition amongst children [33]. Several solutions have already been identified and implemented in order to reduce noise levels at day-care centres: cutting down on the number of children at each location, placing tennis balls under chair legs and modifying the choice of activities and toys. [33, 36] Another way of reducing noise in day-care centres is by adding absorbent acoustic materials. Presented persuasive mobile device sound sensing system could be utilized as a functional solution to reduce the noise in day-care centres by utilizing children's own motivation to keep quiet. The motivation is triggered by the system and its well-being.

To be even more persuasive, a class of students could receive stamps or awards if they succeed in noise prevention. Another option would be for classes to compete against each other. If students behave properly the application could offer tips and facts for learning besides the sound monitoring. The character could also encourage the students.

Another point of view is that there could be this kind of mascot or game in a classroom that could be taught by the students. The mascot could learn and acquire wisdom based on the classroom noise prevention results. Similarly, Ketamo and Suominen [44] have presented teachable octopuses based on geometry tasks. Basically, students would learn by teaching. Classrooms could compare and compete on their sound environments against other schools, even schools in other countries.

A more advanced version of the application could include a feature of the automatic recognition of different sounds. Selin et al. [45] have studied how harmonic and transient bird sounds can be recognized efficiently. The results indicate that it is possible to recognize bird sounds of the test species using neural networks with only four features calculated from the wavelet packet decomposition coefficients. A modified version of this kind of automatic sound recognition system could also be used in learning environments. It is likely that student voices could be recognized in a similar way to birds.

It should be taken into account that it was not studied during this pilot how noise affects to learning. There were no comparisons between silent and noisy classes or with and without the application. In this case, supposition that disturbing background noise has negative effect to learning is based on actualized references. Studies indicate that learning and teaching in classrooms is often impeded by noise and reverberation. Noise exposure in schools is often above reasonable limits for children. Learning in loud classrooms is especially impeded by poor speech intelligibility and decreased performance of learning. [46, 47, 48, 49] Furthermore, noise especially effects on cognitive performance, motivation and annoyance [50].

Even though there is evidence that noise of classroom decrease performance of learning, Lawrence [51] reported that certain kind of background music (for example classical) in a classroom could allow students perform manual tasks effectively and efficiently. The application presented could be enhanced with three types of background music based on three noise categories; classical music, pop music and hard rock. This might have an affect to the persuasiveness of the application.

As mentioned earlier, during this study, teachers were told to conduct as ordinary a lesson as possible and they did not tell the students what was going on. However the application could be utilized more pedagogically. A teacher could point out events of the application and highlight the persuasive meaning. The teacher and students could discuss together about noise; sources of noise, why there is noise and its affects to learning. In addition, the application could be utilized as a problem-solving task for students. They should understand how to succeed in this persuasive game and furthermore they should realize why it is important for them.

VII. Future Work

In addition to further research, the next version of the presented system will work wirelessly using socket[6] server solution (Figure 7.). It would be similar to that presented by Perttula et al. [52], but instead of WLAN[7], networking will be conducted via the Internet. The installation and configuration will be more straightforward without a local network. In this second version, it would be possible to place the mobile phone wherever it is the most suitable. This solution also enables multiple phones to perform as a collective sound sensor. A computer or a device with Internet connection is needed to view the visualizations presented on a web page.

The advanced version will also include an automatic sound pressure level limit measurement option by recording average levels for a given period. In practice, a user does not manually set noise limits as presented in Figure 5. Determining the proper settings was time-consuming during the pilot study because the selected readings were not comparable with decibels as mentioned earlier.

The next study will be conducted in day-care centres. They will probably benefit from it as discussed in previous chapter. According to Kyriacou [43] coping strategies

Figure 7. The system architecture of the planned second version

[6] A socket is an endpoint of an inter-process communication flow across a computer network.

[7] WLAN stands for a wireless local area network.

can be divided into direct action and palliative techniques. Direct action refers to strategies teachers can do to eliminate sources of stress. For instance, if time pressures and deadlines are creating stress, a direct action to reduce the problem would be to seek a time extension, or to seek a change in deadline. Palliative techniques do not deal with the source of stress itself, but focus on reducing the feelings of stress from those sources. The presented system basically offers both of the coping methods since it can be used just to reduce the current noise in the room time to time/when needed or it can be used with the perspective that it eventually will drop the noise levels permanently when used regularly in daily basis.

The next study will also pay attention to pedagogically approach described in previous section. Teachers' role will be more active and thus benefits or disadvantages of the application will be studied more comprehensively. Also noise affects to learning will be studied and analyzed. Comparisons to previous studies' results of noise and learning will be conducted.

VIII. Conclusions

Noise can negatively affect children's learning and language development, can disturb children's motivation and concentration and can result in reduced memory and reduced ability to carry out more or less complex tasks [13]. To reduce noise in a classroom, this article presented a collective sound sensing system in the emerging fields of mobile learning, mobile persuasion and mobile phone sensing. The solution utilizes one mobile phone for sound monitoring and one external display for visualization of the suitability of the current soundscape in a classroom. The custom-made application with entertaining graphics can be considered as a persuasive game, data visualization or interactive teaching material.

Based on the results, the application managed to persuade students to behave more appropriately to create a suitable learning atmosphere in the classroom by reducing noise. As Lane et al. [31] have stated, understanding which types of metaphors and feedback are most effective for persuasion goals is still an open research problem. Building mobile phone sensing systems that integrate persuasion requires interdisciplinary research, combining behavioural and social psychology theories with computer science [31]. The use of large volumes of sensor data provided by mobile phones presents an exciting opportunity and is likely to enable new applications that have promise in enacting positive social changes in mobile learning over the next few years.

The combination of large-scale sensor data together with accurate models of persuasion could revolutionize how we deal with other learning issues in addition to noise in the classroom. Can we create new kinds of learning applications using mobile sensor networks and could this be the way we should proceed in developing learning applications for the 21st century? According to the present study, sensors provide new possibilities for learning applications. In addition, the understanding of sensor-based contextual data and the ability to apply this data to learning could also be considered a new kind of 21st century skill.

Acknowledgment

The authors would like to thank all the students who participated in the study. We also gratefully acknowledge teachers from Meri-Pori Upper Level and Meri-Pori Upper Secondary schools for providing expertise feedback.

References

[1] D., Mwanza-Simwami, "Concepts and methods for investigating learner activities with mobile devices: an activity theory perspective," in I. Arnedillo-Sánchez, M. Sharples, G. Vavoula (Eds.), *Beyond Mobile Learning Workshop*, Trinity College Dublin Press, 2007, pp. 24-25.

[2] L., Naismith, P., Lonsdale, G., Vavoula and M., Sharples, "Literature review in mobile technologies and learning," *NESTA Futurelab Series*, Report 11, Bristol, UK: NESTA Futurelab, 2004.

[3] M., Nichols, "A theory for eLearning," *Educational Technology & Society*, 6(2), 1-10, 2003.

[4] K., O'Brien, "New questions over Google's Street View in Germany," *The New York Times*, April 29, 2010, Available at http://www.nytimes.com/2010/04/30/technology/ 30google.html

[5] M., Oliver and C., Pelletier, "Activity theory and learning from digital games: developing an analytical methodology," in D. Buckingham and R. Willett (Eds.), *Digital Generations, Children, Young People, and New Media*. Mahwah, New, Jersey, London: Lawrence Erlbaum Associates, Publishers, 2006.

[6] M., Sharples, J., Taylor and G., Vavoula, "Towards a theory of mobile learning," in proceedings of *mLearn 2005 Conference*, Cape Town, South Africa, 2005.

[7] A., Trifonova and M., Ronchetti, "Where is mobile learning going?," in A. Rossett (Ed.), proceedings of *World Conference on E-learning in Corporate*, Government, Healthcare, & Higher Education, Chesapeake, VA: AACE, 2003, pp. 1794-1801.

[8] R., Yeh et al., "ButterflyNet: a mobile capture and access system for field biology research," in R. Grinter, T. Rodden, P. Aoki, E. Cutrell, R. Jeffries, & G. Olson (Eds.), proceedings of the *SIGCHI conference on Human Factors in computing systems (CHI '06)*, New York, NY, USA: ACM, 2006, pp. 571-580.

[9] A., Kansal, M., Goraczko and F., Zhao, "Building a sensor network of mobile phones," in proceedings of the *6th international conference on Information processing in sensor networks (IPSN '07)*, New York, NY, USA: ACM, 2007, pp. 547-548.

[10] R., Bose, "Sensor networks motes, smart spaces, and beyond, pervasive computing," *IEEE* 8(3), 2009, pp. 84 – 90.

[11] M., Sharples, Foreword, "Education in the wild: contextual and location-based mobile learning in action," , in E. Brown (Ed.) *Nottingham: Learning Sciences Research Institute*, University of Nottingham, 2010, pp. 4-6.

[12] J., Voogt and N. J., Roblin, "21st century skills," discussion paper, *Enschede: University of Twente*, 2010.

[13] M., Bistrup, M., Haines, S., Hygge, D., MacKenzie, S., Neyen and C., Petersen, C., "Children and noise – prevention of adverse effects," *National Institute of Public Health*, Denmark, ISBN: 87-7899-056-4, 2002.

[14] P., Lonsdale, C., Baber, M., Sharples and T. N., Arvantis, "A context-awareness architecture for facilitating mobile learning," in J. Attewell & C. Savill-Smith (Eds.), *Learning with mobile devices, Research and development*, London: Learning and Skills Development Agency, 2004, pp. 79-86.

[15] A., Syvanen, R., Beale, M., Sharples, M., Ahonen and P., Lonsdale, "Supporting pervasive learning environments: adaptability and context awareness in mobile learning," *IEEE International Workshop on Wireless and Mobile Technologies in Education*, WMTE 2005, 2005.

[16] H-C., Chu, G-J., Hwang, Y-R., Shi, C-I., Lee and W-W., Chien, "A conceptual map-oriented mindtool for conducting collaborative ubiquitous learning activities," in S. C., Kong et al. (Eds.), proceedings of the *17th International Conference on Computers in Education*, Hong Kong: Asia-Pacific Society for Computers in Education, 2009, pp. 559-563.

[17] J., Yau and M., Joy, M., "A mobile and context-aware adaptive learning schedule framework from a usability perspective – a 'diary: diary-questionnaire' study," in S. C., Kong et al., (Eds.), proceedings of the *17th International Conference on Computers in Education*, Hong Kong: Asia-Pacific Society for Computers in Education, 2009, pp. 512-519.

[18] B. J., Fogg, "Increasing persuasion through mobility," in *Mobile Persuasion – 20 Perspectives on the Future of Behavior Change* (Eds. B. J., Fogg and D., Eckles), Persuasive Technology Lab., Stanford University, 2007.

[19] T., Malone and M., Lepper, "Making learning fun: a taxonomy of intrinsic motivation for learning," in *Aptitude, Learning, and Instruction* (Eds. R. E., Snow and M. J., Farr), Hillsdale, N. J., Lawrence Earlbaum, 1987.

[20] S., Santini et al., "First experiences using wireless sensor networks for noise pollution monitoring," *European Conference on Computer Systems: proceedings of the workshop on Real-world wireless sensor networks (REALWSN'08)*, ACM, 2008, pp. 61-65.

[21] N., Maisonneuve, M., Stevens, M., Niessen and L., Steels, "NoiseTube: Measuring and mapping noise pollution with mobile phones," in *Information Technologies in Environmental Engineering, Environmental Science and Engineering*, DOI 10.1007/978-3-540-88351-7_16, Springer-Verlag, Berlin, Heidelberg, 2009.

[22] E., Kanjo et al., "MobGeoSen: Facilitating personal GeoSensor data collection and visualization using mobile phones," in *Personal Ubiquitous Computing Journal*, vol. 12, no. 8, November 2008, published online: August 2007, pp. 599–607.

[23] P., Mohan, V., Padmanabhan and R., Ramjee, "Nericell: rich monitoring of road and traffic conditions using mobile smartphones," in Proceedings of the *6th ACM Conference on Embedded Networked Sensor Systems (SenSys 2008)*, ACM, 2008, pp. 323–336.

[24] E., Miluzzo et al., "Sensing meets mobile social networks: the design, implementation and evaluation of the CenceMe application," in proceedings of the *6th ACM Conference on Embedded Networked Sensor Systems (SenSys 2008)*, New York, NY, USA: ACM, 2008, pp. 337–350.

[25] A., Misra, G., Essl and M., Rohs, "Microphone as sensor in mobile phone performance," in proceedings of the *8th International Conference on New Interfaces for Musical Expression (NIME 2008)*, Genoa, Italy, June 5-7, 2008.

[26] S., Eisenman, E., Miluzzo, N., Lane, R., Peterson, G., Ahn and A., Campbell, "The BikeNet mobile sensing system for cyclist experience mapping," in proceedings of the *5th International Conference on Embedded Networked Sensor Systems*, Sydney, Australia, 2007, pp. 87 – 101.

[27] M., Klatte, J., Hellbrück, J., Seidel and P., Leistner, "Effects of classroom acoustics on performance and well-being in elementary school children: a field study," in *Environment and Behavior*, September 2010, vol. 42, no. 5, pp. 659-692.

[28] *IEEE Standard 100 Dictionary of IEEE Standards Terms*, Seventh Edition, The Institute of Electrical and Electronics Engineering, New York, 2000, ISBN 0-7381-2601-2, page 288.

[29] S., Deterding, D., Dixon, R., Khaled and L., Nacke, "From game design elements to gamefulness: defining gamification," in proceedings of the *15th International Academic MindTrek Conference: Envisioning Future Media Environments*, Tampere, Finland, 2011.

[30] J., Scheible and V., Tuulos, *Mobile Python: Rapid prototyping of applications on the mobile platform*, Wiley, ISBN: 978-0-470-51505-1, 2007, 348 pp.

[31] N., Lane, E., Miluzzo, H., Lu, D., Peebles, T., Choudhury and A., Campbell, "A survey of mobile phone sensing," in *IEEE Communications Magazine*, September, 2010. pp. 140 – 150.

[32] N., Lazzaro, "Why we play games: 4 keys to more emotion," in *Game Developer's Conference*, San Jose, CA, 2004.

[33] A., L'Espérance, A., Boudreau, F., Gariépy and P., Bacon, "Noise reduction in day-care centres by reducing reverberation time: analyses and case studies (Réduction du bruit dans les CPE par la réduction du temps de réverbération : analyses et études de cas)," in *Studies and Research Projects / Report R-463*, Montréal, IRSST, Téléchargement gratuity, 2006, 69 pages.

[34] C., Truchon-Gagnon and R., Hétu, "Noise in day-care center for children," in *Noise Control Engineering Journal*, 30, 1987, p. 57-64.

[35] M., Picard and C., Boudreau, "Characteristics of noise found in day-care centers," in *J. Acous. Soc of Am.*, Abstract, 105(2), 1999, p.1127.

[36] M., Picard, "Revisiting speech interference in classrooms," in *Audiology*, 40, 2001, p. 221-244.

[37] A., Moodley, "Acoustic conditions in mainstream classrooms," in *J. Br. Assoc. Teachers Deaf*, 13, 1989, p. 48-54.

[38] E. Sala et al., "Vocal loading among day care center teachers," in *Logoped Phoniatr Vocol*, 27, 2002, pp. 21-28.

[39] W. P., Schaufeli, M. P., Leiter, C., Maslach and S. E., Jackson, *Maslach Burnout Inventory Manual*, California: Consulting Psychologist Press, 1996.

[40] C. L., Cordes and T. W., Dougherty, "A review and an integration of research on job burnout," in *The Academy of Management Review*, 18(4), 1993, pp. 621-656.

[41] J., Blase, "A Sosial psychologian grounded theory of teachers stress and burnout," in *Educational Administration quarterly*, 18 (4), 1982, pp. 93-113.

[42] C., Kyriacou, "Teacher stress: past and present," in J., Dunham and V., Varma (Eds.), *Stress in Teachers, Past Present and Future*, 1998.

[43] C., Kyriacou, "Teacher stress: directions for future research," in *Educational Review*, 53, 2001, pp. 28- 35.

[44] H., Ketamo and M., Suominen, "Learning-by-teaching in educational games," in J., Luca and E., Weippl (Eds.), proceedings of *World Conference on Educational Multimedia, Hypermedia and Telecommunications*, Chesapeake, VA: AACE, 2008, pp. 2954-2969.

[45] A., Selin, J., Turunen and J., Tanttu, "Wavelets in recognition of bird sounds," in *EURASIP J. Appl. Signal Process*, 1 (January), 2007, pp.141-141.

[46] M., Klatte, M., Wegner and J., Hellbrück, "Noise in the school environment and cognitive performance in elementary school children, part b – cognitive psychological studies," in *Forum Acusticum*, 2005, pp. 2071 – 2074.

[47] D., Shendell, C., Barnett and S., Boese, "Science-based recommendations to prevent or reduce potential exposure to biological, chemical, and physical agents in schools," in *Journal of School Health*, vol. 74, no. 10, 2004, pp. 390 – 396.

[48] P., Nelson, K., Kohnert and S., Sabur, "Classroom noise and children learning through a second language," in *Language, Speech, and Hearing Service in Schools*, vol. 36, July 2005, pp. 219 – 229.

[49] P., Miller and K., Shannon, "Young children's understanding of the effect of noise and interest level on learning," in *Genetic Psychology Monographs*, vol. 110(1), August 1984, pp. 71-90.

[50] M., Haines et al., "Qualitative responses of children to environmental noise," in *Noise & Health, A Bimonthly Interdiciplinary International Journal*, vol. 5, issue 19, 2003, pp. 19 – 30.

[51] D., Lawrence, "Using music in the classroom," in *Advanced Brain Technologies*, 2001, pp. 9.

[52] A., Perttula, P., Tuomi, M., Suominen, A., Koivisto and J., Multisilta, "Users as sensors: creating shared experiences in cocreational spaces by collective heart rate," in proceedings of *14th International Academic MindTrek Conference: Envisioning Future Media Environments*, Tampere, Finland, October 6th-8th, 2010.

AUTHORS

Arttu Perttula (arttu.perttula@tut.fi) is a researcher at the Tampere University of Technology in Pori, Finland (Tampere University of Technology, Pori Unit, P.O.Box 300, FI-28101 Pori, Finland). He got his Master of Science (M.Sc.) from the Tampere University of Technology in 2007. His research is related to mobile human computer interaction. Perttula has been working in different research projects concentrating mainly on mobile social media, mobile learning and mobile games. In the year 2008, Perttula spent six months with Fuji Xerox Palo Alto Laboratory, Inc. as an intern. In the year 2010 Perttula was a visiting researcher at Stanford University, H-STAR Institute (three months).

Jari Multisilta (jari.multisilta@helsinki.fi) is the director of Cicero Learning Network in University of Helsinki (University of Helsinki, P.O. Box 9, FI-00014, Helsinki, Finland), Finland, professor of multimedia in Tampere University of Technology, Information Technology at Pori, Finland, Nokia Visiting Professor and Visiting Scholar at Stanford University. He got his M.Sc. from the University of Tampere in 1992 in Mathematics and his Dr. Tech. at Tampere University of Technology in 1996. The title of his doctoral thesis was "Hypermedia Learning Environment for Mathematics". Prof. Multisilta has studied learning and modern communication and information technologies and has taken part in several research projects on this area. Currently, his research interests include networked and mobile learning, mobile social media, mobile storytelling, and mobile social video applications, especially panoramic video. Professor Multisilta has published over 100 international conference papers and journal articles on his research area. He was a Visiting Fellow at Nokia Research Center on 2008-2009. Multisilta has also been Visiting Scholar at Stanford University, H-STAR Institute for 2007 (2 months), 2008 (1 month), 2010 & 6 months) and 2012 (2 months).

Pauliina Tuomi (pauliina.tuomi@tut.fi), Master of Arts, is a PhD student in Digital Culture at the University of Turku and works as a researcher at the Tampere University of Technology (Tampere University of Technology, Pori Unit, P.O.Box 300, FI-28101 Pori, Finland). Tuomi has been working in different research projects concentrating mainly on mobile learning and game-based learning innovations. Her doctoral thesis concentrates on interactive TV and it is set to define the different characteristic elements of participatory and intermedial TV culture in the 21st century. Her research interests include iTV-entertainment, participatory media culture and mediated communication.

6

Pro-Active Multi-Dimensional Recommender System using Multi-Agents

H. Al Tair[1], M. J. Zemerly[1], M. A. Al Qutayri[1] and M. Leida[2]
[1] Khalifa University, Sharjah, UAE
[2] Etisalat-BT Innovation Centre (EBTIC), Abu Dhabi, UAE

Abstract—**recommender systems currently used in many applications, including tourism, tend to simply be reactive to user request. The recommender system proposed in this paper uses multi-agents and multi-dimensional contextual information to achieve proactive behavior. User profile and behavior get implicitly incorporated and subsequently updated in the system. The recommender system has been developed and applied to the tourism domain. It was tested and evaluated by relatively large set of real users The evaluation conducted shows that most of the users are satisfied with the functionality of the system and its ability to produce the recommendation adaptively and proactively taking into considerations different factors.**

Index Terms—**Multi-agent, multi-dimensional rating, proactivity, recommender system.**

I. Introduction

Recommender systems are normally used to filter huge amounts of information in order to provide users with recommendations pertaining to products/services. These recommendations are usually filtered based either on the information content or ratings by other users. Systems using the former approach are called Content Based Filtering (CBF) while those that utilize the latter are called Collaborative Filtering (CF) [put refs]. These two approaches are viewed as the traditional recommender systems since they were introduced first and have been applied in many recommender systems by product/service providers such as Amazon[1] and Netflix[2]. Each of the said traditional approaches has its own limitations as will be discussed in Section 3 [1]. The traditional approaches are not suitable for use in context aware situations. In such situations recommendations depend on several different conditions that can vary from one a case to another. For example, in the tourism field, the target users of the recommender system would have different preferences and it is not necessary that they all would be in the same touristic areas that may be recommended [2]. Hybrid recommender systems overcome such shortcomings of the traditional approaches [1]. Knowledge-based is an example of the hybrid recommender system types [3]. This approach works based on previously defined knowledge or rules. The benefit of having rules is that they can be used for critical situations and hence have the tasks done more effectively and in harmony with the user context.

To enhance the performance of two dimensional recommender systems a number of researchers proposed integrating them with multi-agent systems [4] and [5]. A major feature of agent technology is that it is based on a collaborative model, especially among agents, with other components of the system. This aspect can be used to facilitate the process of arriving at appropriate recommendations. Another property of agents is that they can be designed in order to be pro-active. This means they have the capability to initiate and execute tasks autonomously.

Most of the above recommender systems are considered as two dimensional (the user and the product/service). Multi-dimensional (MD) recommender systems are used in order to incorporate more dimensions such as contextual information (the user, the product, the time, the weather... etc). All these dimensions can be useful for the process of decision making that subsequently results in recommendations.

Beside the attempts the researchers have done to apply MD and use the contextual information using a hybrid recommender system, we have decided to enhance this approach of MD recommender system [6]. This MD recommender system [6] was tested based on movies datasets from which the traditional two-dimensional CF approach was applied. Reduction based theory has been used in order to reduce the multi dimensions to two dimensions thus allowing to use contextual information as additional dimensions. This technique is elaborated more in the following section. However, to the best of our knowledge, there has not been any attempt so far to use this MD with hybrid recommender system (Knowledge-base) as its base and where multi-agents are deployed.

In this paper we propose a multi-agent recommender system that is based on multi-dimensional rating approach using the Knowledge-base Hybrid recommender system. The Knowledge-base which will be using rules in order to provide relevant recommendations to the users based on their profiles and contextual information. The system is also supported by a set of rules along with the contextual information. The recommendations are presented to the user in a pro-active fashion.

Following this introduction the paper is organized as follows. Section 2 presents an overview of the technology used for this work. Section 3 reviews significant research related work. Section 4 discusses the requirements of the design and then describes the system design architecture along with the inference engine of the system. Section 5 discusses the implementation of the system. Section 6 presents the system testing and evaluation. Finally section 7 concludes the paper and highlights the future work of this research.

[1] www.amazon.com
[2] www.netflix.com

II. Overview of the Technology Used

A. Recommender Systems

The internet is now recognized as the ultimate repository of information. However, this overabundance of information causes a problem to internet users when searching for specific information or services [7]. In order to solve this problem of filtering the information coming from the internet and to meet the user's needs so that those services are recommended frequently, the concept of Recommender System was introduced [8]. This kind of systems is capable of predicting users' preferences and provides products or/and services based on previously rated items [1].

According to Figure 1 recommender systems can be classified based on the recommendation approach used in [9], the algorithmic technique that is applied in the recommender system [10] or the personalized services provided by the recommender system [2]. As for the first classification, there are three types of recommender systems: the Collaborative Filtering (CF) where the popular items with higher ratings are recommended to other users with similar profiles, the Content-Based Filtering (CBF) where recommended items are similar to the items that the user has selected in the past, and hybrid where the first two types of recommender system are combined in order to avoid their limitations.

Recommender systems can also be classification according to [2] into: Service based on the preferences stated by the user explicitly [11], Services based on the location of the user (location-based) [12], and Services based on the user context (context-aware) [13].

The context-aware recommender system can be considered as the most suitable to be deployed in certain application areas such as tourism, education, and health. For example, in the tourism domain the contextual information is useful to be considered when providing a user with a set of recommendations based on time, season, and his/her exact location.

In order to handle multiple contextual information, some two-dimensional algorithms can be extended to support multiple dimensions as discussed in [6]. In that work the authors propose an approach using reduction based theory to implement a multi-dimensional recommender system.

The reduction-based approach is combined with a traditional two-dimensional estimation method (CF). The main advantage of using reduction-based is by reducing the problem of the multi-dimensional recommendations to the two-dimensional recommendation space so that after the reduction almost any two dimensional technique can be used [6]. For instance a recommendation R of a movie that the user is going to watch during the weekend can be represented as:

$$R^{D}_{User\times Content\times Time}(John, Action, Weekend) \quad (1)$$

where D is the set of existing ratings which contains records <user ,content, time, rating> for each of the user ratings. The overall set of ratings for the user John will first eliminate the time as the contextual additional information by selecting all weekend ratings for movies. After

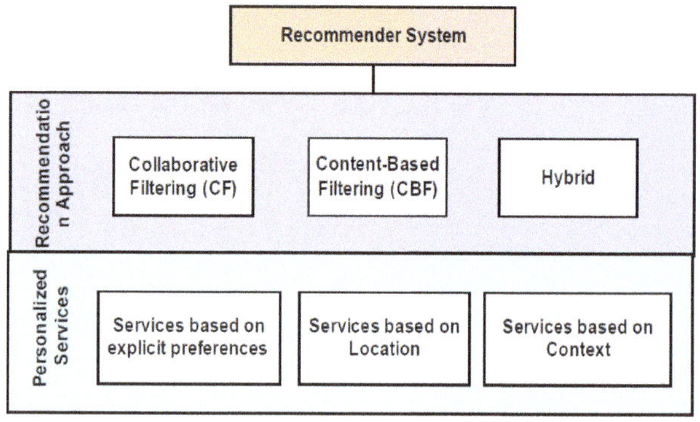

Figure 1. Classification of recommender system

that, it will become a traditional two-dimensional (User x Content).

III. Related Work

In order to discuss the related work we emphasis on the issues of the traditional recommender systems as it was discussed in [1].

A. Issues of Content-based Filtering (CBF)

- New user problem: This depends on the user ratings. The less the number of ratings available by the user, the less accurate recommendations s/he will receive.
- Overspecialization: Only similar items to what the user has rated earlier will be recommended. However, users sometimes may want to change or try new things that they have not selected before.
- Limited Content Analysis: Similar features of two items might make them indistinguishable [14].

B. Issues of Collaborative Fieltering (CF)

- New user problem: The less number of ratings there are, the fewer amount of preferences are defined for a user and hence the recommender system provides less accurate recommendations.
- New item problem: The item which is rated the most by users will be highly recommended. Yet, it happened that there are new items and they are interesting but because they are not rated before, they will not be recommended to users.
- Sparseness: This refers to the number of people who rate an item compared to the ratings themselves. Some items are rated highly but very few people have made those ratings.

Felfernig et al. in [15] argue that collaborative filtering used by, for example, Amazon, cannot be used for tourism applications. The reason being is that collaborative filtering performs best when there are already significant numbers of rated items since the classifications of books for example are known. Moreover, some books are classic in a way that even though they were published 40 years ago, a group of people can still buy such books. As for tourism the personal trip planning is less frequent compared to buying a book. It is rarely the case that tourists will be interested in the exact same events. As it was stated earlier, that collaborative filtering particularly was not sufficiently used yet when it comes to the tourism domain [16]. Tourist mobile guides usually tend to provide a tourist with recommendations that are closer to his/her preferences, which allow more personalization of the suggested recommendations. Yet, these guides fail to use what other tourists might like and highly rate to be recommended [2]. Additionally, the recommendations are normally fired based on an explicit request from the user. However, this is not suitable especially when sudden events occur like change of locations and activities. This can be applied for emergency situations or even when a tourist for instance is on a trip where specific festival or concert is running at the same time and place s/he is in. The sections below explore some solutions to find a suitable recommender system approach for the tourism domain.

C. *Available Solutions*

The available solutions of most of the issues mentioned earlier are:

1) *Hybrid recommender system*

A hybrid approach attempts to overcome the limitations of CF and CBF techniques by combining them to provide more accurate recommendations than those only based on CBF or CF [9, 17]. Based on how the two approaches are combined, one of the classifications of hybrid recommender systems is the unified model which has properties from both content-based and collaborative approaches. There are many models that have been designed and implemented under this category, which mainly uses methods from the content-based and the collaboration approaches with other techniques and methods. One of the techniques that have been used is the knowledge-based as in [18]. The knowledge-based technique improves the accuracy of the recommendations brought by hybrid recommender systems and deal with the limitations of recommender system approaches (cold start) [1]. The cold start concept refers to the problem the recommender system is facing when there is a new user or a new item as stated earlier [19]. The knowledge-based recommender systems are applied in the tourism domain which proved to have significant improvement. An example is the tourism recommender system [3].

2) *Multi-agent recommender system*

Choosing to utilize a multi-agent in a recommender system, especially in the tourism domain, came mainly from the need to facilitate the recommendation process to serve a larger scale of tourists with different services. Agents were used particularly to simulate the human intelligence by using a set of logic rules that each agent will have in order to perform certain tasks on behalf of the users [20]. An example of task-based recommender system is [5]. The tasks assigned to the agents are usually designed to solve smaller problems for more complex and larger problems [21]. Agents have the ability to be social and communicate with other agents and other components in the systems [22, 23]. The interaction between the agents can be done automatically in an intelligent way to pass the required information [24]. Moreover, agents are able to communicate any results directly to end user [25]. Agents are also able to monitor the users' behaviors in order to build their profiles, keep the history, and then take the feedback implicitly [26].

3) *Multi-dimensional rating approach*

Adomavicius et al. [27] proposed a multi-dimensional approach that allows using contextual information about the user such as time and location beside the basic two dimensions that is used by most of the recommender systems (user and item).

Recommender systems were used originally in the tourism domain to enhance the process of arranging a trip independently with most of the services required, such as places to visit, restaurants, hotels and flights bookings. The previous solutions can be combined in order to form a recommender system. The multi-dimensional approach uses the contextual information to provide more accurate recommendations to the users, which is suitable for the tourism domain. The base of the multi-dimensional approach can be hybrid recommender system to avoid the limitations of CF and CBF. The multi-dimensional approach can be developed to be more pro-active and has a more efficient inference engine through integrating it with multi-agent systems. The following section describes the proposed multi-dimensional multi-agent system architecture for a tourism application.

IV. System Architecture

A. *Description of the system architecture*

In order to realize the system, the architecture is divided into four layers as illustrated in Figure 2.

They are as follows:

1) *Data Layer*

This layer is responsible for storing the data such as the profile of the users and the information about the services that are brought from the web which is cached and kept up-to-date to avoid network-induced bottlenecks.

2) *Agent Layer*

This is the core component of the system that contains the inference engine and the environment of the agents. The main parts of this layer are the agents themselves which are created by the Agents Engine Factory. Each agent is specialized in performing one task or one type of the three services provided by the system. In the case of the tourist prototype system implemented in this paper the services are the request of a hotel, a restaurant and/or an event. Each agent's intelligence will be defined as a set of rules. The agents use the multi-dimensional inference engine (MDIE) component in order to store the ratings of different recommendations which will lead to anticipating recommendations to users as it will be explained in the next section. Finally, there are the Information Retrieval Manager and the Profile Manager components. The former manages the information retrieved in the data layer while the latter manages the profiles of the users which are stored in the data layer.

There are four main agents that the system creates for each user based on his/her requirements of services. Their functionalities are explained below:

- Schedule Agent: The schedule agent is created after the minimal information required is collected from the user. Each user will have his/her own schedule agent that will be responsible of creating a schedule that will be filled later with trip activities. The schedule agent will be in charge of sending requests to the agents' engine to create other agents when needed. Moreover, the schedule agent will also be in charge of sending reminders of the activities to the users. Once the user makes a selection, feedback will be taken implicitly and passed to the profile manager to build/update the profile of the user.
- Restaurants Agent: The restaurants agent is created by the agents' engine to bring recommendations of

restaurants to the user. This agent will first request the profile manager to bring specific information that is useful such as the type of trip and check if there is history or not. The information pulled from the database will then be passed to the MDIE entity to predict ratings of recommendations to be brought by the restaurants agent. The recommended restaurants will be passed to the information retrieval manager to store them in the retrieved information database so it can be used later on.

- Hotels Agent: the hotels agent will be created by the agents' engine before even the schedule agent fills the schedule with the activities. This agent will be created based on the need for a hotel or not. Based on the trip dates that the user would specify at the beginning, it will be decided whether this hotel agent should be triggered or not. For example, if the user has inserted the start date of the trip to be equal to the current date or after, there might not be a use for recommending a hotel since the user has already started the trip.
- Events Agent: the events agent will be created based on the need to provide events recommendations. It will access the profile and check if a history exists. The information found will be passed to the MDIE where the recommendation ratings will be predicted and provided to the user.

3) Control Layer

This layer facilitates the transition of the information from the user terminal to the agent's layer and vice versa.

4) View Layer

This layer provides a friendly user interface that a traveler would interact with through the system.

B. Multi Dimensional Inference Engine

In this section we will discuss how the inference engine works for one service which is the restaurants yet it works the same for the other services.

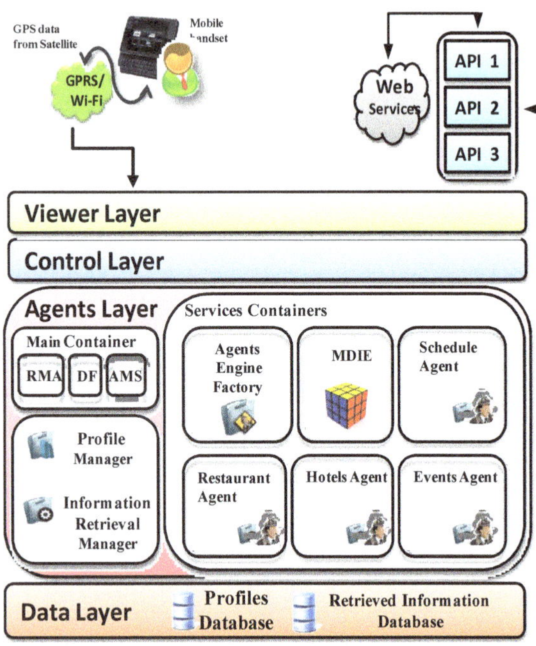

Figure 2. Design Architecture

1) Multi-dimensional recommendations cube

As discussed in [28] implicit feedback can be taken from the behaviors of the users, e.g.: selections, duration, and deletion. However, since the interaction between the user and the system will be reduced to the absolute minimum, in our case only the selections of the recommendations proposed to the user are considered. Therefore, the ratings of the recommendations are taken implicitly by observing the user's behavior and the selections a user makes. The ratings are going to be saved in the database. It can be represented in the recommendations cube as it is the case of the multi-dimensional approach [6] where Adomavicius et al. have followed the multi-dimensional data model of [29].

There are 6 sides for one cube. Each side is considered as a dimension and represented in tables. Each table contains other useful information too. In our proposed architecture we propose to have three different multi-dimensional cubes in a way that each cube is assigned to a service since we have three services provided by the system. There is a slight difference among the three cubes based on the service dimension. We are going to go through all cubes to show how the combination of the dimensions can be used to store ratings.

Figure 3 illustrates the cube of recommendations ratings for the restaurants service. We have decided to use four sides and assign four dimensions, one for each one of them. The dimensions are: user, restaurants, time, and trips. A rating from these dimensions can be represented as the following:

$$R : USER \times RESTAURANT \times TIME \times TRIPS \rightarrow Rating \qquad (2)$$

For instance, as shown in the Figure 3, a user *Ahmed* has been on a family trip in which he stayed in Abu Dhabi. In a weekday he had a meal at an Italian restaurant also in Abu Dhabi. The rating that was given to this combination of these dimensions is 3 according to the user's previous selections and preferences.

2) Recommendation probability

Conditional probability was used since the expected properties of the different recommendations provided to users can be used with different contextual information. The previous ratings which are saved in the ratings cube are going to be the base for getting the preferences of the

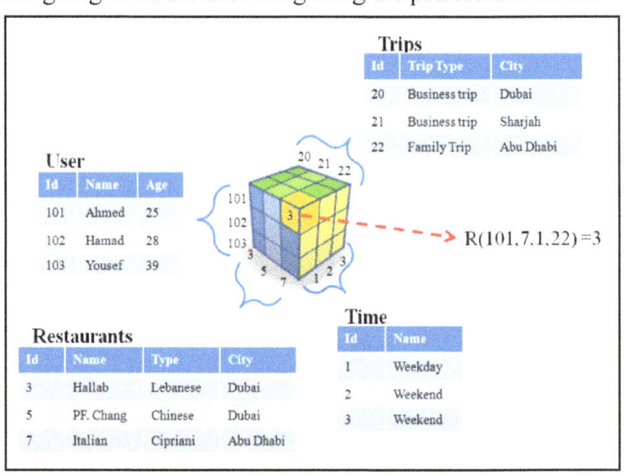

Figure 3. Multi-dimensional model for ratings of restaurant recommendations

users, and sorted according to their priorities. To explain this process we will demonstrate an example involving restaurants. The main attributes of the restaurants are the city and the type of food.

The ratings of the restaurants that a user has been to should be brought from the recommendation cube, as was explained previously. Table 1 shows all the existing ratings which were given by the system to the restaurants properties to the same user. By using a set of rules that will be explained later, along with the reduction based theory, the system would check the current location of the user, which for the example here is assumed to be *Abu Dhabi*. The first step is when the system would use the conditional probability to check the probability of which type of food the user would prefer on this visit, since the type of food is the first attribute.

Note that if the user has no such history random selection will be given to him.

The probabilities of each type of restaurants in Abu Dhabi are going to be calculated as the following:

$$P(Type / AbuDhabi) = \frac{A}{B} \quad (3)$$

where *A* is the rating given to specific type of restaurants in *Abu Dhabi* and *B* is the ratings that were given to restaurants in *Abu Dhabi* regardless of its type. Therefore, when calculating we have the following probabilities:

P(Indian/Abu Dhabi) = 19/24 = 0.79

P(International/ Abu Dhabi) = 2/24 = 0.083

P(Lebanese / Abu Dhabi) = 3/24 = 0.125

According to Figure 4 Indian food has the highest probability that the user would like to have if was in Abu Dhabi. The first attribute that is the type of food will be assigned to be Indian food.

After this step we have the two attributes of the restaurants which are the city *Abu Dhabi* and the type of the food which is *Indian* as the highest priorities to the user. The process of selecting the attributes of hotels and places to visit will be applied in a similar way. The attributes of the hotels are: the number of stars and the city, while the attributes of the places to visit are: the category of place and the city.

In the following section we are going to apply these attributes of the same example on the available restaurants to see what recommendations to send to the users.

3) Multi-attribute theory

The multi-attribute theory is actually an evaluation methodology that is widely used [30] to assess and rank alternatives significantly [31]. Therefore, we have decided to use it in order to assess the recommendations before providing them to the users.

The attributes priorities that were shown previously are going to be used and applied using the multi-attribute theory. The multi-attribute choices are represented in Table 2 where each of the attributes will be assigned a weight according to the priority of the attributes (exact =2, was chosen before =1, never =0). The restaurants list would be brought from the web along with their attributes, food type and in which city they are located. For example, *Restaurant 1* serves Indian food and it is in Abu Dhabi. Therefore, the food type and the city are exactly matching user's preferences and thus both attributes were given the weight of 2. *Restaurant 2* serves Chinese food which the user has never experienced before so it would be given a weight = 0. Yet, city was for example Dubai that he has visited it, but it was not the highest when the probability was calculated. Weights for *Restaurant 3* are similar to Restaurant 1 since both are in Abu Dhabi and serves Indian food. Total of the weights are calculated then and the highest are going to be recommended to the users. In this example *Restaurant 1* and *Restaurant 3* are going to be recommended since they are the most suitable for the user.

The recommended services are going to be kept in the information retrieval database in case another user has common preferences and context can reuse these services or sometimes the same user would like to select another recommendation when making changes to his/her trip schedule

TABLE I.
RATING OF RESTAURANTS THAT A USER HAS BEEN TO

ID	City	Type of restaurants	Rating
1	Abu Dhabi	Indian	15
2	Abu Dhabi	Indian	4
3	Abu Dhabi	Lebanese	3
4	Abu Dhabi	International	2
5	Dubai	Indian	5
6	Dubai	Lebanese	3
7	Dubai	International	2
8	Dubai	International	1
9	Sharjah	Lebanese	1

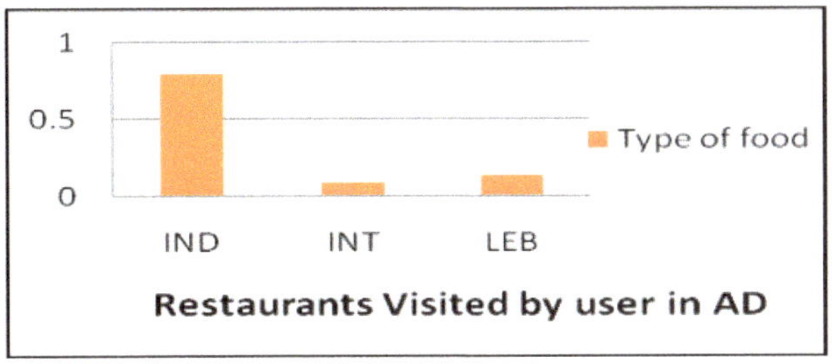

Figure 4. Probability of different types of restaurants

TABLE II.
APPLING THE MULTI-ATTRIBUTE THEORY TO RECOMMENDED RESTAURANTS

Attributes / Recommended restaurants	Food type	City	Total
Restaurant 1	2	2	4
Restaurant 2	0	1	1
Restaurant 3	2	2	4

V. TESTING AND EVALUATION

The purpose of the system evaluation is to validate the system requirements according to the users' experience when using the system. Moreover, the system evaluation is conducted to basically measure the accuracy of the recommendations brought to users according to different profiles and contextual information. The system has been evaluated, also, against other similar systems.

A. Evaluation

The system was implemented and tested by real users as tourists in UAE. There are three different types of experiments that can be considered when evaluating a recommender system [19]: offline, online experience and user studies. The last type which is the user studies is what we have decided to use since it is the most suitable for our system at this stage since it is a prototype. The evaluation of the simulated prototype of the system was carried out by 71 volunteer who managed to do 220 test cases with different scenarios. They are from different educational backgrounds experiences, nationalities, ages, and genders. The system performed well and gave expected recommendations most of the time according to the evaluation conducted.

There are a number of criteria to evaluate a recommender system as in [33]. Based from the evaluation conducted we are going to check whether the system has satisfied these criteria.

- *Transparency:* The transparency is achieved partially by showing the users the recommendations as reminders and notifications prior the activity actual time. Any action taken is notified to the user so s/he can take it and hence update the schedule. The transparency can certainly be enhanced in the future by showing the user's previous selections and reasons of why the recommendations were suggested the first place. A number of users mentioned that in the comments.
- *Scrutability:* The scrutability is achieved by allowing the users to make changes to the trip schedule either manually or automatically. Manual schedule modification can be done by using the edit and delete buttons. On the other hand, the automatic schedule modification is done by the system by firing reminders and notifications. They allow the users to either accept or reject the recommendations. This what most of the users found it interesting about the system that it allows them to make the changes to the schedule either by themselves or by the system in case of rejecting a recommendation. Based on the activities in the schedule, the current location of the user will be taken and compared to where s/he should be. After that, the reminders will be sent to users. In case the location of different the system will sent the reminder as a notification of whether the user is willing to stick to the schedule or not. In case the user decided to not to stick to the schedule, new activities will be recommended to users in his/her current location. The schedule will be updated accordingly automatically. The selections the user made will also be taken into account since the ratings of the recommendations are taken implicitly.
- *Trust:* The trust is achieved in term of how the user is confident about the system starting with its recommendations and ending with how user friendly is the interface design. As shown in Figure 5 it turned out that 87% of the users prefer to receive recommendations based on their interests, 64% of them prefer to try new things very often, while 57% like to have same food types and visit places of their interest. The majority of the users (94%) agreed that the system learns their preferences with time and 76% of them stated that the system brought to them the kind of recommendations that they were looking for while 21% found that it does some times. In spite of that, 66% of the users believe that recommendations can be improved and that other choices can be added to the system which is reasonable suggestions due to the limitations of the resources used for this prototype. It was a delight that 93% of the users showed that they are willing to use this system in the future, since they believed that such a system will be very helpful that they can trust when they plan for trips.

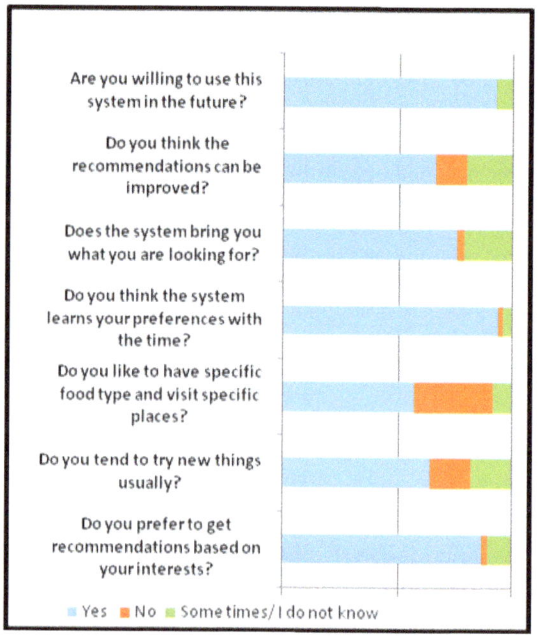

Figure 5. How users trust the system

- *Effectiveness:* As for the system's prediction of recommendations for every user, it was found that 82% of the users have got what they have expected to find of recommendations which are similar to their preferences. Moreover, 93% of the users found that the system brings more accurate and better recommendations with time.
- *Persuasiveness:* this aspect is achieved by comparing the recommendations predicted with what users have in mind, and whether the user accepted the recommendations or not since in our system the users' ratings are taken implicitly and we cannot compare to actual users ratings. 80% of the users showed satisfaction about the recommendations while 17% of them ensured that it does some times. Moreover, 51% of the users rated the recommendations brought as excellent while 49% as medium which is also reasonable and fair since there is a number of users whom services ratings were showing a decrease and fluctuation.
- *Efficiency:* this is achieved by measuring the time the system takes in order to build a proposed plan full of restaurants and events/places recommendations. At the current stage with the limited number of resources we are using for the recommendations, the time they take to be calculated, checked, brought and filled within a schedule is almost 13 seconds at

maximum. The time is taken to bring a list of hotels that are according to users' preferences, is about 20 seconds at maximum.

- *Satisfaction*: 69% of the users found the system not demanding, and 87% agreed that the system is not noisy as well. Moreover, over half of the users (56%) believed that it requires less interaction than other systems. The other half found that it requires constant interaction all the time, or sometime. The reason behind this percentage is how the reminders and notification were designed to be fired in this prototype. They found that it was asking questions all the time. In the prototype the reminders of the schedule activities were designed to come one after the other because it is only a simulation. In real system a user might have one to two reminders/notifications in a day which cannot be considered noisy. Besides that, the system does not require attention from the user all the time, and it does not require a lot of steps in order to fire the recommendations. Furthermore, all of the users found the system easy to use and they only had to explore it for the first time only, and after that they did not take time to recall how to navigate through the system. Over 90% of the users found the interface design user-friendly indeed (Figure 6).

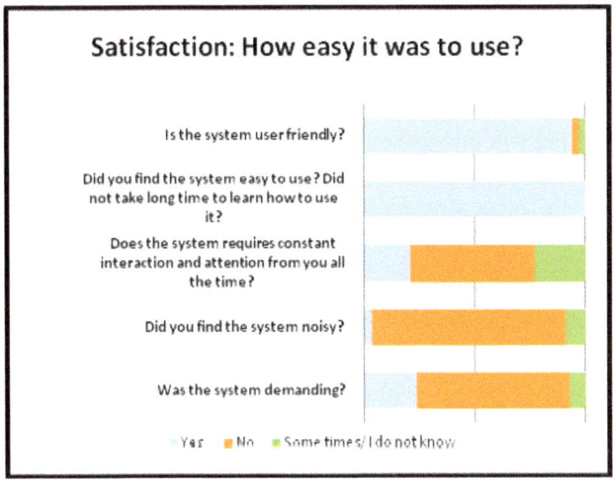

Figure 6. How users are satisfied

B. Comparison between available Systems

Since the architecture is supposed to maximize the pro-activity of a recommender system, there are a number of criteria that our system has met compared to the other projects as shown in Table 3. Creating a dynamic schedule trip is not provided by the other systems. Moreover, the contextual information is not always considered in most of the systems. On the other hand, the systems which use the contextual information mostly do not consider the accuracy of the recommendations that are provided to every user. The limitations of these systems give our proposed system the advantage of the pro-activity and portability of the system, and accuracy and dynamicity of the recommendations. The criteria are:

TABLE III.
EVALUATION OF THE PROPOSED SYSTEM AGAINST OTHER SYSTEMS

Criteria \ Projects	Hybrid Recommender System integrated Multi-Agent [4]	Hybrid Recommender System integrated Multi-Agent [5]	Multidimensional Recommender System [27]	Our Proposed System
Least user effort /interaction	√	√	√	√
Implicate Feedback	x	x	x	√
Estimating Ratings	√	x	√	√
Using the contextual information	√	x	√	√
Ability to handle changes initiatively	N/A	√	x	√
Pro-activity	√	√	x	√
Build/update profile implicitly	x	x	x	√
Usability of the system	√	N/A	√	√
Accuracy of recommendations for every single user	√	√	√	√
Portability/ Heterogeneous	√	√	√	√
Create proposed trip schedule	√	x	N/A	√
Dynamicity of recommendations	√	√	√	√

- *Least user effort/ interaction:* The complexity of any system can be measured by the number of steps a user has to go through in order to reach the result. The less the steps the easier the system to use.
- *Implicit feedback:* The feedback of the recommendations in any recommender system can be taken either implicitly by watching a user's behavior or explicitly by entering the feedback. Deciding to take the feedback implicitly affects the number of steps as described earlier.
- *Estimating Ratings:* The effectiveness of any recommender system can be measured by its ability to predict the ratings of items provided to users based on their preferences.
- *Using the contextual information:* Examples of the contextual information are location and time. Using such information can help providing more accurate and suitable recommendations to users.
- *Initiative:* The system can initiate actions and offer recommendations to the user and internally among the agents or the other components of the system.
- *Build/update profile implicitly*: This feature states that the user will have the ability to create his/her schedule. In case the user did not create his/her profile, the system will do it. Also, updating the profiles will be done on behalf of the user.
- *Usability of the system*: The system usability measures how user-friendly the system is and how easy the user interface is.
- *Accuracy of recommendations for every single user:* This indicates whether the system is accurate when sending recommendations to users depending on each user's preferences and context.

- *Portability/Heterogeneity:* This measures whether the system works in heterogeneous environments
- *Create proposed trip schedule:* Some of the systems allow the users to create a trip schedule while others prepare this schedule automatically.
- *Dynamicity of recommendations:* The dynamicity of the recommendations indicates how they change according to user's profiles and context. They do not remain the same.

VI. Conclusion and Future Work

Current recommender systems approaches have several limitations which include new user problem, overspecialization, limited content analysis, new item problem, and sparseness. Four types of hybrid recommender systems were introduced in order to overcome these limitations. Integrating the multi-agent system with the Knowledge-Based (KB) recommender systems has been proved to enhance its effectiveness. This is due to the agent technology ability to provide the pro-activity feature and the collaboration among the system components.

The main contribution of this paper is enhancing the multi-agent hybrid recommender system by applying the multi-dimensional approach so that it allows the system to consider more factors.

Architecture of a multi-agent recommender system that is based on multi-dimensional rating approach was proposed. It used KB hybrid recommender system as the base of the multi-dimensional rating since there are many benefits of this type especially when using rules for different cases. Such architecture can be applied to different scenarios, yet for this project it was applied in the tourism domain. The system will provide users with trip schedule within seconds. The system has the ability to build the profile of the user from scratch based on user selections. The contextual information is used to provide suitable recommendations of hotels, restaurants and events/places to visit. Agents are specialized in each of these services. The system behaves in a pro-active manner when providing the recommendations in a dynamic schedule. Thus, the interaction between the user and the system is reduced considerably. The feedback of the recommendations is taken implicitly from the user behaviour. The feedback is also used to update the profile of the user to store the preferences which will be used to provide more accurate recommendations in the future.

The system was evaluated by real users using the user studies as the evaluation experimental type. The evaluators performed 220 test cases in which different trips scenarios have been conducted. It was found that the system has satisfied the recommender system criteria.

The system was evaluated also against 3 of the related projects based on technical features like the implicit feedback and how heavy the interaction is.

As for future work, we are planning to maximize the system pro-activity by detecting the trips planned from the user's calendar. Currently since it is a web-based application it is not possible to access a device's calendar. However, if the system is implemented as a mobile application this can be done. Moreover, the mobile application can use the GPS system in any device to provide real time testing.

The inference engine of the system can be enhanced in order to understand the users type (which group) and hence bring more accurate and suitable recommendations. Furthermore, there are some services that can be added to the system to enhance the intelligence and help in bring more accurate and suitable recommendations such as: the weather, currency exchange, cinemas shows timing, and organized trips available with details and their prices.

Modeling different types of user behavior (such as irregular type) within the recommender system will satisfy the majority of users as currently the system only models regular behavior i.e. (the future looks like the past). This will require significant effort but will strengthen the system considerably.

The security of the communication between the server and the user could be improved so that the confidence of the users in the system can be increased. This will require the incorporation of some of the security strategy that is applicable to mobile agent systems.

References

[1] G. Adomavicius and A. Tuzhilin. Toward the next generation of recommender systems: A survey of the state-of-the-art and possible extensions. *IEEE Transactions on Knowledge and Data Engineering*, 17(6):734-749, 2005.

[2] M. Kenteris, D. Gavalas, A. Mpitziopoulos, "A mobile tourism recommender system", IEEE Symposium on Computers and Communications (ISCC), Riccione, Italy, 22-25 June 2010, pp. 840-845.

[3] T. Berka, and M. Plössnig, "Designing recommender systems for tourism", *Proceedings of The Eleventh International Conference on Information Technology in Travel & Tourism, ENTER 2004,*Cairo, Egypt, January 2004.

[4] A. Casali, A. Von Furth, L. Godo, C. Sierra, A Tourism Recommender Agent: From theory to practice, WASI-CACIC 2007 (ISBN 978-950-656-109-3) pp. 1548-1561, Corrientes, Argentina, 2007.

[5] F. Lorenzi, F. Correa, A. Bazzan, M. Abel, F. Ricci. A Multiagent Recommender System with Task-Based Agent Specialization. Proceedings of the 10th Workshop on Agent-Mediated Electronic Commerce (AMEC-X). IFAAMAS, pp. 103-116, Estoril, Portugal, 2008.

[6] G. Adomavicius, R. Sankaranarayanan, S. Sen, and A. Tuzhilin. Incorporating contextual information in recommender systems using a multidimensional approach. *JTOIS*, 23(1), 103-145, 2005.

[7] P. Maes. Agents that reduce work and information overload. *Commun. ACM*, 37(7):30-40, 1994.

[8] P. Resenick, N. Iacovou, M. Suchak, P. Bergstrom, and J. Riedl. Grouplens: An open architecture for collaborative filtering of netnews. In *Proceedings ACM Conference on Computer-Supported Cooperative Work*, North Carolina, USA, 1994, pp. 175-186.

[9] M. Balabanovic and Y. Shoham, "Fab: Content-Based, Collaborative Recommendation," *Comm. ACM*, vol. 40, no. 3, pp. 66-72, 1997.

[10] J. S. Breese, D. Heckerman, and C. Kadie, "Empirical Analysis of Predictive Agorithms for Collaborative Filtering," *Proc. 14th Conf. Uncertainty in Artificial Intelligence*, July 1998, pp. 43-52.

[11] D. Kenteris M. Gavalas, D. Economou. An Innovative Mobile Electronic Tourist Guide Application. Personal and Ubiquitous Computing, 13(2):103-118, 2009.

[12] S. Berger, H. Lehmann, F. Lehner. Location-Based Services in the Tourist Industry. Information Technology & Tourism 5(4):243-256, 2003.

[13] A. Hinz, G. Buchanan. Context-Awareness in Mobile Tourist Information Systems: Challenges for User Interaction. *Proceedings of the Workshop on Context in Mobile HCI*, Austria, 2005, pp. 257-266.

[14] U. Shardanand and P. Maes, "Social Information Filtering: Algorithms for Automating 'Word of Month'," *Proc. Conf. Human Factors in Computing Systems,* 1995.

[15] A. Felfernig, S. Gordea, D. Jannach, E. Teppan, M. Zanker: "A short Survey of Recommendation Technologies in Travel and Tourism"., OEGAI Journal 25 (7), Oesterreichische Gesellschaft fuer Artificial Intelligence, pp. 17-22, 2007.

[16] J. Herlocker, J. Konstan, L. Terveen, J. Riedl. "Evaluating Collaborative Filtering Recommender Systems. ACM Transactions on Information Systems", 22(1): 5-53, 2004.

[17] M. Pazzani. A framework for collaborative, content-based and demographic filtering. *Artificial Intelligence Review*, pp. 393-408, 1999.

[18] R. Burke, "Knowledge-Based Recommender Systems," *Encyclopaedia of Library and Information Systems*, A. Kent, ed., vol. 69, Supplement 32, Marcel Dekker, 2000, pp. 175-186.

[19] G. Shani, A. Gunawardana, " Evaluating Recommender Systems", Microsoft Research Report MSR-TR-2009-159, Nov. 2009, http://research.microsoft.com/apps/pubs/default.aspx?id=115396, Retrieved 15, April, 2009.

[20] W. Bing, W. Fei, Y. Chunming, "Personalized Recommendation System Based on Multi-Agent and Rough Set,"*2nd International Conference on Education Technology and Computer (ICETC)*, vol. 4, pp. 303-307, Shanghai, China, July 2010.

[21] V.N. Marivate, G. Ssali, and T. Marwala. "An intelligent multi-agent recommender system for human capacity building", *Proc. 14th IEEE Mediterranean Electrotechnical Conference,* pp.909–915, Ajaccio, France, 2008.

[22] M. Wooldridge. An Introduction to MultiAgent Systems. John Wiley and Sons, 2004.

[23] I. Rudowsky. "Intelligent Agents". Communications of the Association for Information Systems, Vol.14, pp.275-290, 2004.

[24] S. Macho, M. Torrens, and B. Faltings, "A multi-agent recommender system for planning meetings," *In Proc. Of the Agents'2000 workshop on Agent-based recommender systems (WARS'2000)*, Barcelona, 2000.

[25] H. Zhu, X. Song, S. Zhang, Y. Gao, P. Zhang, "Personalized Recommendation Algorithm Rsearch Based on Multi-Agent," *International Seminar on Business and Information Management 2008, (ISBIM'08)*, Wuhan, pp. 177-180, June 2009.

[26] Y. Blanco-Fernandez, J.J. Pazos-Arias, A. Gil-Solla, M. Ramos-Cabrer, B. Barrangans-Martinez, M. Lopez-Nores, J. Garcia-Duque, A. Fernandez-Vilas, R. P. Diaz-Redondo, "AVATAR: An Advanced Multi-agent Recommender System of Personalized TV Contents by Semantic Reasoning,", *In Proceedings of the Fifth International Conference of Web Information Systems Engineering(WISE 2004), LNCS (3306)*, Brisbane, Australia, pp.415-421,2004.

[27] G. Adomavicius, A. Tuzhilin, "Extending Recommender Systems: A Multidimensional Approach", IJCAI-01 Workshop on Intelligent Techniques for Web Personalization (ITWP'2001), Seattle, Wshington, 2001, pp. 4-6.

[28] C. G. Harrison, D. M. Chess, and A. Kershenbaum, "Mobile Agents: Are they a good idea?", Technical Report, IBM Research Division, T. J. Watson Research Center, 1995.

[29] S. Chaudhuri, And U. Dayal. An overview of data warehousing and OLAP technology. *ACM SIGMOD Record*, 26(1):65-74, 1997.

[30] R. Sch¨afer. "Rules for using multi-attribute utility theory for estimating a user's interests", *In Online Proceedings of the 9th GI-Workshop: ABIS-Workshop 2001*, pages available online at http://www.kbs.uni-hannover.de/~henze/ABIS_Workshop2001/final/Schaefer_final.pdf, Dortmund, Germany, 2001.

[31] J. S. Dyer, P. C. Fishburn, R. E. Steuer, J. Wallenius and S. Zionts. Multiple Criteria Decision Making, Multiattribute Utility Theory: The Next Ten Years. *Management Science* **38** 645-654, 1992.

[32] R. Campbell Ranganathan. "An infrastructure for context-awareness based on first order logic", Personal and Ubiquitous Computing, Volum 7 Issue 6, pp. 353-364, December 2003.

[33] N. Tintarev, J. Masthoff: A survey of explanations in recommender systems. In: ICDE Workshop on Recommender Systems & Intelligent User Interface, 2007.

AUTHORS

H. Al Tair received her B.Sc. in Information Technology majoring Software Engineering from College of Information Technology, UAE University in 2008. She received her M.Sc. degree in Computer Engineering from Khalifa University in 2011. Currently, she is working towards her Ph.D. degree at Khalifa University (e-mail: hend.altair@kustar.ac.ae).

M. J. Zemerly obtained his PhD from The University of Birmingham, UK, in 1989. Since then he worked at various universities in UK such as UCL, Warwick and Westminster. He then joined Khalifa University in year 2000 as an Associate Professor where he is currently the Computer Engineering Program Chair and chairs the eServices and Networks Research Group. His research interests are in ubiquitous computing, and context aware mobile systems. Dr Zemerly published more than 70 journal and conference papers as well as many book chapters. He is also the co editor-in-chief of a journal and currently co-program chair of the ICITST conference series. Dr. Zemerly is a senior member of the IEEE (e-mail: jamal@kustar.ac.ae).

M. A. Al Qutayri received the B.Eng degree from Concordia University, Montreal, QC, Canada, the M.Sc. degree from University of Manchester, U.K., and the Ph.D. degree from the University of Bath, U.K., all in electrical and electronic engineering in 1984, 1987, and 1992, respectively. From 1992 to 1996, he was a Senior Lecturer at De Montfort University, Leicester, U.K. He subsequently joined Khalifa University in the U.A.E., where he is currently a Professor in the Department of Electrical and Computer Engineering. He has published numerous technical papers in peer reviewed international journals and conferences in his fields of research, which include reconfigurable computing, embedded systems design and security, multi-agent systems and applications, and test and design for testability of mixed-signal integrated circuits. Dr. Al-Qutayri is a senior member of the IEEE, member of the IET, U.K., and holds the Chartered Engineer status from the U.K. Engineering Society. (e-mail: mqutayri@kustar.ac.ae).

Marcello Leida received an M.Sc. degree in Computer Science from the Universita' degli Studi di Milano, Italy and subsequently a Ph.D. in Computer Science from the Universita' degli Studi di Milano as part of a research collaboration programme between the Intelligent Systems Research Centre in BT and the Knowledge Management Group of Universita' degli Studi di Milano. During his Ph.D. he did research on Data Integration architectures, Ontology applications, Ontology Matching and uncertainty information management. He joined EBTIC in August 2009 and he is doing research on Enterprise Applications and Business Process analysis. His interests are mainly focused on semantic web, knowledge management and artificial intelligence; he published several papers in international conferences, workshop and journals on these topics. He has been program committee and referee for several conferences and member of IFIP W.G. 2.6 on database semantics. (e-mail: marcello.leida@kustar.ac.ae).

Learner Open Modeling in Adaptive Mobile Learning System for Supporting Student to Learn English

Viet Anh NGUYEN and Van Cong PHAM
University of Engineering and Technology, VNU, Hanoi, Vietnam

Abstract—**This paper represents a personalized context-aware mobile learning architecture for supporting student to learn English as foreign language in order to prepare for TOEFL test. We consider how to apply open learner modeling techniques to adapt contents for different learners based on context, which includes location, amount of time to learn, the manner as well as learner's knowledge in learning progress. Through negotiation with system, the editable learner model will be updated to support adaptive engine to select adaptive contents meeting learner's demands. Empirical testing results for students who used application prototype indicate that interaction user modeling is helpful in supporting learner to learn adaptive materials.**

Index Terms—**m-learning, context-awareness, personalized learning, open learner modeling**

I. INTRODUCTION

Because of its portability, mobile technology is a growing trend in a wide range of activities in modern life such as: communication, entertainment, banking and education. Therefore, mobile learning is also emerging as important research in e-learning field. One of the benefits of mobile learning (m-learning) is the ability to provide and access learning materials anytime in anywhere. For two decades, Adaptive Hypermedia (AH) systems have been developed to provide the learners with adaptive learning materials based on their demands through evaluating learner model. Most AHs are designed for the personal computers, so it requires a definite location and time. Having restricted location and time, the learners find it difficult to approach the learning systems whenever they need. Consequently, the most recent generation of mobile learning research focuses on context-aware mobile learning application. With adaptive engine usages, the learners can easy browse the adapted course content as they want.

Our research addresses the context-awareness adaptation in mobile learning that aims to support Vietnamese students to use the mobile devices such as mobile phone, Personal Digital Assistant (PDA) to learn English in order to prepare for TOEFL test. We are interested in the learner modeling as well as the context factors that affect the students. In addition, we take into account open learner modeling to obtain user information as get the learner to talk the system what they need to know. An improved prototype of our model, CAMLES [1] system also described.

The main contribution of our work is personalized learning materials by using open learner modelling to support the learners can edit their model through negotiation with the system. The rest of this paper is structured as follows: First, we review the related researches on context-aware location dependent learning. In the next section, the context factors using in our model to adapt course content for each student is introduced. As for the fourth section, we represent our context-aware mobile learning, the CAMLES system that focuses on representing how to manage editable learner model and content model as well as the system design and architecture. System implementation with our experiments will also be described in section five. Finally, the discussions and conclusions are summarized.

II. LITERATURE REVIEW

Our literature review presents recent context-aware m-learning applications for learning language. Especially, those support students to learn foreign languages. These applications can be classified into two categories: context-aware location-independent learning and context-aware location-dependent learning. Learners can use the former anywhere that is not restricted in any specified locations. The later application, through location-tracking technologies such as GPS or WLAN, which can automatically identify the learner's location as selecting appropriate learning resources for them, is especially basic. Now, we focus on several typical applications:

- CAMCLL [2], context-aware location-independent learning, teaches Chinese to the students whose language levels are not enough to make conversations in Chinese by supporting appropriate sentences to different learners based on contexts. The CAMCLL context includes time, location, activities and learner levels. Adaptive engine of CAMCLL is based on ontology and rule-based matching.

- TenseITS [3] teaches English language to foreign students through meeting their demands. Learner model is designed based on four context factors: location, interruption/distraction, concentration and available time. Appropriate learning materials for different learners are selected based on the information represented in learner model.

- LOCH [4], context-aware location-dependent learning, supports students to learn Japanese while involving in real time situations. By monitoring the positions of the learners, teachers can establish the communication with the students and g uide them. The context factor in LOCH system is location.

- English vocabulary learning [5] recommends vocabulary for different learners based o n their location, time for their learning and individual abilities. This system uses WLAN to identify learner's position. In addition, it uses some techniques such as maximizing information strategy, evaluating the score of t ime characteristics and estimating the amount of learning words to select suitable vocabulary for different learners.

- TANGO [6] supports Japanese students to identify English words with physical objects via the use of mobile devices through RFID tag reader/writer. TANGO includes six modules to select appropriate English words based on learner models.

- MESLL [7] is designed to support Japanese learners to learn Kanji or Chinese as a second language via SMS function or e mail. The learne rs send an e mail to the system in order to request a test. The system composes a test and feedbacks to them including adaptive English words as well as example sentences.

- PALLAS [8] is desi gned to support to a mobile language learner by providing personalized and contextualized access to learning resources which is considered as a part of contextualization. Personalization of the system has been defined based on learning style preferences, learner's objectives, and current learner's knowledge.

III. CONTEXT FACTORS TO ADAPT

Context is an y information that can be used to characterize the situation of an entity such as a person, place or object that is considered relevant to the interaction between a user and an application [9]. Meanwhile, according to B.Hu, in m-learning, context is the set of suitable environmental states and settings based on situated roles between a learner and a tutor [10].

In our personalized m-learning model, it is suggested that context is the information that has impact on learners in learning activities. We assume that there are several factors having influence on adapting course materials in each learner. Location, tim e, manner, and learne r's knowledge are context factors taken into account in our model. Firstly, location allows information and services to localize. In our model, location allows adaptation system to situated place where learners participate in the course. As S.Cui proposed in TenseITS [3], location is a sp ecial place where students use m obile devices to learn such as home, bus terminal, hotel, etc. Secondly, time refers to the instantaneous time of the day. Specially, the interval that the learner interacts with the system is important for an amount of course materials requiring the learners to learn. Thirdly, the manner of learning is considered as a factor of context using for adaptation. It mentions learner's attitudes such as concentration, interest level when they take part in the course. Finally, learner's knowledge is regarded as an oriental factor to determine what course content should be learned in the next stage.

Context-awareness describes a process in which context factors are us ed to target the provision of adaptive learning materials for the learner in interactive systems based on location, learner's preferences as well as learner's knowledge. This process includes two principal functions: 1) context interpretation and 2) con text implementation [11]. The f ormer collects the learner's input data. Following context processing, the later issues the output that is personalized according to the information reflecting learner modeling.

Reichenbancher [12] noted that there are four di fferent levels of adaptation: information level, technology level, user interface level and pres entation level. Focusing on information level, our model aims to adapt learning materials according to context factors mentioned above. In the next section, we will p resent our personalized mobile learning framework in deeply.

IV. CONTEXT AWARE MOBILE LEARNING ARCHITECTURE

In order to select personalized mobile learning materials based on the c ontext as well as learner' s preferences, we propose architecture with their layers described in Figure 1.

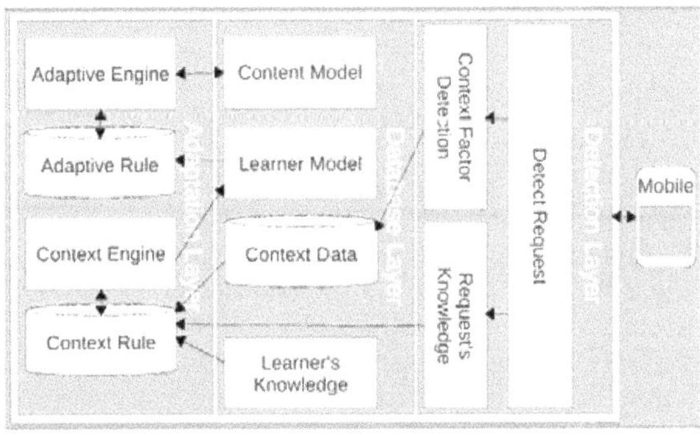

Figure 1. Context-Aware mobile learning architecture

A. *Detection Layer*

The function of the detection layer is to define the requirements of learners, in cluding the normal acade mic requirements or requests to change the assessment system for learners. Requirements normal learning behaviour of learners choose topics, the context in which the system offers, then the system will generate learning content compatible with the choices av ailable to the learners. For the normal academic requirements, the request will b e determined through the identification of contextual factors that the user input options, such as lo cation, time and learning topics. These factors made the user interface to the learner can choose. T he elements are selected according to the person standi ng position and the ti me he or she can learn. This requi rement is applicable to learners, when they choose new t opic or i n the new context. Besides, requests to change the assess ment system for learners are required to change indicator of the level of understanding of the learner after the implementation of the normal acade mic requirements of learner completed. That is, it was onl y made when learners complete their lear ning content through the test questions. For this requirement, students must pay through an examination and assessment of the learning process in the system. Request is the response of learners to assess learning outcomes after the end of l earning content that the system made based on normal academic requirements above. This requirement is a change in the knowledge of the topics that the learners who participate. Therefore, when learners make this re quest means that the learner wants to change the level of knowledge in the database to achieve a high level of knowledge than reality. In addition to giving context and gat her feedback from learners, detection layer includes a t est for eval uation of t he first learn to participate in the system.

1) Detect Request

This component is responsible for cl assifying requests that learners interact with the system. The system is divided into two types of entry requirements:

- Firstly, the requirements in terms of context, when the learner needs in terms of learning content, the request will be passed to the component Context Factor Detection for processing.

- Secondly, the requirements in terms of changes in the level of knowl edge of learners for each topic, which the system was ra ted after the y complete learning content. Learning content i s delivered by systems through data collection requirements on the context in which learners choose.

In addition, the data request is processing the request on a proficiency test, like tests of assessm ents of the candidates qualifications.

2) Context Factor Detection

For a syste m of learning English in context, the determination of the input context elements is a first step in the processing system. Context Factor Detection is part of the Detection Layer, its function is to identify the contextual factors whi ch the user pr ovides the system to start their learning process. In this context is defined, including the location of the user, the time that they can be used to complete the required course content and i ts concentration in the learning process. Each person can learn at the sa me location or different. Similarly, they can have time to learn the same or di fferent. The concentrations are separate. In general, we can say that each individual has a separate context, most of the m are different. Therefore, the composition Context Detection Factor was born with the specific task of determining the value of t he components that the context user su pplied. These values are i nput to the system's data layer. This component works only when there is interaction in terms of context between users and syste ms. That is, in the process of learning, learner does not have any changes in terms of l ocation, duration and concentration, the components will not make it to redefine these elements. Then, the de fault factors pr ovided the context for the system is the contextual factors that considered as users' first choice in the process of using the system. In ot her words, the component Context Detection Factor referred to as user req uirements change in terms of location, time or concentration.

3) Request's Knowledge

After each learning content, learners must complete a test. The syste m will asses s understanding of content learned through the course of th e test results. Results of evaluation of the system can not satisfy learners. The learner might think that their level of understanding of that field as well, wh ile the system is rated at normal levels. Therefore, they will not agr ee with this assessment, they want to change this assessment, they want their system to assess the level of unde rstanding of content. Then, t hey asked to change their level of u nderstanding about that field. When learners have this requirement, the Data Request to transfer the request to the Request Knowledge. This component is responsible for processi ng the request to change the understanding of a topic in which the learner has participated. This component is re quired to receive input from learners on updating the assessment of t heir knowledge about the topic just to participate. The update is a change in their data in the database system. It involves elements corresponding to the context that they are topics to change l evels of underst anding. Hence, it will be combined with contextual elements that learners initially are available to the system. This combination will be held in Adaptation Layer. Th us, the composition Request's Knowledge update request to change data on the level of understanding of the learner to provide treatment system. When learners participate in the system and make learning content, the composition Request's Knowledge can be used several times, repeated until the learner to accept the assessment of the system on t he level of knowledge of person for each subject participated.

B. Database layer

Database layer is a component of the system where data is stored and provided to the system. This layer consists of four subcomponents: context data, learner's knowledge, learner model and co ntent model. The com ponents are closely linked together. Firstly, the context is separated into two co mponents, including data about the context (location and time) and data about the learner's knowledge before. The separation into two components will support the processing system to update the knowledge required from the user feedback better. The system will only have to update the knowledge level of users wi thout having to change the context of their pr evious selections. Learner's knowledge, which stores information about the assessment system for learners. This is the level of knowledge learned at the end of his studies. Secondly, the learner m odel and content model are two ingredients which are base d on rules adapted from that extract the relevant content, stored in a database, users and interactive learning.

1) Context data

Context information is the information obtained from the learner's request such as location, interval of time to learn and concentration. These factors require the learners to fill in before they participate in the course. In this model, we define location as a place where the learners use mobile devices to take part in the course. It is no t a specific common place such as home, bus terminal, hotel, etc. Each location is described by a corresponding discrete value in Table 1. This represents the factors that impact on learning activities such as co ncentration level, the frequency interruption as well as av ailable time to learn. The lower va lue indicates that the location affecting context factors is higher, whereas the higher value indicates that impact is lower.

TABLE I. THE VALUE OF LOCATION FACTOR

No	Location	Value
1	Bus terminal	1
2	Restaurant	2
3	Outing	3
4	Campus	4
5	Home	5

Interval of time is available time that the learner will spend on learning. In terms of time limit in using mobile device, we use four options of interval of time for choosing the time to learn. These are 15, 30, 45, and 60 minutes. Similarly, we use discrete values to identify the level of concentration. The learner can choose one of parameters before participating in the course. Those values are only used to assume the concentration of learner because selection cannot guarantee for that the learners will concentrate as they do.

The concentration parameter is designed to determine the learner's requirements of concentration on learning while student uses mobile device to browse the course. Three concentration levels are low, medium and high. Each of them also is described by discrete value that is 1, 2 and 3 respectively.

2) Learner's Knowledge

Learner's Knowledge is the user's knowledge of the topics given in the system. This knowledge is the result obtained by two methods:

- Firstly, it is the result of the learning process. This result is determined by: after each lesson content, the system will give a certain number of questions. Then the user will complete the questions. The system is based on the amount of questions that users complete right to make the rate of assessment for each topic that the learner has chosen. This rate will be reported to the learner. If the learner does not accept the rate of assessment, or it is the knowledge that the system offers, the user may change that assessment in accordance with their capabilities. The system will rely on feedback data and the previous context, generates a set of questions for users to complete. The system will re-evaluate and inform students. Notification process - feedback loop is done, only when the learner accepts the level of knowledge that the system evaluated.

- Secondly, the initial test of their knowledge of learners without their contents must complete a course before. A set of questions will be generated automatically and randomly covers the most basic themes that initially the user needs to achieve . The completion of these questions and received feedback is determined similarly to the above method.

Knowledge of the learner after the unification and was saved in the database will be divided into five basic levels: poor, average, good, very good, excellent. Each level also is described by discrete value as showed in Table 2.

TABLE II. LEVEL KNOWLEDGE LEVEL

No	Learner knowledge level	Value
1	Poor	1
2	Average	2
3	Good	3
4	Very good	4
5	Excellent	5

3) Content Modelling

We describe the course content as the tree structure with hierarchical nodes that describe topics. They consist of several child nodes. The leaf is a node without child nodes. These contain topic content in detail. Each node includes some attributes to distinguish and they are the basis for adaptation processing. The learner model decides whether node is chosen for different learner or not. It not only decides the numbers of nodes needed to learn but also decides the depth of the tree content that learners are suggested to travel. There are some reasons why the course content is represented as tree structure instead of knowledge graph that modelled in our recent study, ACGs model [13, 14]. Those are: (1) The content of our scenario, the learning topic test support is hierarchical structure, (2) The content adaptation for different learner is to select suitable topics from the course so that it examines the tree processing to select nodes required to learn.

We denoted T (Topic) is the subject study, in which T_i (i = 1...n) is the subject of the T. Similarly, T_{ij} (j = 1...m) is the child of T_i. The topics are arranged under a tree from top to bottom according to the content of the topic. Each topic is a node of the tree. The topics above (as in the general topic) have content that covers the content of the child (a subject in the details.) The child node will inherit the content of the topic at parent level. But it is only reflected in the general level, not goes into detail on each issue reflected in the topic. It foc used on the content corresponding to its position. This raises the problem of how that can be determined in accordance with the contents of that topic. Because threads are arranged according to each topic, tree should have different altitudes. Depending on such topics as wide or narrow, there are many issues of concern which branches child was born. The principal topics with content relevant to general users have average knowledge on that topic. Learners can choose which topics to be able to absorb knowledge in accordance with their capabilities. At the higher topics the content more detail and depth. To be able to learn the content in these topics, the system requires students to understand well the content of lower-level topics. This requirement is entirely accurate, because the topics at high levels are inherited from the subject at a low level, learner may want to learn and understand the need to have certain knowledge of the problem. This knowledge was assessed through the learning process of users in low-level topics.

4) Learner modelling

One of the most important information in this layer is learner model that is basic to select adaptive course content for different learner. It is designed from context factors as well as learner's knowledge. Because all context factors are represented by discrete values, the learner model also is described by them. In this model, we assume that learner model depends on context factors and learner knowledge. With context factors, we designed learner model whose value is calculated by value of location, concentration and time to learn as showed in Table 3. At this stage of the model, we assume that the value of learner model is aggregation of all of context factors. Therefore, there are ten models of learners with values from 3 to 12 respectively.

TABLE III. THE VALUE REPRESENTS LEARNER MODEL BASED ON CONTEXT FACTORS: LOCATION, CONCENTRATION AND TIME

Low(1)				Medium (2)				High (3)			
15	*30*	*45*	*60*	*15*	*30*	*45*	*60*	*15*	*30*	*45*	*60*
1	*2*	*3*	*4*	*1*	*2*	*3*	*4*	*1*	*2*	*3*	*4*
3	4	5	6	4	5	6	7	5	6	7	8
4	5	6	7	5	6	7	8	6	7	8	9
5	6	7	8	6	7	8	9	7	8	9	10
6	7	8	9	7	8	9	10	8	9	10	11
7	8	9	10	8	9	10	11	9	10	11	12

Five rows in Table 3 represent the value for location factor, the first row denotes location at Bus terminal which has minimum value and fifth row denotes location at Home which has maximum value. For instance, for the learner who is at home with low concentration level and time to learn is 45 minutes, the learner model value represented in Table 3 is value 9 (row 5th and column 4th).

As mentioned above, based on learner's knowledge factor, we define learner model as the aggregation of learner model that is based on context and knowledge as shown in Table 4.

TABLE IV. LEARNER MODEL IS COMBINED CONTEXT FACTORS AND LEARNER'S KNOWLEDGE

Learner Model	Learner's knowledge				
	1	*2*	*3*	*4*	*5*
1	2	3	4	5	6
2	3	4	5	6	7
3	4	5	6	7	8
4	5	6	7	8	9
5	6	7	8	9	10
6	7	8	9	10	11
7	8	9	10	11	12
8	9	10	11	12	13
9	10	11	12	13	14
10	11	12	13	14	15

There are fourteen models of learner based on learner's knowledge level and context factors. These models are the basis for adaptation layer to select adaptive course content for different learners. For example, if the learner who can be at Home, concentration level is Medium, time to learn is 30 minutes and knowledge level is Good (This value is evaluated through the test question when learner participates the course), the learner model value is 11.

C. *Adaptation layer*

Adaptation layer includes some functions designed to adapt learning materials for each learner. Based on the results of test as well as learner's background, Learner's Knowledge evaluating component used to identify how learner's knowledge level is. Learner modeling component is constructed to determine all of the context factors such as location, time to learn, and learner's knowledge of different learners affecting to adaptation. The heart of this layer, learning resource selection component, is used to select appropriate adaptive learning content for each learners according to their learner modeling. We designed several rules to choose learning resources from content model as travelling of tree nodes. The child node describes detailed information about parent node. Therefore, if learner travels the tree deeply, the content obtained is more detailed.

Learning material is adapted to different learners in two ways. The first way is that when learner selects one topic from suggested list, the content belonging to this topic is adapted based on learner model of different learners. The second way occurs when the learners finish a test, the system recommends one or more topics that students need to learn.

We classify student into fourteen categories in order to adapt the course content.

The excerpt we used to select learning resources in this model is if - then rules. The rules as described in the Table 5.

Defended on learner model, the adaptive rules include three elements such as height of tree, number of topic and number of test question.

TABLE V. TABLE 5. ADAPTIVE RULES ACCORDING TO LEARNER MODEL

No	Learner model	Rules		
		Height of tree	*Number of topic*	*Number of test question*
1	LM1	1	1	5
2	LM2	1	2	5
3	LM3	1	2	5
4	LM4	2	3	5
5	LM5	2	3	5
6	LM6	3	3	10
7	LM7	3	4	10
8	LM8	3	4	10
9	LM9	4	4	10
10	LM10	4	4	12
11	LM11	4	5	12
12	LM12	5	5	12
13	LM13	5	5	12
14	LM14	5	5	12

The height of tree informs that how information is in details. The number of topics denotes the number of child nodes or sub topics of determined topics. Having several sub topics, the number of topics will decide how many topics are supplied to different learners. Similarly, the number of test questions denotes how many test questions will be required to take after different learners browsing the definite topics.

Adaptation Layer is composed of four main components: Adaptive Engine, Adaptive Rule, Context Engine and Context Rule.

1) *Context Rule*

This component is a set of rules for handling data and the context in which the Database Layer and Knowledge's Request components provide. The use of rules for the context will depend on the requirements of the learner provides. If the request is to provide learning content or assessment tests, the composition will receive data from the two components Knowledge's User and Context Data from the Database Layer provides. Correspondingly, if the request is to change the level of knowledge, the composition will receive data from the Context Data from Database Layer and Knowledge's Request from Detection Layer. After receiving the corresponding data from the components, data and the rules which were built before, is executed in the Context Engine component.

2) *Context Engine*

Context Engine is the execution of laws on the data context associated with it. The result of this process is to model learning to the learner. The calculation is done as in Table 4.

3) *Adaptive Rule and Adaptive Engine*

Adaptive Rule of rule is adapting to each learner model, which is provided by the Database Layer. If the learner model is formed from the required content and knowledge which learners already saved previously, then the law will be applied and processed by the Adaptive Engine components as in Table 5. On the other hand, if the learner model is based on a request to change the assessment of knowledge on the subject studied, the content model will be developed based on the corresponding learner model and the number of random course questions. The number of random questions given corresponds to the model of learning and topics that learners want to change the level of assessment. Then, the content model will not include the learning content and height of tree. To determine the degree of difficulty of questions, with it, the determination of the number and level of difficulty of the questions will be based on previous context models and the level should change. Specifically, the context model will prescribe the number of questions that the system will require the learners to answer; specified levels should change the level of difficulty of questions. The number of questions will be determined by the system default. Meanwhile, the level of difficulty of questions is determined by the ratio between the number of wrong answers the question and the number of respondents answered the question.

$$L = \frac{n1}{n2} * 100\%$$

In which:

- *L: Level of difficulty of a question*
- *n1: Number of wrong answers that question*
- *n2: Number of people answered that question*

For example: With a question has:

n1 = 200; n2 = 500; Then, L = (200/500)*100% = 40%.

Because the level of need to change covers only the value: Poor (1), Average (2), Good (3), Very Good (4), Excellent (5), we will determine the level of difficulty corresponding to the ratio between the Poor [0%, 20%), Average [20%, 40%), Good [40%, 60%), Very Good [60%, 80%), Excellent [80%, 100%]. Specifically, we have the following Table 6:

TABLE VI. CONTENT MODEL AS A CHANGE OF LEARNER

Learner Model	Number of test question	Knowledge's Request				
		1	*2*	*3*	*4*	*5*
1	10	[0%,20%)	[20%,40%)	[40%,60%)	[60%, 80%)	[80%,100%]
2	10					
3	10					
4	10					
5	10					
6	10					
7	10					
8	10					
9	10					
10	10					

The questions will be taken randomly in the ratio range them. For example: Originally, the user selects the context values: Location (home), time (30 minute), concentration (medium) and Topic is Noun. Then, the user completed and evaluated: level of knowledge is Average. The user wants to change this rating to Good. Meanwhile, the number of questions the system offers is 10 and the difficulty of the questions are in the interval [40% -60%).

V. SYSTEM PROTOTYPE IMPLEMENTATION

We implemented CAMLES prototype based on J2ME technology. Therefore, mobile phone needs to support java program as well as GPRS or 3G. In order to use CAMLES, the learners need to download and install application themselves in their mobile phone. At this stage, we develop content model which consists of five main topics: Noun, Verb, Adjectives, Adverbs and Prepositions. Those are considered as parent topics for the entire contents of the system. Under each topic, there will be a corresponding child topic, for example, the child of Adjectives topic has eight children: Manner, Place, Time, Frequency, Sentence, Degree, Interrogative and Relative.

The learner inputs context parameters via mobile interface. The topic content was adapted him. Finishing this topic, the system suggests some question test to evaluate learner's knowledge of the topic and shows the test results as well as recommendations in next screen.

For example, first, students will choose the context, including Time (10 minute), Location (Home) and Concentration (High). Then they will select Topic (Noun). Based on the context and topic, the system will give back to learning content, respectively, as Figure 2.

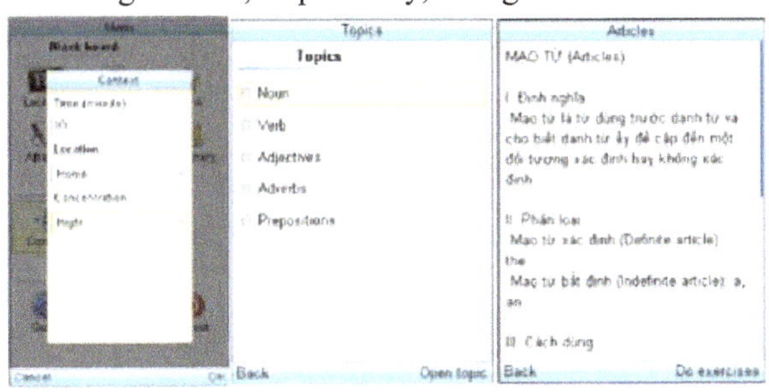

Figure 2. Learner inputs context parameters and adaptive content showed

After that, learners were unsure about their knowledge of course content that the system offers, learners will have to complete the questions that the system provides, as Figure 3.

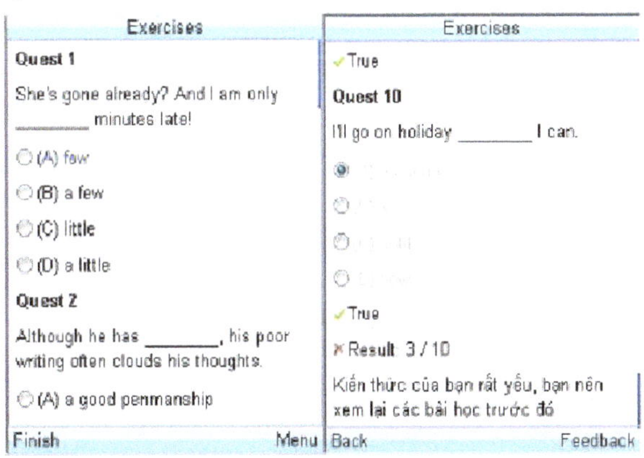

Figure 3. Test questions for evaluating learner's knowledge

End of test questions evaluating the system requires learners to perform, the system will evaluate and respond to learners. The study could change the evaluation of the system on their own knowledge, as in Figure 4.

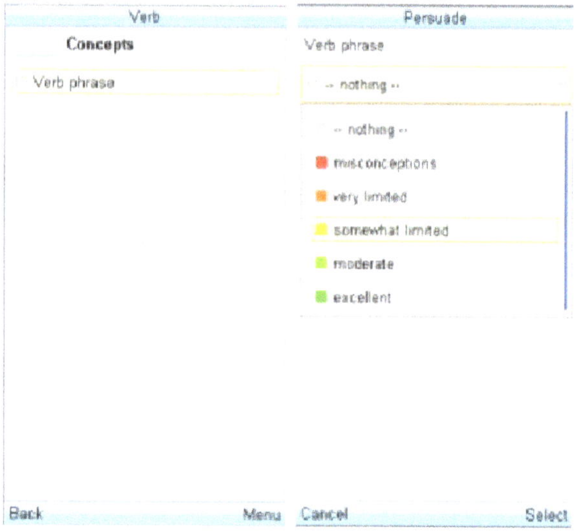

Figure 4. The learner evaluated changes in the knowledge that the system was evaluated

After the learner has cha nged the knowledge level assessment system that has been evaluated, the syste m will provide the learner some questions to confirm that the change of the course is true. Therefore, students will have to complete all the questions. If learn ers complete the right amount of questions in the rate that the system can accept, then students will be accepted that the knowledge itself has changed. Conversely, if learners do not complete the questions, the system will re-evaluate the level of knowledge of the learning results that learners have achieved. The change in the evaluation and testing will stop when they have ac cepted the assessment of knowledge that the system offers, such as in Figure 5.

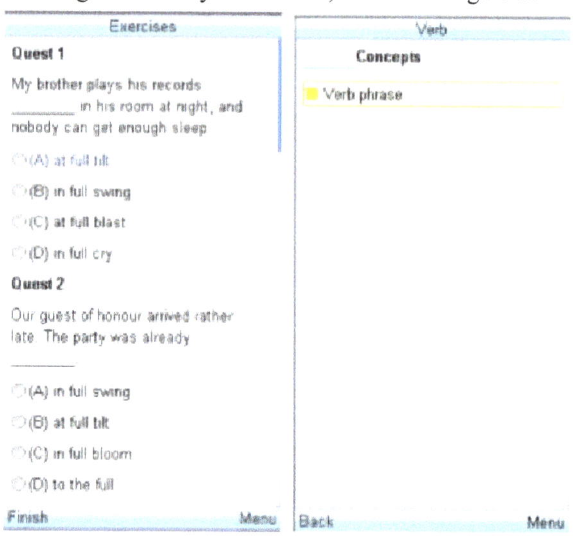

Figure 5. Questions that learners should complete the questions and assessing knowledge levels were changed after completing the questions

To examine our experimentation, we desi gned a questionnaire which includes six questions in order to survey 35 students who used CAMLES system with their mobile phones which support GPRS or 3G to connect to Internet. In order to evaluate our system, students check to one of from 1 to 5 values that 1 was the lowest and 5 was the highest. We classify students into three categories: group one includes students who ha ve never taken the TOEFL test before, group two contain students who have taken TOEFL test and have got below 450 score (paper test), and group three are students have received above 500 score. Table 7 shown average results of the questionnaire for each group.

TABLE VII. RESULTS OF QUESTION-NAIRE

No	Question	G1	G2	G3
1	Do you think the system was easy to use?	3.5	4.0	4.0
2	Would you like to use the sy stem again?	4.5	4.0	3.5
3	Do you think the test question is appropriate for you?	3.0	4.5	4.0
4	The topic that syste m selects is appropriate for you?	4.5	4.0	3.5
5	Did you choose context factors as you in?	3.0	4.5	5.0

According to Question 1 and Question 2, the students were satisfied with system and would like to use the system again. Results of Question 3 showed that the students who have never taken TOEFL test before did not satisfy because the test questions, which we used in prototype, are not easy. The results of this question also denote that the students, who have had high test scores before, satisfied with system. Average score of Question 4 is 3.5 denoting that the topic selected for such students is not good enough because our content model does not have more topic as well as topic content is not in details to support them. Question 5 surveys learners who choose the context whether true as they in or not. For instance, the learners can choose their location at home while they are at Bus terminal. Problem of how to locate learner's location will be resolved in the next stage through location base services. As you see, in Group 1 result, students who have never take the TOEFL test before are interested in our system. However, Average score of Question 5 is 3.0 showing that they often choose the context which is not true as they in. For example, they choose Restaurant location while they are in class.

VI. DISCUSSIONS

Our target users are graduate students who intend to take TOEFL test. However, this approach can be applied to general learners to study English as a foreign language. Our model, context-aware location-dependent learning, adapts learning content according to context as well as learner's knowledge background. To find interests in our system, we compare it with early systems.

In TenseITS [3], learner's knowledge parameter is only calculated at current stage, so if the learner, from second time, backs to the system with the same context factors as he/she inputted previously, the adaptive contents are similar. In our model, learner's knowledge background is stored and is evaluated after the students finish the topic. The results are basic for re calculating value of learner model for next time when learners use system.

The CAMLL [2] is also based on learner level to adapt suitable sentences, however, how the learner's level updating learning progress has not been specified.

At this stage, our learner model is still not distinct for all context cases. Therefore, there are several different contexts having the same value in learner's model. In the future work, we will consider refining the content model as well as adaptive engine in order to match the learner's requests. One notable problem of how to fragment content to display in accordance with the size of the mobile phone is also considered. In addition, we will improve user interface to meet demands of new users. We intend to deploy a web application version of this model because of the disadvantages of the stand-alone application. The web application easily supports different models of mobile phones.

VII. Conclusions

This work represented a personalized context-aware mobile learning architecture. By using open learner modeling technique, this model allows learners to interact with the system in order to get accurate actual learner's demands. In order to do that, we also proposed an adaptation mechanism based on evaluating learner's knowledge to select adaptive learning materials meet the learners. Besides, prototype of use was presented to illustrate the potential of applicability of our system.

VIII. References

[1] N. V. Anh, P. V. Cong, and H. S. Dam. A Context – Aware Mobile Learning Adaptive System for Supporting Foreigner Learning English. In Proceedings of IEEE-RIVF 2010 International Conference on Computing and Telecommunication Technologies, 2010.

[2] K. Al-Mekhlafi, X. Hu, and Z. Zheng. An approach to context-aware mobile chinese language learning for foreign students. In Proceedings of the 2009 Eighth International Conference on Mobile Business, pages 340–346, Washington, DC, USA, 2009. IEEE Computer Society.

[3] Y. Cui and S. Bull. Context and learner modelling for the mobile foreign language learner. volume 33, pages 353 – 367, 2005.

[4] R. G. J. Paredes, H. Ogata, A. Nobuji, Y. Oishi, and T. Ueda. Loch: Supporting informal language learning outside the classroom with handhelds. In WMTE'05, pages 182–186, 2005.

[5] C.-M. Chen, Y.-L. Li, and M.-C. Chen. Personalized context-aware ubiquitous learning system for supporting effectively english vocabulary learning. In ICALT'07, pages 628–630, 2007.

[6] H. Ogata and Y. Yano. Context-aware support for computer-supported ubiquitous learning. In WMTE'04, pages 27–34, 2004.

[7] [6] M. LI, H. OGATA, S. HASHIMOTO, and Y. YANO. Adaptive kanji learning using mobile-based email. pages 520–526, 2009.

[8] S. A. Petersen, J.-K. Markiewicz, and S. S. BjØrnebekk. Personalized and contextualized language learning: Choose when, where and what. Research and Practice in Technology Enhanced Learning, pages 33–60, 2009.

[9] A. K. Dey. Providing architectural support for building context-aware applications. Atlanta, GA, USA, 2000. Georgia Institute of Technology.

[10] B. Hu and P. Moore. "smart context": An ontology based context model for cooperative mobile learning. In CSCWD (Selected Papers)'06, pages 717–726, 2006.

[11] G. D. Abowd, A. K. Dey, P. J. Brown, N. Davies, M. Smith, and P. Steggles. Towards a better understanding of context and context-awareness. In HUC'99, pages 304–307, 1999.

[12] T. Reichenbacher. Adaptive methods for mobile cartography. ICC 2003, (August):10–16, 2003.

[13] N. V. Anh and H. S. Dam. Acgs: Adaptive course generation system - an efficient approach to build e-learning course. In Proceedings of the IEEE Sixth International Conference on Computers and Information Technology, pages 259–265, 2006.

[14] N. V. Anh, N. V. Ha, and H. S. Dam. Developing adaptive hypermedia system based on learning design level b with rules for adaptive learning activities. Journal of Natural Science, Vietnam Nation University, 25(1):1–12, 2009.

Authors

Viet Anh NGUYEN is with the University of Engineering and Technology, VNU, E3, 144 XuanThuy, CauGiay, Hanoi, Vietnam (email: vietanh@vnu.edu.vn).

Van Cong PHAM is with the University of Engineering and Technology, VNU, E3, 144 XuanThuy, CauGiay, Hanoi, Vietnam (email: vcong.pham@gmail.com).

Interdisciplinary Approaches at Institutions of Higher Education: Teaching Information Systems Concepts to Students of Non-Computer Science Programs

Roland Schwald
Albstadt-Sigmaringen University, Sigmaringen, Germany

Abstract—**The aim of this paper is to present a curriculum development concept for teaching information systems content to students enrolled in non-computer science programs by presenting examples from the Business Administration programs at Albstadt-Sigmaringen University, a state university located in Southern Germany. The main focus of this paper therefore is to describe this curriculum development concept. Since this concept involves two disciplines, i.e. business administration and computer science, the author argues that it is necessary to define the roles of one discipline for the other and gives an example on how this could be done.**

The paper acknowledges that the starting point for the development of a curriculum such as one for a business administration program will be the requirements of the potential employers of the graduates. The paper continues to recommend the assignment of categorized skills and qualifications, such as knowledge, social, methodological, and decision making skills to the different parts of the curricula in question for the development of such a curriculum concept. After the mapping of skills and courses the paper describes how specific information systems can be used in courses, especially those with a specific focus on methodological skills. Two examples from Albstadt-Sigma-ringen University are being given.

At the end of the paper the author explains the implications and limitations of such a concept, especially for programs that build on each other, as is the case for some Bachelor and Master programs. The paper concludes that though some elements of this concept are transferable, it is still necessary that every institution of higher education has to take into consideration its own situation to develop curricula concepts. It provides recommendations what issues every institution should solve for itself.

Index Terms—**Business education, enterprise information systems, interdisciplinary approaches, university curriculum development.**

I. Introduction

In today`s economic environment, characterized especially by globalization and the digitalization of the economy, hardly any modern higher education program, for example in the field of engineering, life sciences, business or education will be able to offer its students an education completely without computer systems content.

The reasons are obvious: Modern-day computer information systems are ubiquitous at many workplaces. Professionals from many different areas, from, say, architecture to zoology work regularly using all types of computer information systems. In some cases, these non-IT professionals have to be proficient even with the conceptual side of information systems.

As an example, one can take a marketing manager who has to oversee the implementation of an electronic-business (e-business) solution, a potentially strategic project for a company. This e-business project may include for example a new or improved online shop or an online marketing system.

Or one can think of the VP (Vice President) Logistics of a manufacturing company. Optimizing his area he may well be using the latest release of a major supply chain management system, for example from a company like Infor or Oracle. With such a system he could improve managing the changing demand cycles in his company, for instance.

Or, as a third example, imagine the CEO (Chief Executive Officer) who is using business intelligence software from - for example - Microstrategy to discover trends and interrelationships in his business. Business intelligence systems attempt to improve a company's business performance based on analysis and making use of information. Gartner, a Stamford, CT based information technology research and advisory company defines business intelligence (BI) as follows: "An inter-active process for exploring and analyzing structured, domain-specific information (often stored in data ware-houses) to discern business trends or patterns, thereby deriving insights and drawing conclusions. The business intelligence process includes communicating findings and effecting change. Domains include customers, suppliers, products, services and competitors." [2]

A. Computer Science at Institutions of Higher Education

Universities from around the world have realized the importance of IT concepts for students and have included computer information systems content into their curricula. Computer science courses in business administration programs were included in the early 1990s when microcomputers became popular. The wide availability of the Internet in the late 1990s saw another wave to include more computer science content in non-computer science curricula [3].

The question that arises here of course is how, especially from a conceptual point of view, this can be managed, i.e., how can the manager of a study program at an institution of higher education proceed in a structured way if he or she wants to adapt the program so that it contains the best possible content for the students.

Of course one - possibly the easiest - option for the program manager would be to take a laissez-faire approach and do nothing at all, i.e., leave information systems out of strategic curriculum planning. However, with this approach, an opportunity to design and / or maintain the curriculum in a way that caters to the interests of their future graduates would be missed.

B. Defining the Roles

Since the role of information systems can vary from one study program to another, one should reflect on the particular roles information systems play for students and graduates of the specific program in question. Computer information systems is such a comprehensive discipline in science that it is often taught in dedicated schools or departments. Business administration on the other hand is also a very large field. A generic bachelor of business administration program will probably consist at least of the core concepts of the business administration field, such as accounting, finance, operations management, marketing, human resources and so on. Unlike pure computer science programs, the main subject of business administration programs is companies and not information systems and information technology in general. Therefore the main focus is primarily on the application of information systems and not so much on their design and development. Information systems in business administration programs hence become important if they support business administration processes. In other words: information systems can be seen as tools to support the business in general.

In a very first step to develop a curriculum that includes information systems content in a systematic way one will have to define the role of IT in the university program being planned. For a business administration program for instance it therefore might be deemed essential for example that a student understands the value of information systems for companies from a strategic as well as from an operational point of view. Also, they have to understand information management in general. Teaching information systems in business programs must enable students to wisely use enterprise information systems.

One effect that can be observed is that the regular usage of information systems in classes increases the acceptance of IT among business administration students. The more students see how a very large portion of concepts from "non-IT" modules or courses such as Human Resources, Finance and Accounting, Marketing and so on are supported by information technology, the more they understand the importance of information systems for today's (and probably tomorrow's) companies.

I argue that it isn't necessary to include as many information systems into the courses as possible but to rather introduce relevant types of systems to the students in a coherent, logical manner.

TABLE I.
ROLE OF INFORMATION SYSTEMS IN BUSINESS ADMINISTRATION PROGRAMS

Conception	Practice	Job related
• Value of Information Systems	• Know how to model business processes	• Prepare for upcoming internships
• Experience complexity of large enterprise systems	• Use enterprise information systems	• and for the job market after graduation
• Understand information management		

It is also important information systems in university courses prepare students for internships and for the job market in general. Students should experience the complexity of ERP (Enterprise Resource Planning) and other large enterprise systems to distinguish them from other types of information systems. Also, we take the view that one should teach students how to model information flows and business processes, activities that are often necessary to implement large IT systems in organizations.

See also table I for possible roles information systems can play in business administration programs.

C. Background: Albstadt-Sigmaringen University

I will base my suggestions for a concept to teach information systems to students of non-computer science programs on the situation of the business administration programs at Albstadt-Sigmaringen University. Though this concept is based on a business administration program, the general approach could be similar for some, possibly many, other programs.

Since the examples in this paper are based on the situation at a specific university I will shortly give some background information about this university so one can better assess the concepts proposed in this document.

Albstadt-Sigmaringen University is a public university in the German federal state of Baden-Württemberg (Southern Germany). It has two campuses, one in the city of Albstadt the other in the town of Sigmaringen. Both cities are about 30 kilometers (19 Miles) apart from each other. The university offers Bachelor and Master level courses in the areas of Engineering, Life Sciences as well as Business and Computer Sciences. Currently (Spring/Summer 2010) the university has around 2.600 enrolled students.

There are two pure Business Administration programs available at Albstadt-Sigmaringen University: A Bachelor of Science and a Master of Science program. Students in the last year of the Bachelor program can choose between three specializations: Marketing, Operations Management, and Financial Controlling. The Master of Science program intentionally has a general management focus; therefore there are no specializations available to graduate students.

Both programs therefore have no specialization in MIS (Management Information Systems) or similar. Teaching information systems thus has to happen primarily within the mandatory courses of the programs.

Another particularity here is that the Bachelor of Science in Business Administration program consists of seven semesters with the fifth semester being a mandatory internship semester. One ramification is that students should be familiar with some major business information systems before they start their internship. Another particularity is that the Master of Science in Business Administration and Management program is a so-called consecutive Master program, i.e., is a direct continuation of the previous Bachelor program.

II. CONCEPT

The basic idea of the concept found at this university's business administration programs are the close integration of the skills and qualifications the students should have upon graduation with the courses being taught and the information systems being used in those courses. The students acquire their skills by participating in courses. Depending on the skills emphasized in a course, information systems may or may not be used for teaching purposes.

An important step during the development of a curriculum in general are identifying the relevant qualifications and skills the Business Administration student should have at the time he or she graduates from university. At Albstadt-Sigmaringen University's Business Administration and Computer Sciences Department they are classified into four core skill areas, especially relevant to the Business Administration programs in question:

A. Methodological skills:

Business administration students have to acquire problem solving skills. As they will become managers they will have to have excellent troubleshooting as well as systematic problem solving skills in order to be successful. On a more basic level, methodological or "how to do" skills also include knowledge about how something can be done. For example, in the case of teaching information systems to business administration students, this could mean letting Bachelor students complete a standard sales transaction with ERP (Enterprise Resource Planning) systems such as SAP ERP or Microsoft Dynamics NAV. By completing such a standard transaction they learn that a business process needs different resources, documents, and organizational units. They will experience the complexity of large enterprise information systems. Also, they will see what skills employees should have when working with such systems and programs.

Another example for acquiring methodological skills would be solving a business problem such as tax calculations with a spreadsheet tool. The use of Microsoft Excel to solve simple to intermediate business problems in the first semester of the Bachelor of Science in Business Program for example will – amongst others - enhance the students' acceptance of information technology so that it will be easier in subsequent semesters to introduce them to more advanced programs and systems.

B. Knowledge:

In this context this implies acquiring in-depth, explicit knowledge of relevant subjects, in other words acquiring "hard skills". Unfortunately this often implies learning (and sometimes simply memorizing) facts and information. The providers of such facts are usually experts and other knowledgeable professionals.

However, in a higher education course this also means that students (and lecturers of course) can and should critically analyze those facts and information, even if they come from experts. In the case of information systems, this for example could mean teaching information technology architectures. Students should recognize that there are different ways to design and develop information system structures in a company. The students should understand the advantages and disadvantages of the different types of IT architecture and they should be able to discuss what type of IT architecture is useful for what type of organization for example.

C. Social skills:

They include for example the capacity for teamwork, oral presentation skills and interactive skills. At Albstadt-Sigmaringen University for example, for the duration of a semester, students have to work on a fictitious information systems project as part of a project team. Recently (Summer Semester 2010), a cooperation agreement with a regional IT company has led to the situation where the students can work on real life IT projects as well. The details of this sub concept of the curriculum design will be presented later in this paper.

D. Decision-making skills:

Managers have to make decisions in a wide range of areas. Decision making can take place in a systematic or in a non-systematic way. Systematic decision-making approaches have the advantage of potentially resulting in more sophisticated, well-thought-out decisions.

In the area of information systems, this could for example mean selecting a sales force management system based on previous requirements analysis techniques. As one can infer from figure 1 which shows what areas of skills are emphasized at what level of study in the Bachelor of Science Program at Albstadt-Sigmaringen University, applying decision-making skills makes only sense once one has acquired expertise / knowledge, methodological skills and soft skills. Decision-making skills depend on these other skill types.

Of course there are some other skill areas, such as learning skills, emotional skills, and intercultural skills and so on. However, it appears to be practical and purposeful to have a fairly short list of skill areas. Some of these other skill areas can be in some way included in the

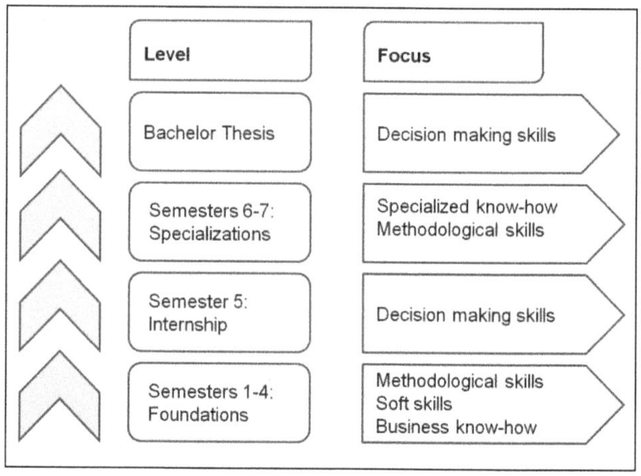

Figure 1. EMPHASIZED SKILL AREAS ON DIFFERENT LEVELS OF THE BACHELOR OF SCIENCE IN BUSINESS PROGRAM

four skill areas defined here (e.g., intercultural skills or communication skills could both be defined in a way so that they can be included in the social skills area; similarly, critical and analytical thinking skills could be included in the knowledge / hard-skills area); other skills might be only to a lesser degree relevant for business administration graduates.

Though all four relevant skill areas mentioned here have connections to information systems, one can also see that especially when it comes to methodological skills it makes sense to work with actual information systems. Therefore, in the next steps, when we identify the skill areas in the curriculum, we have to have a closer look at courses where methodological skills are in the focus.

Those four core skill areas in reality often overlap. As was stated further above, especially the decision making skills depend on the other forms of skills. For example, a manager who wants to effectively select one information system over the other, he or she has to have at least a general knowledge of the architecture of such systems. In practice, there are often no clear borders between the four core skill areas. Figure 2 illustrates this.

E. *Expectations of Potential Employers*

Since these skill areas are indispensable for the development of a curriculum, one has to find out what skills potential future employers of the university's students consider essential. While there seems to be at some institutions a lack of strategic interaction between business schools and the business community [4], this seems to be not a serious problem in the case of Albstadt-Sigmaringen University where there are regular contacts between enterprises and the university where one gets feedback on what skills are demanded from the graduates. Feedback is given for example by internship companies and as well by the advisory board of the Master program that mainly consists of members from the business community.

Also, on the university's supervisory board, there are members from the business world that give valuable feedback regarding what skills are demanded from graduates in the workplace and what the expectations of future employers will be. In addition to that, some visiting lecturers work regularly for local companies and are aware of the skill sets that graduates should possess. Other information sources to see what companies demand from graduates are job advertisements, whether from employment websites, regional and national newspapers, or job openings sent directly to the university.

The university is small enough so that this kind of highly valuable information will be disseminated within the organization and can actually be used.

F. *Identifying the Relevant Employers*

The Business Administration programs at Albstadt-Sigmaringen University are mainly targeted to the needs of SMEs / SMBs (small and medium enterprises, respectively small and medium sized businesses), especially manufacturing companies in the region surrounding this institution. That does not mean of course that students cannot or will not work for larger companies in the end. However, the focus of the Business Administration programs in question is on educating and preparing future managers of SMEs / SMBs.

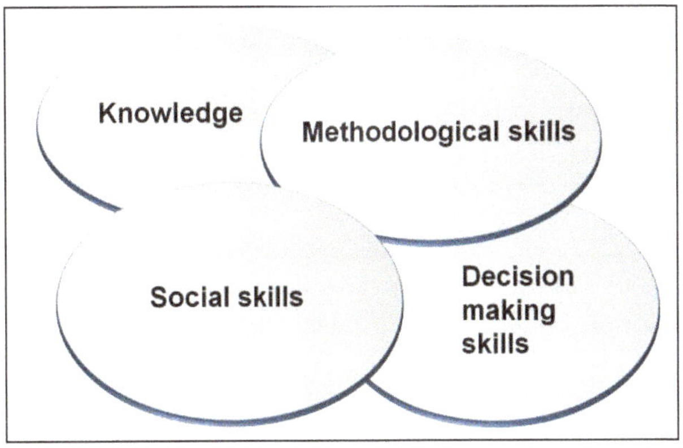

Figure 2. OVERLAPPING OF THE FOUR CORE SKILL AREAS

TABLE II.
COMPARING LARGE COMPANIES WITH SMALL AND MEDIUM-SIZED BUSINESSES

Large Sized Enterprises	SMBs
Company and brands are worldwide known.	Company and brands often little known outside regional markets.
International production sites.	No or little international exposure.
Many employees with international experience.	Few internationally experienced employees.
Access to international financial markets.	No similar financing options available.
High research and development capability.	Often only very specialized research and development.
Potential to temporarily offset losses caused by risky international business activities.	Limited potential to spread risks within the organization.

It is important to define the type of companies the programs want to address with their graduates because it is a big difference whether one manages a large or a medium-sized company. There are several very general factors that differentiate small and medium-sized businesses from large companies [5].

Apart from having less employees and a smaller revenue base, smaller companies can have several typical characteristics that distinguishes them from large sized companies. For example, sometimes they have a smaller know-how with regard to international markets. They are often more dependent on a small number of large customers. Also, they are typically not that well versed with respect to management techniques and they are often short on financial as well as human resources.

Despite those rather negative characteristics, one can also find some potential advantages a smaller company might have over a bigger company. For example, they often have a production and sales program that is more specialized and therefore suitable to use market niches. Often, the motivation of the employees can be very high, due for instance to the short information and communication channels in these types of companies. Often SMEs / SMBs are more customer friendly and have a higher customer proximity. Smaller organizations are often very flexible compared to larger companies and have a shorter reaction time when it comes to adapt to new challenges.

TABLE III.
MAPPING SKILLS AND COURSES

	Metho-dological skills	Know-ledge	Decision-makings skills
Information Systems I	◆		
Internship Preparation			◆
Accounting I		◆	
Business Statistics	◆		
Intro. Business Law		◆	
Operations Management		◆	
Marketing Seminar			◆
Quant. Applic. in Business	◆		
Bachelor Thesis			◆
...			

G. *Mapping Skills and Courses*

After having defined the role of IT in the Business Administration program under review and if one also has specified what type of companies one targets in the sense that students will primarily seek work there after graduation, then, in the next step of this concept, these skills and qualifications are being mapped with the courses offered for the business administration program in question.

Since there are only two dimensions involved in this step, it is an option to create a simple table highlighting those skills that are particularly important for a specific course.

Table III is an example how such a table can look like. It is based on the Bachelor of Science in Business program at Albstadt-Sigmaringen University. The two-dimensional table for the Master of Science in Business and Management program is similarly arranged but also includes a fourth column for "soft skills". In order to keep this paper clear and brief it is not included in this document.

In the case of IT systems used for teaching, especially those courses with a strong focus on methodological skills are identified. In the above example there are three courses that stress methodological skills. In the subsequent step, those courses where it makes particular sense to use information systems are identified.

To better assess what IT / IS content should be allocated to which course one must also have a thorough look at the sequence of the relevant courses identified in the previous step.

Finally, in this concept one investigates possible information systems that could be used in those courses. At the university on which this example is based on, it was for example decided to use Microsoft SQL Server for the Data Analysis course and SAP Business Intelligence for the Data Warehouse course. During the Project Management course for instance, students will work with Microsoft Project Management. Those systems were selected because of their wide usage amongst decision makers.

H. *Managing Necessary Resources*

To implement such a concept one needs resources, i.e., human resources, information systems, and available teaching facilities such as computer labs. In the case of Albstadt-Sigmaringen University's Business Administration programs, there are eleven tenured professors, nine non-tenured faculty / staff members, and a number of external lecturers available. Two out of the eleven tenured professor positions have a specific IT focus while five out of the nine non-tenured staff members take care at least to a certain degree of information systems.

Among the problems one might encounter when including concepts from one discipline, such as computer information systems, into a university program of an essentially different discipline, such as business administration, is that there might not be sufficient – with respect to quantity and quality – resources available. The available full-time professors of business administration programs might not be familiar with relevant enterprise information systems. In that case it will be necessary to either create such a position within the department or make use of computer information systems faculty from other departments. However, by using faculty members from other departments it might be that they have only a very technical view of information systems and cannot adapt their teaching style and content to the requirements of a business administration program. The optimum would be to have someone teach computer information systems who has knowledge in both disciplines, i.e., IT and business and has experience teaching business administration students. Lecturers with that type of interdisciplinary skills however are not abundant at smaller universities such as Albstadt-Sigmaringen University.

If for whatever reason it is not possible to get adequate help from other departments it is an option to create IT dedicated positions within the business administration department. By offering IT-oriented faculty positions within the business administration department it is at least possible that this will especially attract instructors with an affinity toward business. That type of instructor can be potentially more useful than a pure IT professional with a limited ability to think outside of the box.

One way of working around the possibly arising problem that there might not be enough qualified tenured faculty available is to use scientific coworkers or assistant professors who do the teaching part especially when it comes to offer in-class exercises using real enterprise information systems. Of course this also works only if these staff members on their part are computer literate enough.

III. EXMAPLES

The following two examples show how such a concept can be implemented, based on two courses from Albstadt-Sigmaringen University.

A. *Enterprise Information Systems*

As Bachelor of Business Administration students at Albstadt-Sigmaringen University have to complete a compulsory internship in their 5th semester of studies, it will be of potentially great benefit to them if they had the opportunity to work with modern enterprise information systems beforehand. An often stated optional or mandatory requirement one can read in internship offers is that applicants should have at least a basic knowledge of information systems. Often it is stated in those offers that it would be an advantage for an applicant to be familiar with programs such as Microsoft Office but also with larger enterprise information systems such as SAP ERP. However, for internship positions it is normally not the case that applicants are required to have much more than good navigation and general application skills in large enterprise software let alone deep, specialized knowledge in certain system modules such as finance or logistics.

As a typical intern will work with enterprise information systems such as ERP systems or data base management systems, the teaching of that type of systems therefore has to take place before the internship semester. According to the course scheme at Albstadt-Sigmaringen, a course called "Enterprise Information Systems" will take place in the fourth semester of studies, i.e., the semester before the mandatory internship semester. As during the internship semester it will be especially the "to-do" skills, i.e., the methodological skills that come into play, the course "Business Information Systems" (4th semester, Bachelor of Science in Business Program) should have a focus on methodological skills as well.

Since SAP ERP and Microsoft Dynamics NAV are often cited enterprise information systems in job offers and also because students who have completed their internships already often report having worked with those systems during their internships, those two systems were selected to become part of the "Enterprise Information Systems" course. The university has contractual agreements with both companies, i.e., Microsoft and SAP AG. This allows the university to use their systems for either free or at least at educational prices. Apart from the systems themselves, included in those contractual agreements are usually also training material and training for the teaching and research staff.

In this business information systems course, the students get an introduction to both systems. Thereafter they learn how to navigate and complete some simple business processes such as the procurement of material and so on. There are 12 semester hours reserved for exercises with the systems. The maximum group size usually doesn't exceed 25 students. The students have access to the systems for the duration of the semester so that they can do the exercises again at their leisure or explore the systems on their own. There are also staff members available if questions arise.

B. Project Enterprise Software Systems

In the 7th semester of the curriculum for the Bachelor of Business Administration program appears a course called "Enterprise Software System Project". This course is supported by a regional IT services company that amongst others is a Microsoft partner and offers the ERP system "Microsoft Dynamics NAV." The support of that company consists for one in supplying real projects, i.e., they present actual internal projects such as the development of a new module for an ERP system. The company also provides two of their staff members to facilitate the students' project work, i.e., provide help if the students encounter problems and so on.

The students work in groups of seven to nine persons. The groups each elect a project manager who is responsible for the organizational issues of the project. At the start of the course, the students get an introduction to the business cases they have to solve. They will get some information on how to approach this problem in general, e.g.,

TABLE IV.
IT CURRICULUM AT ALBSTADT-SIGMARINGEN UNIVERSITY

Sem.	Courses with IT content	IT tools / systems used
Bachelor of Science in Business Administration		
1	Introduction to Business Information Systems	MS Excel
2	---	---
3	Intermediate Business Information Systems	MS Approach, SAP BI, MS Visio etc.
4	Enterprise Information Systems	SAP ERP, MS Dynamics (NAV), ARIS Toolset
5	Internship	Typically various
6	---	---
7	Project Enterprise Software Systems	MS Project, MS Dynamics (NAV), SAP BI etc.
Master of Science in Business Administration and Management		
1	Data Warehouse Systems Data Analysis	MS SQL Server, SAP BI, Cubeware, Microstrategy
2	E-Business Management	---
3	---	---

they are told to do a market analysis, to define the modules' minimum and maximum functions and so on.

If the project involves an information system the students are not yet familiar with, the functionality of this system will be demonstrated by someone who is in charge of the course and the students will also have the opportunity to do some exercises on the system itself.

At the end of the semester, the student groups present their findings and make recommendations. During the whole semester the external IT staff as well as the professors in charge at the university support the groups and give the groups directions and feedback. The IT services company then decides whether and to what extent they will make use of the projects' outcomes.

A rough overlook of the sequence of the major contents of the mandatory IT courses offered at Albstadt-Sigmaringen University is given in the following table [6]:

IV. CONCLUSION

In this paper I have explained a possible curriculum development approach for teaching information system content to business administration students at a university in Europe, based on an example from Albstadt-Sigmaringen University.

While in recent years it has become a matter of discussion whether IT still creates value and adds to the competitiveness of a business [7], the adaptation of non-computer science curricula to include information systems content has progressed and flourished.

An important point for the development of a curriculum are the relevant qualifications and skills the students should have at the time he or she graduates from university, such as methodological skills, a working knowledge of relevant systems, social skills and decision-making skills. A practical approach to teaching business information systems is to accompany lectures with realistic exercises on major information systems.

A. *Benefits of the Concept*

As stated earlier in the text, one option for the business administration program manager would be not to take any action regarding the subject of information systems and technology at all, i.e., to disregard this issue from a conceptual point of view. The consequence of such a non-managed approach would be that information systems topics will be more or less randomly instructed, without any prior systematic planning of the IT content. As a result, every instructor in the program teaches information systems at will or – maybe even worse – not at all. In a university program with a high portion of mandatory courses as is the case at Albstadt-Sigmaringen University's Business Administration programs, this will lead to an incoherent study plan. To avoid these problems it therefore makes sense to decide to take a managed, systematic approach.

There are several, however largely intangible benefits to a managed, systematic approach. E.g., in the case of Albstadt-Sigmaringen University, the implementation of the concept took into consideration particularities of the institution and the academic programs concerned. The Business Administration programs are targeted at the needs of future mid-level managers at mid-sized companies / SMBs and the IT content too is geared toward this group. So when entering the job market, the graduates have a fairly good sense of how information systems can help such a business being more successful.

The mapping of skill areas and courses allows identifying modules where the use of information systems is particularly favorable. This will lead to a better, more efficient allocation of resources. Information systems will only be used where they add value to the education of the students. At the same time when it does make sense to use information systems in a specific course, the system used will benefit the learning objectives of the university program.

Also, this method forces the program manager and his or her collaborators in the curriculum planning process to think about the skills and qualifications the graduates should have from a strategic and tactical point of view. By using some sort of matching table to map skills and courses one gets a clear, transparent overview of the distribution of the skill areas within a university program. Done thoroughly, it will lead to a well balanced distribution of the different skill areas over the course of a university program.

To sum it up, with this structured yet flexible enough concept it was possible to systematically drive the curriculum and its content toward the program manager's stated goals.

B. *Limitations of the Concept*

On the other hand, there are also some problems this approach could not completely solve. Especially the interface between the undergraduate and the graduate programs is difficult to handle. Though being a consecutive Master program, a considerable number of students have not been enrolled in the previous Bachelor in Business Administration program. Therefore, those students may well have a different skill-set when they start their Master studies at Albstadt-Sigmaringen University than those students who had been enrolled in this university's undergraduate program.

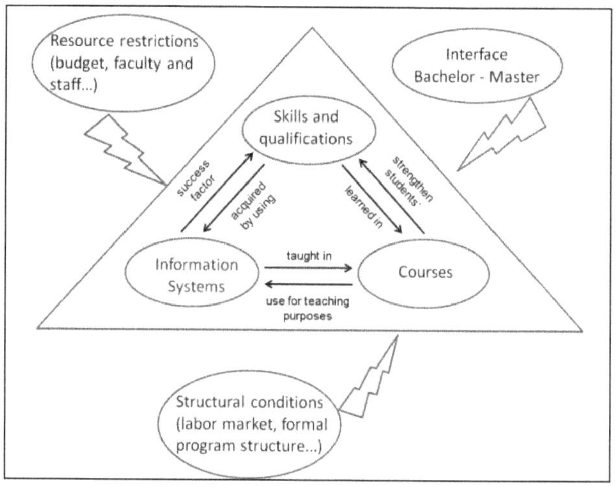

Figure 3. ELEMENTS OF AN INFORMATION SYSTEMS CONCEPT TO BE MANAGED

Another problem is that the IT content is only partially reflected in the study regulations. That means that to some degree the university is dependent on the support of the dean of studies. He or she plays a vital role in the design not only of a curriculum but also how that curriculum is actually filled with content.

And another restriction can be seen in the teaching faculty. If they are not computer-literate, it is difficult to include computer information systems in the courses. One workaround can be for example to use teaching assistants or external lecturers.

In addition to skills and qualifications requirements from potential future employers, designing a curriculum should also consider what competing universities are offering their students. Inferring from discussions with students it becomes clear that they do compare different institutions before deciding to enroll in a specific one. That implies that a university should offer a program that is at least as competitive as those of institutions of higher education one usually competes with for the best talented students. Admission interviews for the Master of Science program often reveal the desire of prospective graduate students to learn more about analytical decision making in organizations for example. For the curriculum that means that business intelligence and similar subjects have to be included in the offering.

An overview of some of the elements of an information systems concept for a non-computer science course is given in figure 3.

One aspect one can deduce from this figure is that the environment surrounding the university and its relevant programs plays a big role in the development of curriculum concepts.

C. *Recommendations*

Therefore, though this concept theoretically could be adopted by similar universities, one should point out that the curriculum design should certainly take into account the particularities of every individual institution. A mere takeover of concepts, including those presented in this paper, without adapting them to the realities of the relevant institution of higher education will probably not be crowned with success.

Based on the above analysis, among these particularities that one consequently has to take into account during the curriculum development process are:

- The class sizes and the number of places per computer lab. For courses where one uses software systems there should be enough computers available in a classroom or computer lab so that every student in that class can do the exercises on the system directly. At a typical undergraduate Business Administration course at Albstadt-Sigmaringen University, the number of students attending will be somewhere between 30 and 55. The largest computer lab has 35 workstations; therefore one usually divides one particular semester into two groups, one morning group and one afternoon group.
- The qualifications of the available tenured and non-tenured teaching staff. The scale of the planned IT content depends significantly on how many faculty members are available either inside or outside the department, and what their qualifications are.
- The requirements of employers. Starting point for the development of a curriculum in general and the IT content of a university program in particular in the concept presented in this paper are the requirements of potential future employers of the students. If a program manager intends to prepare the students for the relevant job market, he or she should exactly know what qualifications and skills potential employers expect when hiring graduates of that university.

As there are very few publically available documents and other resources on the subject of information systems teaching concepts, I suggest that further research should be carried out in order to assess the effectiveness of various curriculum concepts at universities. One way this could be done is by asking alumni of institutions with different curriculum concepts if and how information systems concepts they have learned during their studies have helped them in their career. One research goal would be to assess the effectiveness of various curriculum concepts. Precondition for this however would be to learn more about the existing curriculum concepts at institutions of higher education. It would be especially helpful if we could learn more about the process how universities include computer information systems in the curricula of non-computer science programs.

References

[1] R. Schwald, "Incorporating Information System Content into Non-Informatics Curricula: An Example from a Business Administration Program", in *Proceedings of the IABE 2010 Bangkok Summer Conference*, vol. 7 (1), G. Savatsomboon, M. D. Gavriletea, Eds. IABE, 2010, pp. 45-49.

[2] Gartner (2010, April 30): *Business Intelligence* [Online].

[3] F. G. Cauley, F. D. Aiken, and L. K. Whitney, "Technologies Across Our Curriculum: A Study of Technology Integration in the Classroom", *J. of Edu. for Business*, vol. 85 (2), pp. 114-118, Nov./Dec. 2009.

[4] R. K. Plice, B. A. Reinig, "Leveraging Alumni and Business Community Relations to Assess the Information Systems Curriculum", *J. of Edu. for Business*, vol. 84 (3), pp. 142-150, Jan./Feb. 2009.

[5] J. Gutmann, R. Kabst, *Internationalisierung im Mittelstand*, Wiesbaden, Germany: Gabler, 2000, pp. 10-11.

[6] R. Schwald, "Curriculum Development for Teaching Business Information Systems to Business Administration Students – An Example from a University of Applied Sciences", in Multikonferenz Wirtschaftsinformatik 2010, M. Schumann et al, Eds. Universitätsverlag Göttingen, Göttingen, Germany, 2010, pp. 339-340.

[7] N. G. Carr, "IT Doesn't Matter", *Harvard Business Review*, vol. 81 (5), pp. 41-49, May 2003.

Author

Roland Schwald is a scholar with the Department of Business and Computer Science at Albstadt-Sigmaringen University, Germany. He holds a Bachelor of Science degree in Business Administration with a specialization in Marketing. Also, he earned an MBA degree from the University of Miami at Coral Gables with specializations in Finance and International Business. He has taught courses in systems analysis as well as some others. His main interests are in global enterprise information system management, business education, and international business management. E-mail: Schwald@hs-albsig.de

9

Aspects of Content, Context and Adaptation Modeling in Mobile Learning Application Design

N. Uday Bhaskar[1], U.S.N Raju[2], P.Govindarajulu[3], B.V.Ramana Reddy[4]
[1] Govt. Degree College, Pattikonda, A.P., India
[2]NIT Warangal, A.P., India
[3] S.V. University, India
[4] MERITS, Udayagiri, A.P., India

Abstract—**Mobile leaning application development has been influenced by the e-learning models, methods and a variation of these for the better understanding and accommodation of additional elements that prop up in m-learning scenarios and applications. Here uniquely designed models for various aspects of m-learning scenarios and situations by concentrating on Learner Styles, Contextual Features, Learning Content, Learner Behavior, and Adaptation methods are proposed.**

Index Terms—**Adaptation, Content, Context, Mobile Learning.**

I. Introduction

Learning has gone through major changes from its inception, from the basic form of teacher-centric to class room learning to distance learning to e-learning and now, to present m-learning , i.e., mobile learning. Some call m-learning as the form of e-learning with the support of mobile devices and mobile technologies. But mobile learning is more than e-learning through mobile devices, as it is highly influenced by the various elements that come into play when the learner is mobile in nature with his mobile device. Mobile learning can be distinguished "by rapid and continual changes of context, as the learner moves between locations and encounters localized resources, services and co-learners"[10], and these different situations are described by different learning contexts. Nowadays, time is the most valuable, and people tend to be too busy and are spending more time in traveling and the inability to be at one place brings the need for supporting learning activities at any time, place and through different devices; this is the origin of mobile learning [20], whose main purpose is to allow high degree of mobility for learning.

In Human Computer Interaction (HCI), a context feature is any information that can be used to characterize and interpret the situation in which a user interacts with an application at a certain time. The context aware applications are described as "more attentive, responsive and aware of their learner's identity and their learner's environment [40] and hence can derive the learner's needs implicitly in the context which surrounds him/her at any point in time. The task of a context aware mobile learning (CAML) application is to sense the mobile environment and adapt to the changing context during a student's learning process [10].

Characteristics [12] of mobile learning include: a) Urgency of learning need, b) Initiative of knowledge acquisition, c) Mobility of learning setting, d) Interactivity of the learning process, e) Situatedness of instructional activities and f) Integration of instructional content. One of the most straight forward application for the usage of mobile devices as educational supporting tool is messaging where in the learning content is delivered to the learners is implemented in many research experiments [47][48][44][23][16]. Y.S. Chen et. al. [11] used content-based butterfly image learning (BWL) system.

The capabilities that mobile learning possesses made it widespread, and learners are now able to learn at any time and any place with the support of mobile technologies. With context aware mobile learning, many different learning contexts emerge because of the dynamic and continuous change in the learning settings of the learner's mobile learning environment. Mobile learning can be distinguished "by rapid and continual changes of context, as the learner moves between locations and encounters localized resources, services and co-learners" [10]. Context is a key in the design of more adaptive mobile learning systems [35] and context awareness must be integrated within the systems in order to be truly effective [31].

There are many proposed definitions and taxonomies of context. Shilit [51] divided context into three categories: Computing Context, User Context and Physical Context. Chen and Kotz [22] extended it by adding a time context. Dey, Abowd and Salber [2] gave four categories: identity, location, status and time. The importance of obtaining the actual location of the learner helps to identify the contextual features surrounding the location that affect/support the learner's ability to study. Schmidt et al. also found that location is the most often used context parameter and assert that mobile applications could benefit from a broader notion of context. This work investigates the use of multiple sensors to gather a wider range of context information [43]. The systems [9] also includes two location-specific features which their activities adapt to accordingly, namely the concentration level of the learner and the frequency of interruption at a specific location.

From an informational perspective context aims to provide an information space that can be "adapted according to the user's context", and complemented by tools/processes that promote the socially situated construction and sharing of context [39]. Wang's six dimensions of the M-learning context: 1) Identity 2) Learner 3) Activity 4) Collaboration 5) Spatio-temporal and 6) Facility [52]. The first four of these would establish the situational context of M-learning, and the last two would be associated with the environmental context. It seems very likely the future context aware applications will capture a rich variety of context information and also exhibit a rich variety of context-aware features than the present situation. Providing support for mobile learning application designers in the integration of a wider range of context into the applications will encourage the development of useful and compelling mobile learning applications for mobile devices of the mobile learners and tutors.

Learner's needs are mostly intangible which in most of the cases affected by habit, self-image and even issues of motivation [39] (e.g., a person might be more active in the morning than in the evening). The design of a system must focus on reducing communication barriers by analyzing what can be known about a learner and how to support that information with task, learner and system models. As a rule learner must play an active role in the definition of the context about which the system must be aware of [39]. In addition to being able to obtain context information, applications need to have some "intelligent" component which functions as a predictor of a user's intentions. Developers can intelligently use context information in four primary ways [42]: 1) resolving references, 2) tailoring lists of options 3) triggering automatic behaviors and 4) tagging information for later retrieval.

The fact that different learners have distinct needs, preferences of personal features has been considered with adaptation [38] [17] and recommendation [37] purposes. In the context of mobile learning, the contents also need to be adapted to different devices [3]. A classification of the characteristics that can be used with adaptation purposes in mobile learning environments is presented in [1]. In [34], Yudelson, et al. used techniques for interpreting student behavior data and constructing student models can be classified into three categories: formal, semi-formal (heuristic) and informal (ad-hoc). Formal approaches use methods either from cognitive sciences or from artificial intelligence. Traditional symbolic AI techniques like semantic networks, rule-based reasoning tend to be replaced by non-symbolic techniques like machine learning, neural networks, genetic algorithms, Bayesian models; while symbolic AI techniques try to infer knowledge about the user based on each of his/her actions at click level, non-symbolic techniques have the advantage of extracting information about the user from his/her entire navigation path, viewed as a whole.

For MOBILearn project at the University of Birmingham, an interactional model of context has been developed, having advantages like [4]: it ensures that context is much more than location; it can be used to guide effective choices and propose future actions, rather than simply acting as a filter on information. The University of Tampere for MOBILearn project has developed an adaptive user interface system [4] in relation with the context awareness subsystem. The adaptive user interface subsystem received data from the context awareness subsystem and this contextual, presence and device information was utilized to optimize the user interface for a mobile phone and a PDA.

II. Content-Context-Adaptation Cycle

For a given or available learning scenario, the basic information to acquire is content design and development, context identification and adaptation mechanisms to place. These three elements and their corresponding requirements are drawn from the learning scenario's objectives and implementation. Among the three, the content design and development is primary as it is the knowledge to be delivered to m-learner through an m-learning application. When the m-learning application becomes context aware, then comes the role of context into play. Next, the adaptation mechanisms or strategies that vary according to the contexts identified and modeled. It is this adaptation which brings completeness to a context aware adaptive m-learning application to fulfill all m-learning activities and achieve m-learning objectives.

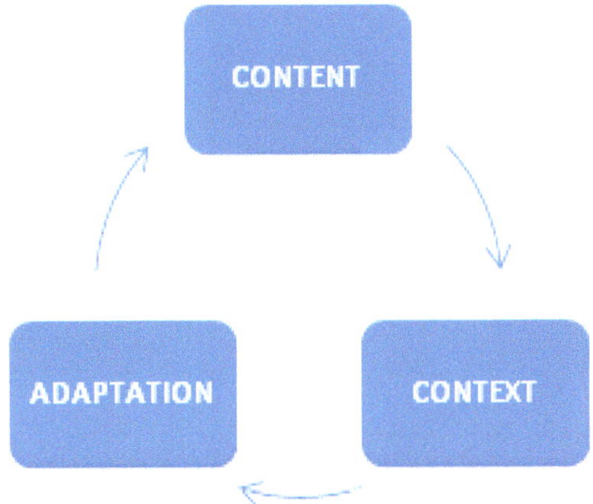

Figure 1. Content-Context-Adaptation Cycle

Given an m-learning scenario, the learning objectives determine the content or knowledge to be delivered to m-learner. Context brings in the ability to impart the content to m-learner so as to make the m-learning activity more appropriate, timely, useful resulting in achieving the learning objective in the most effective manner. Adaptation mechanisms understand and identify the effect of context and content on the m-learner's learning activities and tries to improvise for better learning outcome in the future learning activities. Hence, these three elements are neither static nor fixed. These elements evolve as learning scenario and its constituent learning activities are better understood and achieve higher learning objectives.

Any context aware and adaptive m-learning application development process will be more efficient if it includes this cyclic nature of content-context-adaptation in its process iterations. The cyclic nature is inherent in the relationship between these three elements. Content has to be modeled according the various contextual elements and their features. Adaptation mechanisms primarily depend on the contexts and their values for better adaptation. Depending on the results of adaptation strategies implemented, the content has to be remodeled to achieve better outcome of m-learner's learning activities using context aware m-learning application installed in the mobile device of the m-learner.

III. Content Modeling

The various views of researchers on mobile technologies and its content are worth to consider. Mobile content is any type of media structured for mobile devices, like graphics, tones, films and games [50]; that is delivered through mobile internet [30], using interactivity [27]. In relation to the context of mobile learning, it can be said as "... any service or facility that supplies a learner with general electronic information and educational content that aids in acquisition of knowledge regardless of loca-

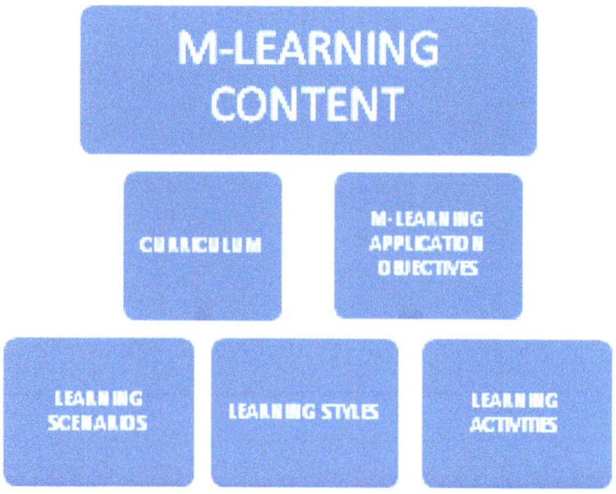

Figure 2. M-learning Content Aspects

tion and time..." [32]. Few of the m-learning projects have addressed the problems of adaptation of learning tasks and personalization of course content based on student's model, learning styles and strategy [8][19]. Norbayah et. al. [36] measures the potential of mobile content in learning to increase learners' tacit knowledge development. At Ultra lab m-learning project, the team produced m-learning materials for people with literacy and numeracy problems [14].

Unlike regular learning content developed for any e-learning application or to any specific requirement, the learning content for mobile learning systems is influenced by a good number of elements, having impact on the content creation and delivery. The very nature of the mobile learning brings these elements into its fold for successful implementation and execution. This motivates the m-learning content creators to consider these aspects in the content development process for modeling the learning content that best suits to a given situation or application. Figure 2 shows these aspects at a glance.

The most influential aspect of m-learning application is said to be the learning styles, for an effective impact of the learning content over the learner in a learning activity. The learning styles designate everything that is characteristic to an individual when he/she is learning, i.e., a specific manner of approaching a learning task, the learning strategies activated in order to fulfill the task [18]. Coffield [13] has identified 71 models of learning styles, among which 13 were categorized as major models according to their theoretical importance, their widespread use and their influence on other learning style models. Whether stable or flexible, genetically determined or experience related, psychological traits or strategies, all categories of learning styles have been proven to exert an influence on learning [18]. In the model [25] by I-Hsueh Tsai et.al., for the development of mobile learning curriculum model, proposed 6-stages; among these, the emphasis was on the design and development of mobile learning content as an integral part for completeness. Personalization of the teaching material has been studied and evaluated in the area of psychology of learning and teaching methods [49][33][7].

The awareness of learning context is important. A learning system that examines the learning context shall adapt learning process with respect to context change [53]. This learning process is enriched by the learning content it presents to the learner making his/her more involved in the learning activities. Anna Trifonova highlights the need for the dependency of the content which could be relative to location context (i.e., the system knows the location where the learner resides and adjusts to it), temporal context (i.e., the system is aware of time dependent data), behavioral context (ie., the system monitors the activities performed by the learner and responds to them adjusting its behavior) and interest specific context (i.e., the system modifies its behavior according to the learner's preferences); and a mix of these is possible and likely. Sharifi et.al.,[45] argued that mobile applications must have context-awareness and personalization as integral parts of the application, and adaptive user interfaces must be generated for the learner, firstly, to maximize learning potential at different locations, and secondly, to decrease the limitations of mobile devices such as the usability of the small screens [5] [41].

In the Electronic Guidebook [24], mobile web content was specifically created for the Exploratorium museum (an interactive science museum) in San Francisco. It was noted that the purpose of context-awareness is to facilitate learning on mobile devices [5]. In the project [28] at Tampere University of Technology, the study-content is presented in the form of a game and the electronic device is used to measure the average students' knowledge level and to adopt the speed of presenting new material to the learners. The Kidsroom project [6] at MIT Media laboratory provided a room that guides children through an interactive storytelling game. Navigation of learning materials is adapted according to the activity context.

The learning content development is a crucial phase and cannot be interpreted as minor task of the entire m-learning application development process. The basic materials that form the low level content has to be transformed into higher forms of content by following a process that treats low level materials against various aspects like syllabus, structure & scope, learning scenarios, learning styles, various contexts. This needs to be carried out in a well defined fashion. A five stage process model for learning content modeling and development in which all the above said aspects are considered is presented here. In the generic process for content modeling presented in the figure 3, every stage will result in intermediate learning content model or design that is specific to that stage-level and can be utilized in an m-learning application considering only those aspects of that stage for its execution. For example, if a preliminary form of m-learning application is used to deliver learning content to the learners' mobiles according to their course specifications, then such an m-learning application can pause at this stage before entering to stage two and use the intermediate learning content model or design produced by the stage one of this process. Similarly, each stage of this process will enhance the design of the learning content by considering various aspects of m-learning needs, and this style gives scope to applications for future scalability options from learning content perspective.

The basic form of learning content is a set of raw materials such as lecture slides, images, clips, animations, selected-passages, FAQs and so on. These materials are created by various tutors and instructors using several authoring tools of their convenience and importance. These materials are very basic to use for any particular application or situation or scenario so as to produce good results. These materials need to be given more appropriate form that enables the tutors to combine or relate or mix various levels of instructions to the students.

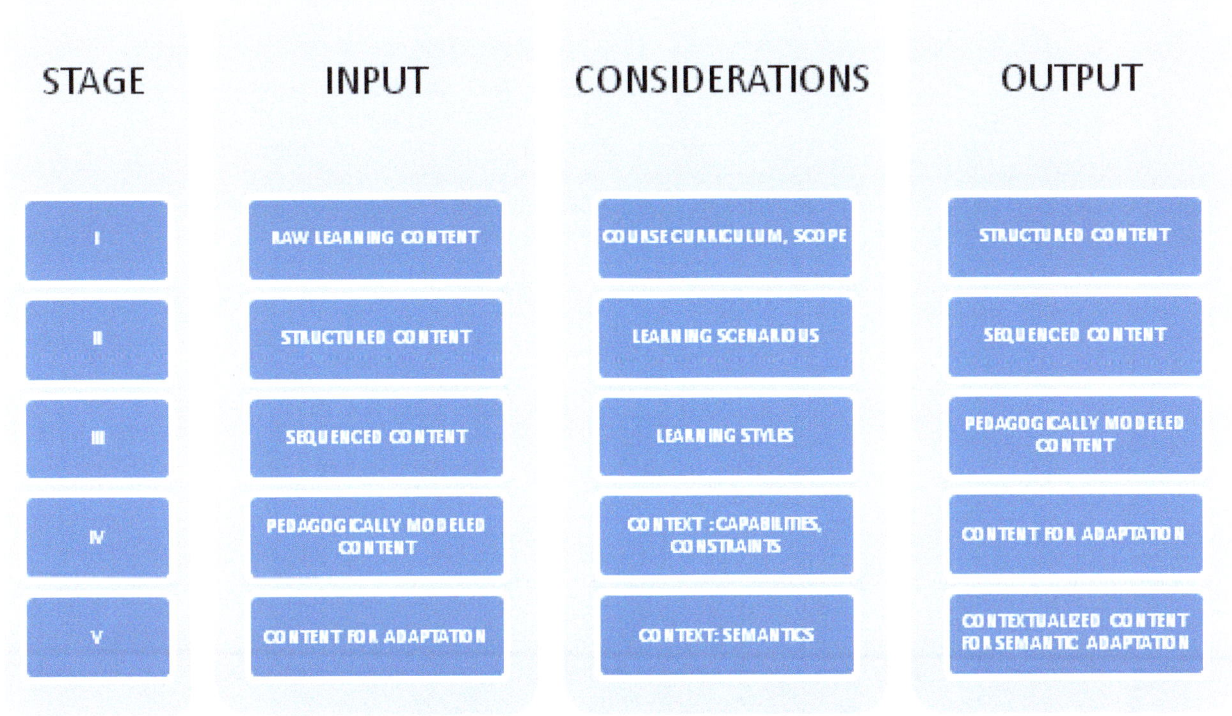

Figure 3. Generic Process Framework for Content Modeling

IV. Context Elicitation

Given the diversity of contextual information, it is useful if categorized, to make it easier and understand in a systematic fashion. The context categories identified can be utilized in a number of forms by context aware mobile learning applications. Context can be said to be as comprising of various components that are interdependent in different ways. To understand and identify its relevance for mobile learning one need to scrutinize it in different perspectives and ways in which mobile technologies can usefully support it. One way to do this is to take as many forms of it which we call context exploration and make the stakeholders aware of them when they are involved in the development of context aware mobile learning application.

The following forms of context are explored to atomic levels and serve as a starting point for eliciting contextual requirements in a different manner that can help in focusing the design and development process for CAML applications in more effective fashion.

More clarity to the requirements elicitation is brought by identify and representing the most comprehensible properties of the various contextual elements that are visualized for the context aware mobile learning application. This information makes the application designers more comfortable when designing the contextual elements implementation aspects of the application. The table 1 gives a view of the various properties of the contextual elements.

Contextual element position could be continuous or discrete. In continuous positioning service, a learner's position is known at all times as he/she moves through the space/area covered by the location service. In this kind, application is provided with more information. In a discrete type, the learners must actively provide the system with the information about their position. As the application uses a wider range of context inputs from various sources and sensors; the complexity for the application designers increases. With the higher-level context abstractions (formed by the low level context inputs) the handling of contexts becomes ease to the designers and developers. Dey et al., assert that using higher-level context abstractions that sit on top of individual component abstractions should make it easier for application designers to use context. The higher the level of the context, it is easier to use, at the expense of some information hiding [15].

Given the diversity of context information, it is recommended to categorize and make it comprehensible in a systematic manner. Another contextual element property is shared and individual. Individual context possess information relevant to the learning activity and interaction between the learner and the m-learning application. Shared context possess information relevant to collaborative learning activities or tasks where in a workgroup or a defined set of learners share common goals, interests and experiences.

Given the learning activity that is carried out by the learner with the application comes the contextual element in two variations in terms of their participation in the learning activity. The set of contextual elements that are relevant to the current learning activity are said to be Active and those contextual elements even with value are not relevant to the learning activity currently executed by the learner are said to be Passive. Understanding this variations help the designers to concentrate on the contexts that are required in any given learning activity's design and development.

TABLE I. CONTEXTS

CONTEXT	CHECKLIST TO ELICIT CONTEXT REQUIREMENTS
LEARNER CONTEXT	Preferences, Profile, Calendar, Goals, Contacts, Competences, Practices, Cognitive Abilities, Intentions, Tasks, Results, Needs, e.t.c.
ACTIVITY CONTEXT	Subject, Purpose, Objectives, Expected Outcomes, Pedagogical Theory, Participants, Teams, Achievements & Results, e.t.c.
CONTENT CONTEXT	Content Type, Size, Presentation, Delivery Mode, Storage Options, Content Structure, Content Sequence, e.t.c.
LOCATION/PLACE CONTEXT	. Naming every location such as Home, Library, Lab, Class Room, Hostel, Canteen, e.t.c. . Categorizing such as Private, Public, On Campus, Off Campus, e.t.c.
TIME CONTEXT	. Exact Timing Options . Breaking the day into sessions like Early Morning, Morning, Forenoon, Afternoon, Evening, Night, Late Night, e.t.c. . Time Interval or Time Latencies
ENVIRONMENT CONTEXT	. Weather Context: Sunny, Cloudy, Rainy, Snowfall, e.t.c. . Temperature Context: Very Cold, Cold, Tepid, Warm, Hot, or setting up temperature units. . Noise Context: No Sound, Quiet, Normal, Louder, Hurting or setting up noise levels units.
DAY CONTEXT	. Like Working Days, Holidays . Like Week Days, Weekends . Like On Vacation, Expedition, Festival, Project Period, e.t.c. . Like days of the week, Monday, Tuesday,…,Sunday
MOVEMENT CONTEXT	. Structuring like Slow, Still, Fast, e.t.c. . Structuring like Standing, Walking, Driving, e.t.c.
COMPUTING CONTEXT	CPU power, Display Size, Color Resolution, Input Method, Output Method, Memory Available, Battery Power, Learner Settings.
HISTORY AS CONTEXT	Previous Activities, Outcomes from Database Logs.
USABILITY CONTEXT	Frequency of Access, Usage Statistics, Usage patterns, e.t.c.
CUSTOM CONTEXT	Derived from the Learning Scenarios and Learning Activity Needs.

TABLE II.
PROPERTIES OF CONTEXTUAL ELEMENTS

PROPERTIES OF CONTEXTUAL ELEMENTS	
NATURE	Natural Artificial
STATE	Active Passive
FORM	Individual Shared
SOURCE	Software Hardware
BEHAVIOR	Static Dynamic
COMPUTATION	Computable Non Computable
RELEVANCE	Instantaneous Historical
ADAPTATION	Adaptable Non Adaptable
FLOW	Continuous Discrete
REPRESENTATION	Low Level High Level

The contextual element participation in a given learning activity or task is not sufficient unless their evolution is known. The contextual elements that are of interest to the current learning activity, if undergoes change during the activity's execution are said to possess Dynamic behavior. The contextual elements which participate in the learning activity but remain unchanged are Static in their behavior. The finer details of the contextual elements and their properties make the designers more comfortable in their task execution.

V. ADAPTATION MECHANISM

Mobile Learner is highly mobile in nature and this poses a major challenge in m-learning scenario in which the learning content has to be delivered to the learner based on his/her current context. There are systems or architectures [21] [26] [46] [29] that deliver learning content by adapting to the learner's context. All of them take the context (Physical, Device), the learner style, learner preferences, learner's knowledge, learner's experience or a combination of some of these as a means to adapt and deliver the appropriate content. So, an adaptive delivery of the learning content to the learner based on his place, time, device, physical activity is done. But this doesn't necessarily mean that they are interested in it or in any associated learning activity.

There is a need for identifying the interest of the learner on the delivered learning content by modeling and understanding the learner's actions against the learning content. In this work a design methodology called Acceptability Analyzer for identifying learner's interest or acceptance level on the delivered learning content by modeling his/her actions against the learning content with an m-learning application supporting the learning content is devised.

The design methodology is based on the notion that the learning content delivered to the learner has attributes specific to it and its type. The learning content is accessed by the learner with the support of an appropriate application that supports the learning content type and the application has its own specific actions that are used by the learner to interact with the application in the learning activity process.

Figure 4 shows a general schema for a context aware adaptive m-learning system which includes the Acceptability Analyzer module supporting the Adaptation module by sending the information regarding the learner's acceptability levels for the learning content delivered in a context. This is used by the Adaptation module for future decision making or evaluation purposes.

Learning content to be delivered to the learner is modeled and stored in a Learning Content Repository. It is connected to the adaptation module for access in taking decisions to deliver appropriate learning content. The History Log, contains details about learner's experiences and interactions of the past, is also connected to the adaptation module for decision making purpose. The Learner Model consists of details about the learner profile which includes learner preferences, learning style and learner schedule that are accessed, used and updated by the adaptation module. By providing this (learner's) information initially, the system will be able to automatically determine the learning preferences and contextual features of the learner at a given place and time. This saves time and effort for the learner while he/she is on the move. The learner can override a particular learning preference at any time and or location. The learner model is consulted by the adaptation module to draw inferences and take decisions.

M-Learning Application is a mobile software application that is used by the learner in his mobile device to carryout the learning activity with the learning content provided to him/her by the adaptation module taking into all the contextual elements into consideration. The Acceptability Analyzer in figure 5 is the centre of the methodology where in all the concentration regarding the learner's Action analysis is carried out. The source of information for this component are the learner's actions against the m-learning application, learning content presented to the learner and the Action Specific Model which gives information about what to be understood or inferred for the Actions recorded by the Action Recorder module.

The factors that cause interruption in the leaning activity are termed as Acceptability Affecting Factors (AAFs) and play major role in the learning activity's final state. The AAF Categorizer gives information about the factors that cause interruption and their impact on the learning activity flow to the Acceptability Engine. The Action Sequence Log stores all the sequence of actions against the application by the learner in a specific format which ensures the storage of corresponding contextual element values for the recorded actions.

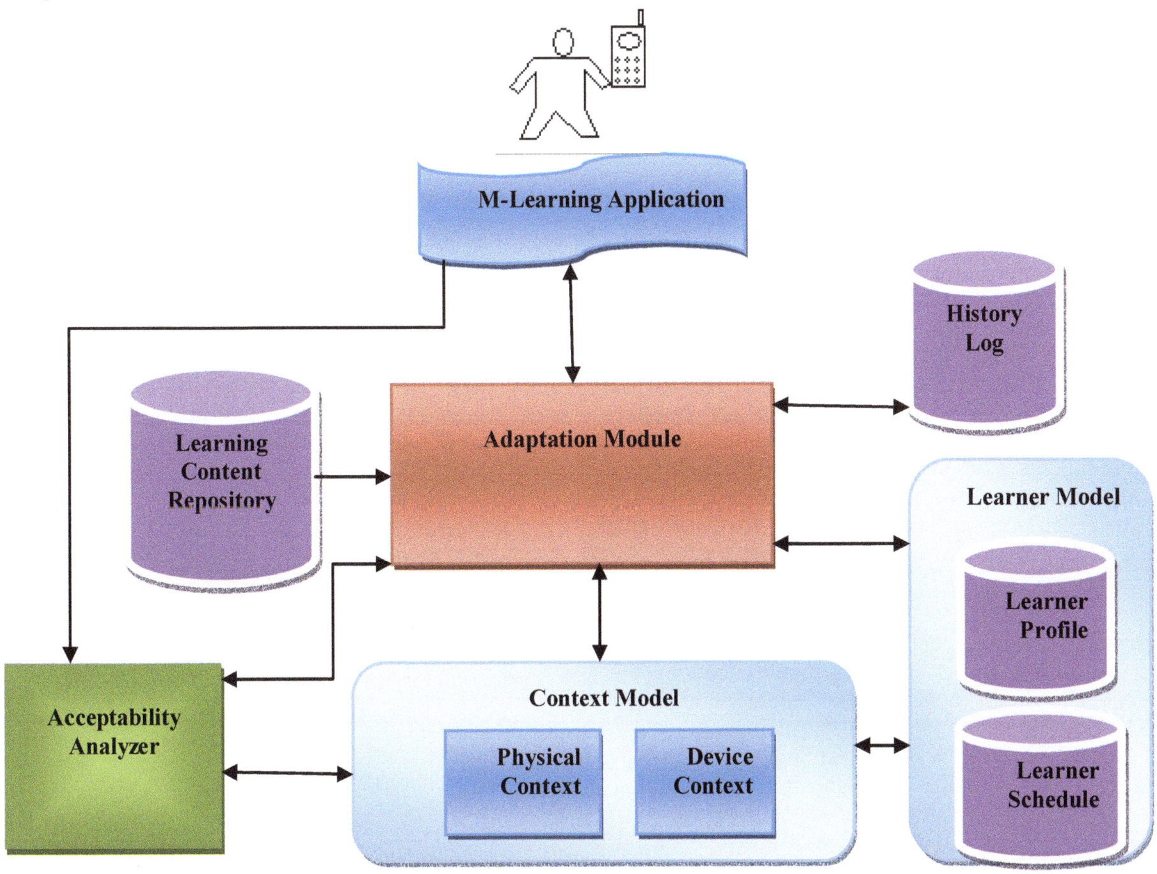

Figure 4. General Schema

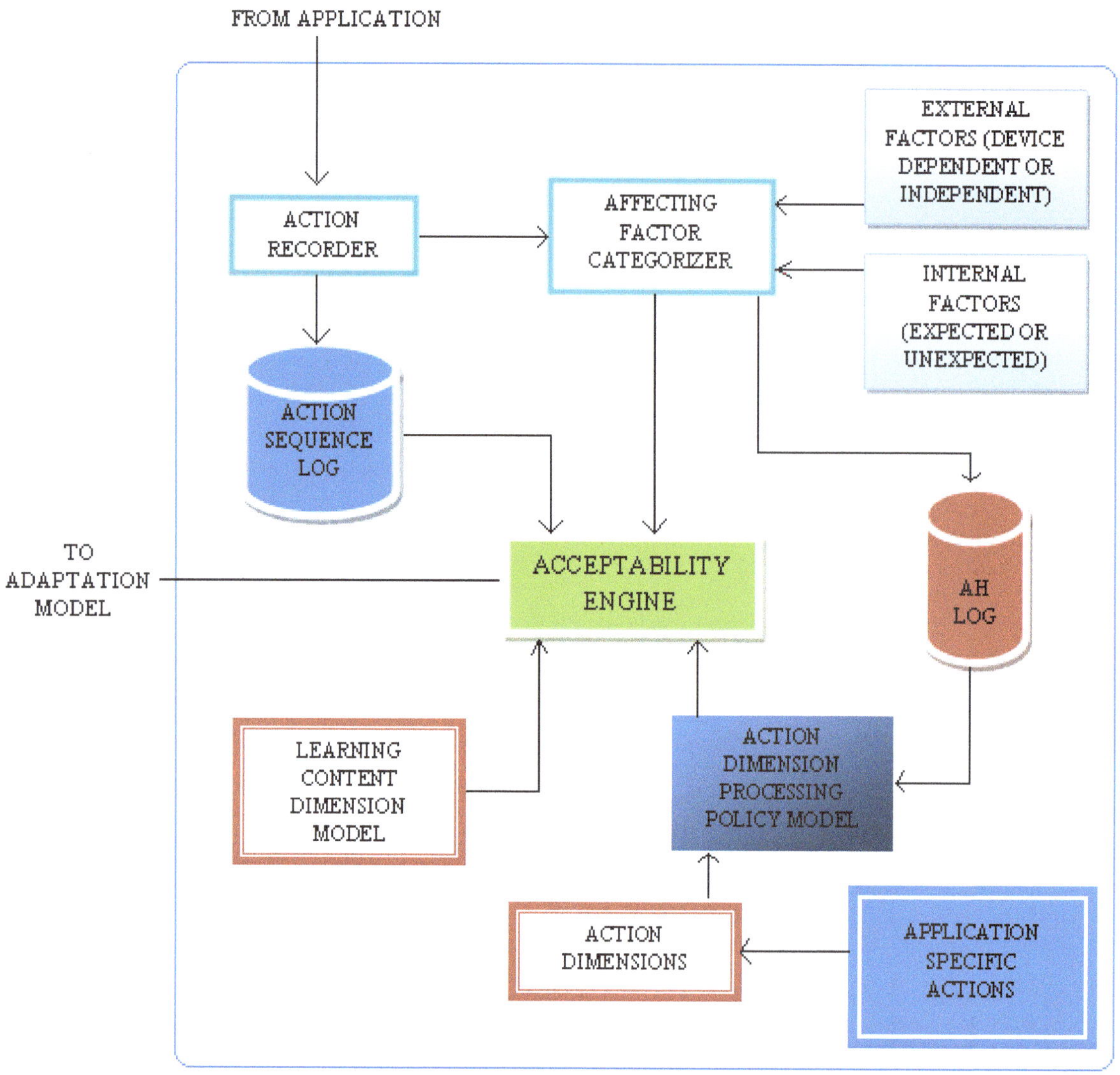

Acceptability Analyzer

The Action Dimension Model feeds information to the Acceptability Engine regarding what to infer for each action selected by the learner against the application in terms of Learning Content Dimensions, which are given by the Learning Content Dimension Model. The Learning Content Dimensions are used by the Acceptability Engine to infer how the Content Specific Attributes and Application Specific Actions are related to each other and their effect on the decision making process of acceptability level of the content by the learner. The Acceptability Engine can be implemented by using any intelligent techniques for inference available in the literature.

VI. Conclusion

Presented are various context aware and adaptive methods for m-learning and also elicited their functioning for better understanding. These various designs, models and methods concentrate on only either adaptation mechanisms or context awareness or learning styles and are subjected to a particular problem and its solution. There is a need for better adaptation in context aware m-learning applications' design to integrate various contextual elements and identify their relationships with regard to the m-learning application under development. This also feeds the adaptation process with finer information about contextual features and learners responses to the learning content under these various contextual features, their values by analyzing the learner's actions or interactions that he/she carries out in the learning activity provided by the m-learning application. This requirement sows seed to various models in the coming future.

References

[1] A. Jappinen, J. Nummela, T. Vainio, M. Ahonen, "Adaptive Mobile Learning Systems – The Essential Questions from the Design Perspective", In the Proceedings of MLearn 2004, Roma, Italy, pp. 109-112, 2004.

[2] A. K. Dey, G.D. Abowd, D. Salber, A conceptual framework and a toolkit for supporting the rapid prototyping of context-aware applications, Human Computer Interaction, 16, pp. 97-166, 2001.

[3] A. Zimmermann, M. Specht, A. Lorenz, "Personalization and Context Management", User Modeling and User-Adapted Interaction, 15, Springer-Verlag, pp. 275-302, 2005.

[4] Antti Syvanen, Russell Beale, Mike Sharples, Mikko Ahonen and Peter Lonsdale, "Supporting Pervasive Learning Environements: Adaptability and Context Awareness in Mobile Learning", In the Proceedings of the 2005 IEEE International Workshop on Wire-

less and Mobile Technologies in Education.

[5] Beale R and Lonsdale P, Mobile Context aware systems: the intelligence to support tasks and effectively utilize resources, Mobile HCI, 2004.

[6] Bobick A F, Intille S S, Davis J W, Baird F, Pinhanez C S, Campbell L W, Ivanov Y A, Schutte A, and Wilson A, The KidsRoom: a perceptually-based interactive and immersive story environment, MIT Media Laboratory perceptual Computing Section Technical Report No.398, November 1996, revised June 1998

[7] Brusilovsky P : Adaptive and Intelligent Technologies for Web-based education. In C. Rollinger and C. Peylo (Eds.), Special Issue on Intelligent Systems and Teleteaching, Kiinstliche Intelligenz, vol.4, pp. 19-25.

[8] Brusilovsky P, Adaptive navigation support in educational hypermedia: The role of learner knowledge level and the case for meta-adaptation. British Journal of educational Technology, vol.34(4), 2003, 487-497.

[9] Bull S, Cui Y, Robig H and Sharples M, Adapting to Different Needs in Different Locations: Handheld Computers in University Education, WMTE, pp. 148-152, 2002.

[10] Chan T., Sharples M, Vavoula G and Lonsdale P,(2004) Educational Metadata for Mobile Learning. International Workshop on Wireless and Mobile Technologies in Education.

[11] Chen Y S, Kao T C, Yu G J, and Sheu J P, A Mobile Butterfly-Watching learning system for Supporting Independent Learning, IEEE WMTE 2002.

[12] Chen Y S., Kao T C, Sheu J P and Chiang C Y (2002). A Mobile Scaffolding-Aid-Based Bird-Watching learning System. In M.Milrad, H.U. Hoppe and Kinshuk(Eds.), IEEE International Workshop on Wireless and Mobile Technologies in Education pp. 15-22, Los Alamitos, USA: IEEE Computer Society

[13] Coffield F, Moseley D, Hall E and Ecclestone K, Learning styles and pedagogoy in post-16 learning. A systematic and critical review, Learning and Skills Research Centre, UK, 2004.

[14] Collett M, Stead G, : meeting the Challenge: producing MLearning Materials for Young adults with Numeracy and Literacy Needs, Proceedings of the European Workshop on Mobile and Contextual Learning, pp. 61-62, UK, June, 2002.

[15] Dey A.K., Abowd G.D. & Salber D., A Conceptual Framework and a Toolkit for Supporting the Rapid Prototyping of Context-Aware Applications, Human Computer Interaction, 2001, Vol 16, pp. 97-166.

[16] Divitini M, Haugalokken O K, Norevik P : Improving communication through mobile technologies : Which possibilities?, Proc. Of IEEE International workshop on Wireless and Mobile Technologies in education, pp 86-90, Sweden, August, 2002.

[17] E. alfonseca, R.M. Carro, E. Martin, A. Ortigosa, P. Paredes, "The Impact of Learning Styles on Student Grouping for Collaborative Learning: A case study", Special Issue of User Modelling to Support Groups, Communities and Collaboration, Kluwer Academic Publishers, Accepted for publication.

[18] Elvira Popescu, Philippe Trigano and Costin Badica, Towards a Unified Learning Style Model in Adaptive Educational Systems, In the proceedings of ICALT 2007.

[19] Esmahi L & Lin F., A multiagents framework for an adaptive e-learning system. Book chapter, to appear in Designing Distributed Learning Environments with Intelligent Software agents, Lin Fuhua, Editor, Idea Group Inc. (IGI), 2004.

[20] Estefania Martin, Nuria Andueza, Rosa M. Carro (2006), "Architecture of a System for Context-based Adaptation in M-Learning", In the Proceedings of the Sixth International Conference on Adavanced Learning Technologies (ICALT'06).

[21] Estefania Martin, Rosa M. Carro and Pilar Rodriguez, "A Mechanism to Support Context-Based Adaptation in M-Learning", EC-TEL 2006, LNCS 4227, pp. 302-315, 2006.

[22] G. Chen, D. Kotz, A survey of context-aware mobile computing research, Technical Report, Dartmouth Computer Science TR2000-381, 2000.

[23] Garner I, Francis J, Wales K : An Evaluation of the Implementation of a Short Messaging System (SMS) to Support Undergraduate Students, proceedings of the European Workshop on Mobile and Contextual Learning, pp. 15-18, UK, June, 2002.

[24] His S : The Electronic Guidebook : A Study of User Experience using Mobile Web Content in a Museum Setting, in the proceedings of IEEE International Workshop on Wireless and Mobile Technologies in Education, pp. 48-54, Sweden, August, 2002.

[25] I-Hsueh Tsai, Shelley Shwu-Ching young & Chia-Hang Liang, Exploring the course development model for the mobile learning context : A preliminary study, in the Proceedings of ICALT 2005.

[26] Jane Yau and Mike Joy (2007). Architecture of a Context-aware and Adaptive Learning Schedule for Learning Java. ICALT ,2007

[27] Jared Bernstein (2006). Born Digital, Not Yesterday: Next-generation Web User Seeks Interactivity [Online].

[28] Ketamo H : mLearning for kindergarten's mathematics teaching, Proceedings of IEEE International Workshop on wireless and Mobile Technologies in Education, pp. 167-170, Sweden, August 2002.

[29] Larbi Esmahi and Elarbi Badidi, "An Agent-Based Framework for Adaptive M-learning", IEA/AIE 2004, LNAI 3029, pp. 749-758, 2004.

[30] Laura Marriott (2006). What Services Are Driving Mobile Internet Usage? [Online].

[31] Lavoie M (2006) I, Mlearning: Identifying Design Recommendations for a context-aware mobile learning system. IADIS International Conference Mobile Learning.

[32] Lehner F and Nosekebel H (2002). The Role of Mobile Devices in E-learning- First Experience with a E-learning Environment. In M.Milrad, H.U. Hoppe and Kinshuk(Eds.), IEEE International Workshop on Wireless and Mobile Technologies in Education pp. 103-106, Los Alamitos, USA: IEEE Computer Society.

[33] Litchfield B.C., Driscoll M.P., & Dempsey J.V :Presentation sequence and example difficulty: Their effect on concept and rule learning in computer-basedinstruction.Journal of computer – based instruction, vol.17, pp. 35-40.

[34] M. Yudelson, T. Gavrilova, and P. Brusilovsky, "Towards user modeling meta-ontology", Proc. of 10th International User Modeling Conference, Springer Verlag, pp. 448-452, 2005.

[35] Malek J, Laroussi M and Derycke A (2006) A Multi-Layer Ubiquitous Middleware for Bijective Adaptation between Context and Activity in a Mobile and Collaborative learning. ICSNC 2006.

[36] Norbayah Mohd Suki, Ahmad Rafi M. Eshaq and Radha Krishna Rao G S V, A Study on Edutainment Framework for M-learning, in the proceedings of ISCIT 2006.

[37] O.R. Zaine, "Building a Recommender Agent for e-Learning Systems", In the Proceedings of International Conference on Computers in Education, 2002.

[38] P. Brusilovsky, "Adaptive hypermedia", User Modelling and user Adapted Interaction, Vol. 11, pp.87-110, 2001.

[39] Patrick Brezillon , Using Context for Supporting User Efficiently, In the proceedings of the 36th Hawaii International Conference on System Sciences(HICSS'03).

[40] Prekop P and Burnett M , Activities, Context and Ubiquitous Computing, Computer Communications 26, pp. 1168-1176, 2003.

[41] Rodden T, Chervest K, Davies N, Exploiting Context in HCI Design for Mobile Systems, GIST Technical Report G98-1, 1998.

[42] S. A.N. Shafer, B. Brummitt, and J. Cadiz, "Interaction issues in context-aware intelligent environments", Human Computer Interaction, 16:363-378, 2001.

[43] Schmidt A, Beigl M & Gellersen H, There is More to Context than Location, Computer & Graphics 23(6) Decemeber 1999, pp. 893-901.

[44] Seppala P : Mobile Learning and Mobility in Teacher Training, Proceedings of IEEE International Workshop on wireless and Mobile Technologies in Education, pp. 130-135, Sweden, August, 2002.

[45] Sharifi G, Deters R, Vassileva J, Bull S and Robig H, Location-aware Adaptive Interfaces for information access with Handheld

Computers, Adaptive Hypermedia, Springer-Verlag, Berlin Heidelberg, pp. 328-331, 2004.

[46] Soo-Joong Ghim, Yong-Ik Yoon, and Ilkyeun Ra, "A Context-Adaptive Model for Mobile Learning Applications", WISE 2004 Workshops, LNCS 3307, pp.102-113, 2004.

[47] Stone a, Briggs J : ITZ GD 2 TXT – How to use SMS effectively in M-Learning, Proceedings of the European Workshop on Mobile and Contextual Learning, pp. 11-14, UK, June, 2002.

[48] Stone A, Briggs J, Smith C : SMS and Interactivity – Some Results from the Field, and its Implications on effective uses of Mobile Technologies in Education, Proceedings of IEEE International Workshop on Wireless and Mobile Technologies in Education, pp. 147-151, Sweden, August, 2002.

[49] Tennyson R. D & Christenson D. L: MAIS: an intelligent leaning system, In D.H. Jonassen (Eds.), Instructional Designs for microcomputer courseware. Hillsdale: N.J Erlbaum (1988).

[50] The free Disctionary (2006). Mobile Content [Online]. Available: http://encyclopedia.the freedictionary.com/Mobile+content

[51] W. N. Schilit, System architecture for context aware mobile computing, Ph.D. Thesis, Columbia University, May 1995.

[52] Y. Wang, Context Awareness and Adaptation in Mobile Learning, presented at the 2nd International workshop on Wireless and Mobile Technologies in Education, 2004.

[53] Yuan-Kai Wang (2004), "Context Awareness and Adaptation in Mobile Learning", In the Proceedings of the second IEEE International Workshop on Wireless and Mobile Technologies in Education (WMTE'04).

AUTHORS

Dr. N. Uday Bhaskar, Assistant Professor of Computer Science, Govt. Degree College, Pattikonda, A.P., India. He can be reached by udaynagella@gmail.com.

Dr. U.S.N.Raju, Assistant Professor of CSE, NIT Warangal, A.P., India. He can be reached by usnraju@yahoo.com.

Dr. P. Govindarajulu, Professor, department of computer science, Sri Venkateswara University, Tirupati, India. He can be reached by pgovindarajulu@yahoo.com.

Dr.B.V.Ramana Reddy, Professor & Principal, MERITS, Udayagiri, A.P., India. He can be reached by busireddy@gmail.com.

10

Accessible Internet-of-Things and Internet-of-Content Services for All in the Home or on the Move

Dimitrios Tektonidis and Adamantios Koumpis
Research Programmes Division, ALTEC Software S.A., Thessaloniki, Greece

Abstract— Public authorities and national healthcare service providers across the European Union and beyond are increasingly under pressure to deliver better services with diminishing resources. Users, quite rightly, demand a high standard of service and increasingly also convenience focused on the needs of the users rather than how the services are organised to deliver them. Users, on the whole, also prefer to be supported in their homes, and be free to move as much as possible. While users without impairments and generally well provided for, there are increasingly also users who have special needs, including the aged and those with a range of physical and mental disabilities who are unable to make effective use of such services from their homes due to the lack of proactive and accessible services.

The idea we promote in the chapter is to provide better support to users with disabilities and impairments from the comfort of their home by means of providing them with a set of scalable services which can be either offered for free or purchased through some central form of a marketplace repository.

***Index Terms*—Internet of Things, Internet of Content, smart home, smart hospital, smart city, service interoperability, service connectivity, dependable services.**

I. Introduction

The arrival of Smartphones and Tablet PCs has created the Post-PC era and led to a development of IOS and IOT. However people with disabilities cannot fully take advantage of these benefits. Our vision is to provide better support to users with disabilities and impairments from the comfort of their home by providing:

- *Advisory services* to help manage their health and social care needs and ensure they can effectively access relevant services.
- *Monitoring services* that help users manage on their own and remind them to about essential aspects such as taking their medication and monitoring their condition and generally look after themselves properly.
- *Alerting services* when the users situation suddenly deteriorates and urgent help is needed.
- *Dependable services* that will check whether services are functioning correctly and notify service provider when the connection to special care users is lost unexpectedly.
- *Accessible Interfaces* using suitable interface technology and user interfaces that are appropriate to the special needs of users.
- *Smart Interfaces* that are sensitive to the context and current needs and that ensure appropriate services are made use of.
- *Orchestration Services* that make sure that services are orchestrated effectively where needed to deliver a seamless service at the point of service delivery.
- *Third party services* when the users may want to use services such as online grocery shopping or public transport or use social networking facilities.

The concept is built on the fundamental principle that users should be supported as much as possible by accessible services to maintain independence through monitoring and advice and involve available services when needed to address emerging needs.

The services can be provided by both public and private providers (and even include wireless connection to devices such as blood pressure monitors, flow meters, glucose level monitors and the like) to ensure that the user is appropriately supported in managing by themselves while ensuring they receive emergency services if needed. This should include other services required for daily living and to facilitate the user managing on their own and giving them the freedom of independent living. It is also envisaged that access to social networking services be provided to mitigate social exclusion suffered by housebound users who live on their own.

We believe that making this vision (we call it "IOTC4ALL" – Internet of Things and Content 4 (: for) ALL) reality will have a significant impact on social inclusion and supporting disadvantaged users to lead independent lives with the assurance of support where needed. The vision is to develop an open service model that allows new service providers to enter and which can easily be extended to different member states independent of the technology used.

The aim is to enable to a large variety and diversity of users with special needs the usage of complex services through devices that although are considered smart they are difficult in terms of use for several people. Also a key aspect of the aspired architecture is the inclusion of people with disabilities to execute these services: though it is about services, the architecture requires an enriched Service Oriented Architecture by components that will enable and facilitate the discovery and execution of services from smart and mobile devices.

In practical terms we need to offer a platform where service providers can:

- Design services
- Implement Services
- Deploy
- Publish

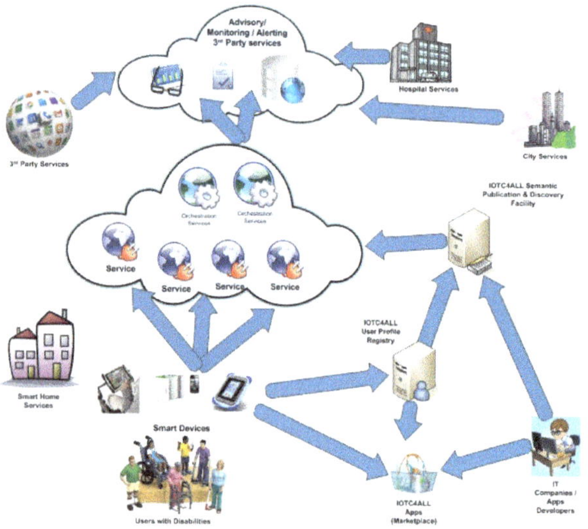

Figure 1. Overall conceptualisation to support the IOTC4ALL vision

And where the providers of devices and third party service providers can:

- Identify relevant services
- Compose and configure services for the specific user
- Configure interface devices (tablet, PDA, smartphones, web cams)
- Add further home-based healthcare devices where required (blood pressure meters, glucose meters etc)

To achieve these goals, we have to deal with the following issues:

- *Service Orientation:* Given that we need to get away from the website orientation to a service orientation how services can be exposed so that they can be accessed by special devices for impaired and disabled users that aim to make them accessible to those with special needs (and potentially combine them with other related services to provide a user centric overall service).
- *Discoverability:* Given a variety of services from a variety of providers and the needs and entitlements of individual users, how the services a user can access be discoverable by assistive devices or special interfaces users will use.
- *Service Entitlement:* Given that services are selective in terms of users meeting certain criteria it should be possible to determine the subset the user is entitled to given their personal circumstances and not to confuse users with irrelevant services and rather work with those they can use at that point in time.
- *Service Composition and Integration:* Services should be fashioned around the user rather than the user having to adapt to the services. So for the user the division between services should be transparent; some services may be provided in collaboration or based on joint or shared responsibilities by different organisations (healthcare, public services etc) and need to integrate seamlessly at the point of delivery to the user. This will require suitable approaches for service composition and integration.
- *Context sensitivity:* Given that the user will only requires some of the available services at any particular point in time it will be important to focus on the context and the immediate needs of the user to increase the assistive powers and not overburden the user who may be bewildered by the sheer amount of services available and what to use in any given situation given on the stated or inferred needs (the latter if some home based measuring devices are used for example)
- *Accessibility:* Given that users will not be IT literate or have special requirements with respect to interfaces in term of presentation, navigation and interface type (especially for blind and with motor disabilities) so that users can effectively use the services. This may also have to be combined with a number of special devices for input/data collection or as output devices.

We envisage three main provider communities regarding the proposed service execution platform:

- *Service providers* (public authorities and public services) aiming to manage their services (healthcare, social care and public services) delivery to the service users (disabled or aged) integrate them for a seamless service provision to the end user,
- *Services consumers* will be (disabled or aged) customers that use services through apps from their home using mobile devices (e.g. smart phones, tablets etc.) or stationary devices (PC or PC enabled TV) directed at IT illiterate or inexperienced individuals without any technical knowledge.
- *Third Party Providers* - IT solutions and value added service who are involved in the service design, composition (Service designers) and IT service provision. If the original service providers are unable to provide interfaces third party service providers will take on the role of generating devices and interfaces and additional services.

In the next section we present a set of informative scenarios of use that can be easily adapted and customised to fit the realities faced by a broad category of consumers and citizens today. Though emphasis is given to categories facing the risk of exclusion due to age or some type of disability, it is obvious that the same scenarios are applicable to the average European citizen and service customer.

II. User Scenarios

The aim of the IOTC4ALL vision is to assist users with disabilities and impairments in their quest for independent living. The goal is to support them in their independence while offering them support services through accessible smart interfaces so that they can effectively receive the services they need at a given point in time. These services are to adapt to the users current needs and thus be user centric rather than service provider centric (which is what typically happens currently). Orchestrating services around the users' current needs using accessible devices and interfaces is the main goal of such a research effort. In the following we discuss a range of typical scenarios that we support:

Scenario1: A blind user who lives on her own and who needs help with managing aspects daily living, managing to get around on foot and with public transport. May also need help with online shopping to reduce the need to get out of the house unnecessarily also during bad weather periods. Also as living on their own needs alerting facility in case they suddenly get unwell.

Scenario2: A user with mental impairment who can generally manage living on his own but needs reminding about important medication they have to take regularly and as forgetful to manage aspects of daily living. He may also need help from doctors/nurses/heath visitors to make sure he receives help if needed and to monitor that he is still safe looking after himself. The user may also get socially isolated and may want wanting to interact with others in self-help groups or with friends and family.

Scenario3: Elderly user with mobility and ongoing health management problems. She needs to monitor her condition and manage her health. At time she may need to be taken to places such as shops and doctors/hospital and would benefit from help with home shopping but is not computer literate.

Scenario4: User with motor problems who can view screens but needs voice activated or simplified input devices as has motor impairment and consequently problems operating keyboards. He also needs special transport as wheelchair bound and needs to receive prepared meals.

Scenario5: An obese young patient, by using smart devices (wireless weight scale or heart-rate measurement device), can keep contact with physicians sending to the Smart Hospital System all his physical parameters such as weight and the daily amount of physical exercise. Plus, the System can go back to the child suggesting to do more physical exercise and, eventually, giving him a target to reach by exercise games or by walking with a pedometer.

Scenario6: A diabetic patient can find useful, in addition to those already described, also a wireless glucose meter which sends detected data to the System, so the medical staff is always updated about the state of the patient.

Scenario7: An elderly person who has to follow a complicated medication plan can use a Smart Cabinet to store medicines: automatically the System can warn the user if he's going to take a medicine that it shouldn't be taken with certain foods or at the same times as other medications they are incompatible with (e.g.: an old man who gets confused by his wife's medicines) and needs reminding about the medication he has to take regularly.

What unites these scenarios s the users have some form of disability or impairment which makes it difficult to manage everyday activities and manage their health and social needs. IT is in principle possible to service all of these needs but this may be too demanding for the users in question who would still want to manage on their own as far as possible while having the assurance of help when needed. Through the use of smart assistants with accessible interfaces a number of available services can be managed on a context sensitive basis and tailored to the current needs of the user. These can be categorised into 3 main areas:

Disability Related: Users with disabilities and special needs will want to be able to live on their own rather than being in the looked after system, but to do this effectively it is important that they receive support services that allows them to manage and that keeps them safe. Therefore the delivery of services and the integration of the user with these services in the home are needed.

For users with disabilities (including aged users who are not computer literate) this presents a challenge. Their disabilities make it difficult for them to effectively use services, because they are provided in a way that is difficult for them to use or confusing or because their impairments such as blindness makes it difficult to navigate complex services. They need suitable interface to overcome accessibility problems and with assistive function to manage and reduce complexity. There is a concrete need for an end-to end solution to provide accessible interfaces on suitable devices, and accessible services that need in some cases to be integrated for a more complete solution which requires in some cases some form of orchestration.

Mobility Related: Users with disabilities or impairment may need help concerning their specific impairments and their ability to function in an urban environment and find available services whilst on the move and this scenario aims to provide location based services and direction to those services the user may require in terms of finding them or using transport to get to them from their home and on the way back. This may also include the ability to call upon help in case the user suddenly gets unwell and needs attention.

Apart from mobility aspects this also includes aspects of being able to access benefits and other public services the user may be entitled to. Thus it should be possible that these services are tailored to the individual and their circumstances and become immediately usable in an easy-to-use fashion.

Healthcare related: Users may have specific healthcare needs but will want to be able to stay in their homes and be supported in their homes rather than occupying hospital beds but which needs more assistance and monitoring at home to keep them safe and avoid the healthcare system being clogged up with unnecessary hospital stays. Technology for this exists but we shall aim to provide an integrated workable and usable solution for the inferm or frail to be supported using available services more efficiently and in a more user focused fashion.

The basic driver for this scenario is that a smart hospital is not necessarily an accessible one. So, this second use case in the IOTC4ALL vision aims to offer accessibility to a smart hospital features. Sensor devices enable function centered on patients, and in particular on diagnosing patient conditions, providing real-time information on patient health indicators. Application domains include different telemedicine solutions, monitoring patient compliance with medication regiment prescriptions, and alerting for patient well-being. Heterogeneous wireless access-based remote patient monitoring systems can be deployed to reach the patient everywhere, with multiple wireless technologies integrated to support continuous bio-signal monitoring in presence of patient mobility

What is needed is some form of intelligent user interface that:

- adapts to the specific form of disability

- is able to determine the changing needs and wishes of the user
- is able to identify relevant services
- is able to support the user to access and use these services
- is able to orchestrate services for a given goal where needed
- reminds the user to complete routine tasks.

All then scenarios require a common set of functionalities:

- Special visual displays/output devices (large screens, voice output, braille etc)
- Easy input devices (predictive text, restricted options, voice command, point and click)
- Intelligent Interfaces that adapt to needs (using semantics to analyse current situation and context and structure services accordingly, also remind users where needed)
- Ability to match needs with available services (using semantics to select services on user need, entitlement and user wishes)
- Ability to orchestrate services for a given goal (using an orchestration service to orchestrate services towards a common goal – e.g. book doctors appointment and book ambulance or taxi or advise which bus and when)
- Ability to interface with special local devices (e.g. blood pressure meter to monitor condition and advise to take medication or seek medical advice or assistance)

There are some common set of requirements across several scenarios. These are to be met by a smart assistant that provides an interface the users are able to manage effectively and that assesses the current context to provide easy access to and orchestration of services to meet the users' needs and wishes:

- Manage medical conditions through monitoring and advice
- Manage balance between supporting independence and calling for help when needed
- Manage transport needs
- Manage support needs in the home
- Finding services suitable for the user
- Reducing complexity
- Online shopping and Home delivery of goods to avoid unnecessary trips to shops.

III. Accessibility Matters: Opening Service Access to All

Accessibility is increasingly becoming and issue as well as assistive technology for both the able bodied and impaired users. Anti-discrimination legislation now requires businesses and public authorities alike to support disabled or disadvantaged users alongside their able-bodied counterparts. This is starting to attract attention also from the commercial community and with patent applications in these areas becoming more frequent, but where the EU is decidedly lagging behind as compared to the US as can readily be seen from the statistics readily available from ISO and EPO. A number of EU projects have begun to address aspects of accessibility, where movement and mental disabilities were more important, and to a lesser degree the sensory ones (including visual and hearing). Current EU-wide disability and accessibility-applicable research address Web, Sensors, Robotics and Brain-Machine Interfaces, as well as motor-related, mental-, and cognitive disabilities.

In a recent survey carried out by the authors it came to light concerning the user needs, that respondents rated cognitive, mental and multiple disabilities as important and that health and employment are areas where new eAccessibility solutions will be needed most. An initial Trend Analysis revealed that standard technology for devices and connectivity were less of an issue, as these are now relatively widely available, while virtualisation, suitable HCI Interfaces, Avatars and Robots as companions were identified as potential weak areas in need of further work. The time horizon for those is considerably longer making them less of an easy picking for researchers and industry alike. IOTC4ALL will be addressing healthcare needs of users with disabilities and making them substantially more accessible than what is available at present through much improved HCI interfaces.

Among the areas identified in need of development in the survey are solutions for increased independence of older people and active ageing with expectations from the community to meet the rights to participation of people with disabilities. With growing health care expenditure, Growing expenditure on social protection benefits, countermeasures to address fragmentation of (ICT) and develop affordable solutions with significant benefits are needed.

What is needed and where IOTC4ALL will make a concrete contribution to novelty will be by developing new models where users have more control over services they consume, New model for evaluating and implementing accessibility. This will promote the business value of investing in accessibility and raise awareness of accessibility issues in the industry.

Among the needs that can be identified in this area and which IOTC4ALL will be supporting:

Advanced active interfaces: Dialog interfaces are a reality in many situations nowadays. However, they usually work with specific limited vocabulary and in a specific context. We aim to develop proactive interfaces that do not necessarily wait for the user to start the interaction but to prompt and remind users based on an analysis of context. This will be achieved by leveraging semantic technologies for assessment of context and accessing relevant services at that point in time so that the user receives focused assistance. IOTC4ALL will create innovative solutions here that are currently not available in the market nor have been reported in the academic literature.

Automatic adaptation of multimodal interfaces: Modern devices such as tablets and smart phones can contain a lot of information about users, such as documents, music, preferences for various applications and so on. So, having a personal profile on a mobile device seems quite natural for the future of ICT. In addition, mobile phones are enhanced with connectivity sensors which allow them to interact with other machines through WiFi, Bluetooth, NFC to include additional monitoring devices such as blood pressure meters, glucose meters and pulse meters. Interfaces should be smart and work out the context and adapt to the needs and preferences of the user and who can interact with remote services to support the user. This goes in tandem with advanced active interfaces and will generate interfaces that adapt to the users disability problems and then actively seek to assist the user to assess their current needs and locate and access relevant services making also use of local devices where appropriate to take readings. Again there are no solutions of this type currently available that cover all these aspects even if devices exist we can draw on to develop the proposed innovative service.

Promoting discoverable and integratable services: Ubiquitous connectivity and network services together with cloud based services and technologies mean that services can be accessed remotely but the way services are made available (usually using standard websites and applications) which are largely provider centric and the way they are presented often make it difficult to consume and integrate and orchestrate them so that users with disabilities can be readily and effectively be supported. A new service model is needed that does allow them to be discovered and accessed as a service so that appropriate smart interfaces can connect to them remotely and make the services available alongside each other and combinable and orchestratable at the interface level. We create innovation and make a concrete contribution in this area.

IV. Setting up a Solution

The proposed Service Execution Environment based on the vision and the scenarios previously presented aims to provide a new approach that will improve the use of services from user groups that have accessibility problems. The definition of a service will be enriched with all the required features and parameters that are related to the accessibility of the service. Therefore the proposed solution will investigate and develop the following:

- An enhanced Service Model that will include additional specifications related to the e-accessibility issues.
- Semantics that will enrich existing services in order to increase their accessibility to groups of users that this is required.
- A Personalised Application and Service Marketplace where developers will be able to publish their applications and users will be able to find services and applications based on their profile.

A. The IOTC4ALL Service Model

The provision of the services for IOTC4ALL, will follow the life-event model that has been implemented successfully in the past for e-Government Services1. The life-event approach facilitates discovery of the service because it bases the services discovery in the user profile and on WHAT the user wants to do ("I want to go to the hospital", "I want to buy groceries"). The execution of these services will be based on the user profile based on his/her needs. Services execution will be differentiated based on the disabilities of the user ("What is the preferable means of transport for a visual impaired person", "How can he/she input the necessary data for executing the service")

The user will be able to download the application to his/her device from the IOT4CALL marketplace according to his/her needs. The services will be installed as Apps (Mobile Applications) to his/her mobile device (Tablet/Smartphone). The installed services will use information from the IOTC4ALL User Profile Registry. The execution of the service will involve the IOT4CALL Semantic Discovery Facility that will be able to locate and retrieve information from available information links. We use the term "information link" as in Linkeddata2 to define every available system or service that can be accessed through the web.

The IOT4CALL Marketplace will provide a variety of applications that will make accessible a large variety of services capable to cover the scenarios and improving the lifestyle of the people with disabilities.

B. Enhancing services components to enrich their accessibility

Today as the number of available services grows the percentage of people with disabilities that can actually access these services is reducing. Web Accessibilities guidelines are applied to many web portals (especially governmental). However these guidelines cannot be applied to mobile applications. Therefore the design for the usability of such applications is limited to the specifications of the mobile device (i.e size to the screen) and not to who is going to use is. The problem is not only to the user interface but also to the service that the mobile application is using. Of course, the cost for reengineering the existing service is huge therefore IOTC4ALL propose an approach that will enhance the existing creating wrappers that will include all the additional features that are related with accessibility.

For IOTC4ALL the usage of a service breaks down into components that are independent and can be enriched without changing its structure and operation. To facilitate the enrichment of the services components, IOTC4ALL will provide to the service providers semantic descriptions to enrich their service. Following the approach of Linkeddata where software systems are defined as Information Links, IOTC4ALL will use Semantic Web Technologies to define the requirements of enhancing existing services without changing their structure.

C. The IOTC4ALL Marketplace

Marketplaces are a common practice from mobile applications and in the future for software in general. Existing Marketplaces (iTunes, Android Marketplaces etc) gather information from the user like the device he/she is using, location and other to suggest applications that may be in his/hers interest. For the applications developers, marketplaces enable them not only to sell their application but also to evolve providing new versions and communicating with their customers.

In IOTC4ALL we can make use of a Marketplace that will also gather information based on accessibility providing intelligent filtering for the available apps. In addition it will provide all the necessary infrastructure to facilitate both the users (customers) and the developers.

---

[1] OneStopGov project http://islab.uom.gr/onestopgov/

[2] http://www.linkeddata.org

D. Smart Devices UI

IOTC4ALL' goal is to enable to a large variety and diversity of people with special needs the usage of complex services through devices that although are considered smart they are difficult in terms of use for several people. To give better supports to users with disabilities and impairments from the comfort of their home or in mobility it is crucial make the above devices fully accessible by means of the definition and implementation of novel interface solutions which are appropriate to the special needs of people.

In particular, a holistic approach in defining an adaptive smart interface should be adopted to pursue a global optimization by considering not only the different kind of devices but also their sensitiveness to the context and people needs, in so ensuring services are appropriately used.

In order to be useful, the design of user interfaces should consider many aspects of human behaviours and needs. The complexity of the extent of the involvement of a human in interaction with a machine is sometimes invisible compared to the simplicity of the interaction method itself.

The existing physical technologies for Human Computer Interaction (HCI) essentially can be categorized by the relative human sense that the device is designed for. These devices are basically relying on three human senses: vision, audition, and touch (Te'eni et al. 2007).

Input devices that rely on vision are the most used kind and are commonly either switch-based or pointing devices (Myers 1998)The switch-based devices are any kind of interface that uses buttons and switches like a (virtual or physical) keyboard (Shneiderman 1998). The pointing devices examples are mice, joysticks, touch screens, graphic tablets and pen-based input (Murata 1991). The output devices can be any kind of visual display or printing device (Nielsen 1994).

The devices that rely on audition are more advanced devices that usually need some kind of speech recognition (Rabiner 1993). Nowadays, all kind of non-speech and speech signals (Brewster 2003) and messages are produced by machines as output signals. Beeps, alarms, and turn-by-turn navigation commands of a GPS device are simple examples.

Recent methods and technologies in HCI are trying to combine former methods of interaction together and with other advancing technologies such as networking and animation. These new advances can be categorized in three sections: wearable devices (Barfield and Caudell 2001), wireless devices (Yacoub 2007). The technology is improving so fast that even the borders between these new technologies are fading away and they are getting mixed together. Few examples of these devices are: radio frequency identification (RFID) products, personal digital assistants (PDA), and virtual tour for real estate business. Current directions and advances of research in user interface design include intelligent and adaptive interfaces as well as ubiquitous computing.

1) Intelligent and Adaptive HCI

As mentioned before, it is economically and technologically crucial to make HCI designs that provide easier and more pleasurable experience for the users. To realize this goal, the interfaces are getting more natural to use every day. Evolution of interfaces in note-taking tools is a good example. First there were typewriters, then keyboards and now touch screen tablet PCs that you can write on using your own handwriting and they recognize it change it to text and if not already made, tools that transcript whatever you say automatically so you do not need to write at all.

One important factor in new generation of interfaces is to differentiate between using intelligence in the making of the interface (Intelligent HCI) (Maybury and Wahlster 1998) or in the way that the interface interacts with users (Adaptive HCI) (Kirlik 2006).

Intelligent HCI designs are interfaces that incorporate at least some kind of intelligence in perception from and/or response to users. A few examples are speech enabled interfaces that use natural language to interact with user and devices that visually track user's movements or gaze and respond accordingly.

Adaptive HCI designs, on the other hand, may not use intelligence in the creation of interface but use it in the way they continue to interact with users. An adaptive HCI might be a website using regular GUI for selling various products. This website would be adaptive -to some extent- if it has the ability to recognize the user and keeps a memory of his searches and purchases and intelligently search, find, and suggest products on sale that it thinks user might need. Most of these kinds of adaptation are the ones that deal with cognitive and affective levels of user activity.

Another example that uses both intelligent and adaptive interface is a tablet PC that has the handwriting recognition ability and it can adapt to the handwriting of the logged in user so to improve its performance by remembering the corrections that the user made to the recognised text.

Finally, another factor to be considered about intelligent interfaces is that most non-intelligent HCI design are passive in nature i.e. they only respond whenever invoked by user while ultimate intelligent and adaptive interfaces tend to be active interfaces. The example is smart billboards or advertisements that present themselves according to users' taste.

2) Ubiquitous Computing and Ambient Intelligence

Ubiquitous computing (Ubicomp), often used interchangeably by ambient intelligence and pervasive computing, refers to the ultimate methods of human-computer interaction that is the deletion of a desktop and embedding of the computer in the environment so that it becomes invisible to humans while surrounding them everywhere hence the term ambient.

The idea of ubiquitous computing was first introduced by Mark Weiser during his tenure as chief technologist at Computer Science Lab in Xerox PARC in 1998. His idea was to embed computers everywhere in the environment and everyday objects so that people could interact with many computers at the same time while they are invisible to them and wirelessly communicating with each other (Riva et al. 2005).

Ubicomp has also been named the Third Wave of computing. The First Wave was the mainframe era, many people one computer. Then it was the Second Wave, one person one computer which was called PC era and now Ubicomp introduces many computers one person era.

3) From unimodal to multimodal interfaces

An interface mostly relies on number and diversity of its inputs and outputs which are communication channels that enable users to interact with computer via this interface. Each of the different independent single channels is called a modality (Jaimes and Sebe, 2007). A system that is based on only one modality is called unimodal. According to the nature of different modalities, they can be divided into three categories:

Visual-Based: Facial Expression Analysis, Body Movement Tracking (Large-scale), Gesture Recognition, Gaze Detection (Eyes Movement Tracking)

Audio-Based: Speech Recognition, Speaker Recognition, Auditory Emotion Analysis, Human-Made Noise/Sign Detections (Gasp, Sigh, Laugh, Cry, etc.), Musical Interaction

Sensor-Based: Pen-Based Interaction, Mouse & Keyboard, Joysticks, Motion Tracking Sensors and Digitizers, Haptic Sensors, Pressure Sensors, Taste/Smell Sensors.

Basic visual user interfaces in smart devices are the standard graphic based WIMP (window, icon, menu, and pointing device). Smart devices also have multiple modalities and are able to interpret information from various sensory and communication channels. For instance, smartphones uses touch based interactions and some (e.g.LG Optimus) allow users to use personalized gestures as shortcuts to commands. Multimodal interfaces process two or more combined user input modes (such as speech, pen, touch, manual gesture, gaze, and head and body movements) in a coordinated manner with multimedia system output (Oviatt 2003). They are a new class of interfaces that aim to recognize naturally occurring in the forms of human language and behavior, and which incorporate one or more recognition-based technologies, e.g., speech, pen, vision. Thus, they can provide richer and more natural methods of interactions, e.g., gesture, speech, and even all the five senses.

The term multimodal refers to combination of multiple modalities. In MMHCI systems, these modalities mostly refer to the ways that the system responds to the inputs, i.e. communication channels. The definition of these channels is inherited from human types of communication which are basically his senses: Sight, Hearing, Touch, Smell, and Taste. The possibilities for interaction with a machine include but are not limited to these types.

Therefore, a multimodal interface acts as a facilitator of human-computer interaction via two or more modes of input that go beyond the traditional keyboard and mouse. The exact number of supported input modes, their types and the way in which they work together may vary widely from one multimodal system to another. Multimodal interfaces incorporate different combinations of speech, gesture, gaze, facial expressions and other non-conventional modes of input. One of the most commonly supported combinations of input methods is that of gesture and speech.

An interesting aspect of multimodality is the collaboration of different modalities to assist the recognitions. For example, lip movement tracking (visual-based) can help speech recognition methods (audio-based) and speech recognition methods (audio-based) can assist command acquisition in gesture recognition (visual-based).

Some examples of applications of multimodal systems are Smart Video Conferencing, Intelligent Homes/Offices, Driver Monitoring, Intelligent Games, E-Commerce and of special interest for our project Helping People with Disabilities. A good example of Multimodal System for disable people is to address and assist disabled people (as persons with hands disabilities), which need other kinds of interfaces than ordinary people. In such systems, disabled users can perform work on the device by interacting using voice and head movements.

4) Smart devices today

In the past mobile devices were difficult to use and had limited user input methods (number keyboard or stylus) with small displays. Thus advanced usage was limited to business applications and platforms like Symbian and Blackberry. With the release of the iPhone this suddenly changed. Multi-touch interaction on a large colourful display enabled a world of new mobile applications affordable und manageable for everyone. The new interaction paradigm was adopted by all players in the market including Google, Nokia and also Microsoft. A few years later with Apple's iPad this concept has evolved into a new kind of personal computer, the tablet PC. While UMPCs and Netbooks appeared too bulky and complicated, the tablet PC brings the simplified mobile touch interaction of the iPhone into new interaction situations: touchscreen and overall user interface design are intuitive and visually accessible. The much larger screen should improve accessibility for people with low vision and other partial vision impairment.

Furthermore, the capacitive touchscreen is more accessible to those with dexterity limitations by introducing an electronic stylus; while those with limited finger movement and/or sensation in their fingertips have greater success using resistive touchscreen devices because of the ability to use a stylus. But, the electronic stylus for the capacitive touchscreen interfaces levels the playing field.

Cameras are increasingly used in conjunction with GPS functionality by software applications for augmented reality, mixed reality, and location-based services. Additionally, cameras are becoming an essential feature for devices used in the growing area of remote health monitoring. Being able to take a picture of a wound, infection or other visible condition will enhance the ability of people (with or without disabilities) to avoid the disruption and considerable effort of going to the doctor's office.

In addition, easy input devices have been developed, such as predictive text, restricted options, voice command, point and click.

V. CONCLUSIONS

Smart house – smart hospital – smart city: are there any common threads amongst them? One may think that each one of these application areas is worth a dedicated examination. In a house we live for years so it is easy to see why a person may need to invest on making it (more) accessible. While in a hospital people stay (fortunately) only for short (and usually unexpected) periods of their life – and again there smartness expected from a hospital may be of totally different nature than the smartness we expect from a home to exhibit to its inhabitants. Finally, for the smart city case, the requirements are also different: people may need to "connect" to different types of information provision services and infrastructures and two basic service types are: (a) what services are offered to me <*here*> and (b) how can I get <*there*>.

What we can consider as common denominator of all these are:

- The users have access to their mobile phone (smartphone, Android or iphone) so the unique access point to information from the environment (house, hospital or city) comes through their mobile phone.
- It is through their mobile phone that they may recognize the various addressable entities i.e. the "things" (of the Internet of Things) that are surrounding them and which can be tagged with QR codes / RFID tags or be accessible through NFC, and
- It is again through their mobile phone that they will be informed about the availability of the various content elements (of the Internet of Content) that are retrievable and which can be tagged with QR codes / RFID tags or be accessible through NFC

What is our understanding of a service in the context of the IOTC4ALL vision? This comes up to a series of highly connected links and ideas:

- Any retrieval of location-based (house / hospital / city) content is a service (though a basic one – could we come up with more sophisticated ideas?)
- Any access to a "thing" for getting information about its status or for manipulating is also regarded as a service (again this may be regarded as basic ones).
- More sophisticated or complicated services may base on the composition of more than one basic services – this way we may support different families of scenarios to be realized.
- What we promote as an idea and as a role is this of industrial enterprises (like the ones participating in the project) which will invest on offering such types of composite services to people. So given one house / hospital / city setting, user X may decide to buy the service bundle by value added service provider A, or B or C, or, in case s/he wants may still access 'manually' each of the offered 'content' and 'things' manually. More specifically:
    - User X may decide to buy a basic service from value added service provider A that enables organized scenario selection according to a (simple) thematic organization.
    - Or spend some more money and buy the more expensive service B, where there is also increased intelligence i.e. the selection of a particular scenario follows the mood of the user and the user does not have to choose one. This case is of interest to categories like older people or bipolar patients who may be unable to make a decision on their own and so this extra degree of automation helps their independent living. Finally,
    - User X may choose to buy a premium service from provider C that offers him/her with an advanced level of customer service and care that extends beyond basic service activities offered at level A and slightly more sophisticated or automated of level B, to include a more personalized and customized approach that create a customer-centric atmosphere. In this case, there may be increased interoperability offered for services to bridge gaps between the various environments that the user is changing – i.e. from the home through the city to the office and then back to home. This extra level of semantic processing that helps information flow between the various smart settings (house / city / hospital) is offered if the customer agrees to pay extra money.

Below we briefly summarise each of the supported scenario clusters:

Smart home scenario: A smart home in the context of this paper is considered as an augmented environment with the ability to consolidate information ("content") embedded in appliances, as well as micro-systems and multi-modal sensors to offer users enhanced access to information and functionality and an ability to control and mange the environment at a higher level of abstraction. So why shouldn't the home take initiative to either understand (proactively) the user's mood and offer the most appropriate setting adjustment or react to user's wishes (user chooses to watch TV, light settings are adapted and phone calls, etc.)

Smart hospital scenario: What is our idea for a smart hospital? In the context of this paper is mainly related with enhancing accessibility of to its patients (as well as the medical staff working). The users in this pilot may be able-bodied patients as well as visually / cognitive / age challenged persons. The idea of accessibility in the hospital context relates with helping patients take their drugs on time (so all blister packs and other drugs' packages may contain RFID labels) and facilitating their access to medical equipment that embeds RFID tags. The doctors, nurses, caregivers and other staff members wear a "smart badge" storing information (besides their employee ID number) that may help the patient ask for advice or guidance. Same way as doctors, on arrival, each patient receives a wristband with an embedded RFID tag storing a unique identifier, and some information about his / her profile that besides medical information may include also information significant for the types of interaction that the particular patient is capable to perform with services / "content" / "things" available within the hospital environment.

Smart city scenario (mobile phone as a "white cane"): Consider a visually impaired person travelling through a city. Similar to a foreigner in a hostile place feeling lost and requiring guidance, the use of a mobile phone to provide such assistance is ideal. The primary aim of this scenario is to enable visually / cognitive / age challenged persons to acquire the necessary guidance and assistance by capturing geo-tagged information using their mobile phone, and return any information, such as a specific pharmacy, the local police station, or even points of interest. Additionally, the GPS capability of the phone is exploited to enhance the locally available services recognition process (service spots).

As far as the links with the other two scenarios are concerned we should notice that with the use of the mobile phone, the same way that a visually / cognitive /

dementia challenged person is enabled to get guidance and assistance for the urban environment for short-range navigation and exploration in the home and hospital settings.

REFERENCES

[1] W. Barfield and T. Caudell, Fundamentals of Wearable Computers and Augmented Reality, Lawrence Erlbaum Associates, Mahwah (2001)

[2] S. Brewster, "Non speech auditory output", in J.A. Jacko and A. Sears (eds), The Human-Computer Interaction Handbook: Fundamentals, Evolving Technologies, and Emerging Application, Lawrence Erlbaum Associates, Mahwah (2003)

[3] A. Jaimes and N. Sebe, "Multimodal human computer interaction: a survey", Computer Vision and Image Understanding, 108(1-2), pp 116-134 (2007)

[4] A. Kirlik, Adaptive Perspectives on Human-Technology Interaction, Oxford University Press, Oxford (2006)

[5] M.T. Maybury and W. Wahlster, Readings in Intelligent User Interfaces, Morgan Kaufmann Press, San Francisco (1998)

[6] A. Murata, "An experimental evaluation of mouse, joystick, joycard, lightpen, trackball and touchscreen for Pointing - Basic Study on Human Interface Design", Proceedings of the Fourth International Conference on Human-Computer Interaction 1991, pp 123-127 (1991)

[7] B.A. Myers, "A brief history of human-computer interaction technology", ACM interactions, 5(2), pp 44-54 (1998).

[8] J. Nielsen, Usability Engineering, Morgan Kaufman, San Francisco (1994).

[9] Oviatt, S. L. Advances in Robust Multimodal Interface Design." IEEE Computer Graphics and Applications, 23 (2003): 52-68

[10] L.R. Rabiner, Fundamentals of Speech Recognition, Prentice Hall, Englewood Cliffs (1993).

[11] G. Riva, F. Vatalaro, F. Davide and M. Alaniz, Ambient Intelligence: The Evolution of Technology, Communication and Cognition towards the Future of HCI, IOS Press, Fairfax (2005)

[12] B. Shneiderman, Designing the User Interface: Strategies for Effective Human-Computer Interaction (3rd edition), Addison Wesley Longman, Reading (1998)

[13] D. Te'eni, J. Carey and P. Zhang, Human Computer Interaction: Developing Effective Organizational Information Systems, John Wiley & Sons, Hoboken (2007)

[14] M.D. Yacoub, Wireless Technology: Protocols, Standards, and Techniques, CRC Press, London (2002)) and virtual devices (K. McMenemy and S. Ferguson, A Hitchhiker's Guide to Virtual Reality, A K Peters, Wellesley (2007)

AUTHORS

Dimitris Tektonidis holds a BSc in Software Engineering and a PhD in Computer Science for research in the area of Enterprise Application Integration. His main fields of research are Enterprise Application Integration (EAI), IS interoperability, Service Oriented Architectures (SOA), Web Services, Semantic Web, e-Commerce, e-Government and Enterprise Systems (ERP, CRM, SCM). (e-mail: dte@altec.gr).

Adamantios Koumpis heads the Research Programmes Division of ALTEC S.A., which he founded at 1996 (then as independent division of Unisoft S.A.). Adamantios holds a PhD degree from the University of Kingston, UK and a Bachelor degree from the University of Crete, Greece. (email: akou@altec.gr).

11

On the Development of a Web-Based M-Learning System for Dual Screen Handheld Game Consoles

Hend S. Al-Khalifa
King Saud University, Riyadh, Saudi Arabia

Abstract—**This paper presents our experience on the design and development of an M-Learning web-based system for the Nintendo DSi game console. The paper starts by addressing the difficulties that emerged from the lack of resources on design guidelines for dual screen devices also the absence of adequate techniques and methods to support the design decisions. Then it explains how we overcome these challenges by adopting a design decision suitable for the screen requirements of the Nintendo DSi console. Finally, we present the components of our M-Learning system and the results of a preliminary usability evaluation.**

Index Terms—**Handheld Game console; Mobile Learning; Nintendo DSi; Interactive classroom.**

I. Introduction

Designing interactive Web-Based Mobile Learning (M-learning) systems is a highly demanding task that depends on the capabilities of the target handheld device. Given the different abilities and sizes of handheld devices (e.g. Pocket PCs, Mobile Phones and Game consoles), design guidelines and techniques were proposed to help designers on developing applications and websites that adapt to their specifications. One example of such guidelines is the Mobile Web Best Practices (MWBP) from W3C [1], which includes considerations for: multiple screen sizes and densities, performance optimization, touch targets, gestures, actions and device capabilities.

However, with the recent proliferation of ubiquitous handheld game consoles that comes with dual screens (i.e. two adjacent screens) and unique interaction modalities, a set of additional challenges are introduced in the design process. As an example, the Nintendo DS series of handheld game consoles, which comes with dual screen and touch screen features using stylus and finger, has introduced a new challenge for web-based applications developers. This challenge has originated from the interface constraints inherited in the Nintendo DS consoles, which incorporates both touch and dual screen capabilities.

Therefore, in this paper we report our experience in designing, developing and evaluating a web-based M-Learning system designed for the Nintendo DSi game console. While designing the system, we applied touch and small screen interface design recommendations previously reported in the literature, besides we used our ad hoc design techniques to utilize the dual screen feature in the DSi console.

II. Motivation

Handheld game consoles are widely used by the young generation for entertainment purposes. Little has been reported in their educational use. Given their manageable prices, wide spread, ease to use and varied interaction modalities have made them candidate technologies to be utilized for improving the educational process.

In fact, the pedagogical expectations of using such a technology in the classroom, will greatly impact students' learning and kick start a new learning experience.

Therefore, our system objectives can be summarized as follows:

- Improve and enhance the communication between teachers and students in the classroom.
- Utilize the technology (i.e. Nintendo DSi) to improve the educational process.
- Provide a cheap alternative technology with powerful capabilities to help perform Rich-Interactions in the classroom.
- Abandon the traditional method of classroom interaction.

III. Handheld Game Consoles

A. Overview

A handheld game console is "a lightweight, portable electronic device with a built-in screen, games controls and speakers" [2]. Most new handheld game consoles come with a built-in web browser that can connect to the Internet using Wi-Fi connection and interact with the user using stylus or touch. Sony PSP and Nintendo DS are among the dominant game consoles in the market, with Nintendo DS becoming the fastest-selling handheld gaming device.

The Nintendo DS, in particular, has the most interactive features that are not found in other portable game consoles, such as dual screen, touch screen, stylus, built-in wireless connection and built-in microphone. The Nintendo DS has three types of consoles: Nintendo DS, Nintendo DS Lite and Nintendo DSi. In addition to the previous features, two built-in cameras have been added to the DSi model.

Handheld game consoles are designed to a very limited and specific range of usage and applications, most of the applications designed for these game consoles are entertainment-oriented with a few relevant to education and learning. However, in recent years many research were conducted to explore the potential of these consoles in teaching and learning e.g. [3]. Other potential applications envisioned for the dual screen handheld consoles are: reading eBooks, displaying and interacting with large maps, and using it as an assistive technology as reported by [4].

B. Application Design

Handheld game consoles usually have their own proprietary software (called software development kits) used to develop applications working on their environment. However, designing web pages developed for desktop computers cannot be displayed the same on handheld consoles. This is because the screens of handheld consoles are typically smaller than their desktop counterparts, besides they have less processing power and memory specifications. Thus, designing web pages that adapt to handheld consoles with small screens size and touch capabilities is different compared to laptop/desktop computers that operate using basic pointing devices.

Overall, we have not found any guidelines proposed specifically for user interface design targeted to handheld consoles; on the other hand, user interface design guidelines for mobile device have been widely adopted to cope with their limitation [5]. In particular, small-screen research has been limited to web page design considerations for mobile devices [6]. Also, we have not found specific guidelines for designing web pages usable and viewable by web browsers in game consoles, especially for those which operate with dual screens. Based on the limited resources, we opted to benefit from user interface design guidelines for mobile devices in designing interfaces for handheld game consoles.

So, to display web content properly on mobile devices, some basic design principles need to be followed, among them [5]:

1. Develop applications that provide direct, simple access to focused valuable contents;
2. Trim the page-to-page navigation down to a minimum;
3. Adapt for vertical scrolling or reducing the amount of vertical scrolling by simplifying the text to be displayed; and
4. Reduce the number of users' keystrokes

In the educational domain, Churchill and Hedberg [7] provided a set of design recommendations for displaying learning objects in small-screen handheld devices. Some of these recommendations include: design for one step interaction, minimize scrolling, design for short contact time and design to match the task. Similarly, other guidelines have been suggested by [8] to create highly efficacious, user friendly and usable mobile interfaces. These guidelines are based on user analysis, interaction and mobile learning interface design. The aim of the guidelines is to use them as benchmark for interface designers when designing usable mobile learning portals.

IV. Our Practice

In this section, we describe our experience of developing the Nintendo DSi Interactive Classroom System (NDiC). NDiC is a Web-Based M-Learning system designed to create an interactive classroom, and promote understanding among all students using game console devices. The system consists of two components: an authoring tool and a classroom management system.

Before shedding the light on the system components we first need to explain the underpinning design approach used to design the interactive assessment sheets that will be displayed on the web browser of the Nintendo DSi game console. Afterwards, we will present the implementation of the system with a preliminary usability evaluation.

A. The Nintendo DSi Web Browser

The Nintendo DSi encompasses two LCD screens, each have 256 x 192 screen resolution with 3.25-inch wide, to show different information. The top screen is for display only while the bottom one is a touch screen that supports both stylus-operated and finger-operated interactions. Stylus operated touch screens support interactions of type: click, drag, drawings, gestures and handwriting.

The Nintendo DSi uses Opera web browser that is developed specifically for it. Opera is designed to utilize the capabilities of the DSi console such as dual screens and touch interactions. It also uses a virtual keyboard for texting purposes.

There are two ways of viewing web pages in the Opera web browser, the default setting is the Overview Mode, which shows on the top screen a zoomed out overview of the page the user is currently browsing. The user can switch between showing the zoomed out overview on the top screen or the bottom screen. However, in the Column Mode, all page content are compressed into a single column that shows in both screens, which makes reading long texts easier. Moreover, file formats supported in the Opera web browser are: HTML, XML, JPG, PNG, GIF, CSS and JS.

Given the previous features of the Opera browser, yet, it has many limitations [9] [10] such as:

1. It does not support plug-ins e.g. Adobe Flash.
2. It does not support font styles.
3. It does not store any cookies when power goes off.
4. Using some Cascading Style Sheets (CSS) properties slows the browser.

B. The Design Approach

Design recommendations from the literature were taken into consideration in our current system design. In fact, we break down the recommendations into three design perspectives, namely: (1) the touch aspect that requires interaction with a stylus and/or finger (2) the small screen size and (3) the dual screen.

For the touch interaction, the Nintendo DSi supports touch using finger or via a stylus, however a recent study by McKnight and Cassidy [11] revealed that finger is not a preferred method for interaction compared to stylus. This is because children fine motor skills are not sufficiently developed to enable them to point their finger accurately on small targets. Besides using stylus increases the pointing accuracy of children when smaller buttons are used [11]. Based on the findings of McKnight and Cassidy and on the fact that web controls (such as radio buttons and checkboxes) appear small when displayed on the Nintendo DSi Opera browser, we decided to design the interaction to accommodate the stylus operations.

For the small screen size, we need to create web pages with simple and organized designs that limit horizontal scrolling [12]. This can be done by splitting web pages into smaller, logically related units that can fit onto a handheld device's screen. We also need to choose legible fonts, icons and controls [6]. Basically, the controls size can be small since the stylus has high pointing accuracy compared to the finger. However, to display large images on the web page, their sizes must be reduced to fit the screen size of the DSi screen and also saved in a format supported by the Opera web browser.

As for the dual screen design recommendations, we did not find any clear guidelines in this respect; so we overcome this shortcoming by developing our own approach based on cycles of trails-and-errors. Our approach suggests that static elements of the user interface are placed on the top display and elements that require interaction are placed on the bottom display, forming a web page that is executed across both screens. So, when designing the web page, the important information (in our case the assessment question) is displayed at a fixed location on the screen, i.e. in the top of the page so that it occupies the top screen, while the answer area is displayed in the bottom screen. The distance between the question area and the answer area is fixed based on the Nintendo DS screen size.

Finally, we used open source web standards to design and build our web application. The elements of a web page must leverage the capabilities supported in the browser, so we used HTML for structuring the content, CSS for styling the content, and JavaScript for the interaction.

With these design considerations in mind, we next introduced our Web-Based M-Learning System that used the above guidelines for authoring interactive web pages that can be displayed legibly on the Nintendo DSi screen.

C. *The System Architcture*

The major modules of the NDiC M-Learning system are:

- **The authoring tool**: used by the instructor to author assessment sheets in the form of web pages.
- **Classroom Management system**: used by the instructor to publish assessment sheets, track students' progress and manage users' accounts.

D. *The Authoring Tool*

Figure 1 shows the user interface of the authoring tool. This tool allows the instructors to author interactive assessment sheets that have one or more types of questions: MCQ, Hot Map (where an image is decomposed into a set of clickable regions using radio buttons), Fill in the blank and True/False. The instructor has the ability to also save the assessment sheet for later alteration.

Once an assessment sheet is finalized, the instructor can save it for the web using two methods:

1. Active tracking, which tracks students' answers immediately. This allows the instructor to view students' progress while answering the assessment question(s) and show whether they are doing well or not.
2. Passive tracking, where the instructor chooses to disable the students' tracking feature. This can be done when the teacher wants the students to solve the questions as an exercise. The embedded JavaScript in the web page will automatically check the answers for the students and provide them with the appropriate feedback.

The authoring tool was developed using Visual Basic 2005.

E. *Classroom Management System*

The classroom management system is a web-based (NDiC URL: http://www.ndicproject.com) application used to handle the administrative details of the classroom and publish assessment sheets to the students.

The classroom management system consists of two major subsystems: classroom subsystem and management subsystem.

The classroom subsystem helps the instructor to upload assessment sheets to the server so that the students can access them, retrieve statistics regarding students' answers and create polls. The system also has a tracking module to track students' progress. The tracking module aids the instructor on tracking students' progress in real-time while they are solving the assessment sheet. It can also show different statistics about the students progress (when sheet tracking is set to active). The instructor can also create polls and receive the students' responses immediately.

The management subsystem (Figure 2) is for the system administrator to manage instructors and students' information such as adding, editing, or deleting. Also, the administrator can manage classrooms information.

The classroom management system was developed using PHP for server–side scripting and MySQL for database running on Apache web server.

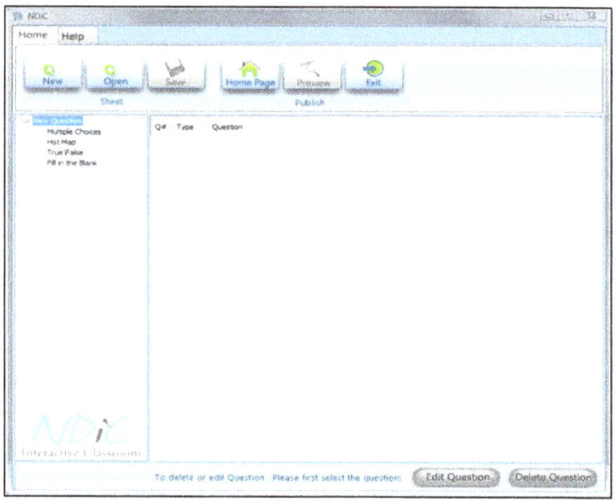

Figure 1. Authoring tool user interface

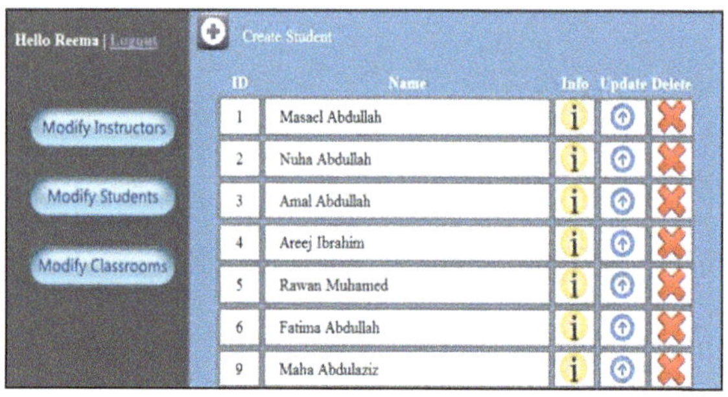

Figure 2. The modify students page management subsystem

V. System Demo

In order to use the NDiC system, a student must point his/her DSi browser to the login page of the NDiC web application. Once the student is logged in, the system records the student attendance.

In the student's log in page a list of assessment sheets are displayed which are ready to be selected. Once the student selects an assessment sheet the assessment questions are displayed to the student one by one, as shown in Figure 4.

Using our system 255 students can be supported at most in a single class with the presented demo system. Also unlimited number of classes can be created and supported.

If the active mode was setup for the assessment sheet, the tracking module in the web application will track the logged in students and monitor their progress while solving the assessment sheet, as shown in Figure 3.

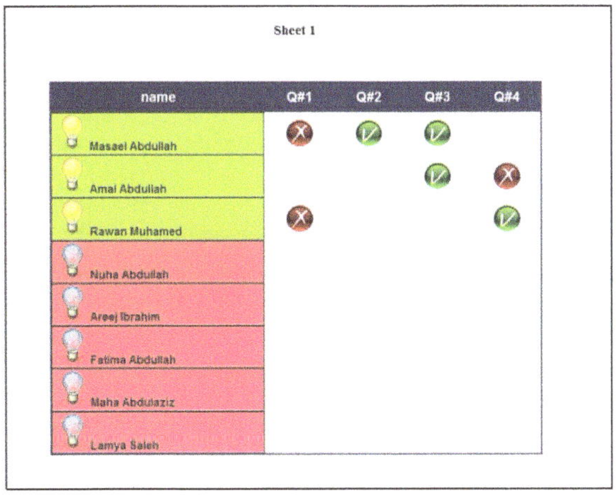

Figure 3. View of the tracking system showing students' who is currently solving 'sheet 1' assessment sheet, the names with green backgrounds and illuminated light bulbs mean logged in students.

VI. Preliminary System Evaluation

Evaluation of applications designed for the use by handheld devices is a challenging task as reported by [13]. Different evaluation techniques from the domain of Human-Computer Interaction (HCI) can be considered under certain context; however, usability evaluation is considered a first step for evaluating any interactive system.

Using usability evaluation, we evaluated the NDiC system by a group of elementary school students and teachers in a small-size math/geography class. The teachers used the system to author and publish assessment sheets, and the students used the system to solve these sheets.

We then used the System Usability Scale (SUS) questionnaire, a well-known usability instrument, to measure the usability and user satisfaction of the NDiC system. Question items included in the SUS survey has a five-scale Likert scale ranged from strongly disagree (1) to strongly agree (5). The items are:

1. I think that I would like to use this system frequently.
2. I found the system unnecessarily complex.
3. I thought the system was easy to use.
4. I think that I would need the support of a technical person to be able to use this system.
5. I found the various functions in this system were well integrated.
6. I thought there was too much inconsistency in this system.
7. I would imagine that most people would learn to use this system very quickly.
8. I found the system very cumbersome to use.
9. I felt very confident using the system.
10. I needed to learn a lot of things before I could get going with this system.

The questionnaire was distributed among the students (age 9-13) and instructors. The students were asked to rate their agreement to the statements from strongly disagree (1) to strongly agree (5). In summary, the mean ratings of students' satisfaction were high; in general it was 93%.

On the other hand, the mean ratings of instructors' satisfaction were moderate, it scored 76%. The items affected by the instructors' ratings were items 7, 8 and 9. By further asking the instructors about their reasons of the selected

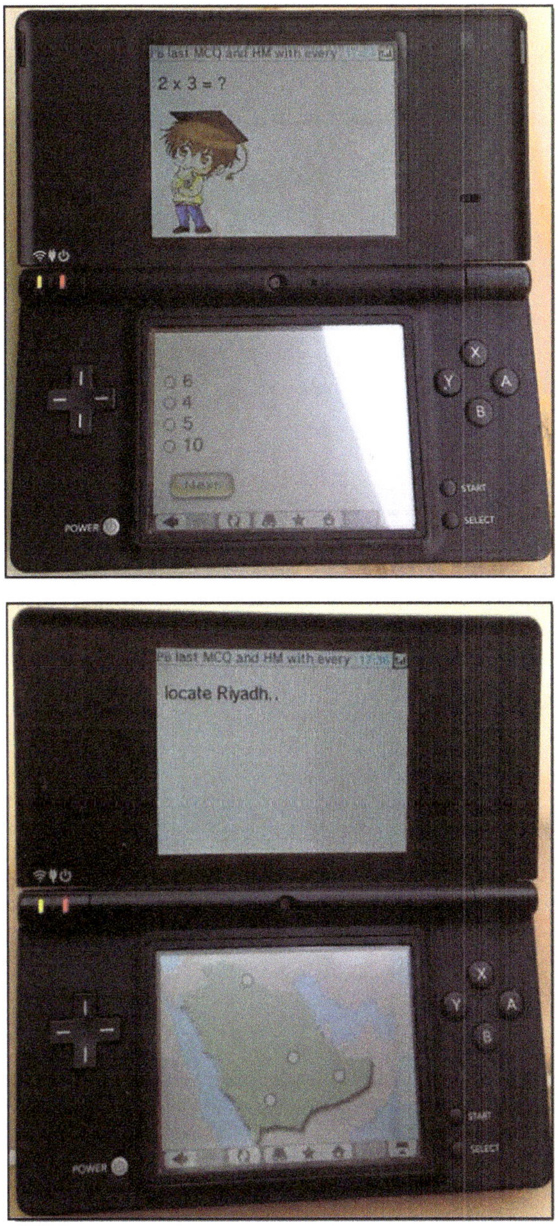

Figure 4. A picture of an MCQ question (top) and a hot map question (bottom) on the NINTENDO DSi console

ratings, they reported that the system does not support Arabic language which is their native language and their English is not so good, which actually hindered the system use. Also the instructors reported that they lack some technical skills, which implies requiring some training to use the system.

VII. Conclusion and Lessons Learned

With the wide spread of game consoles among the young generation, provided by their outstanding capabilities, a shift toward developing M-Learning systems that utilizes the capabilities of these consoles has increased.

In this paper we showed how we designed a low cost web-based interactive classroom system that uses a well-know game console for aiding student interaction in the classroom. Our designed system took into consideration the capabilities of the game console and the design recommendation reported in the literature. A preliminary evaluation of the system that was carried by a number of students and instructors revealed encouraging results.

A future expansion of this project will include publishing assessment sheets that consider different screen resolutions e.g. the new Nintendo DSi XL.

Key lessons learned when designing web pages with interactive controls for the display on handheld devices with dual screen can be summarized as follows:

1. Use web controls that benefit from the stylus supported functionality.
2. Place interactive elements at the bottom screen and leave the top screen for display only.
3. Take care of the aspect ratio of the dual screen size, to avoid the horizontal and vertical scrolling effects.
4. Avoid overriding the default font settings (size and family) that are set by the handheld browser.
5. Limit the number of characters displayed in one page to around 250.
6. Avoid using controls with more than 4 options (e.g. drop-down list, radio buttons).

Finally, we hope that what have been reported in this paper can be considered as a starting step for developing a framework for designing web-based learning applications designated for dual screen handheld game consoles.

Acknowledgment

The author would like to thank her undergraduate project students for developing the system.

References

[1] W3C, "Mobile Web Best Practices 1.0," 29-Jul-2008.

[2] "D 4.1 – Standards and technology monitoring report," University of Maribor, 24-Apr-2007.

[3] J. Kirriemuir, "The relevance of video games and gaming consoles to the higher and further education learning experience," The Joint Information Systems Committee, TSW 02-01. Retrieved June, vol. 2, p. 0201, 2002.

[4] A. Pulman, "Does a handheld gaming device make an effective assistive technology tool?," presented at the ALT-C 2007, 2007.

[5] L. Qiao, L. Feng, and L. Zhou, "Information Presentation on Mobile Devices: Techniques and Practices," in Progress in WWW Research and Development, vol. 4976, pp. 395–406, 2008.

[6] D. Zimmerman and T. Yohon, "Small-screen interface design: Where are we? Where do we go?," in Professional Communication Conference, 2009. IPCC 2009. IEEE International, pp. 1-5, 2009.

[7] D. Churchill and J. Hedberg, "Learning object design considerations for small-screen handheld devices," Computers & Education, vol. 50, no. 3, pp. 881-893, Apr. 2008.

[8] D. S. K. Seong, "Usability guidelines for designing mobile learning portals," in Proceedings of the 3rd international conference on Mobile technology, applications & systems - Mobility '06, p. 25, 2006.

[9] "Web Specification Support in Opera 8."

[10] S. McManus, "Webdesign for Nintendo DS Opera web browser," Webdesign for Nintendo DS Opera web browser.

[11] L. McKnight and B. Cassidy, "Children's Interaction with Mobile Touch-Screen Devices:," International Journal of Mobile Human Computer Interaction, vol. 2, no. 2, 2010.

[12] J. Cartman and R. Ting, Strategic Mobile Design: Creating Engaging Experiences, 1st ed. New Riders Press, 2008.

[13] K. Huang, "Challenges in Human-Computer Interaction Design for Mobile Devices," in Proceedings of the World Congress on Engineering and Computer Science, 2009.

Author

Hend S. Al-Khalifa is with the Information Technology Department, College of Computer and Information Sciences, King Saud University, Riyadh, Saudi Arabia.

Enhanced Practical Wireless Communications Education via Blended Instructional Tools

Ashraf Tahat[1], Ahmad Jamal[2], Mohammad Kalbouneh[1]
[1] Princess Sumaya University for Technology, Amman, Jordan
[2] Umniah Mobile Telecommunications, Amman, Jordan

Abstract—**Today's graduating telecommunications engineers need marketable skills that are typically developed beyond the traditional classrooms. These skills set can be acquired by experimenting for hands-on experience that resembles the professional field workplace. In this paper, we address blending of methodologies, applications, and the use of the Internet for delivery of knowledge, and instruction in an undergraduate wireless communications engineering course. This solution combines several different delivery methods, such as face-to-face, team collaboration, field measurements, in conjunction with high-end professional simulation software. Two examples of handover in a dual-mode cellular network and radio frequency (RF) signal propagation modeling are considered. Demonstrations and verification of theoretical models are presented based-on blending of instructional means. Lessons are learned to enhance hands-on experiments beyond traditional laboratories to strengthen in-depth analysis, learning skills of students, in addition to promoting proper team collaboration on projects.**

Index Terms—**Blended learning, handover, hand-off, hands-on laboratory, m-learning, 3G, GSM.**

I. Introduction

Revelutionary advances in technology have driven the academic field to undergo a variety of corresponding significant changes to cope with the demands of the modern industry and workplace. Hence, classroom lectures on a particular advanced technical concept, such as wireless radio frequency fundamentals; need to be complemented at times by experimental illustrations. Because wireless communications related courses include topics that constitute techniques used in industries such as voice communications, data telecommunications systems [1], security systems, telemedicine and eHealth applications [2], ECO driving in conjunction with automotive maintenance [3], and defense systems, it is imperative for those industries that technical staffs such as engineers and technicians have the practical hands-on experience in addition to the theoretical background in wireless communications. Although laboratory experimentation form an invaluable supplement to theory in training of engineers and technicians, it falls short in many live field scenarios where in-depth understanding of some fundamental principles is vital for practical real-life network designs. Moreover, in most of educational institutions and training schools, there are several wireless communications courses for teaching the theoretical concepts, but no or not enough laboratory equipment to demonstrate or verify the application of the theory [4]. The main reason is that the equipment required in high-frequency wireless or radio communications laboratories are of significantly complex technology, and expensive. Hence, small to mid-size educational institutions cannot afford to obtain such equipment, and establish those hands-on experimental laboratories [5].

Undoubtedly, some of the countless opportunities modern technologies provide lie in the effective use of applications software and the Internet as potent instructional tools [6]. A simulation software-based laboratory, with experiments carried out on professional high-end planning and simulation software (e.g., ASSET3G by Aircom International) [7], could be used alone to replace field experiments by creating virtual simulations of desired scenarios within the capabilities of the deployed software. However, in the context of mobile RF field environment, simulation-based experiments always present an approximated and idealized result that can have a significant discrepancy from actual real-life field set-ups that exist in a live network.

A blending of a variety of low-cost learning tools and different delivery methods (such as collaboration in experimental field campaigns, simulation software, Internet-based applications, and m-learning [8], in addition to traditional instructional means) could provide for means to alleviate the previously described issues [6]. This has great resemblance to what is termed "*blended learning*," which refers to the combination of various event-based learning activities, including face-to-face classrooms, live e-learning, team collaboration, and asynchronous self-study [9].

In this paper, we consider two example fundamental concepts from a wireless communications engineering course, namely *handover* and *wireless signal propagation* [10], to illustrate the concrete application of the proposed learning approach [6]. This is carried-out as follows: Firstly, students learn the theory and engineering models of the above wireless phenomena through face-to-face instruction. Secondly, they form teams to go on field measurement campaigns to collect relevant data from three cellular sites (urban, suburban, rural) utilizing low-cost mobile application installed on their smart-phones. Thirdly, the collected data is exported into a special wireless network simulation and design software in a laboratory to perform simulations [11]. Finally, analysis and comparisons are performed to verify the studied fundamental theoretical concepts.

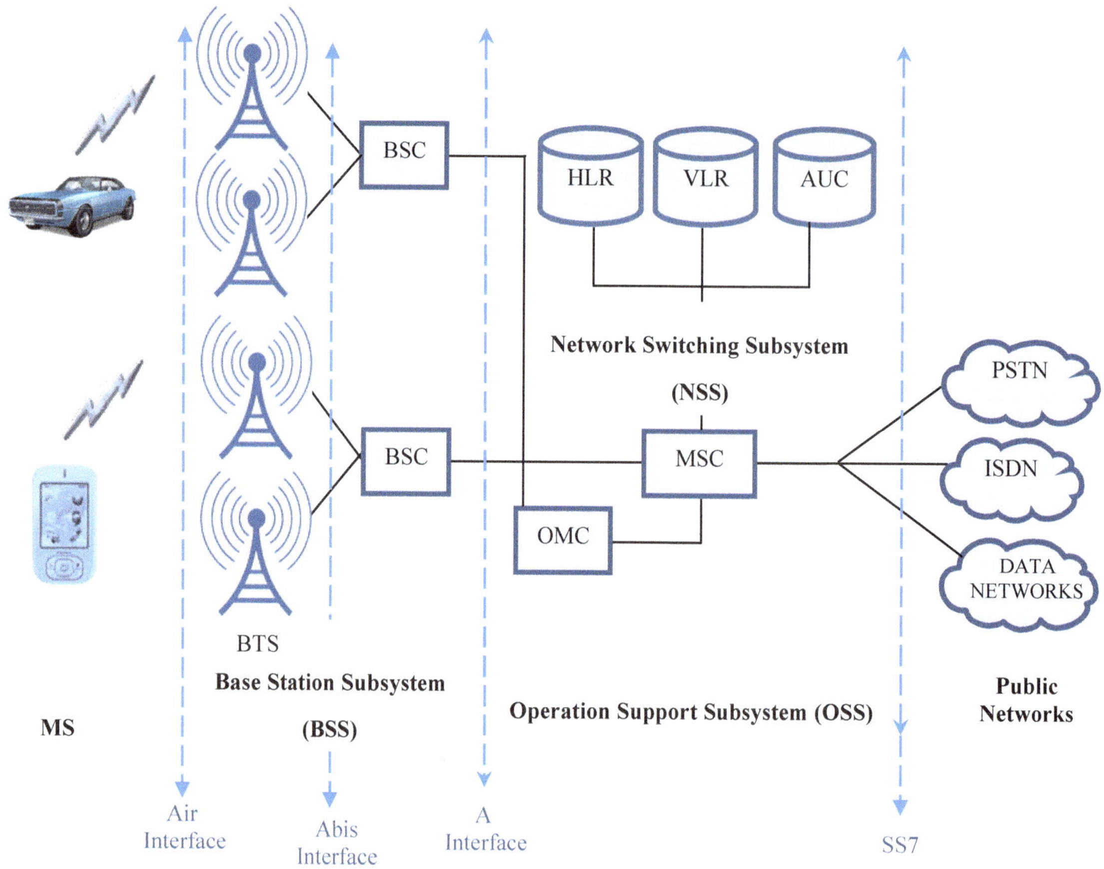

Figure 1. GSM network architecture

This paper is organized as follows: in section II, we review relevant cellular (wireless) concepts and models, and in section III we present the utilized measurement and visualization tools. In section IV, we describe simulation set-up and analysis. Finally, we conclude in section V.

II. Theoretical Background

Dual-mode WCDMA-GSM mobile terminals require an interworking mechanism between the WCDMA and GSM technologies [10], [12]. For instance, if the user has established a voice call using WCDMA technology, and then moves outside of WCDMA coverage, the voice call needs to be handed over to GSM without any perceived disturbance during handover using dual-mode mobile station (MS) in a live network. Handover (*hand-off*) plays an instrumental role in ensuring interworking between WCDMA and GSM technologies.

A. GSM Mobile System Architechure

GSM (Global System for Mobile communication) is a digital mobile telephony system that is the most widely used mobile (cellular) radio communications standard in all parts of the world [13]. GSM uses a variation of time division multiple access (TDMA) where it digitizes and compresses a subscriber's data, then sends it down a channel with other streams of other users' data, each in its own time slot. A 200 kHz carrier is divided into timeslots for individual phones to use. This allows eight full-rate or sixteen half-rate speech channels per carrier radio frequency. These eight radio timeslots (or eight burst periods) are grouped into a TDMA frame. The channel data rate for all 8 channels is 270.833kbit/s, and the frame duration is 4.615ms. GSM is typically deployed at either the 900 MHz or 1800MHz frequency bands. While in North America, the 850 MHz and 1900 MHz bands were already allocated instead [10]. And in rare cases the 400 and 450 MHz frequency bands are assigned in some countries because they were previously used for first-generation systems. Regardless of the frequency selected by an operator, the transmission power in the MS is limited to a maximum of 2 watts in GSM850/900 and 1 watt in GSM1800/1900 phones.

The GSM network is structured into a number of discrete sections as depicted in Fig. 1: a) The Base Station Subsystem (BSS) (the base stations and their controllers). b) The Network and Switching Subsystem (NSS) (the part of the network most similar to a fixed network). This is sometimes also just called the core network. c) The operations support system (OSS) for maintenance of the network. d) The GPRS Core Network (the optional part which allows packet based Internet connections).

B. 3G (WCDMA) Mobile Network Architechutre

Third-generation mobile services are still now being deployed throughout the world [13]. Wideband code-division multiple access (WCDMA) technology is an excellent framework for providing multimedia services, since it meets the performance demands of the mobile Internet, including Web access, audio and video streaming, ing, and video and IP multimedia calls [10].

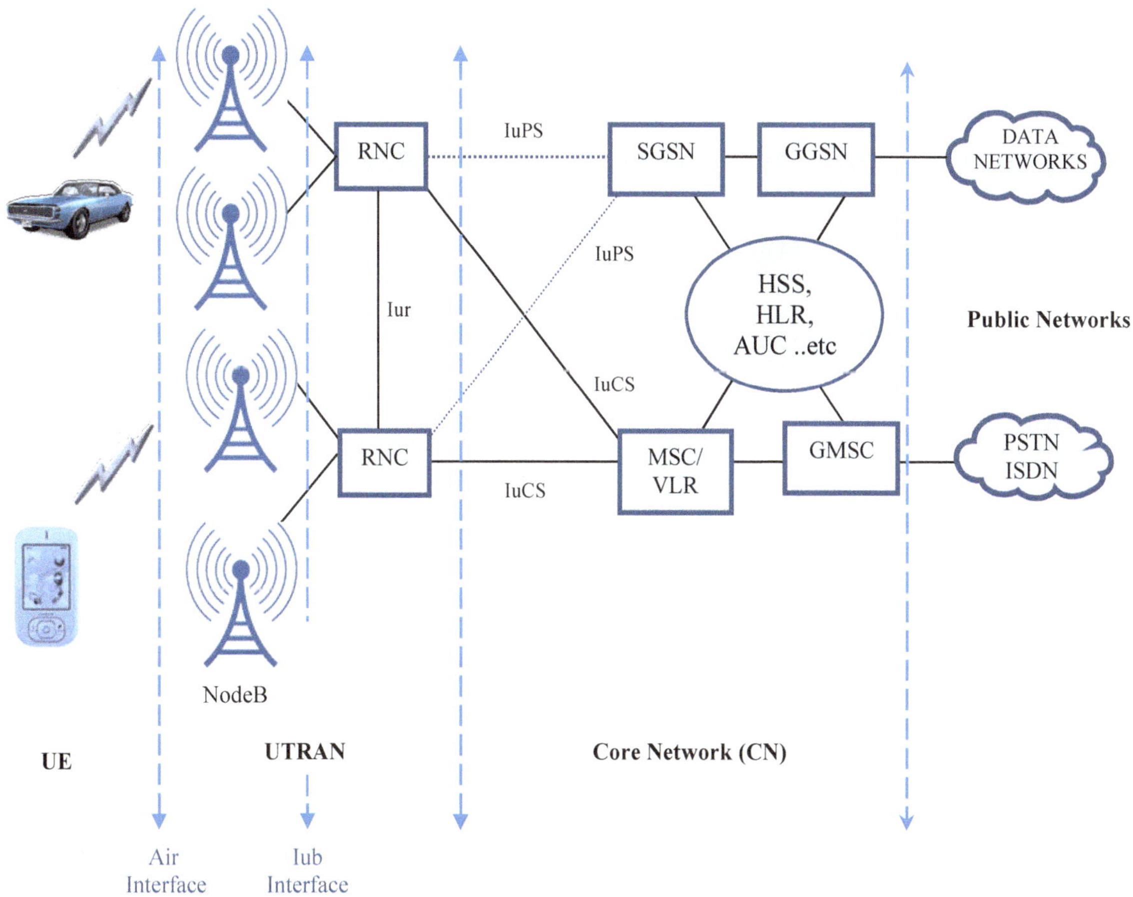

Figure 2. 3G UMTS network architecture

WCDMA transmits on a pair of 5 MHz-wide radio channels and uses direct sequence CDMA transmission technique unlike GSM that uses TDMA.

Services advertised as 3G are required to meet IMT-2000 technical standards. Recent 3G releases, often denoted 3.5G and 3.75G, provide mobile broadband access of several megabits per second to smartphones and mobile modems in laptop computers. While most WCDMA technology will initially only be deployed to cover urban areas, GSM networks have a nationwide or global footprint that provides access to mobile services, such as voice, data, short message service (SMS) and multimedia messaging service (MMS). Therefore, operators who already have a GSM network want to capitalize on their investments when migrating to WCDMA technology. A third-generation MS equipped with both WCDMA and GSM technology would put the end-user in contact with seamless, (practically) worldwide, mobile service. As depicted in Fig. 2, a UMTS network consist of three interacting domains: a) Core Network (CN) which contains the databases and network management functions b) UMTS Terrestrial Radio Access Network (UTRAN) c) User Equipment (UE). The basic Core Network structure is similar to the GSM network with GPRS. Base Station is referred to as NodeB, and control of NodeB's is performed through the Radio Network Controller (RNC).

C. Handover Between 3G (WCDMA) and GSM Networks

Handover between WCDMA and GSM allows the GSM network to be used to give fallback coverage for WCDMA technology. This means that subscribers can experience seamless services in an initial commercial launch. Although the network solely communicates with the mobile terminal using one access technology (GSM or WCDMA) at a time, the mobile terminal needs to perform measurements on GSM while communicating in WCDMA and vice versa [12].

The dual-mode mobile terminal reselects a GSM cell when that cell is ranked higher than the current WCDMA cell or any other WCDMA cell. WCDMA and GSM cells are ranked together according to signal strength. This same type of ranking applies in GSM. When performing cell re-selection in WCDMA, the mobile terminal either measures GSM cells continuously or when the quality of the serving WCDMA cell falls below a given threshold. The mobile terminal is solely allowed to select a new WCDMA or GSM cell when the average received quality and average signal strength exceed a minimum threshold [12]. The minimum-quality threshold (signal-to-noise ratio) ensures that the mobile terminal can receive the information transmitted by the potential target cell. This criterion also takes into account: a) The maximum transmit power that the mobile terminal is allowed to use when accessing the cell. b) The maximum radio frequency (RF) output power that the mobile terminal can transmit.

D. *Wireless Signal Propagation Modeling*

One of the main important parameters required to ensure service is the received signal power. During planning, propagation models are utilized to predict path loss and received signal power. Propagation Models are mathematical attempts to model the 'real' radio environment as closely as possible. Most propagation models need to be tuned or calibrated by being compared to measured propagation data; otherwise you will not be able to obtain accurate coverage predications [14].

Several models have been developed to meet the requirements of realizing the propagation behavior in different conditions [10]. These models have been traditionally applied to frequencies below 2 GHz. According to Jordan's Telecommunications Regulatory Commission, the used propagation model in Amman, and Jordan in general, is COST 231 Hata for the frequency bands of 800MHz to 2100MHz. UMTS site density is generally greater than GSM site density; hence accuracy in the first hundreds of meters from the base station is of high importance [14].

We used ASSET3G by Aircom International [7] to predict signal strength values of received power and to provide coverage maps based on these results. Base station configuration parameters and GIS digital map including clutter, elevation and vectors are needed for this process. The digital map of Amman, Jordan we have used had a 20 m resolution. We used the Standard Macrocell Model in ASSET3G [11] for estimating path loss. This is a slightly modified Okumura-Hata model [10]. The path Loss formula for this model is as follows [11]

$$PL_{MC} = K_1 + K_2 \log_{10}(d) + K_3 H_{ms} + K_4 \log_{10}(H_{ms}) + K_5 \log_{10}(H_{eff}) + K_6 \log_{10}(H_{eff}) \log_{10}(d) + K_7 (Diffn) + C_{Loss}$$

(1)#

where d is the distance from the base station (BS) to the mobile station (MS) in kilometers, H_{ms} is the height of the mobile station above ground in meters, H_{eff} is the effective base station antenna in meters, *diffn* diffraction loss calculated using the Epstein-Peterson, Bullington, Deygout, or Japanese Atlas knife edge techniques. Table 1 lists the default *K*-parameters values for popular standard cellular frequencies. These values represent typical starting values based on urban environment. C_{Loss} is clutter specifications such as heights and separation are also taken into account in the calculation process.

It is vital that the model has to be accurately calibrated (tuned) to correspond to real network environment to obtain more accurate prediction values for a real live project [14]. To do this, we need to compare predicted propagation data with student team survey data to produce an accurate propagation model that functions correctly and to be utilized by in-laboratory simulations.

TABLE 1
DEFAULT PROPAGATION MODEL PARAMETER VALUES

Parameter	900 MHz	1800 MHz	2000 MHz
K_1	150.6	160.9	162.5
K_2	44.9	44.9	44.9
K_3	-2.55	-2.55	-2.55
K_4	0.00	0.00	0.00
K_5	-13.8	-13.8	-13.8
K_6	-6.55	-6.55	-6.55
K_7	0.7	0.8	0.8

Figure 3. Photo of a GSM BS site location near QAIA road.

III. FIELD MEASUREMENTS AND ANALYSIS TOOLS

Firstly, the student measurement campaign was sent to identify areas near city limits were 3G coverage starts to diminish and handover to a GSM BS with adjacent cell coverage becomes necessary. This utilized test-bed in a live network was set-up with a 3G system operating in the 2100 MHz frequency band and GSM system operating in the 900 MHz frequency band. The measurements were performed with one dual-mode mobile terminal connected to one BS at a time. The transmitted power for each BS was 33 dBm for 3G, and the transmitted power for each GSM BS was 41 dBm.

Secondly, the signal propagation measurements campaign has collected readings at three different locations representing urban, suburban, and rural environments. Some measurements were close to a BS and some were near cell borders, taken in a gradually increasing manner. Selection of locations was based on various criteria, where distance, elevation, and line-of-sight (LOS) capabilities were the main differing attributes at each location.

Physical recorded measurements included the Received Signal Strength Indication (RSSI), GPS location coordinates, and a photo of the BS tower site. Fig. 3 shows a photo captured at the considered GSM site near Queen Alia International Airport (QAIA) road. GPS coordinates were tracked at each BS location, thus a terrain profile could be constructed to fully determine the obstacles that may interfere with LOS. Mobile application software CellTrack [15] version 9.1 was used on a Nokia X6 phone in this drive test procedure to take the measurements.

A. CellTrack

CellTrack is a mobile application software utilized to extract and collect some phone information about the serving cell it is connected to [15]. Fig. 4 shows the CellTrack application software interface running on the Nokia X6 phone. CellTrack provides the following parameters: *Cellid* the ID from the actual cell. *LAC* the Location Area Code from your cell. *Net* the network you are moving in (e.g. 416 for Jordan and 77 for Orange Jordan). If a WCDMA/UMTS cell is present, it is also shown here. *Name* the name of the cell. This is shown by an algorithm when you select 'Cell Name from ID' in the settings, or from the cell broadcast when you have turned-on *'Cell Name from CBS'* and have defined a service number. CBS overwrites the generated name. *Signal* shows the quality of the receive signal. The in a percentage value is taken from the phone API itself, while strength is displayed as a dBm value. *Battery* should show the capacity of your battery. *Description* shows information from the database. *Diagram* shows the dBm values as green and the percentage as a black line. The red lines represent cell changes. The equipment reported the RSSI values, which represent the received signal strength in dBm..When analyzing signal strength it is useful to relate the received signal strength to the distance between the BS and the SU. This section analyses the signal strength for the Down Link (DL). The physical measurements included only the RSSI. Sample measurements listed in Table 2 here were performed at each of a three selected locations by the subscriber station.

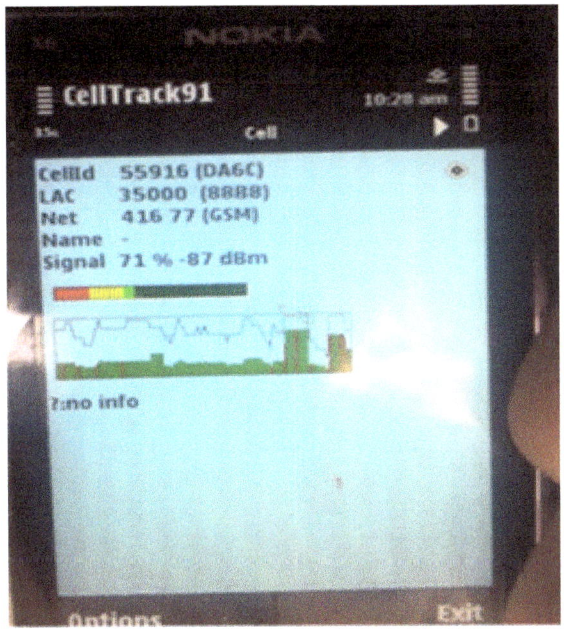

Figure 4. CellTrack mobile application software interface

The collected measurements for each site were tabulated in Excel® spreadsheets and then plotted. Fig. 5 shows a sample plot of measured data that depicts handover near the QAIA 3G and GSM sites indicating where handover has occurred from 3G to the GSM site. We will show later generated plots of received signal power

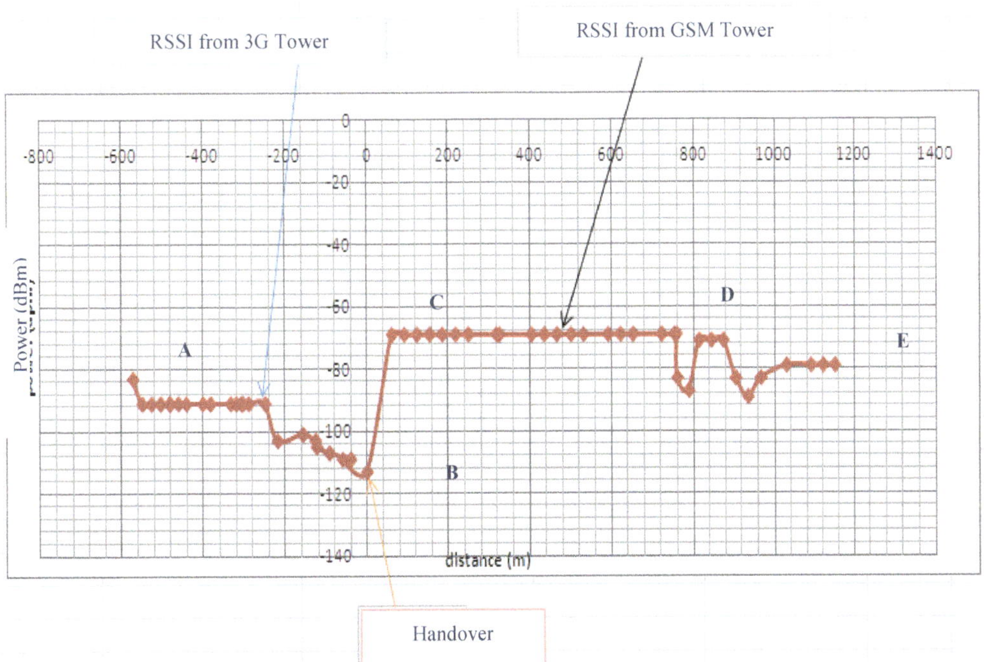

Figure 5. Plot of recorded measurements path between 3G and GSM BS for QAIA locations.

TABLE 2

A SAMPLE OF RECORDED MEASUREMENTS AT TEST SITES

Site Readings– 3G and Header File – 3G			
POINT	Longitude	Latitude	RSSI
1	35.88472222	32.0702777	-79
2	35.88333333	32.07083333	-85
3	35.88277778	32.07138889	-79
4	35.88222222	32.07166667	-75
5	35.87805556	32.07222222	-85
6	35.87777778	32.07222222	-89
7	35.87111111	32.06944444	-69
8	35.87722222	32.07222222	-99
9	35.87694444	32.07194444	-89
10	35.87666667	32.07166667	-105
11	35.87666667	32.07166667	-110
12	35.87638889	32.07138889	-105

(RSSI) and predicted values super imposed on the same plot to emphasize the value of this approach and appreciate the methodology in bringing a real feel of the physical phenomena.

B. Google Earth

Google Earth® is a virtual globe, map and geographical information software that was acquired by Google in 2004 [13], [16]. It maps the Earth by the superimposition of images obtained from satellite imagery, aerial photography and GIS 3D globe. Fig. 6 shows a sample Google Earth view of one of the considered test paths, QAIA, surrounding area and measurement locations superimposed using tags.

Figure 6. A Google Erath view of measurement locations.

Google Earth® is currently available for use on personal computers running Windows, Mac OS, or Linux operating systems, and is also available for mobile viewers as a free application [16]. It displays parts of the surface of the Earth in addition to availability of 3D images of terrain and buildings. It uses digital elevation model (DEM) data collected by NASA's Shuttle Radar Topography Mission (SRTM).

Figure 7. 3D earth view of the 3G BS site location near QAIA road.

Google Earth can provide a lot of information about a location. However, the information is stored in layers, which can be turned on or off. Layers include roads, border labels, parks, …etc. Two layers are useful for creating a more three dimensional globe: Terrain layer simulates the elevation levels, and the 3D Buildings layer lets you zoom through cities, and between buildings. 3D views of many mountains have been improved by the use of supplementary DEM data to fill the gaps in SRTM coverage [13]. Google Earth is also available as Web Map Service client. It supports managing three-dimensional Geospatial data through Keyhole Markup Language (KML). It has the capability to show 3D structures, which consist of users' submissions using SketchUp, a 3D modeling program software. This means one can view the whole earth in three dimensions. Fig. 7. shows a 3D view of the 3G BS cell site near QAIA road.

C. *ASSET3G by Aircom International*

Several groups of measured data sets were imported into ASSET3G by Aircom International [7] using the InSignia standard file format [11] to calibrate relevant propagation models, predict signal strength values of received power, and to provide coverage maps based on these results and relying on base station configuration parameters in conjugation with GIS digital map as shown Fig. 8.

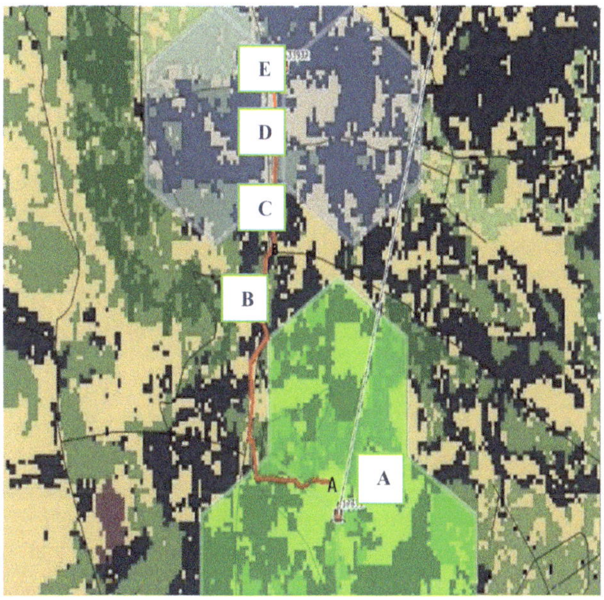

Figure 8. Measurement path between BS locations for QAIA sites in ASSET3G software.

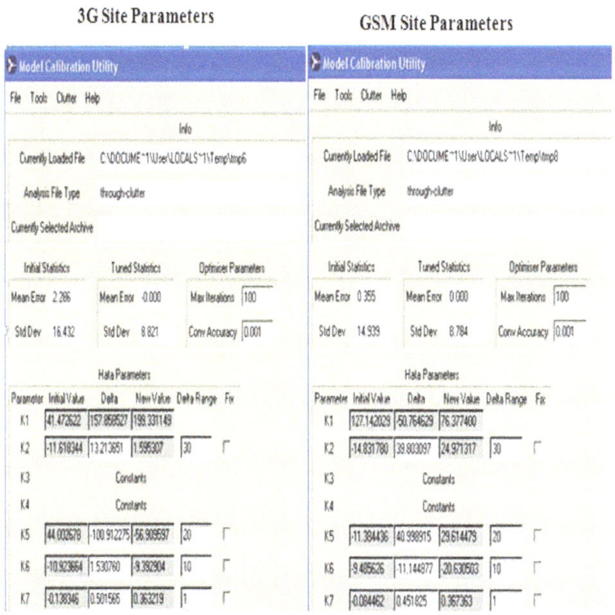

Figure 9. Propagation model tuning (calibration) in ASSET3G software.

ASSET3G enables you to auto-tune your propagation models, using analysis-information from a comparison between your model's path-loss parameters predictions using default *K* parameters of Table 1 and 'real-world' measurement data to find new *K* parameters appropriate to each site's geographic area as shown in Fig. 9.

We show in Fig. 10 a plot of the actual measured received power in the coverage area of the GSM site. On the same graph, we plot the predicted received power using the ASSET3G planning and simulation software before propagation model tuning. Then superimposed on the same graph we plot the predicted received power after tuning the propagation model using the measurements campaign actual values. It is evident how the calibration process relying on field merriments is vital in obtaining predictions with significant accuracy for wireless network planning and optimization.

Now that we have calibrated (tuned) propagation models conforming to the environment we are working in (thanks to experimental data files), coverage footprints for each site is created after running the path-loss predictor to produce the coverage footprints for each considered site. Fig. 11 shows coverage map (RSSI) for the QAIA 3G and GSM sites.

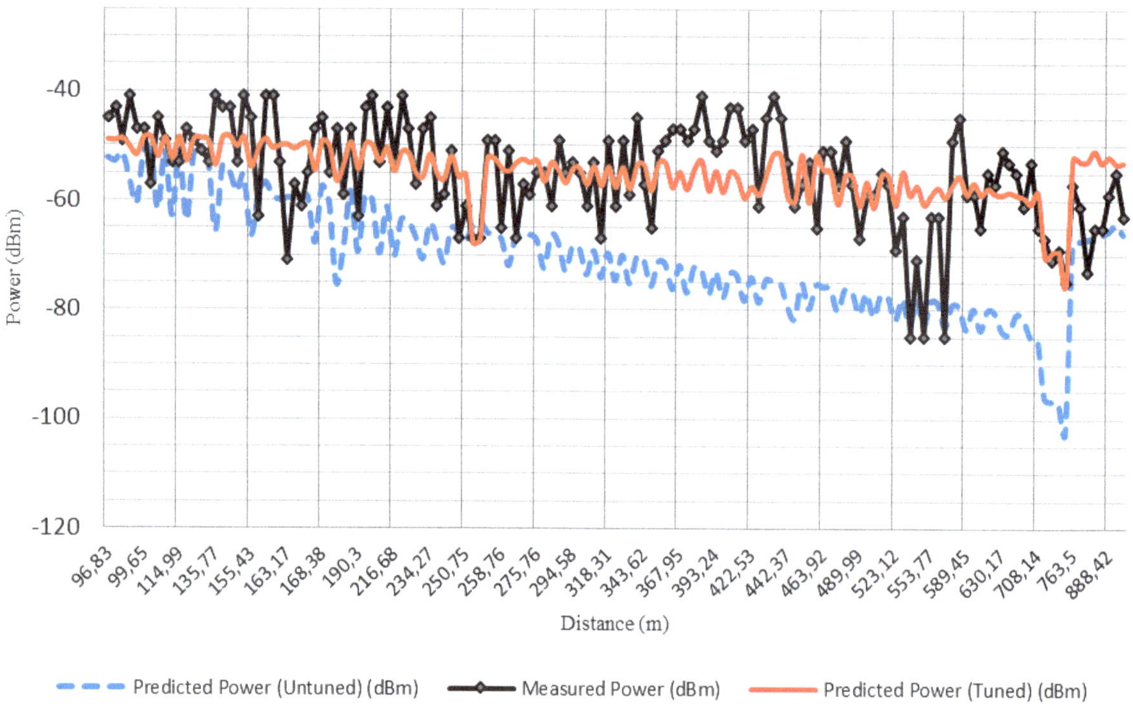

Figure 10. Measured and predicted RSSI for GSM 900MHz site before and after tuning near QAIA.

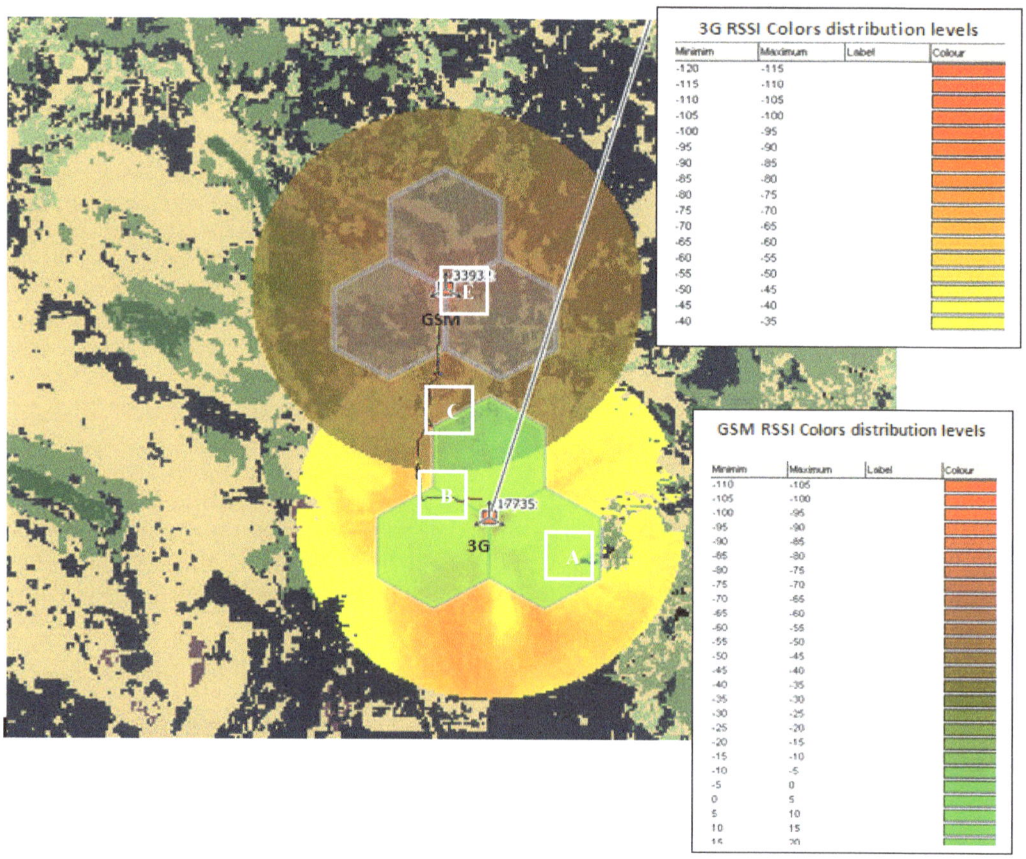

Figure 11. Simulated coverage of 3G and GSM sites in ASSET3G software near QAIA.

IV. NEIGHBORING ANALYSIS IN ASSET3G

Handovers can be classified into three categories [12]: *Hard* handovers which are intended to be instantaneous in order to minimize the disruption to the call. A *hard* handover is perceived by network engineers as an event during the call. When the mobile is between base stations cell borders, then the mobile can switch to any of the base stations, so the base stations bounce the link with the mobile back and forth. This is called ping-ponging. *Soft* handover is one in which the channel in the source cell is retained and used for a while in parallel with the channel in the target cell. The interval, during which the two connections are used in parallel, may be brief or substantial. Soft handovers may involve using connections to more than two cells (connections to three, four or more cells can be maintained by one phone at the same time). *Softer* handover occurs when a call is in a state of soft handover, the signal of the best of all used channels can be used for the call at a given moment or all the signals can be combined to produce a clearer copy of the signal. The latter is more advantageous. Such combining is performed both in the downlink (forward link) and the uplink (reverse link) in a *softer* handover scenario.

There are two ways to define neighboring cells "neighbors" to a certain base station, *manual* and *automatic* [11]. ASSET3G refers to the *manual* method with "Database", and for the automatic method with "Plan". The neighbor analysis feature in ASSET3G software can be used to view all the relationships currently existing in the Site Database.

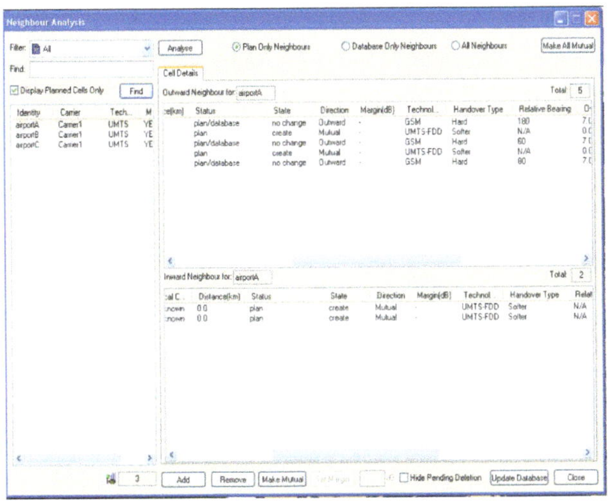

Figure 12. Neighboring analysis window in ASSET3G

You can also examine the differences, on a cell-by-cell basis between the database and any new potential neighboring data that has been generated or loaded. As shown in the Fig. 12, at point B, QAIA sector A, performed a hard handover at distance 700 m away from the 3G site to the GSM site.

Referring to Fig. 5, which represents a plot of the actual observed received power (RSSI) from either of the 3G or the GSM BS that the MS is currently connected to, indicates that a handover has occurred at approximately 1.3 km away from the 3G site to the GSM site. This is in agreement with what is predicted by ASSET3G neighbor analysis feature.

V. CONCLUSIONS

This paper has presented a blending methodology for a variety of learning (instructional) tools to provide for an inexpensive practical and hands-on learning approach in wireless communications course. This has proved to be effective for verification and for illustration of fundamental but hard to perceive concepts in mobile communications. Two examples, handover and RF signal propagation modeling, were presented and verified. In addition, these approaches emphasize students' collaborative work with their peers and active participation within educational framework.

REFERENCES

[1] A. Tahat, M. Khalaf, and O. Elmuhesen, "A solar energy water heater remote monitoring and control system," *in Proc. International Conference on Electronic Devices, Systems, and Applications (ICEDSA)*, pp. 98-103, April 2011.

[2] A. Tahat, A. Sacca, and Y. Kheetan, "Design of an integrated mobile system to measure blood pressure," *in Proc. IEEE 18th Symposium on Communications and Vehicular Technology (SCVT)*, pp. 1-6, November 2011.

[3] A Tahat, A. Said, F. Jaouni, and W. Qadamani, "Android-based universal vehicle diagnostic and tracking system," *in Proc. IEEE 16th International Symposium on Consumer Electronics (ISCE)*, pp. 137-143, June 2012.

[4] A. Kara, E. Aydin, R. Oktem, and N. Cagiltay, "A remote laboratory for training in radio communications: ERRL," *in Proce. The 18th IEEE International Symposium on Personal, Indoor and Mobile Radio Communications (PIMRC'07)*, pp. 1-6, August 2007.

[5] N. Cagiltay, E. Aydın, R. Oktem, A. Kara, M. Alexandru, and B. Reiner, "Requirements for remote RF laboratory applications: an educators' perspective," *IEEE Tran. on Education*, vol. 52, pp. 75-81, February 2009.

[6] A. Tahat, A. Jamal, M. Kalbouneh, and H Jaber, "Blending of learning tools for enhanced practical wireless communications education," *in Proc. 6th Interactive Mobile & Computer Aided Learning International Conference (IMCL)*, pp. 1-6, November 2012.

[7] Aircom International, http://www.aircominternational.com/, July 2012.

[8] J. Attewell, and M. Gustafsson, "Mobile communications technologies for young adult learning and skills development (m-learning)," *in Proc. IEEE International Workshop on Wireless and Mobile Technologies in Education (WMTE '02)*, pp. 158-160, August2002.

[9] D. Garrison, and H. Kanuka, "Blended learning: Uncovering its transformative potential in higher education," *The Internet and Higher Education*, vol. 7, pp. 95–105, 2nd quarter 2004.

[10] A. Molisch, *Wireless Communications*, 2nd ed., West Sussex, England: Wiley & Sons, Ltd., 2011.

[11] ASSET User Reference Guide, Version 6, AIRCOM International Ltd, Leatherhead, Surrey, UK, 2007.

[12] G. Alsenmyr *et. al.*, "Handover between WCDMA and GSM," Ericson Review, No. 1, 2003.

[13] Wikiepedia, http://en.wikipedia.org/ July 2012.

[14] A. Tahat, and M. Taha, " Statistical tuning of Walfisch-Ikagami propagation model using particle swarm optimization," *in Proc. 19th IEEE Symposium on Communications and Vehicular Technology (SCVT) in the Benelux*, pp. 1-6, November 2012.

[15] CellTrack, http://www.afischer-online.de/sos/celltrack/, May 2011.

[16] Google Earth, http://www.google.com/edu/teacher/ , July 2012 .

AUTHORS

Dr. Ashraf Tahat received the B.S., M.S., and Ph.D. degrees in Electrical Engineering from the Illinois Institute of Technology (IIT), Chicago, Illinois, USA, in 1992, 1993, and 2002, respectively. He is an Associate Professor with the Department of Communications Engineering at PSUT in Amman, Jordan where he was the Department Chairman. He is currently a Visiting Professor in the Department of Electrical and Computer Engineering at McGill University in Montreal, Quebec, Canada, working on digital signal processing for modern communications systems such as LTE-A. In 2002, he worked at the ECE Department, IIT, as a Post-Doctoral Fellow. From 1998 to 2000, he was with 3Com Corporation working on voice-band PC modems. In 1997, he joined Lucent Technologies (AT&T), Inc. (currently Alcatel-Lucent) as a consultant Research Engineer to assist in the development of DSP-based ADSL modems. Dr. Tahat is a senior member of IEEE. (e-mail: aat@ieee.org).

Ahmad Jamal received the BS degree in communications engineering from Princess Sumaya University for technology (PSUT) in Amman, Jordan in 2012. He is currently working at Umniah mobile communications copmany in Amman, Jordan as a Network Operator Engineer. He is responsible for command line configuration for Huawei/ZTE 3G & 2G equipment, in addition to analysis and investigation of KPIs. (e-mail: ahmad_89@hotmail.com).

Mohammad Kalbouneh received the BS degree in communications engineering from Princess Sumaya University for technology (PSUT) in Amman, Jordan in 2012. He worked at Umniah mobile communications company in Amman, Jordan as a technical support engineer. (email: mohammad314278@yahoo.com)

13

New Learning Spaces? M-Learning's, in Particular the iPad's Potentials in Education

György Molnár Dr.
Budapest University of Technology and Economics, Budapest, Hungary

Abstract—**The iPad is a very popular touchscreen tablet computer, which was introduced approximately two, two and a half years ago. It runs the MAC operating system making it suitable for different general and special uses. The paper focuses on the fact how the use of iPads can help education, make teaching and learning more efficient. In addition to the technical, infrastructural parameters, which are essential as much as the user's experience and it usability are concerned, the paper provides a few concrete applications as well. Today, the use of iPads is general either in everyday life, or in public education or tertiary education. The author analysis his experiences from the viewpoint of students, but also mentions applications useful for teachers. Students can use iPads with the appropriate applications in any stage of the learning process, e.g. using notes, course materials, schedules and assignments, cooperation, group assignments by collaboration. The paper also describes such solutions or respectively suggestions by means of which the monitoring of shared content and data can be realized by the teacher. A few professional applications (e.g. used by students of medicine, engineering and natural sciences) are also listed.**

Index Terms—**Learning support ICT devices; M-learning; IPad; Iphone; iOS; tablet PC**

I. Introduction

The paper discusses a technological stage of the development of mobile communication having a significant influence on the future of our learning and knowledge, whose focus is a new dimension of informal learning. The new learning environment can be characterized by m-learning, which means a novel impulse in LLL forming a learning potential. iPads – a very rapidly developing family - have been chosen for presentation out of the mobile communication devices usable for supporting learning. The first part of the paper describes the physical, technical parameters of the device, then a few good practices of general and learning-oriented use. In the last part a few Hungarian and international practical examples in the field of concrete application are presented. The use of the iPad is the most widespread among teachers from the Y generation, but is relatively common among those from early X generation and previous generations. The statistical data showing the fast spread of Apple IOS devices as opposed to Microsoft Windows devices prove this fact.

The use of smartphones is now on everyday basis. The devices besides they use as phones function as multifunctional tools which can be helpful in navigation, internet browsing, global positioning, picture taking and video shooting. They also feature music playing and word processing functions. Most of the new generation phones are shipped with touch screens, GPS, support multimedia and operate on 3G or 4G (LTE) networks. The touch screens apply interactive multi-touch technology which has been developed since the 80' in order to display information and provide opportunities for visual communication. Jeff Han, the researcher of New York University was first to develop a functioning multi-touch screen that could be controlled by multiple fingers. The technology is based FTIR (Frustrated Total Internal Reflection) and computer based image processing.

Nowadays the use of iPad is common in everyday life, elementary and high schools, as well as at universities.

II. Body of Paper

A. *M-learning*

The aim of this paper is to present a new, future dimension of informal learning, which will have a huge impact on education and is based on mobile communication technology.

This new learning space can be described as m-learning. It provides a new impulse to lifelong learning. From all devices applicable in education, this chapter will focus on a significant and fast developing segment, the iPads.

M-learning can be defined as access to course content by means of mobile devices and the related learning activity. Or rather the type of learning in the course of which an online CMS, LMS system can be accessed by means of a networked mobile communication device. This sort of use is provided by smart phones, and tablet PCs. Such a smart phone is Samsung GalaxyTab touchscreen device with Android Browser, which makes it possible to download optional web content and provides full scale multimedia. The intelligence of mobile communication devices limits the circle of downloadable content (web 2.0, flash, audio, and video). The smart phones available today use different operation systems, which makes compatibility a bit hard. Such a system is the long known Symbian, the Windows phone, the new Android, Apple IOS, the Bada and Blackberry platforms. The other main challenge is problem free content managements requires web feature optimized for mobile screens (unique CSS, optimized contents). In addition to these points of view, m-learning possibilities are increasingly used; furthermore, a great part of the developments focus on this direction, e.g. the official Moodle mobile client will soon be available to complete the well-known Moodle LMS system. The paper and our research work focuses on such a mobile communication device, the iPad, which is suitable for supporting m-learning [1], [2], [3].

The author primarily reflects his impressions from the students' points of view. The same he alludes to applications available to the teachers. The iPads preinstalled with the proper applications can be used by the students during the period of the learning process. Such are notes, excess to the learning materials, agendas and tasks, collaboration and group tasks which involve teamwork. The author introduces solutions and suggestions which allow the teacher to monitor content and information shared by the students. The author also reveals some applications used by different professions, i.e. medicine, engineering, technology.

B. The iPad as a device

The general characteristics and physical parameters of iPads are the following:

- 9.7" touch screen monitor (1024x768), in case of iPad3 with HD resolution up to 3.1m pixels
- 10-hour battery
- Thinness: 8.8 mm
- Weight: 613 g
- Built-in Wi-Fi, 3G, and ultra-high speed4G LTE
- 16, 32, 64 GB inner memory
- MAC OS X operation system
- Built-in camera, in case of the latest, 3rd generation a 5MP camera
- Phone function, SIM card slot
- Types: iPad 1., 2., and 3.
- Fast booting and standby time
- 1080p HD video recording

The use of iPads has been increasing rapidly as opposed to the use of Microsoft-based systems in everyday, work and learning environments as well. MAC operation systems used by iPads are absolutely different from Microsoft platforms. In spite of their openness the file system of the operation system cannot be accessed to directly and fully, thus the direct basic file operations cannot be realized, which is common in the case of MS platforms. Special applications need to be used, such as iTunes, dropbox and teamviewer. So the MAC based iPad platform is not compatible with the most commonly used Microsoft operation system, which means that iPad applications cannot be used on MS platforms. However, Microsoft applications or respectively their equivalents can be run on iPads. After registration, the users can download iPad applications by means of the App Store using their Apple ID usernames. This way, applications can be upgraded, one account can be used for upgrading five iPads. This is similar to that of the android smartphones, in the case of which a Google ID is needed to be able to use the Market. Each iOS operating system is connected to an Apple ID, which can be used for the iTunes Store, iCloud and the App Store [4].

In the case more than one device is connected to the same Apple ID, then the formerly purchased applications can be download to other iOS devices free of charge. Furthermore, they can be set to automatically download the upgraded applications to each device. In addition to the Apple Store a huge collection of music, films, TV programs, books, podcasts and university courses are available for the iOS devices. The other huge repository is the Apple ID iCloud, which provides a storage capacity of

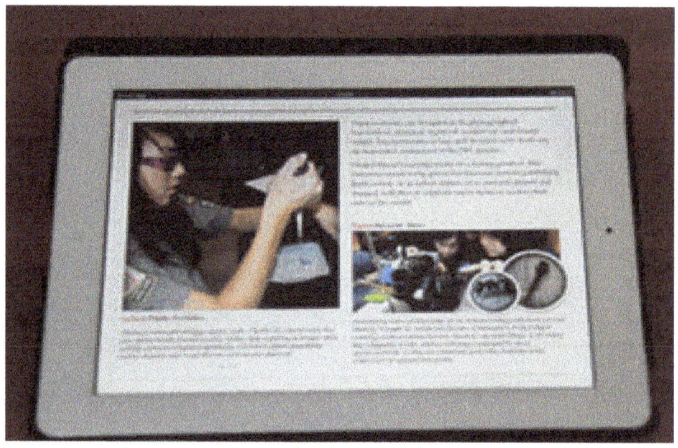

Figure 1. IPad 2, Source: Own figure

five GB for each user, which can also be used for security saving. This process can be automatically algorithmized, so if the given device is connected to a WiFi network, a security copy of its operation system is saved by the iCloud. Consequently, in case of any problem the security copy can be used. These devices can be synchronized with any PC by means of the iTunes application, which is suitable for storing downloaded programs, media content and security savings. E.g. the iPhoto, a part of the operation system, can easily be used on an iPad. When connected to Microsoft Windows, the system recognizes the iPad as a driver and the automatic play discussion panel pops up, by choosing "Importing videos and pictures" photos and videos can be imported by means of the Windows Photo Gallery application. iTunes synchronization can be carried out by means of an USB cable, or Wi-Fi, if the PC and the iOS (iPad) are connected to the same Wi-Fi network. The iPad applications of international importance are the following:

- the iPad as a source of information in addition to its GPS and navigation functions can be used by employees of United and Continental Airline
- the knowledge base and the connected 3D programs of anatomy can be used by employees of outpatient departments, hospitals, certain diagnostic measurements are also supported
- by employee of labour market enterprises

C. Using IPADs in teaching and learning

The adaptation of iPads in education is still in its infancy except for a few developed countries. Their application in education can be realized by means of two methods:

- iPads are distributed to the students before classes, prepared for performing a given task, then they are collected when the class is over. This way the iPad devices can be fully controlled in the school learning environment. The Apple provides a solution for such application: the so called Bretford PowerSync Case (See Figure 2) which can store, upload and synchronize 30 iPads at the same time. This mechanism should be used when pupils/students require more teacher attention. In such a case the same programs, applications run on the iPad.
- The iPads are handed out to the pupils/students for individual use, either at class or after class. In this case each pupil/student has an iPad and an Apple ID of his/her own, and use them as a personal device.

This approach requires a bit more school organization to synchronize the necessary programs and applications. This method can be useful as the pupils/students can learn the use of the device more quickly. This approach should be used in the case of students over 18.

In the following practical educational applications of iPads will be presented and categorized by task type. Electronic textbooks, auxiliary materials, classroom note taking, content and activity management, group work, pair work, online tasks and independent learning are to be used in classroom environments. The kno, inkling and dropbox programs support electronic course materials, while the elements, notes plus and iannotate pdf are used for note taking. The iStudiez pro and iTeacher Book program support activities, iWork and Quick office Pro HD are used for preparing documents, the SketchBook Pro, iMovie and Garageband applications are used for media management. In addition to independent learning, pair work is also supported by means of Skype for iPad, GoToMeeting, iThoughts HD or Share Board. The last can realize working on several iPads at the same time to be shown by means of a Wi-Fi device. In addition of general application, job specific programs can also be downloaded such as software supporting studies in Anatomy, Maths, Chemistry, Mechanics, and Electrical Engineering. The use of blackboard is on no secondary importance either. Figure 3 shows iPad learning management programs. And not neglected the possibility of using the blackboard program either (See figure 4) [5].

D. iPad-related surveys, researches

The statistical data concerning the use of iPads are primarily to be found in professional works and sources. A good example is the statistical data of US Abilene Christian University showing our laptop and iPad uses and frequencies, such as:

- 50% of the lecturers use the iPad every lesson, 84% of them regularly
- The efficiency rate of knowledge transfer has increased to 79% by means of the iPad
- A much more frequent use of the course material was characteristic when these devices were used [6].

In conclusion, it can be stated that the use of the iPad means a bigger experience for the lecturers of the institution, both from the viewpoint of lecturers and students. The devices are said to have been received positively by the students. This urged teachers to recognize the fact that this technology can be used in an efficient way at class. Other statistical data show that educational institutions all over the world are interested in the integration of mobile devices into the teaching-learning process.

In Hungary, a methodological experiment of novel approach was introduced in the academic year of 2011/12 by Károly Eszterházy College and the Teacher Training Primary School of the College. The device in the focus of the research is the iPad. The pupils and teachers of class 8c started the academic year in an electronic learning environment involving iPad table devices and a touch-screen interactive whiteboard.

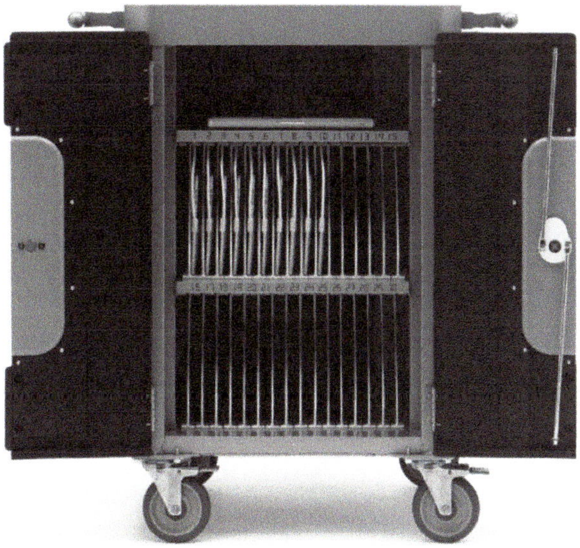

Figure 2. Bretford PowerSync for iPads, Source: Apple iOS Learning Solutions. 2011 cited 2011; Available from: http://images.apple.com/education/docs/

Figure 3. : iPad learning management programs, Source: Own figure

Figure 4. Blackboard application on IPAD, Source: Own figure

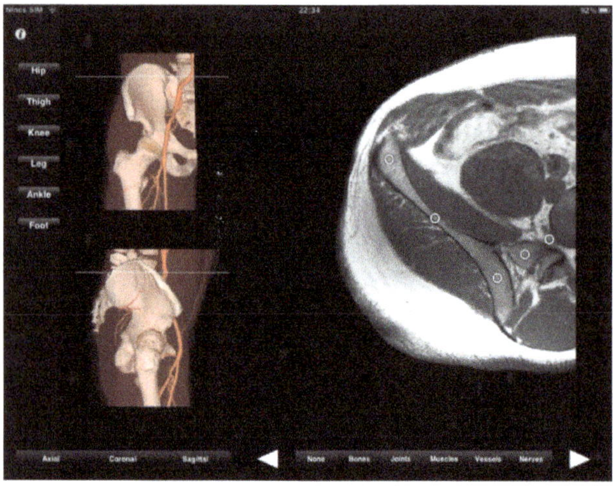

Figure 5. medicine application on IPAD, Source: Own figure

Figure 6. E-learning course on IPAD, Source: Own figure

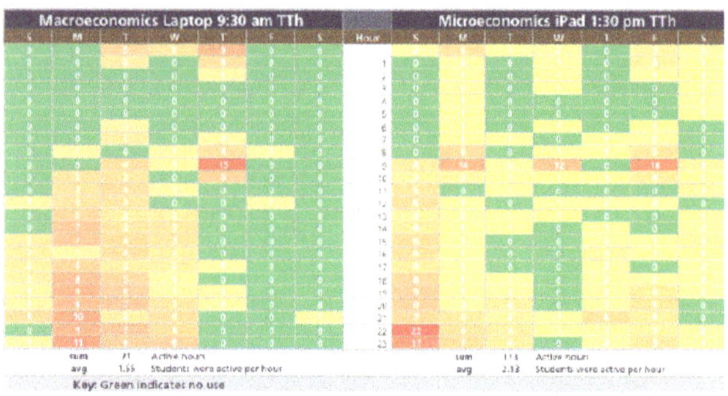

Figure 7. IPAD application in tertiary education, Source: Dr. Ian Sheppard, http://www.acu.edu/technology/mobilelearning/documents/research/ipad-or-ifad.pdf

The course materials – Hungarian Literature, History, Physics, Biology, Maths, Informatics, Geography and English Language - to be used by the table devices were provided by the Mozaik Publisher.

The process of knowledge transfer and acquisition is enriched by new interactive applications in addition to the course materials and the collections of activities in this up-to-date learning environment.

In the 2nd stage of research each student of class 8. D and the members of faculty started the semester in a digital learning environment equipped with an iPad2 tablet and an interactive board The experiment aims to

Figure 8. IPAD application in Primary School, Source: http://www.ipad.ektf.hu/?q=gallery

take advantage of iPad2 devices, which main element is the iBooks Author curriculum editor application.

Starting from this term the methodical emphasis is set on new, different bases. The main concept differs from the previous IT experiments, that not only a device is given to the staff, but a complete edition of textbooks in iBook form is provided both to students and teachers. The experiment is unique both in aspects of revolutionized curriculum and common access to the system.

But there is more to it than the curriculum conversion and digitalization. Taking advantage of iBook Author developer program, the traditional textbooks can be presented with interactive tasks, animations, and self-monitoring interactive tests.

Some experienced teachers took part as mentors in the process and word as a catalysts they editorial work which was done by multiple developer teams, including graphic designers, and IT professionals.

The primal goal of experiment was to study the opportunities provided by the new learning environment. The observation and analysis helped mark the adequate and up to date learning trend.

III. Conclusion

Analyses carried out and experiences acquired so far suggest that classroom-based teaching and learning supported by iPads is more useful and efficient than otherwise. The iPad is an extremely innovative device when its proper services and applications are utilized. Naturally, when such a device is introduced in a school, appropriate technological background, staff and course materials need to be provided. The most recent iPad-based developments have just been released.

They are available here: http://events.apple.com.edgesuite.net/1201oihbafvpihboijhpihbasdouhbasv/event/index.html [7]. Such a new development is the iBooks application, which automatically upgrades the content of the downloaded i-books in the e-library when the authors have changed the content. This program is only available on iPads and i-MACs.

A similar new application is the iBooks author only developed for MACs. Finally, iTunes U application is one of the latest programs being suitable for designing complete courses featuring audio, video, books, and other content by means of e-learning course design.

Recently the latest, 3rd generation iPad has been released. The new model offers better user experience, high resolution, 3.1MP screen, 5MP iSight optics and ultra high speed 4G LTE wireless technology.

All in all, the iPad tablet PCs are perfect for m-Learning learning environments, which cannot only be used by the Y and Z generation learners, but by lecturers, educationalists as well for supporting learning in addition to maintaining motivation.

The iPad inspires creativity and various functions of mobile learning while becoming the students' favorite learning tool. Thanks to the pre-installed applications as AppStore and iTunes U it supports collaborative work on Web 2.0 platform.

It also allows the students to be involved in sharing and editing the content and information. All this is possible regardless the time and space thanks to the iBook textbooks.

The latest generation of iPads opens a new horizon of learning both in formal and informal dimension, and build up a powerful learning potential.

REFERENCES

[1] Albert-László Barabási, Behálózva – a hálózatok új tudománya (Networked – the new science of networks) (Magyar Könyvklub, 2003., ISBN 963547895X)

[2] György Molnár, The requirements and development areas of the ICT aided learning environment, In: András Benedek (ed.): Digital pedagogy – Typotext Budapest 2008., pp. 225-255

[3] Piet Kommers, Piet A.M, ICT as explicit factor in the evolution of life-long learning. International journal of continuing engineering education and life-long learning, 20(1/2010), pp. 127-144

[4] György Molnár, Flashes or steady light? Or the potentials of developing networked learning, In: Miguel Baptista Nunes, Maggie McPherson (ed.): Proceedings of the IADIS International Conference e Learning, IADIS international conference E-learning 2011, Volume II. Rome, Italy, July 20-23, 2011, ISBN: 978-972-8939-38-0, pp. 405-408

[5] György Molnár - András Benedek, The empirical analysis of a web 2.0-based learning platform, In: Constantin Paleologu, Constandinos Mavromoustakis, Marius Minea (ed.): ICCGI 2011, The Sixth International Multi-Conference on Computing in the Global Information Technology, Luxembourg, June 19-24, 2011., ISBN: 978-1-61208-008-6, pp. 56-62

[6] György Molnár - András Benedek and János Cz. Horváth, Jump over the shadow? From the traditional education to the non-typical one or the experiences of an electronic learning environment, In: Miguel Baptista Nunes, Maggie McPherson (ed.): Proceedings of the IADIS International Conference e Learning, IADIS international conference E-learning 2010, Freiburg, Germany, ISBN: 978-972-8939-17-5, pp. 243-247

[7] http://events.apple.com.edgesuite.net/1201oihbafvpihboijhpihbasdouhbasv/event/index.html, last visit 14.05.2012

[8] http://www.ipad.ektf.hu/?q=gallery, last visit 14.08.2012

[9] http://www.acu.edu/technology/mobilelearning/documents/research/ipad-or-ifad.pdf, last visit 14.08.2012

AUTHOR

György Molnár Dr. is with Department of Technical Education, Budapest University of Technology and Economics, Budapest, Hungary

14

The Implications of Mobile Notifications for User Experience of a Social Network Service

Sami Vihavainen[1], Kaisa Väänänen-Vainio-Mattila[2]
[1] Aalto University, Espoo, Finland
[2] Tampere University of Technology, Tampere, Finland

***Abstract*—Smartphones enable an always-on connection to Social Network Services (SNS). A typical way of interacting with SNSs is to access them when the user has a suitable situation to check the status of her social networks or to write an update. One way to enhance the usage of SNSs is to have the service automatically push notifications about events to the smartphone user interface. However, there is no research on how users experience such SNS notifications. We present an explorative field study with 11 participants to assess how users experience mobile notifications compared to reading SNS content manually, initiated by the user. The participants first used Facebook for a month without notifications and then for a month with an application called Socially that sends frequent notifications about Facebook events to the user's smartphone desktop. The participants who kept the notification feature on reported increased reading of Facebook. However, after a while, many were unwilling to receive the notifications, mainly because of lack of control. We report the qualitative findings on user experience, which reveal for example, that the use of mobile notifications decreases interest in Facebook. Notifications limit user control, and using Facebook feels more meaningful when accessed manually. Implications for design are discussed.**

***Index Terms*—Mobile notification, Social Network Service (SNS), Field study, User experience.**

I. Introduction

Social Network Services (SNSs) are increasingly a part of people's everyday lives. They are used to maintain contact with acquaintances, both close ones and more distant ones [1]. Nowadays, SNSs are often used with smartphones. There are mobile SNS applications, for example for Facebook and Twitter, which enable people to follow their friends' doings and socialize with them while on the move.

Features of automation have been developed for SNSs, meaning that some tasks that were previously performed by the users are allocated to the SNS. For example, location sensors of a smartphone perform automatic "check-ins" on Foursquare [2]. This kind of automation reduces the need for users' "manual" or user-initiated interaction with the service while supporting access to a broader set of information.

Recently, automation has been offered also in the form of mobile notifications. Notification systems have been defined as "interfaces specifically designed to support user access to additional digital information from sources secondary to current activities" [3]. Mobile notifications inform the user about happenings on an SNS by pushing information on the smartphone desktop. However, earlier research has shown that from the user's perspective, notifications are a double-edged sword. Notifications are valuable to the user in conveying important or relevant information, but they also come at the cost of interrupting the user [4].

How users experience notifications for SNSs has not been studied. To fill this gap, we studied the user experience (UX) implications for mobile notifications on SNSs in the field. By user experience, we mean users' subjective experiences, which arise from user-system interaction in the context of use. User experience covers both pragmatic (practical, goal-oriented) and hedonic (emotional, non-instrumental) aspects of system use [5]. A field study with end users using a system or a service in real contexts of use is a strong method to gain understanding of the details of user experience [6].

In studying mobile notifications, we focus on how service automatically follows happenings on the user's SNS, analyzes the acquired data, decides what information to deliver, and delivers that information by pushing the information to the user's phone as a notification. Thus, the system works as an assistant to follow activities of other people, which is a principal motivation for using services like Facebook [7]. We consider this to be an important area of research because notifications, if properly implemented, can enhance a user's social interaction by assisting the user in being better aware of the happenings in her social circle. However, it is important to the user experience that the notifications' interruptions are not perceived to be irritating.

To study the user experience of mobile notifications, our main research question was as follows:

How do mobile notifications affect the user experience of social network services?

We conducted an explorative field study with end users using Facebook with an existing, relatively widely-used mobile application called Socially [8]. Socially pushes News Feed updates from users' Facebook contacts to users' mobile phone desktop based on predefined time intervals; 30 minutes was the default setting. It also gives a light sound and vibrates the phone each time a new notification pops up. With Socially, the user is also able to manually read the Facebook News Feed by opening the News Feed screen and looking through the updates. Our approach was to study a group of current Facebook users, introduce them to Socially, and gather and analyze data on how they use Socially in connection with Facebook.

Our aim was to explore the user experience of mobile notifications and provide insight to designers to consider when designing SNS functionality in mobile contexts. From a broader viewpoint, we explored the potential of automatic features in the SNS context, i.e. how tasks can be transferred from a human to a machine in mobile social applications.

In the remainder of this paper, we will first present related research followed by the research methods. We will then present the study process and the results. After discussing the design implications for mobile notifications and automation for social media, we will provide conclusions the contributions of the study.

II. Related Research

In this section we examine research related to notifications in Human-Computer Interaction (HCI). We also briefly present the concepts of automation and perceived control and how these concepts relate to notifications.

Previous research has found that if properly implemented, notifications can support people's awareness of others' social states, actions, and activities [9]. However, notifications generate an interruption, and it is not self evident that they are perceived positively by users. The benefits of interruptions must outweigh the detriments [3]. McCrickard and Chewar [3] argued that users' possible dissatisfaction with notifications is due to designers' failure to estimate user task prioritization. This results in notifications presented at inappropriate times and in an unsuitable presentation style. Earlier research has also found that intensity, which can be "scaled from not notifying at all to trying explicitly to grab the entire user attention" (p. 3), is affected by the presentation modality and the amount of information presented to the user [4].

Notifications can be used also in mobile devices such as smartphones. However, the use contexts of mobile devices are diverse, and the user is normally performing some other primary task. Therefore, the notifications have to compete with the user's environment to get the user's attention [10]. User attention is limited; it has even been proposed that in today's information society, attention is the scarcest resource [11]. Because a notification generates an interruption, a notification appearing at the wrong time and too intensely generates annoyance. It is unwise to get the user's attention at all costs. Several approaches to notifications, such as information filtering, choosing modality, and attentive UI, have been proposed to get a user's attention in a satisfactory way [10]. The mobile context also sets requirements for notifications because the interaction events with mobile phones are often short and fragmented [12], such as checking Facebook happenings while on a bus [13]. In addition users accept notifications more quickly if they are timed to appear at the end of mobile interaction episodes rather than at random times [14].

Recent research has also examined easing the user's burden to keep up with SNS events by aggregating events from several SNSs. For example, Cui et al. [15] developed and studied a LinkedUI application that enables the user to follow several SNSs in one consistent user interface. They found that the user preferred the aggregation approach to accessing SNSs through a traditional mobile web browser. LinkedUI decreased difficulties such as in switching windows and delays loading web pages. Overall, in the mobile and web environment, there are several ways to inform the user about changes taking place on SNSs or other web services. These change indicators can reduce the user's burden in looking for new content [16].

Automation means that some aspects of human activity are transferred to the system. This also means that control is partially transferred from human to machine. Different levels of automation can be built into HCI [17], from full machine control to full user control. Notifications implement a form of automation, which means that the machine (in this case the SNS and/or the mobile device) takes control of when the user should start interacting with the SNS.

Psychology research [18] has investigated the notion of control. Control refers to "the extent to which an agent can intentionally produce desired outcomes and prevent undesired ones. When individuals believe they can do this, they are said to have personal control, perceived control, or a sense of control" (p. 554). In information systems, perceived control has been found to affect users' motivation to use the system. Novak et al. [19] found control to be a major determinant of the flow experience with online environments, which again affects the depth of interaction with the service. They defined control as the "user's perception of her ability to successfully navigate through the Web environment and her perception of how the Web responds to her inputs" (p. 27). On the other hand, lack of perceived control may cause desperation and hopelessness [20]. In the HCI context, it is important to avoid negative emotions, and that user has the impression that she is in control of situations [21]. In addition, people feel happier about the outcomes that they have accomplished themselves compared to similar outcomes that have been accomplished by someone else or by chance [20].

Earlier HCI research has looked into notifications and automation usage in domains other than SNSs. In psychology, perceived control has been explored as a concept that may improve the user's experience with the system. In our study, we investigated how mobile notifications, as a form of automation, affect the user experience, including the user's sense of control.

III. System Description

In our study we used Facebook and Socially to study the user experience of mobile notifications. With Socially, the user is able to read Facebook (and Twitter, Linked In, and Foursquare) by starting Socially and opening Facebook News Feed manually (Figure 1) or by reading notifications that the system pushes from Facebook (Figure 2). The user is also able to write status updates. Socially has other features that were not the focus of this study, including notifications during incoming calls, caller location, and synchronizing Facebook profile photos and birthdays with the phone. In our study, we used version 2.10. Based on the Facebook site, there were about 50,000 active users (27.10.2011), and the application had a 3.9/5.0-star average rating based on 173 reviews on Facebook.

Socially scans the user's social network services (e.g., Facebook or Twitter) and pushes new update notifications as a pop-up to the phone's desktop (Figure 2).

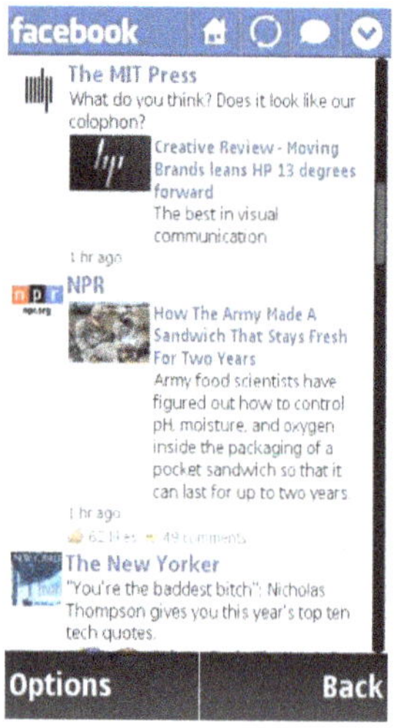

Figure 1. Socially's Facebook News Feed reading view.

Figure 2. Socially pushes recent updates to a phone's desktop at predefined time intervals.

When the new notification pop-up appears on the desktop, the phone gives a short audio alarm and the phone vibrates. The audio alarm is about 0.5 second (knock-knock-knock). Volume follows the phone's general ringing volume settings. Vibration cannot be turned off. It is possible to set the updates to come in time intervals of every 30 minutes, one hour, four hours, once a day, or never. The default setting has the phone push Facebook updates to the phone's desktop every 30 minutes. The updates are the most recent updates that would appear on the user's Facebook News Feed. When the notification appears, the user can scroll through the updates one by one, using the phone's arrow keys or push Read or Exit. The maximum number of new updates to scroll through is 15 per notification. Pushing Read opens a view where the user can read and scroll through the entire News Feed, as shown in Figure 1. Pushing Exit closes the notification. It is also possible to shut the feature off either by turning off Desktop alerts or setting the update time interval to Never.

Pushing the social notifications to the phone's desktop makes the phone work for the user by keeping an eye on what is happening on the user's SNS. This potentially enhances user's awareness of her social surroundings. She does not have to manually check the Facebook News Feed; it is delivered to her phone's desktop.

IV. Field Study

We recruited 11 participants for this qualitative study from the greater Helsinki area in Finland. We conducted a study in the field where we first studied participants' Facebook use strategies and then the use of mobile notifications through Socially in concert with Facebook.

A. Users and Devices

The users were recruited from the SizzleLab [22] living lab environment. There were nine males and two females. All were young adults from 19 to 27 years of age (mean age = 23.3). All had used Facebook in their everyday lives before the study. Of the total, 10 had Nokia N97 phones, which they had used as their primary phones for at least six months before the study. One participant had an HTC smartphone with an Android platform. All the participants also had unlimited data plans. The phones and data plans were provided to the users for free.

B. Data Collection

1) Field Study

Use of an existing application was studied in the everyday life context to better reveal the social and technological issues that unfold only in long-term real life use and might not come up in laboratory and prototype conditions. The field study had two phases. In the first phase (Phase A), people's Facebook use activities and strategies were studied. This phase lasted from 26 to 41 days (an average of 31 days per participant), depending on the users' personal availability for the study. In the second phase (Phase B), we intervened and introduced the users to Socially. We asked the users to use Socially with Facebook. We instructed them to use the application as they wanted but encouraged them to at least try it. This phase lasted from 31 to 46 days, depending on the participant's availability (an average of 41 days per participant).

A pseudo-Facebook profile was established for the study, and all the participants added the pseudo profile as their Facebook contact for the period of the study. This was done to log how often participants updated their Facebook profiles during Phase A and Phase B. Participant privacy was taken seriously, and access to the pseudo profile was granted only to researchers who were part of the project group. All the data related to participants was anonymized for publication.

The participants installed Socially in their phones themselves. Because we wanted them to use the application as realistically as possible, we did not want to force them to use any of its features. However, all participants were introduced to all the features. They were asked to send us an e-mail or SMS check when they had successfully installed the application, and we told them to ask for help if they had any problems. We provided technical support if needed. We did not reveal to them that the focus of the study was notifications. As a reward, each participant was given a 50€ fee at the end of the study.

2) Open-ended Questionnaires

During the field study, two open-ended questionnaires were conducted. The first took place in the middle of Phase A and concentrated on the participants' Facebook use habits and experiences. The second was conducted at the beginning of Phase B and concentrated on first reactions and experiences with Socially and if the participants had had any problems with it. Both questionnaires included real-life examples. At this point, we kept the questions quite general so that our research focus would not be revealed and so that the questions would not direct participants' use of Socially.

3) Final Interviews and Questionnaires

At the end of the study, we interviewed the participants. Six participants were individually interviewed face-to-face. The five remaining participants were not available for interview. However, they were sent open-ended questionnaires, which were formulated based on the already conducted individual interviews. Interviews were semi-structured and consisted of reactions and use of the automated features of Socially. The interview protocol and the questionnaires covered the following areas:

1. Participant's background (e.g., age, technology know-how)
2. General use of the phone (e.g., whether it worked properly)
3. General use of Socially (e.g., perceived usefulness)
4. Experiences with Socially's notifications
5. Experiences reading Facebook News Feed manually
6. Whether they planned to use Socially after the trial and why

C. Data Analysis

Qualitative data from the open-ended questionnaires and face-to-face interviews was analyzed based on grounded theory as defined by Straus and Corbin [23]. This means that the theories or result themes are found bottom-up from the data, and no pre-study hypotheses are formed. Furthermore, qualitative data was analyzed using content analysis [24]. First, the key points related to data were identified. Then the key points were categorized into larger themes. Research questions were used to guide the analysis. Quantitative data on Facebook updates during Phases A and B was analyzed using a paired sample t-test.

V. Results

In this section, we first describe users' interaction with Facebook during Phases A and B (Section A) to provide background on the users' Facebook habits and motivations (Phase A) and general use of the notification feature (Phase B). Section B summarizes the user experiences with Socially. We then present the main result themes that arose from the study data in sections D–G.

A. Facebook Interaction in Study Phases A and B

1) Phase A—Reading News Feed and Hanging Out

For most of the participants, reading was the main activity on Facebook; motivations varied from a need to keep up with happenings to a need to relax.

Many claimed that they use Facebook mainly to follow what their friends are doing or for private chatting and mailing, but they are passive in writing about their doings and whereabouts in the public status box. According to our questionnaire, on average, they read Facebook about 5.1 times per day. However, some were clearly more active than others; the standard deviation was 4.6. As everyone reported that they read Facebook every day, the users were active readers of Facebook.

Regarding Facebook contribution, we counted the times they had written a status update or sent a link (e.g., YouTube video or a web site) to their profile during Phase A. The total number of updates was 86. On average, the participants sent 0.26 updates per day; the standard deviation was 0.38. One of the users (P8) was a very active contributor, sending 1.2 updates per day on average, whereas five participants sent no updates.

Using Facebook seemed more serious to some than to others. For example, one participant described how he needs Facebook to be up to date on his friends' doings and new events.

"[Without Facebook] I would be so clueless about everything what happens." – P2

But many described their use as "hanging out". It was popular among the participants to read Facebook when there was nothing else to do. It was more about relaxing and passing the time than serious objectives. Using Facebook while on the move was popular. Half of the respondents reported that they have a habit to read Facebook while on a bus or a train. Nine participants had used Facebook with their phones before the trial. The majority reported reading as the main activity with the phone. The participants' basic Facebook habits were in line with earlier studies of SNS uses [7] and interaction moments with the service [13].

2) Phase B—Differences in the Use of Notifications

Overall, participants differed on whether they used the notification feature offered by Socially. Three participants (P1, P2, P6) kept the feature on for the entire use period (about 4 weeks) while six (P3, P4, P5, P8, P9, P10) kept it on from one to seven days; two (P7, P11) did not use it at all. Next, we will discuss these differences in the use of notifications. We will also elaborate on the findings by comparing them to earlier research. We will refer to the above-mentioned groups as High-users, Low-users, and Non-users. P1, P2, and P6 belong to the High-users; P3, P4, P5, P8, P9, and P10 belong to the Low-users; and P7 and P11 form the Non-users group. Non-users had problems understanding the logic of how the notifications feature works. P7 accidentally deactivated the notifications after installation. She deactivated the "Desktop Alerts" feature without realizing that it also prevented notifications from showing up. P11 had the feature on but he did not keep Socially running in the background. He did not realize he needed to keep the application running to be able to use automated features.

B. The General Effect of and User Experience with Socially

1) The Main Effect of Socially: Reading Increased But Writing Did Not

Six participants (P2, P3, P4, P5, P6, and P10) reported that Socially increased their Facebook reading but did not have a clear effect on how they updated their status lines. Four (P1, P7, P8, and P11) reported that Socially did not have a significant effect on how or how much they used Facebook; one (P9) reported it actually reduced her Facebook activity while the notifications were on.

The most important reasons participants reported for increasing reading activity was that it was quicker to manually read the Facebook News Feed with Socially than how they had previous read the Feed; and because of the notifications (when turned on), they were persuaded to read the updates more often. In total, 160 updates were sent during Phase B, and only 7 were written using Socially. On average, participants sent 0.35 updates per day, with a variance of 0.50. On average, there were more updates sent per day in phase A (0.26), but the difference was not significant. Based on paired samples t-test, the difference between Phases A and B was not significant; the p-value was 0.078.

2) User Preferences and Experiences with Socially

Most of the participants reported that they liked Socially. Six participants (P2, P5, P6, P7, P10, and P11) planned to continue using Socially after the trial period. Three (P4, P8, and P9) were a bit uncertain, and two (P1 and P3) would not use it after the trial. The main reasons for continuing the use were notifications and ease of use. For three participants, notifications were one of the main reasons they wanted to continue using Socially. However, one (P11) did not understand that Socially had to be kept in the background. After he realized that, he wanted to continue using the system so that he could receive notifications. For three, the main reason was that Socially was easier and faster to use than other ways (web browser or Nokia's Facebook application). With Socially, it was convenient to open the application manually and pull the Facebook updates to the phone.

Most found that from a pragmatic perspective, Socially worked well. It was fast and easy to use. However, from a hedonic perspective, there was more variation among the participants. Many found that notifications were not pleasant and did not suit their Facebook use habits whereas some liked felt they were a bonus.

C. The Main Themes Related to User Experience of Mobile Notifications

During the analysis phase, four main themes related to user interaction with mobile notifications came up: Users had varying control strategies over notification interruptions (1); ease of skipping insignificant messages favored manual reading (2); notifications decreased interest in reading Facebook (3); and users' goals affected the perceived usefulness of notifications (4).

1) Users Have Varying Control Strategies over Interruptions

The effect of practicing control over notification frequency and intensity came up when comparing the High-users and the Low-users. Two from the High-users had made the update interval less frequent than the default setting of "every 30 minutes" (to "every four hours"). For them, notifications were actually one of the main reasons they wanted to continue using Socially after the study. One was bothered by the intensity (alarm) of the notifications. He wanted to keep the notification feature on, but because he could not turn off the alarm, he chose to lower the sound level of his phone. The third user who kept the notification feature on used the 30-minute default interval for the entire period. However, he kept his phone silent all the time. He said that he always keeps his phone on silent because the phone's alarms generally disturb him. For him, the notifications were not as intense as for others. But he did keep the phone's vibration feature on to be aware of incoming messages and phone calls. Of the six Low-users, none had practiced any control over the mechanisms of the notifications. All of them used the 30-minute default until they turned the feature off. Therefore, although the system had a mechanism for the user to adjust the frequency of notifications and to control interruptions based on weighting the value of interruption against the costs, they did not use this mechanism. They also did not control the intensity of notifications by adjusting the phone's sound level like two of the High-users did. The Low-users thus perceived the notifications as more intrusive than the High-users.

In addition to controlling the settings of the notifications, varying levels of behavior control practiced by the participants were reported. Two of the High-users said they read the updates practically every time they showed up unless they were in a hurry. But even they did not actually read the new status lines every time even though they might have checked the pop-up.

"But it's a different thing what I will actually do after I have opened the key lock. Will I push Read or Exit? It depends a lot on what else am I doing at the time." – P2

They managed to control the notification situations with their behavior: They simply did not pay much attention to notifications when they did not want to read them. Even unimportant notifications did not cause interruptions so severe that it would have irritated them as they were in the middle of their primary tasks.

Some of the Low-users related how they at first thought that the feature was pretty cool, but that after a short while they realized they would not need or want notifications about new Facebook updates but wanted to control their Facebook reading manually.

"I tried the feature for about three days but then it started to get on my nerves, when it beeped every day. I thought this is not for me because I like to go to Facebook often and watch the new updates by myself anyway." – P5

"Those kinds of things [notification features] are really neat. I was really excited at first, but then again in the long run it was not that good anyway." – P10

This brings up how the user experience can change over time. After the notifications were first considered useful, withdrawal occurred. Unlike the High-users, most of the Low-users did not report practicing much behavior control. Of the Low-users, five reported that they checked the notifications very often before inactivating them. Only one never really checked the updates but just ignored the update alerts. Low-users said that the alarm sounds Socially gave when sending notifications to the phone's desktop were disturbing. The alarm was too persuasive and one participant mentioned how she reacted physically to the alarms.

"I reacted perhaps quite physically to it. Like, hey! Now it alarms, I have to check it out." – P5

To ensure that the push alerts do not bother the user, Socially automatically stops showing updates at night. This window is from 9PM to 8AM. Although this was a nice feature, it was not enough because it did not adapt to the user's real life well enough. In the following case, the user's strategy to control notifications was eventually to shut off the notification feature.

"I really do not want to wake up during the weekend mornings because some half-acquainted person has added a YouTube-video to her Facebook wall. – P5

Earlier research [4] has shown that the intensity of notifications is influenced by the modality in which the notifications are presented, including whether they are silent or not. Based on the results, frequency seems to be a defining factor of the perceived intrusiveness of the notification system. The control possibilities for the user to adjust the intensity of notifications should not be overlooked or made too difficult. Users do not necessarily explore the control possibilities spontaneously but may turn the system completely off if controlling the notification is not easy enough.

2) Ease of Skipping Insignificant Messages Favors Manual Reading

Many participants reported that they wanted to control their Facebook reading manually by scrolling through the News Feed. This was because they felt that it gave them better control over what to focus on. There were two main reasons notifications were not perceived as a valuable way to follow Facebook. The social importance of one's Facebook contacts varied widely. However, notifications treated all the updates as equally significant. For example, one user had about 1000 Facebook friends and was not interested in getting all of those friends' status updates to his phone; for instance, he only gets one-to-one text messages that are directed only to him. Some participants had blocked some of their Facebook friends' status updates because they perceived that many were useless. The fact that this could not be done with Socially was disturbing. Several participants stated that the greater the user control in defining the filtering principles of Facebook status lines, the more they would like the notification feature. Second, mobile notifications emphasized individual messages whereas the News Feed shows several updates at the same time. With the entire News Feed on display, it is easier to control the unwanted, boring, and useless status updates by just skipping updates. It was considered important to control pre-filtering of updates and filtering during the reading process. Participants thought that there was a lot of unnecessary "noise" on Facebook that was not considered meaningful. Many were irritated that the notifications did not provide a meaningful extent to which user was able to intentionally produce desired outcomes and prevent undesired ones.

Showing one update at a time makes the first update especially important. Reading the first update can be a crucial moment of interaction when the user makes the decision to continue reading Facebook. One of the participants mentioned that when he gets notifications, he checks the first status update and based on that decides whether to explore the updates further or stop the prevailing interaction event.

"If that first one [status update] is something that does not interest me at all, I might push Exit and check the others sometime later just by doing it manually." – P2

Another user had actually never read other than the first one.

"I never continued reading the updates after the first one, because already the first one seemed useless." – P5

We consider this to be a valuable insight for designers. If the first shown status update were the most commented on or most liked, it might trigger a deeper interaction event and persuade the user to continue reading Facebook.

3) Notifications Decrease Interest in Facebook

Losing control of what to read and when decreased user interest in Facebook. For some users, notifications made the content of Facebook updates feel less meaningful. One of the Low-users noted that during the first days of use, she checked the updates every time. However, when reading her friends' status updates from the phone, she realized how irrelevant they were to her current context. After that "enlightenment", she said she used Facebook less than she would have normally because she learned to ignore the notification alarms but did not read Facebook manually from phone or from a desktop computer. She thought that there was nothing interesting on Facebook to read, but when she turned the feature off (after keeping it on for a week), she started to read Facebook from her desktop computer again.

"During that one week I opened Facebook from my computer only like two times. That was because all the updates came to the phone. But somehow I then noticed that if these all are this kind of crap, why do I bother reading them at all. However, after I stopped the push feature I started to use Facebook more often with a computer." – P9

For her, the push alarms were eventually annoying and made her uninterested in Facebook.

"It is weird, that suddenly it is really interesting if you do it by yourself, but not if someone pushes it to you." – P9

Another participant described similar feelings when reading Facebook manually and getting notifications.

For me it feels somehow more meaningful to check the updates manually, than if I am just being told [through notifications]. – P1

Psychology research has produced results similar to these findings. As previously mentioned, people feel happier about outcomes that they have accomplished themselves compared to similar (equally pleasant) outcomes that have been accomplished by someone else or by chance [20]. This might partly explain why reading the News Feed manually was somehow more meaningful than getting the same content delivered through notifications. During the manual reading process, the user accomplishes the interaction event whereas through notifications, something else is the primary agent accomplishing the interaction event.

4) User Goals Affect Perceived Usefulness of Notifications

Participants' Facebook use motivations and the goals they try to achieve by using Facebook affected how useful the notification feature was perceived. The High-users who perceived that the notifications enhanced the user experience of the system were keen on being aware of what their Facebook friends were up to. Two of the High-users discussed how Facebook was all about knowing what friends are doing and what kind of social events have occurred, are occurring, or will be occurring. One was actually surprised when we asked if the notifications had bothered him. The other felt that the notifications were one of the main reasons he will continue using Socially after the trial. The third User (who kept his phone silent) explained that for certain intervals, notifications are a nice bonus. Overall, these users perceived that notifications helped them be more aware of the happenings in their social network. The value was greater than the detriment of notification interrupting the current primary task. Using the user experience related notions presented by [5]; for the High-users notifications enhanced the hedonic quality of Socially. This was because they wanted to be aware of others' social lives, and for this notifications were useful. Notifications enhanced the user experience of the service by explicitly supporting users' hedonic goal of being related to others. These users were more focused on following others than the participants who did not perceive notifications as useful and who used Facebook to pass the time and relax.

Four of the Low-users found that they did not want to enhance their Facebook use in general. Ironically, this is exactly what the notifications try to do. Although all the Low-users used Facebook quite actively in their everyday lives (as discussed in 4.3), it was not a high priority and was used to pass the time or relax. This was a significant reason why many of the Low-users did not like notifications. Therefore, the implementation of notifications was not in line with the users' hedonic goals of passing the time and relaxing. The Low-users perceived that their control of when to read Facebook was too limited and did not feel that automation supported their use goals. One of the Low-users thought that notifications take too much attention from more important tasks like studying.

"If one should be studying, and Facebook is such a temptation anyway and easy to get stuck with, these [notifications] just increase that danger." – P10

VI. Discussion And Design Implications

It can be tempting for designers to add automated features to social media applications. Potentially, automation can increase the amount of social interaction and enhance the user experience. However, the findings of this study show that automation in the social media context is not all about efficiency and productivity. Concerning notifications, the user has to feel that notifications support motivations and goals in using the service and that the benefits overweight the intrusiveness of the generated interruptions. It is also important that the user has control over the intrusiveness of notifications. Careful consideration is required when implementing features that push SNS content to people's personal phones. Next, we will discuss our findings and offer recommendations on how to design automation in a mobile notification context. We will also discuss the broader implications of automation for mobile social applications.

A. Use Goals Are Significant in Defining Social Media UX

People can have various motivations for reading about happenings on SNSs like Facebook. Some might feel it very important not to miss information about their friends' doings whereas others might read SNSs only when they do not have anything else to do or want to take a relaxation break from more important tasks. Notifications made it efficient to be up to date on Facebook happenings. It did assist in the pragmatic goal of getting information. However, the use goals played a significant role in user experience. Designers should thus concentrate on supporting users' hedonic goals and not only pragmatic goals. As a concrete example, the service could have a feature that when first used would ask the user questions concerning her goals related to Facebook use and set up notification settings based on the user's answers.

B. Losing Control May Reduce Motivation to Follow SNS Updates

As the functions are automated, control is transferred from human to machine. This may lower the user's perceived control of following SNSs. Several participants reported that reading the updates manually made the information feel more meaningful, and they felt more motivated to read Facebook manually than by using mobile notifications. With notifications, the trigger for checking SNS updates comes from a machine. For most users, it was better for the trigger to come from the user herself. Notifications can actually decrease reading activity compared to manual reading, as was reported by one of the participants. In information systems, perceived control has been found to affect users' motivation to use the system [19], and the lack of autonomy of actions has been shown to reduce motivation [25]. Research indicates that people prefer accomplishing outcomes themselves to having them accomplished by someone else [20]. So it is important for designers of mobile notifications to consider the importance of perceived control for user experience.

C. Notifications Can Violate the Privacy of Information Input

Users can interpret notifications as an invasion of privacy. This occurs when privacy is understood not only as a possibility for the individual to control what other people know about her (output of information) but also the individual's ability to control the information flow about others to him or her (input of information). This input-output distinction of privacy has been presented by [26]. With notifications, user control of information input might be in danger, and lack of control can be perceived as a violation of privacy. For example, user privacy was violated when she was not able to sufficiently control getting a notification at 9 A.M. on Sunday morning about some half-acquaintance linking a YouTube video to Facebook. Her strategy of controlling her privacy was to shut off the notifications feature completely. We propose that when developing systems that suggest something that is not a high priority for the user, designers should concentrate on situations in which the user already interacts with his or her phone. For example, notifications could show up only when the user actively uses the phone. Thus, a severe interruption could be avoided. However, in these situations, it is important that the notification does not stand out too forcefully so it does not complicate the primary task the user is performing.

D. Activity recognition can support relevance of notifications

Activity recognition has been proposed as a way to automatically find proper moments for interruptions (e.g., [27] and [4]). The fast development of smartphones and ubiquitous computing has enabled smartphones to contain numerous sensors that acquire data on a user's physical surroundings as well as interaction with the phone and phone's interaction with surrounding devices. For example, ContextPhone platform developed in 2005 records data from 16 different types of sensors that could then be used to trigger actions within a service that uses the data [28]. A more recent example is the Android operating system-based Funf [29], which is able to record data from over 30 different built-in mobile phone data probes [30]. It might be useful for developers of notification features to try to use built-in sensors such as an accelerometer, magnetometer, and GPS sensor to model the activities of the user. Such a sensor-based approach could reveal users' activities, such as whether a user is on a bus or cycling. Successful recognition of activities (see, e.g., [31], [32]) might enhance suitable moments for notifications. However, although this kind of adaptive automation might create a better user experience, it is likely that the user control is needed. We still (after tremendous advances in machine learning) seem to be in the situation where humans are vital to systems with automation [33].

D. *Repetition Frequency of an Automated Task Should Not Be Extreme by Default*

Designing automation in the mobile social context is not only about allocating tasks between humans and machines but also about how to inform the user about the possibilities and workings of automation. For example, if the settings of automated features could be changed by the user but the default setting is at one extreme (i.e., very frequent or very rarely performed automated features), the user might not even be aware of the possibility of delegating tasks to the phone, or the user might become so irritated because of intense automation that he or she turns the feature off. This happened with many of the participants — they got irritated partly because the pop-up frequency of notifications was so high. Notifications might have been perceived more positively if the default pop-up frequency had been, for instance, every two hours.

E. *The Time Window for Automation to Assure the User of Its Usefulness Is Short*

It took at most one week for users to decide whether the notifications feature was useful. The majority of the participants were positive about the feature beforehand, and they thought that getting the updates automatically to their phones' desktops was nice. However, after a week, the majority had turned the feature off because it did not adequately accommodate their needs and Facebook use. We believe this window is extremely critical in automated systems that in some way cause interruptions in mobile contexts. Although more data is needed to define a precise time frame during which users either approve or disapprove of the feature, the implication for design is that designer should make the system adapt to a user's usage habits fairly quickly. For example, after noticing that a user does not browse and read the pushed Facebook updates, Socially could automatically suggest alternative notification settings (e.g., lower notification frequency or silent notifications).

F. *Users Must Be Able to Choose the Level of Automation Based on the Significance of Information*

In SNSs, the sizes of social networks and the closeness of the contacts vary. For example, some of the participants had over 1000 Facebook contacts whereas some had about 100. Raising 1000 contacts' Facebook updates almost to the level of personal text messages was not perceived as useful. Only important messages directed specifically to the user were designated to be pushed to the phone automatically. The current iPhone and Android Facebook applications let users manage notifications by selecting what kind of messages the user wants as notifications, such as friend requests, events, or likes. However, the other solution could be to use varying levels of automation depending on the significance of information. Automation does not have to be all or nothing. For example, highly relevant information could include an alarm whereas information of less relevance would not, and the least relevant information would not be pushed as a notification at all.

Another important consideration is of course how to determine what is significant and what is not. One possibility would be that the user selects the automation level for different kinds of messages (e.g., friend requests, likes, etc.). Another possible direction for information filtering could be to combine automatic filtering (e.g., a crowdsourcing approach in which the system automatically filters out updates that have not been read or are not considered important by one's peers) and user control and lowering the level of automation by giving the user the option to include his or her own criteria (e.g., in form of tags) in the filtering algorithm.

Regarding the parts of the selection process where the user is not in control, a discussion about the values of design would be necessary. For example, the developer might want to raise the importance level of updates that contain names of some brands, meaning that notifications could be considered mobile ads. Making these kinds of decisions without informing the user might be ethically dubious and might also lower the user's perceived control of what to read on SNSs.

VII. Limitations of the Study

There are three limitations of this study. First, the number of participants was small. This qualitative study emphasized finding implications for designs using mobile notifications. Our aim was primarily to understand the details of the user experience, so we chose the qualitative approach. With a larger sample size, additional results on the quantitative effect of notifications on usage amounts could have been established. It is also possible that a larger user sample would have revealed further user innovations on how to use notifications to limit excessive Facebook use, such as by reading Facebook only when a notification pops up. Second, the results of the use and the effect of notifications were self-reported. Before the study, we tested and considered a context logging application to run in the background. However, because the Socially application was not our own but a commercial one, we did not have access to the source code, and existing context logging systems were not able to log quantitative data on the use of notifications. These limitations open up an area for future research. Logging the use of notifications of a large group of users could make it possible to quantify their effect in more detail. Third, as Socially's notifications were fairly intense (frequent), they can be considered as a somewhat extreme case of mobile notifications. Thus, many participants stopped using notifications after some days. However, the intensity of notifications as the default enabled us to find results related to user control and experience, which might not have shown with less intense notifications.

VIII. CONCLUSIONS

It is apparent that having enough perceived control over automation-assisted reading of an SNS is essential for a good user experience. Most people who used the mobile notifications reported increased reading of Facebook. However, after using Socially notifications for a while, many were unwilling to receive them and felt the system limited their control too much. Whereas some participants perceived that notifications helped them be better aware of their social surroundings, others noted a decreased interest in Facebook. Having a manual means to control Facebook reading felt more suitable for varied everyday use. We believe that it is valuable for designers to understand that people will practice control over the notification system one way or the other to suit their SNS use purposes. Offering easy-to-use technical ways for the user to adjust the intensity and content of notifications assists the adaptation of the notification system to individual needs. Inaccurate, impractical, or limited technical ways to personify the notification system might make users control them by shutting them down or ignoring notifications completely. In this paper, we suggested design implications for better user experience of mobile notifications. A valuable direction for future work would be to study the best and most convenient ways to ensure that the intensity and value of notifications meets the use practices of an individual SNS user. Considering the rapidly-developing sensing capabilities of modern smartphones, we see activity recognition as an interesting path to study the possibilities of automatically finding the proper contexts for notifications. However, at the same time, the importance of perceived control to a good user experience must be kept in mind in the design process of SNSs.

ACKNOWLEDGMENT

We would like to thank Erkki Kuisma and Hannu Vilpponen from Nokia Research Center for supporting the study, Antti Oulasvirta for support in the early phase of the study design, and Risto Sarvas for comments on the early version of the manuscipt.

REFERENCES

[1] E. Gilbert and K. Karahalios, "Predicting tie strength with social media," in *Proceedings of the 27th international conference on Human factors in computing systems*, 2009, pp. 211–220.

[2] Auto Check In Lite. (2012). *foursquare*.

[3] D. S. McCrickard and C. M. Chewar, "Attuning notification design to user goals and attention costs," *Communications of the ACM*, vol. 46, no. 3, pp. 67–72, 2003.

[4] N. Kern and B. Schiele, "Context-aware notification for wearable computing," in *Proceedings of the 7th IEEE International Symposium on Wearable Computers (ISWC'03)*, 2003, pp. 223–230.

[5] M. Hassenzahl, The thing and I: Understanding relationship between user and product. In: Blythe, M.A., Overbeeke, K., Monk, A.F., and Wright, P.C. (Eds.) *Funology: From usability to enjoyment*. Springer, 2003. pp. 31-42.

[6] D. J. Lazar, D. J. H. Feng, and D. H. Hochheiser, *Research methods in human-computer interaction*. West Sussex, UK: John Wiley & Sons, 2010.

[7] A. N. Joinson, "Looking at, looking up or keeping up with people?: motives and use of Facebook," in *Proceeding of the twenty-sixth annual SIGCHI conference on Human factors in computing systems*, New York, NY, USA, 2008, pp. 1027–1036.

[8] Socially mobile client. (2012).

[9] J. M. Carroll, D. C. Neale, P. L. Isenhour, M. B. Rosson, and D. S. McCrickard, "Notification and awareness: synchronizing task-oriented collaborative activity," *International Journal of Human-Computer Studies*, vol. 58, no. 5, pp. 605–632, May 2003.

[10] P. Isomursu, M. Isomursu, and M. Ervasti, "User Attention in Mobile Devices," presented at the ACHI 2012, The Fifth International Conference on Advances in Computer-Human Interactions, 2012, pp. 161–164.

[11] T. H. Davenport and J. C. Beck, *The attention economy: Understanding the new currency of business*. Harvard Business Review Press, 2002.

[12] A. Oulasvirta, S. Tamminen, V. Roto, and J. Kuorelahti, Interaction in 4-second bursts: The fragmented nature of attentional resources in mobile HCI. In *Proceedings of CHI 2005* pp. 919-927. 2005.

[13] A. Oulasvirta, T. Rattenbury, L. Ma, and E. Raita, "Habits make smartphone use more pervasive," *Personal and Ubiquitous Computing*, vol. 16, no. 1, pp. 105–114, 2012.

[14] J. E. Fischer, C. Greenhalgh, and S. Benford, "Investigating episodes of mobile phone activity as indicators of opportune moments to deliver notifications," in *Proceedings of the 13th International Conference on Human Computer Interaction with Mobile Devices and Services*, 2011, pp. 181–190.

[15] Y. Cui, M. Honkala, K. Pihkala, K. Kinnunen, and G. Grassel, "Linked Internet UI: A mobile user interface optimized for social networking," in *Proceedings of the 12th international conference on Human computer interaction with mobile devices and services*, 2010, pp. 45–54.

[16] K. Väänänen-Vainio-Mattila and M. Wäljas, "How do users find out what's new: a study of change indicators in mobile services," in *Proceedings of the 15th International Academic MindTrek Conference: Envisioning Future Media Environments*, 2011, pp. 201–204.

[17] R. Parasuraman, T. B. Sheridan, and C. D. Wickens, "A model for types and levels of human interaction with automation," *IEEE Transactions on Systems, Man and Cybernetics, Part A*, vol. 30, no. 3, pp. 286–297, 2000.

[18] E. A. Skinner, "A guide to constructs of control.," *Journal of personality and social psychology*, vol. 71, no. 3, p. 549, 1996.

[19] T. P. Novak, D. L. Hoffman, and Y. F. Yung, "Measuring the customer experience in online environments: A structural modeling approach," *Marketing Science*, vol. 19, no. 1, pp. 22–42, 2000.

[20] P. C. Ellsworth, Some reasons to expect universal antecedents of emotion. In P. Ekman & R. J. Davidson (Eds.), *The nature of emotion: Fundamental questions*. New York: Oxford University Press, 1994, pp. 150–154.

[21] S. Brave and C. Nass, "Emotion in human-computer interaction," *The human-computer interaction handbook: fundamentals, evolving technologies and emerging applications*. New York: Taylor & Francis, pp. 81–96, 2003.

[22] SizzleLab. (2012).

[23] A. C. Strauss and J. M. Corbin, *Basics of qualitative research: grounded theory procedures and techniques*, 2nd ed. Sage Publications, Inc, 1990.

[24] Y. Zhang and B. M. Wildemuth, "Qualitative analysis of content," Applications of social research methods to questions in information and library science, Westport, CT: Libraries Unlimited, pp. 308–319, 2009.

[25] R. M. Ryan and E. L. Deci, "Intrinsic and extrinsic motivations: Classic definitions and new directions," *Contemporary educa-*

tional psychology, vol. 25, no. 1, pp. 54–67, 2000.

[26] I. Altman, The environment and social behavior: Privacy, personal space, territory, and crowding. Monterey: Brooks/Cole Publishing Company, 1975.

[27] J. Fogarty, S. E. et al. Predicting human interruptibility with sensors. *ACM Transactions on Computer-Human Interaction, 12*, 2005, pp. 119-146.

[28] M. Raento, A. Oulasvirta, R. Petit, and H. Toivonen, "ContextPhone: A prototyping platform for context-aware mobile applications," *IEEE Pervasive Computing*, pp. 51–59, 2005.

[29] N. Aharony, W. Pan, C. Ip, I. Khayal, and A. Pentland, "Social fMRI: Investigating and shaping social mechanisms in the real world," *Pervasive and Mobile Computing*, vol. 7, no. 6, pp. 643–659, Dec. 2011.

[30] Funf. 2012.

[31] J. R. Kwapisz, G. M. Weiss, and S. A. Moore, "Activity recognition using cell phone accelerometers," *SIGKDD Explor. Newsl.*, vol. 12, no. 2, pp. 74–82, Mar. 2011.

[32] S. Reddy, M. Mun, J. Burke, D. Estrin, M. Hansen, and M. Srivastava, "Using mobile phones to determine transportation modes," *ACM Transactions on Sensor Networks*, vol. 6, no. 2, pp. 1–27, Feb. 2010.

[33] R. Parasuraman and C. D. Wickens, "Humans: Still vital after all these years of automation," *Human Factors: The Journal of the Human Factors and Ergonomics Society*, vol. 50, no. 3, p. 511, 2008.

AUTHORS

Sami Vihavainen is with the Aalto University's Helsinki Institute for Information Technology HIIT, P.O. Box 15600, 00076 Aalto, Finland (e-mail: sami.vihavainen@hiit.fi).

Kaisa Väänänen-Vainio-Mattila is with the Tampere University of Technolgy's unit of Human-Centered Technolgy, department of Pervasive Computing, P.O. Box 589, 33101 Tampere, Finland (e-mail: kaisa.vaananen-vainio-mattila@tut.fi).

15

Guidelines on Implementing Successful Seamless Learning Environments

A Practitioners' Perspective

Noriko Uosaki, Hiroaki Ogata, Mengmeng Li, Bin Hou and Kousuke Mouri
The University of Tokushima, Minami Josanjima, Tokushima, Japan

Abstract—**This paper describes implementation guidelines for ubiquitous seamless mobile learning featuring the learning-log system named SCROLL (System for Capturing and Reminding of Learning Log). This system helps users to share and remind ubiquitous learning experiences. We review recent researches on mobile assisted learning in order to set pedagogically-based guidelines for instructors, learning environment designers and the administrative end or decision makers including ICT coordinator, and local authorities. Main foci in our guidelines are for language teachers to practice SROLL-based mobile learning in their classes for both in-class and out-class learning according to their situation or conditions. Our goal is to help educators to implement an effective seamless mobile learning environment with SCROLL.**

Index Terms—**mobile assisted language learning (MALL), mobile learning, m-learning, seamless learning, task-based learning**

I. Introduction

Pervasion of high-efficiency mobile phones or smartphones has expanded the potential of mobile learning environments. Researches on this area have been very active during this decade [1] [2] [3]. Mobile learning has been recognized as one of the natural directions toward which CALL (Computer Assisted Language Learning) is heading [4] [5]. Reference [6], which indicated that the learners preferred mobile platform rather than PCs, endorsed this trend. Mobile and ubiquitous technologies have been expected to foster shifting from classroom-based learning to the one that is free from time and space boundaries, which is, what they call, mobile cloud-education [7]. It can provide smart anytime-anywhere learning that is customized and adapted to individuals, and delivered via personal portables devices.

In this paper, after overviewing the recent research tendencies on seamless mobile learning, we analyze how seamless the learning could be and categorize it into 7 types according to the facilities being afforded, and then we set implementation guidelines for those who are considering adopting seamless mobile learning in the new era of high performance smartphone.

There have been quite a few technology-driven researches on these fields, but however sophisticated systems (software) and devices (hardware) being developed, unless we have a good guidance so that practitioners can use them in a real educational environment, they might be useless, being left unused. But such guidelines have rarely been found out on any journals so far. Even if they exist, since mobile technology is a rapidly and constantly advancing field, the guidelines which were set in the past are destined to get out of date soon. Therefore it is our aim to produce new guidelines on implementation of an effective seamless mobile learning environment. Though this is especially featuring SCROLL (System for Capturing and Reminding of Learning Log), we believe that it can help implementation of any seamless mobile learning environment.

The rest of this paper includes research updates on mobile learning in Chapter II. Chapter III presents keys to successful implementation of seamless mobile learning, and Chapter IV describes learning scenarios in mobile environment. Chapter V gives conclusions with our plans for further development of seamless mobile environment.

II. SEAMLESS MOBILE LEARNING

A. Mobile Learning

Reference [8] defined "mobile learning" as "any educational provision where the sole or dominant technologies are handheld or palmtop devices". According to [9], "mobile" as in mobile learning encompasses both learning supported by mobile devices and also learning in an era characterized by mobility of people and knowledge. In this paper, "mobile learning" by which we mean "learning with its use of mobile and wireless technologies with the scope of the mobility of the learners and learning situation.

The new generation smartphones as represented by iPhone and the open sourced mobile OS Android provide users with such advanced functions as the multi-touch interface, full browser, and millions of applications. One of the key features of smartphones is that they are equipped with a range of sensors such as the accelerometer, ambient light sensor, GPS, microphone, camera, compass and so on. Aggregation of mobile devices with such multi-sensor systems would enable users to benefit from information gathered via other surrounded devices and users [10] [11]. It will significantly impact the future of education, providing new environment called surround learning or seamless learning and inducing more social interaction and collaborative learning among learners. Thus these mobile technologies will possibly play a critical role in the future CALL activities.

B. *Seamless Learning*

"Seamless learning" is used to describe the situations where students can learn whenever they want to in a variety of scenarios and that they can switch from one scenario to another easily and quickly using one device or more per student ("one-to-one") as a mediator [12]. It allows learners to learn anytime, anywhere, and provides them with multiple ways of learning throughout the day. The term, seamless learning is also used to describe a lifelong learning in various environments across time and location seamlessly through the use of the technology as a mediating tool [13]. In this paper, by seamless learning, we mean learning which occurs with seamless transitions between in-class and out-class learning, between handheld use and desktop use.

How seamlessly we can carry out class (seamless rate) depends on the facility we afford. Fig. 1 shows the correlation between mobility enhancement and seamless rate. Higher the mobility we afford, higher the seamless rate we attain for our learning environment. 'Mobile' in Fig. 1 means a learning environment where one-to-one based internet connected mobile device such as smartphone, tablet, PDA is available. 'Fixed' in Fig. 1 means a learning environment where one-to-one based internet connected fixed computer is available. Learning activities according to each condition will be introduced later in Chapter IIII.

Table 1 shows types of learning available according to each mobile/fixed condition. There are seven types of mobile learning based on the seamless rate: 1) full seamless learning, 2) semi-seamless learning, 3) in-class only non-seamless mobile learning, 4) in-class only non-seamless mobile learning, 5) outside-class only non-seamless mobile learning, 6) in-class only non-seamless fixed learning, 7) outside-class only non-seamless fixed learning.

When mobile is available for both in-class and outside-class, learning can be fully seamless, thus we call it "full

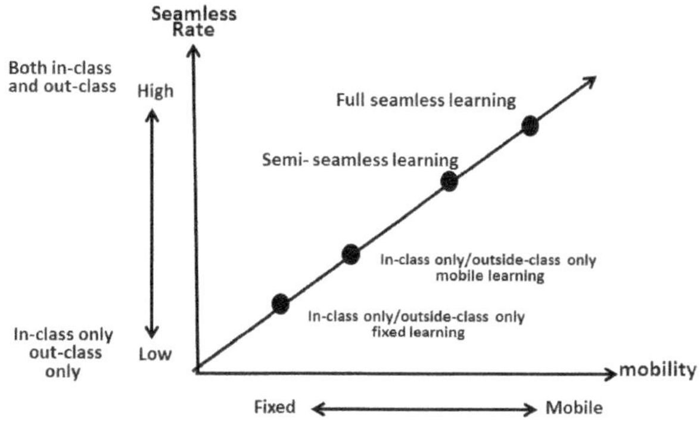

Figure 1. Correlation between seamless rate and mobility

seamless". When either mobile or fixed computers are available for both in and outside-class, we call it "semi-seamless" because even though mobility gets lower, still in- and outside-learning can be connected through either fixed or mobile devices. If there is mobile/fixed available only in-class or only outside-class, it cannot be seamless anymore. Thus we call it non-seamless such as in-class only non-seamless, or outside-class only non-seamless.

C. *SCROLL*

1) *System Overview*

SCROLL is a client-server application, which runs on different platforms such as Android mobile phones and PCs (cf. http://ll.is.tokushima-u.ac.jp/learninglog/signin) [14].

Users can register what they have learned anytime, anywhere. The registered object is called as a ubiquitous learning log object (ULLO). They can take quizzes made from their ULLOs for recalling them (cf. Section 2.C.2)). The System lets learners share their ULLOs. The learners are able to watch other learners' ULLOs. If they like other learners ULLOs, they can "relog" them to their own pages just like "retweet" in Twitter so that ULLOs uploaded by other learners can be their own ULLOs.

TABLE I.
SUGGESTED LEARNING STYLE ACCORDING TO THE SEAMLESS RATE

Seamless rate	In-class		Outside -class		Type of Learning
	Mobile	*Fixed*	*Mobile*	*Fixed*	
High	✔	✔	✔	✔	Full seamless learning --- cf. Chapter IV.A (It covers all learning activities described in Chapter IV)
	✔	✔	✔		
	✔		✔	✔	
	✔		✔		
		✔	✔	✔	Semi-seamless learning---cf. Chapter IV.B, E
	✔	✔		✔	
		✔	✔		
	✔			✔	
		✔		✔	
	✔	✔			In-class only non-seamless mobile learning --- cf. Chapter IV.C
Low	✔				
			✔	✔	Outside-class only non-seamless mobile learning --- cf. Chapter IV.D
			✔		
		✔			In-class only non-seamless fixed learning
				✔	Outside-class only non-seamless fixed learning
					Computer-less learning

Using this system, teachers can grasp what their students have learned outside the class (informal settings). For example, they ask their students to record the words that they have learned into SCROLL as ULLOs. In the next class, they make a reflection using their students' ULLOs. Through this process, they can check whether the ULLOs given by their students are correct or not, and allow their students to share their knowledge. In this way, SCROLL enhances and integrates both formal and informal learning.

2) *Quiz Function in Learning Log System*

According to [15], memories can stay in the short-term "buffer" only for a limited period of time while they are strengthening their associations to be stored in long-term memory. Therefore the System is designed to support retaining vocabulary in a long-term memory by giving them quizzes. The system gives learners quizzes to help them retain their vocabulary. It generates quizzes automatically based on ULLOs registered, reclogged, and viewed by the learners. Quizzes will be generated until they give them correct answers. Four months after they made the correct answers, they will be given quizzes again to make sure if the learners retain them in their long-term memories. After a certain interval, the system gives them quizzes which they have answered correctly to make sure if they are retaining their vocabulary. By using this methodology, it is expected that their short-term memory will be reinforced into long-term memory. Logs of all the quizzes are stored to be analyzed and evaluated. Wrong-answer rates and difficulty-level choices from "too easy" to "too difficult" reflect the quiz generation. Difficulty level adjustment is made in order to facilitate their learning processes. If they challenge quizzes, they can accumulate the scores, which are shown in the right-hand side of the screen: Today's Score and Your All Score. It is expected to motivate them to solve more quizzes.

III. Keys to Successful Implimentation of Mobile Learing

A. *Four Influential Factors for Successful Seamless Learning*

What factors will influence successful mobile learning? In this section, we will pick out four influential factors and show guiding principles to assess the feasibility of successful mobile learning implementation. The assessment checklists for implementation possibility of mobile-based learning are shown in Fig. 2 and Fig.3.

Factor 1: Management and facilities

In order to manage mobile learning initiative effectively, obtaining senior level buy-in is inevitable. As [16] has pointed out, the cost of the technology, the infrastructure and the services and applications are important issues when implementing mobile learning. As a general advice, institutions should try to make use of what is already in place in order to keep costs down. If every student reports smart phone owners, then it is easy to carry out mobile learning without budget difficulties. Since smart phones are prevailing rapidly, it is expected most students in classroom own them. In this case, purchasing some devices and lending them only to those who do not own them might be another possibility with low-cost performance.

Fig. 2 shows you whether your budget and students' device condition fit for mobile-based seamless learning. According to [17], "smartphones account for half of all mobile phones". Although smartphone penetration is incredibly high, there must be some population who do not want to own mobile phones in any time. Thus we cannot expect every student owns one in class. Therefore some budget is necessary to purchase some extra mobile phones to lend the students who do not own them.

Factor 2: Instructors' abilities

For successful implementation of mobile-based seamless learning, one of the most important things is the instructors' motivation whether they are really for mobile-based seamless learning. Without their willingness, it is hardly possible. In fact many instructors are not highly motivated to use ICT technology [18]. Therefore computer literacy of teachers is an important issue to be dealt with. However since our focus in this study is not on enlightening teachers, this problem is to be discussed at another opportunity in the future.

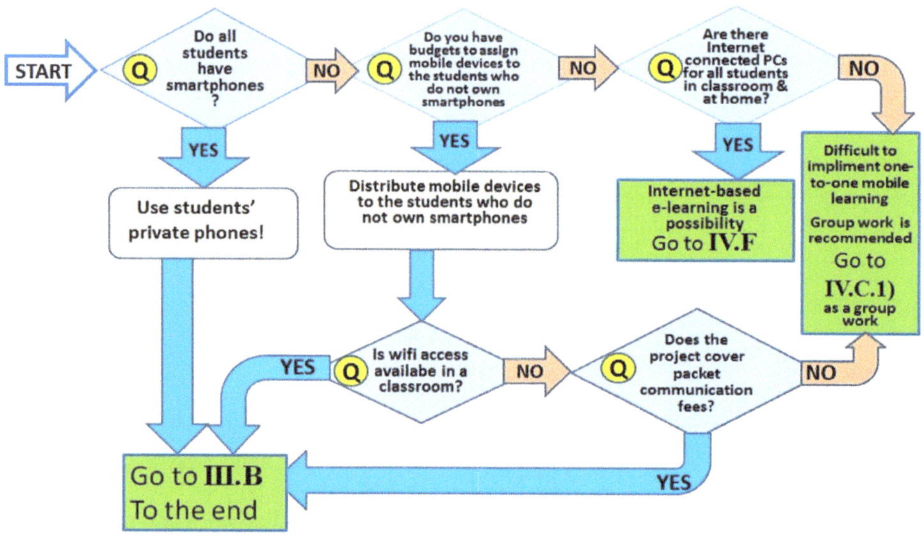

Figure 2. Devices/budget checklist

In order to reinforce entwining in-class learning with outside-class one, SMALL system (Seamless Mobile-Assisted Language Learning Support System) has been developed [19] (cf.III.C.1). Still teachers' active involvement is inevitable. What the students learned outside-class should be introduced during class. For instance, if one of the target words in class is "contemporary" and somebody already uploaded this word to the System, let him talk about how he encountered the word. The word context is important to learn vocabulary as mentioned. Students' presentation in turn as either solo work or group work is recommended as another activity to link outside-class learning to in-class learning (cf. IV.A).

Factor 3: Flexibility of curriculum and class setup

Even though teachers are highly motivated, if they have to follow the strict teaching guidelines including teaching methods, teaching materials, implementation is difficult. They need to have a flexibility in making decisions of what kind of class they run including learning approaches, teaching materials, seating arrangement and so on. Generally mobile learning fits task-based, interactive and collaborative learning. It is reported that traditional seating arrangement (desks all lined up and all the students face the teacher and a blackboard) makes students less active, less cooperative, less engaged in learning. In order to maximize the learning opportunities, traditional lecture style class is not recommendable.

This is endorsed by the study from [20], which reported that only about 5 percent of the information delivered through lecture was retained. Compare that with retention rates at 50 percent for discussion group and 70 percent for practice by doing. Even higher, at 80 percent, was retention by students teaching others. Thus successful mobile learning depends much on instructors' flexibility in running class.

Fig. 3 shows if your condition fits for seamless mobile learning based on your flexibility to choose teaching/learning materials. If you are totally free go to Section III.C.1) and if the flexibility is limited, go to III.C.2) accordingly.

Factor 4: Learners' motivation

Even though schools have budget (Factor 1), teachers are highly motivated (Factor 2) and have a flexibility (Factor 3), we cannot expect effective learning without learners' active participation. Things which hinder their active participation could be their low computer literacy and low motivation.

As for the former factor, according to [21], it is reported "students say that these environments (blended-learning environments) best support how they learn." Compared with the situation in late 20th century when 'computer literacy' is popularly used, students' computer literacy has been dramatically improved and most students have a favorable attitude IT-based learning. Therefore the second factor is most important and most difficult. Learning motivation has been a major research topic since early 20th century in the field of pedagogical psychology.

Reference [22] categorized motivation into two types: intrinsic and extrinsic motivation. It is argued that Deci and Ryan's Self-determination theory (SDT) could describe one's motivation in more understandable way.

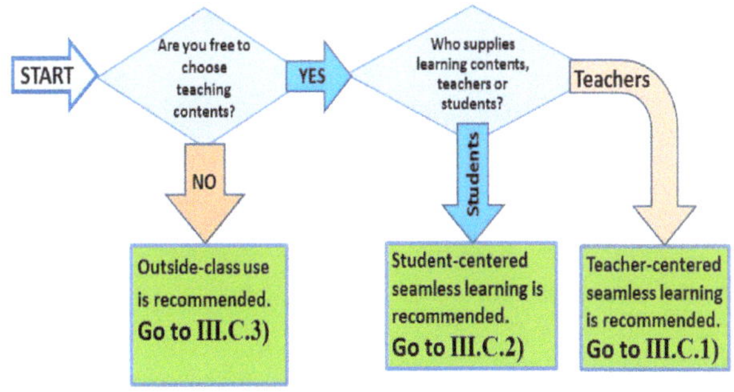

Figure 3. Content checklist

Reference [23] contended that there is a self-determination continuum ranged from non-self-determined to self-determined which is subdivided into three parts: motivation, extrinsic motivation and intrinsic motivation.

It is explained that the more self-determined, the more motivated. Thus, it is recommended to let students think that learning is not forced and that it is determined by themselves. First ask them why they learn and let them make a practical goal setting of their learning (eg. TOEIC 600 points). This process is important. Without active use of mobile learning system, any ideal m-learning environment could be useless. It is also important to let them keenly sense that in order to achieve their goal in-class learning is not sufficient and that outside-class, anytime-anywhere-based ubiquitous learning is inevitable. Generally speaking, students are busy, small-step style where the task can be achieved for a short period of time is recommended. Though it is believed that intrinsic motivation is better than extrinsic one, still it might be necessary to inspire some extrinsic motivation such as introducing a reward system where students' hard works would reflect their grades. Putting these factors into consideration, an effective mobile learning would be realized.

B. Nine Considerations in practice

In order to carry out a successful seamless learning, it important not only to grasp the environment: what is available, what is not available, and learners' computer literacy and mobile literacy, but also to grasp some techniques to conduct a class. This section suggests 10 vital things for practitioners to keep in mind

1) *Conduct questionnaires*

At the beginning of the semester, it is highly recommended to conduct questionnaires to grasp learners' computer literacy and mobile environment. Web tools such as Google document can facilitate this process. The followings are sample questions:

1. Do you have mobile phones? Is it smartphone?
2. Do you have computers at home? Is it internet connected?
3. Do you share computers with your family? Do you have one of your own?

2) *Initial registration*

Create students accounts beforehand and make a user list with the name, email address, password, username, device number because it is so time-consuming that without this process it is difficult to carry out anything in the first m-learning class. Note that you should print out the student account list and cut into each piece and hand them to the students on the first day.

3) *Numbering of devices to be distributed*

If you are going to deliver devices to the students, it is highly recommended to number the devices to be used and to make a list with student numbers, names, device numbers, and seat numbers. It is very convenient to make device numbers and seat numbers identical if possible. If the students are going to use their own devices, skip 4) to 6).

4) *Prepare Lend Out Equipment Forms*

Distribute lending forms when you distribute devices asking the students to fill in the form with their contact information such as their e-mail addresses and cellphone numbers.

5) *Prepare the contact information slip for each device*

It is highly recommended to post or insert the contact information slip with the name of the project leader, his/her e-mail address and cellphone number in case someone lost his/her device. There will be a possibility that the finder may get contact with the project leader.

6) *Charge devices*

If the mobile devices are to be provided to all students, it is important for the practitioners to charge devices before distribution. It means lots of electrical outlets needed. After being distributed, we need to ensure them to charge the device every day. When students use their private smartphones, it will be their routine work to charge them every day

7) *Privacy*

Privacy is an important issue to be considered when taking photos and upload them as ULLOs. It is recommended to advice students not to take a photo with someone's face in it or the ones which are considered too private.

8) *Make use of Description Column or images*

We learn words from the context [24] [25] [26]. When the students upload the words they learn to the system, they should be encouraged to input its context in which the word appears or just take a photo of the article and upload as image. Since students often do not like to be bothered with uploading contexts, teachers should keep encouraging them. It often works if they are informed that context uploading will reflect their grades.

9) *Comment column*

A comment column in each ULLO interface is for asking questions or exchanging information. Therefore let students note that when they write comments on ULLOs, they should not write things off-topic.

C. Creating Contents

1) *Creating teacher-centered contents for inside- and outside-class learning*

SMALL System proposed by [19] is based on SCROLL and linking function between inside- and outside-class learning is newly added (http://ll.is.tokushima-u.ac.jp/ecourse/). Its aim is to entwine in-class learning with out-of-class learning. When learners upload what they have learned outside-class to the System, then it links it with the one used in the textbook and vice versa.

First, teachers upload textbook data (PDF or text) to the system. Students can read textbook files anytime anywhere for previewing and reviewing with mobile devices just like a paper book. .In the electronic textbook, registered words are hyperlinked and when the teacher clicks them, a side bar pops up and it shows names of the students who registered them so that the teacher will be able to know how many students and who have registered them.

The electrical textbook can be accessed through internet-connected PCs and smartphones. They can use during class and at home. Teachers registered the target words of each chapter before class and let them view target words as a preview assignment. Students can "relog" them if they are unknown words and challenge quizzes.

As mentioned, this system aims to entwine out-of-class learning with in-class learning. Fig. 4 shows how in-class vocabulary learning and out-of-class vocabulary learning are linked. When a student registers "credit" during his out-of-class learning, the system shows him the textbook context where it appears as well as Chapter title and line number. Then he can realize that he has already learned it in the textbook during his in-class learning. That way it reminds him of what he had learned but forgotten already. The System also entwines students' self-learning with that of the others.

Fig. 5 shows how the "contemporary" object is connected to those registered by other classmates. When he clicks them, he can learn other contexts where "contemporary" is used. This linking function is significant because, as mentioned earlier, students need to encounter as many different contexts as possible to grasp the true meaning of the word. For another reason, as a general concept, people are likely to forget what they have learned. In fact forgetting learned vocabulary is a serious problem during learning English vocabulary [27]. Even though the student felt "contemporary" was totally new to him, the system lets him know that he has learned it before in the textbook.

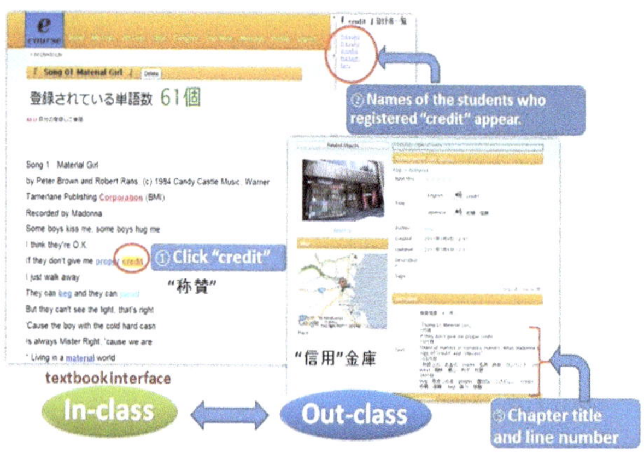

Figure 4. Link between in-class learning and out-of-class learning

2) *Creating student-centered contents for inside- and outside-class learning*

Students can upload contents just as teachers do on the textbook upload page shown. Looking for some articles which they get interested in (any topic will do such as music, sports, movies, TV shows) and upload them so that they can share. They are supposed to present concerning the material they chose and the vocabulary which was linked to the textbook vocabulary during class in turn. It is recommended to do it as a group work.

3) *Creating contents for outside-class informal learning*

When teachers should follow the teaching guideline strictly and do not have a flexibility to choose teaching/learning materials, SCROLL would be mainly for outside-class learning. Either According to [28], 5 to 16 exposures are necessary to fully acquire a word. SCROLL is a good tool to assist learners to reinforce word-meaning connection by giving them a quiz and recommending ULLO where they come across the place where they learned it

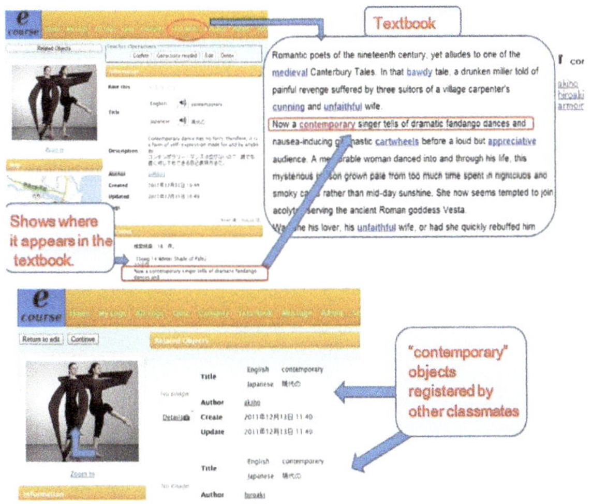

Figure 5. Link between one's learning and other classmates' learning

In order to stimulate students to learn/upload more words, a top ten upload ranking list (names of the learners who uploaded top most to 10th) is shown on each user's top page. As mentioned before, they can accumulate the scores when they challenge quizzes. The score display is expected to motivate them to solve more quizzes. Let them note that their commitment to SCROLL learning reflected their grades.

IV. Learing Designs in Seamless Mobile Environment Featuring SCROLL

Some of the learning scenarios suggested in this section are based on 'experiential learning' proposed by [29]. This learning theory contends that it is effective to make students active and learn by doing, For the sake of the instructors' convenience, lists of things necessary, and things to do before class are provided. In recent years many educational institutes have introduced LMS (learning management system) such as Moodle. Combination of SCROLL and LMS will make seamless mobile learning more effective. If there is no system available, Edmode is a possibility. Edmode, founded by Nicolas Borg and Jeff O'Hara in 2008, is an educational social networking site to connect teachers, students, and parents, connect teachers with other teachers, and connect class with other classes. Teachers can post assignments, create polls for student responses, embed video/sound clips to support students' learning. Facebook, Twitter or Dropbox have similar functions. Therefore, the learning activities introduced below are based on the combination use of SCROLL and LMS or any form of sharable holders.

A. *Full seamless learning*

1) *Full-seamless vocabulary learning with SMALL*

(1) Environment: in-class and outside-class mobile/fixed

(2) Activity outline

Students preview target words in the textbook before class and take quizzes. They receive lecture from a teacher during class. They were assigned to learn vocabulary and uploaded what they have learned out-of-class to the System. Teachers introduce their students' outside learning in class or students present what they have learned outside-class. The vocabulary they learn outside-class will be linked to the vocabulary in the textbook by the system (cf. Fig. 4 & Fig. 5).

(3) Target language skill: reading, vocabulary building

(4) Level: beginner to intermediate

(5) Activity style: solo work

(6) Before class: Prepare reading contents for outside-class learning. Register target words to the System.

(7) Procedures

1) Before class, as a preview, students see "all logs" registered by the teacher, relog them and answer quizzes at home.

2) During class they read and comprehend the contents with the help from the teacher.

3) Students upload words they have learned and answer quizzes both inside- and outside-class. Teachers evaluate their status watching their logs and quiz logs.

(8) Points in practice

1) It is necessary to keep encouraging students to do learn vocabulary outside-class. It sometimes works just to let them keep record of the hour they spent for vocabulary learning (cf. Appendix 1). The System teaches us how long they logged in, but it does not necessarily show the time they spent for vocabulary learning.

2) Reference [19] shows that the group with the system felt more fun in vocabulary learning and learned more words than the group without the System. Therefore informing them of its significant effectiveness will motivate students to use the System.

B. *Semi-seamless learning*

1) *Semi-seamless vocabulary learning with SMALL*

(1) Environment: in-class and outside-class fixed (+in-class mobile)

In this condition, it is less-seamless since there is no mobile device out-side class. It is hard to get access to the System when they are not home nor classroom. Still it is possible to link in-class learning with outside-class one using fixed computers, though it would be less effective.

(2) Activity Outline to (8) Points in practice are the same as A.1).

C. *In-class only non-seamless mobile learning*

Two In-class Task-based Learning activities are introduced in this section. Task-based language learning or TBLL is a popular learning method based on 'experiential learning'. Much emphasis is put on learning by doing.

1) Orienteering on campus using mobile devices

(1) Environment: in-class mobile and PC

(2) Activity outline

Students go outside classroom with mobile devices such as smartphones, tablets and do orienteering-based activities. If the class is not in one-to-one condition, one-to-group work is recommended.

(3) Target language skill: speaking, reading

(4) Level: beginner to intermediate

(5) Activity style: pair work / group work

(6) Before class: Register ULLOs on campus. Register questions on "description" area on the registration interface (cf. Fig. 6).

(7) Procedures

1) Make pairs or groups and let them go to the places instructed by the teacher.

2) Then the students receive push mode messages which inform them that there are some ULLOs around them. Read "Description" and do the task as instructed. They input an answer in a comment column (cf. Fig. 7). Another possibility is to

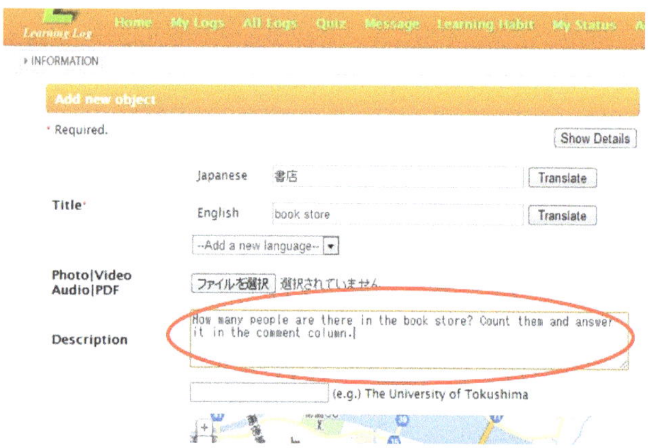

Figure 6. Description area in Learning log upload interface

answer the question orally using voice recorder and upload the sound files to some instructed sites such as LMS, Edmode, and dropbox. Then go to another object that instructs next task. They are not allowed to use their native language. (cf. Appendix 2).

3) The number of the task should be the same with the number of the group members so that every single member could be a leader of at least one task.

4) If they have some problem or trouble, they make a call to the teacher, who gives them some advice or new instruction according to their situation. As a rule they are supposed to use the target language while communicating with their teacher.

5) The teacher checks the e-mails/voice mails sent by the students and give them the next task

6) The pair/group who finished all the tasks first will be awarded and it reflect their grading.

(8) Points in practice

1) Make pairs or groups and let them go to the places instructed by the teacher. It is appropriate to give them four to five tasks during one class. Accordingly, make a group with 4 to 5 students each.

2) Drawing can be one of the tasks.

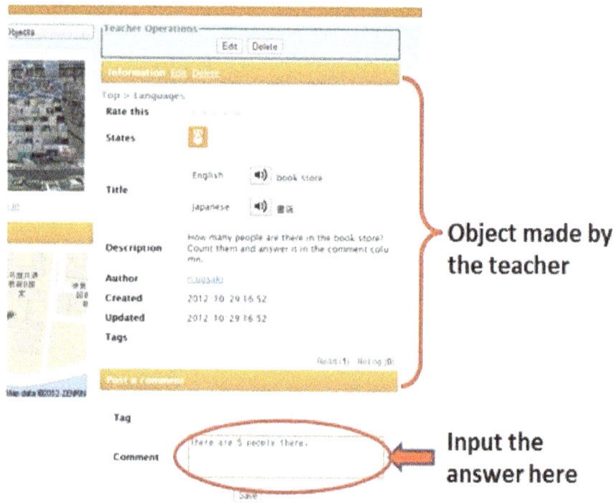

Figure 7. Comment column in Learning Log Object interface

3) Use "describe ...." in instructions so that the students can learn how to describe the situations which they encounter in

4) We can make it a competition-style where the group who finished all the tasks first is awarded.

5) We can make the students create tasks themselves.

2) Interviewing using smartphones

(1) Environment: in-class mobile and PC

(2) Activity outline

Interview activity: Peer-to-peer interviews and if possible, interviews with international students using smartphones. The students uploaded the recorded files of interviews to SCROLL, LMS, or Edmode, and shared the files by listening and made summary reports which were also uploaded to the LMS. (2) Target Language Skill: speaking

(3) Level: intermediate to advanced

(4) Activity style: pair work

(5) Before class: Make pairs.

(6) Procedures

1) Make pairs and let them interview with each other on a given topic. Prepare questions before interview sessions. Make recording interview session using voice recorder function of smartphones and upload the recorded file to the System.

2) After the interview session, let them make a summary of the interview and submit the file to the System.

3) It is easy to share uploaded files by SMALL. So listen to other classmates' interviews and vote for

the best interviewer. The best interview might deserve an add-on to his/her grade.

(7) Points in practice

1) If the students are beginners and feel it hard to make conversation continue, one solution is to deliver them sample dialog video clips (or sound clips) and let them practice in a pair-work style (cf. 'deliver' can mean 'upload files to LMS, Edmode, Google+, Dropbox, Youtube etc.).

2) Delivering useful phrases for interviewing, either it is digital or analog, might help.(cf. Appendix 3).

3) It will stimulate other students to upload best performers' video clips or recorded files to LMS, Edmode, dropbox, or any form of sharable holders

D. Outside-class only non-seamless mobile learning

1) Outside-class Task-based Learning

When teachers do not have a flexibility to choose teaching/learning materials, it is difficult to carry out task-based learning during class. In such a case, task-based learning for outside-class use is recommended. SCROLL is about to add a function to let learners to get involved in a task-based learning so that they can actually experience learning situations such as shopping, asking the way, visiting a doctor and so on (cf. Fig. 8 and Fig. 9).

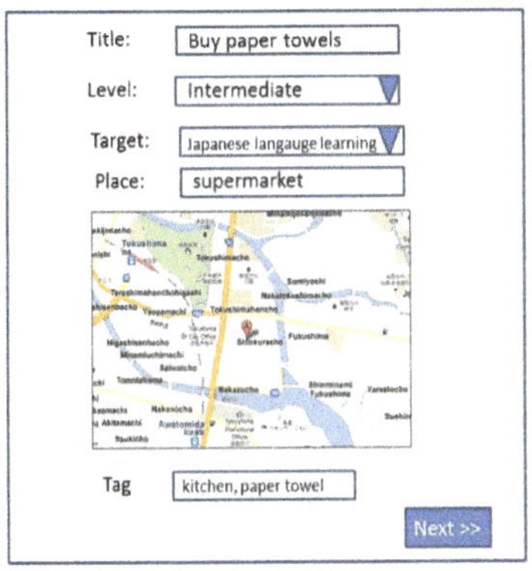

Figure 8. Task-based learning Interface (1)

(1) Environment: out-class mobile

(2) Activity outline

Learners choose the task and proceed as instructed. They can learn useful vocabulary and expressions during the task.

(3) Target language skill: Japanese language learning, speaking, vocabulary

(4) Level: intermediate to advanced

(5) Activity style: solo work / pair work

(6) Procedures

Learners choose the task such as shopping, opening an account at the back and so on as they want. They proceed the task according to the System guidance. It is an informal learning. Therefore basically it is all up to learners' autonomy.

2) Outside-class Context-aware Vocabulary Learning

(1) Environment: out-class mobile

(2) Activity outline

When learners come across the place where an object was registered, SCROLL lets them know and let them learn it.

(3) Target language skill: Japanese vocabulary building

(4) Level: beginner to intermediate

(5) Activity style: solo work / pair work

Figure 9. Task-based learning Interface (2)

(6) Procedures

1) When learners get cross to a registered object (Japanese word), the System lets them know that the learning object is near them and encourage them to learn Japanese vocabulary throughout their daily lives. The System also generates the quizzes to reinforce their vocabulary learning.

2) Reference [30] shows that the learners benefited from the system and the context based recommendation and learning habit based prompting also motivated them to study more. Therefore informing them of its effectiveness will motivate students to use the System.

E. PC-based vocabulary learning using SCROLL

This method is recommended when smartphone penetration rate is low and no budget is available. In this case, it is practically impossible to carry out seamless mobile learning. It should be used only as a last resort.

(1) Environment: in-class PC / outside-class PC

(2) Activity Outline

Students read books and learn vocabulary using SCROLL.

(3) Target Language Skill: reading, vocabulary building

(4) Level: beginner to intermediate

(5) Activity Style: solo work / group work

(6) Before class: Prepare reading contents or just use textbook. Register target words to the System.

(7) Procedures

1) Before class, as a preview, students see "all logs" registered by the teacher, relog them and answer quizzes at home.

2) During class they read and comprehend the contents with the help from the teacher.

3) Students upload words they have learned and answer quizzes both inside- and outside-class. Teachers evaluate their status watching their logs and quiz logs.

(8) Points in practice

1) In order to encourage students to do home assignment, teachers need to use carrot and stick policies: carrot (some rewards or extra points at the end of the semester) and stick (extra assignment or detention). Outside-class learning is important because they are in lack of learning time at school.

2) If the teachers want to encourage students to read what they are interested, "noredink" is recommendable (cf. http://www.noredink.com/). This site aims mainly for practice and master of grammar. When they make an account, students choose their interested fields and then, they can read sentences from their chosen fields.

V. Conclusions and Future Works

This paper has set practice-based guidelines for instructors or any kind of parties who are interested in ubiquitous seamless learning to implement an environment using SCROLL. Some sample learning scenarios for both formal and informal learning have been suggested in order to help teachers to prepare class and to promote learners' autonomous learning.

We insisted earlier the necessity of setting new guidelines for implementation of seamless mobile learning environment by saying that however sophisticated systems and devices being developed, unless we have a good guidance, they might be useless. We are earnestly hoping that our guideline will meet the demand in the age of rapidly-growing mobile and wireless technology and helps implementation of seamless mobile learning environment in any level of education from K-12 to higher education all over the world.

As our future work, utilizing sensor technology, more customized learning recommendation system is under development so that the system can give the learners more recommendation at more appropriate timing and more appropriate places. We believe that the smartphones will be equipped more sophisticated kinds of sensors in the future and the device will know learners better to catch their learning habits more accurately.

References

[1] N. Cavus, and D. Ibrahim, "MOLT: A Mobile Learning Tool That Makes Learning New Technical English Language Words Enjoyable," *International Journal of Interactive Mobile Technologies*, Vol 2, No 4, pp.38–42, 2008.

[2] A. Kukulska-Hulme, and S. Bull, "Theory-based Support for Mobile Language Learning: Noticing and Recording," *International Journal of Interactive Mobile Technologies*, Vol 3, No 2, pp.12–18, 2009.

[3] I. Ismail, R. M. Idrus, A. A. Ziden, and M. Rosli, "Adoption of Mobile Learning Among Distance Education Students in Universiti Sains Malaysia," *International Journal of Interactive Mobile Technologies*, Vol 4, No 2, pp.24–28, 2010.

[4] G.M. Chinnery, "Emerging technologies. Going to the mall: mobile assisted language learning," *Language Learning & Technology*, Vol. 10, No. 1, pp.9–16, 2006

[5] G. Stockwell, "Vocabulary on the move: investigating an intelligent mobile phone-based vocabulary tutor," *Computer Assisted Language Learning*, Vol. 20, No. 4, pp.365–383, 2007.

[6] P. Thornton, and C. Houser, "Using mobile phones in English education in Japan," *Journal of Computer Assisted Learning*, Vol. 21, No. 3, pp.217–228, 2005.

[7] M. Wang, and W. P. Ng. Jason, "Intelligent Mobile Cloud Education: Smart Anytime-Anywhere Learning for the Next Generation Campus Environment," *Proceedings of the 8th International Conference on Intelligent Environments (IE'12)*, pp. 149–156, 2012.

[8] J. Traxler, "DEFINING MOBILE LEARNING," *Proceedings of the IADIS International Conference Mobile Learning*, pp. 261–266, 2005.

[9] M. Sharples, J. Taylor, and G. Vavoula, "A Theory of Learning for the Mobile Age," In R. Andrews and C. Haythornthwaite (eds.) *The Sage Handbook of eLearning Research*. London: Sage, pp. 221–247, 2007.

[10] B. Hou, M. Li, H. Ogata, and Y. Yano, "Supporting Japanese mimicry and onomatopoeia learning using sensor data," *International Journal of Mobile and Blended Learning*, Vol. 2, No. 1, pp.40–54, 2010.

[11] M. Li, H. Ogata, B. Hou, N. Uosaki, and Y. Yano, "Development of personalized and, context-aware model in learning log system," *Workshop Proceedings of the 19th International Conference on Computers in Education*, Asia-Pacific Society for Computers in Education, Chiang Mai, Thailand, pp.321–328, 2011.

[12] T.-W. Chan, J. Roschelle, S. Hsi, Kinshuk, M. Sharples, T. Brown, C. Patton, J. Cherniavsky, R. Pea, C. Norris, E. Soloway, and N. Balacheff, "One-to-one technology enhanced learning: an opportunity for global research collaboration," *Research and Practice of Technology Enhanced Learning*, Vol. 1, pp. 3–29, 2006.

[13] W. Chen, P. Seow, H.-J. So, Y. Toh, and C.-K. Looi, "Extending Students' learning Spaces: Technology-Supported Seamless Learning," *Proceedings of ICLS2010*, Chicago, Illinois: ISLS, pp. 484-491, 2010.

[14] H. Ogata, M. Li, B. Hou, N. Uosaki, M. M. El-Bishouty and Y. Yano, "SCROLL: Supporting to Share and Reuse Ubiquitous Learning Log in the Context of Language Learning," *International Journal of Research and Practice on Technology Enhanced Learning*, Vol.6, No.2, pp.69–82, 2011.

[15] R. C. Atkinson, and R. M. Shiffrin, "Human memory: A proposed system and its control processes," In K. W. Spence and J. T. Spence (Eds), *The psychology of learning and motivation*, Vol. 2, pp.89–195, London: Academic Press, 1968.

[16] G.M. Vavoula, P. Lefrere, C. O'Malley, M. Sharples, and J. Taylor, "Producing Guidelines for Learning, Teaching and Tutoring in a Mobile Environment," *Proceedings of the 2nd IEEE International Workshop on Wireless and Mobile Technologies in Education (WMTE'04)*, p. 173, 2004.

[17] Nielsen's survey in 2012:

[18] J. Jiang, and F. Chen, "The Chinese School Message System: A New Interactive Mode of Promoting Educational Development by ICT," *Proceedings of the 3rd International Conference on Information and Computing (ICIC)*, 4-6 June, Wuxi, China, pp.173 – 176, 2010.

[19] N. Uosaki, H. Ogata, M. Li, and B. Hou, "Towards seamless vocabulary learning: how we can entwine in-class and outside-of-class learning," *International Journal of Mobile and Learning Organization*, Vol.6, No.2, pp.138–155, 2012.

[20] National Training Laboratories Institute for Applied Behavioral Sciences, "*The Learning Triangle: Retention Rates from Different Ways of Learning*," Bethel, Maine, 2005.

[21] ECAR Study of Undergraduate Students and Information Technology, 2012.

[22] K. A. Noels, "New Orientations in Language Learning Motivation: Towards a Model of Intrinsic, Extrinsic, and Integrative Orientations and Motivation," In Z Dornyei & R. Schmidt (Eds.), *Motivation and Second language learning* pp.43–68. Honolulu: University of Hawaii Press, 2001.

[23] R. M. Ryan, and E. L. Deci, "Self-determination theory and the facilitation of intrinsic motivation, social development, and well-being," *American Psychologist*, Vol.55, No.1, pp.68–78, Jan 2000,.

[24] W.E. Nagy, P.A. Herman, and R.C. Anderson, "Learning words from context," *Reading Research Quarterly*, Vol. 20, pp.233–253, 1985.

[25] S. Krashen, "We acquire vocabulary and spelling by reading: additional evidence for the input hypothesis," *Modern Language Journal*, Vol. 65, pp.43–58, 1989.

[26] Y. Sun, and Q. Dong, "An experiment on supporting children's English vocabulary learning in multimedia context," *Computer Assisted Language Learning*, Vol. 17, pp.131–14, 2004.

[27] C-M. Chen, and C-J. Chung, "Personalized mobile English vocabulary learning system based on item response theory and learning memory cycle," *Computers and Education*, Vol. 51, pp.624–645, 2008

[28] I.S.P. Nation, *Teaching and Learning Vocabulary*, New York: Newbury House, 1990.

[29] J. Dewey, *Experience and Education*, Kappa Delta Pi, 1938.

[30] M. Li, H. Ogata, B. Hou, and N. Uosaki, "Context-aware and Personalization Method in Ubiquitous Learning Log System," *Journal of Educational Technology & Society* (in press).

Appendixes

Appendix 1

Daily Vocabulary Learning No. ( ) Name( )

Date	min./total	Date	min./total	Date	min./total	Date	min./total
11/15	min.	11/22	min./ min.	11/29	min./ min.	12/6	min./ min.
11/16	min./ min.	11/23	/	11/30	/	12/7	/
11/17	/	11/24	/	12/1	/	12/8	/
11/18	/	11/25	/	12/2	/	12/9	/
11/19	/	11/26	/	12/3	/	12/10	/
11/20	/	11/27	/	12/4	/	12/11	/
11/21	/	11/28	/	12/5	/	12/12	/
Sign		Sign		Sign		Sign	

Appendix 2: Sample Instructions

1) Go to the gymnasium and describe what people there are doing.
2) Go to the tennis courts and count the number of people who are playing tennis and describe them
3) Go to the roof and describe what you see.
4) Go to the café "XXX", count the number of people who are sitting at tables and report today's special lunch.
5) Find some international students and ask them where they are from and their majors and so on
6) Find at least three things which are orange, and name them.

Appendix 3: Useful expressions for interviewing

1) Where are you from?
2) What is your hometown famous for?
3) What are your hobbies?
4) Do you belong to any club or circle?
5) What kind of sports do you like?
6) What is your favorite food?
7) What kind of music do you like?
8) Do you have your favorite artist or bands?
9) Do you have any plan for next weekend?
10) What did you do during the holidays?

Authors

N. Uosaki is with the University of Tokushima, Faculty of Engineering, 770-0813 Japan (e-mail: n. uosaki@gmail.com).

H. Ogata is with the University of Tokushima, Faculty of Engineering, 770-0813 Japan (e-mail: ogata@is.tokushima-u.ac.jp)

Mengmeng Li is with the University of Tokushima, Faculty of Engineering, 770-0813 Japan (e-mail: limengmeng@iss.tokushima-u.ac.jp)

Bin Hou is with the University of Tokushima, Faculty of Engineering, 770-0813 Japan (e-mail: houbin@iss.tokushima-u.ac.jp)

Kousuke Mouri is with the University of Tokushima, Faculty of Engineering, 770-0813 Japan (e-mail: mohrikohsuke@iss.tokushima-u.ac.jp)

16

Client-Server and Peer-to-Peer Ad-hoc Network for a Flexible Learning Environment

Ferial Khaddage[1], Christoph Lattemann[2]
[1] Deakin University, Melbourne, Australia
[2] Jacobs University, Bremen, Germany

Abstract—**Peer-to-Peer (P2P) networking in a mobile learning environment has become a popular topic of research. One of the new emerging research ideas is on the ability to combine P2P network with server-based network to form a strong efficient portable and compatible network infrastructure. This paper describes a unique mobile network architecture, which reflects the on-campus students' need for a mobile learning environment. This can be achieved by combining two different networks, client-server and peer-to-peer ad-hoc to form a sold and secure network. This is accomplished by employing one peer within the ad-hoc network to act as an agent-peer to facilitate communication and information sharing between the two networks. It can be implemented without any major changes to the current network technologies, and can combine any wireless protocols such as GPRS, Wi-Fi, Bluetooth, and 3G.**

Index Terms—**Peer-to-Peer, client-server, network, mobile learning.**

I. Introduction

Mobile Learning is on the rise with the upcoming of new quasi cost-free wireless technologies (flat-rates); The Importance of mobile learning is increasing not only for off-campus learning environment but also for on-campus as well. Mobile learning has shown to be effective in sharing and transferring learning content among students and is often effective when used to build mobile student groups to form an on-campus mobile learning environment [1].

Mobile Peer-to-Peer network is already offering crucial and unique opportunities for innovations especially at universities, and can advance the delivery mechanism of learning content.This is not only true for off-campus students, but also for on-campus students who require wireless access to course content. Therefore this network can provides easy, reliable and free access to the learning content while on-campus.

With the emerging of high bandwidth wireless channels such as Third Generation Network (3G) widely installed the wireless Local Area Network (Hotspots). Also with the advancements of the new mobile devices and the provision of new business models by the telecommunication providers, which is based on flat-rate, the broad usage of mobile phones for retrieving and accessing learning content at universities is becoming more feasible and is on the rise.

Wireless technologies and mobile networks are constantly changing and they are being developed at a rapid speed, hence, new possibilities are emerging that can be used along the 3G telecommunication network.

The network infrastructure and architecture are normally achieved by designing high-availability networks which take a unique approach to the subject by covering the ideas underlying networks, the architecture of the network elements, and the implementation of these elements are considered crucial aspects. This paper describes mobile network architecture and applied technology that is capable of providing wireless access to learning content within the university campus. Also wireless protocols suitable for mobile devices such as Wi-Fi, Bluetooth and 3G are being explained. Concluding remarks, advantages and limitations will be presented at the end of this paper.

II. Wireless Protocoll for Mobile Phones

The rapid development of mobile devices especially mobile phones for communication purposes, have triggered its application in educational institutions. Also the software trend towards ad hoc and peer-to-peer networks and the ability to target mobile phone devices for networked applications for the delivery of learning content, offers a great opportunity for higher education institutions and in particular for universities.

In the following sections we discuss mobile network architecture which comprises a State-of-the-Art communication technologies and wireless networks for mobile learning purposes. Some of them are already being developed at universities and in particular Deakin University. The main focus lies on the local wireless networks (WLAN), as this is the network which mostly used by student to access learning material and for the delivery of e-learning content. Wireless Local Area Network (WLAN) is normally free of charge especially when accessed by students while on-campus, and it is safe and secure, because the university offers a secured network, and has implemented all security measures to prevent unauthorised login attempts.

In the following sections we discuss the wireless network protocols:

A. Third Generation Network (3G)

Number Smart phones use a mobile network technology to deliver and receive different types of data. Information technologist, system and network developers classify this technology into generations. With the advancement and the rapid development of the mobile phones an advanced protocols were created and they are the 2G and the 3G networks. 3G system must allow simultaneous use of speech and data services, and provide peak data rates of at least 200 Kbit/s, this is according to the IMT-2000 specification. 3G networks consist of cells; each cell contains a base station using a number of radio frequencies or channels. The mobile phone is always communicating with the network via those channels. These channels consist of two types, one to handle verbal data, audio and voice communication, and the other to manage signalling and messaging, therefore users can receive a massage when they are unable to make a call or when the call drops [2] [3].

GPRS (General Packet Radio Service) is a wireless communication service which can be classified as a 2G Technology. This protocol is used mostly for smart phones communication. It is a packet-switched, "always on" connection that remains active as long as the phone is within range of the service. It allows smart phones to do things like run applications remotely over a network, interface with the Internet, participate in unsynchronized communication sessions, and it does act as a mobile modem for a computer and transmit and receive e-mails. GPRS can send and receive data at a rate of 114 kilobytes per second [4]. It provides moderate speed data transfer, by using unused Time Division Multiple Access (TDMA) channels in the GSM network. GPRS offers up to 171.2 Kbps, depending on the network availability, channel coding scheme and terminal capability [4]. However, in practice, the bit rate is 40-80 Kbit/s. But this bandwidth will only provide poor quality to user for real-time streaming and communication.

GPRS is penetrating the mobile network at a rapid speed. Almost every mobile in the market support GPRS and GPRS services like Multimedia Messaging Service (MMS) and Wireless Application Protocol (WAP) are widely used. To use it, it is necessary to re-implement the Wireless Service Subsystem. Point-to-Point (PTP) service will be used to discover an active device and peer-to-peer transfer. Also, the higher the data rate, the lower the error correction capability. Generally, the connection speed drops logarithmically with distance from the base station. This is not an issue in heavily populated areas with high cell density like a university environment, but may become an issue if access attempt are made remotely [5].

B. *The Wi-Fi Technology*

A wireless network uses radio waves that communicate in two ways: *a centralized network* with access point, called infrastructure network and, *a decentralized network* without access point, called *ad-hoc network.* Despite that Wi-Fi power consumption is considered high, still this technology is the most widely used especially at universities. This is because Wi-Fi allows connectivity in peer-to-peer mode, which enables devices to connect directly with each other. Hence, in the context of education at universities, learning content can be wired with higher bandwidth.

The area around a Wi-Fi access point is considered as a hotspot. Universities as well as several companies and institutions have already created their own hotspot, for example Deakin University, McDonald restaurants, Starbucks, to enable users to access the Internet via a partly free of charge Wi-Fi network.

Most of the mobile phones today are embedded with Wi-Fi; therefore student can use their mobile phone to access the Wi-Fi network [4]. This type of connectivity is useful in mobile learning environment.

The Wi-Fi radio transmission happens at frequencies of 2.4 GHz or 5GHz, which is considered very high and allows the signal to carry more data [6]. Wi-Fi uses a license-exempt portion of the radio spectrum so that everyone can set up his own Wi-Fi network without applying for a license or paying for broadcasting rights. Also, Wi-Fi can provide the high bandwidth for good quality audio streaming. There are currently three main specifications 802.11, 802.11a, 802.11b and 802.11g in the family of IEEE 802.113.2 [7].

All three specifications use the Ethernet protocol and Carrier Sense Multiple Access with Collision Avoidance (CSMA/CA) for path sharing.

C. *Bluetooth*

Bluetooth is a specification for a mobile wireless Personal Area Networks (PANs), also known under the IEEE 802.15.1 standard [8]. Bluetooth is a rapidly growing technology [9], focusing on building Personal Area Networks (PANs) and connecting mobile devices cheaply and wirelessly. Bluetooth is available in multiple classes with varying power requirements, and varying range. These specifications make the ideal connection within an ad hoc peer-to-peer networking environment. Fig.1 is an illustration of the location of Bluetooth within the wireless network. It is also a brief illustration of the three main categories in wireless protocols, which are 2G and 3G wireless networks, WLAN and LAN and the location of the Bluetooth technology lies.

Bluetooth can offer many functionalities, it enables different mobile devices to connect and exchange data and information. Bluetooth is designed for low power consumption within a short range of 1 – 100 meters with a data transmission rate of up to 24 Mbit/s. It is based on low-cost transceiver microchip which does exist in almost each mobile device [14].

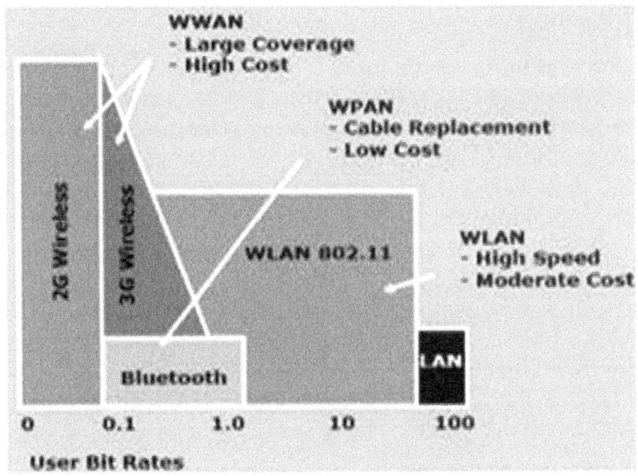

Figure 1. Wireless Protocols Feature [9]

Therefore Bluetooth allows mobile phones to communicate with one another as long as they are within range. Since Bluetooth is a radio communication system, mobile phones do not have to be in line of sight of each other, unlike the requirement for infrared information transfers.

Therefore Bluetooth, if used, will enable the system to be portable, flexible, and cost effective. Cost is an important aspect to consider, because communication via a mobile phone can still be expensive, and charges for downloads by the network carriers are often quite expensive. However, on the other hand communication and sharing resources via Bluetooth doesn't cost the student anything, as it is normally free of charge [9].

Also there are three different connection types available for Bluetooth see table1 below

TABLE I.
TYPES OF BLUETOOTH CONNECTIONS

Connection identifier	Expanded name
Btgoep	General object exchange profile
btl2cap	Logical link control and adaption layer protocol
Btspp	Serial port profile

First, *btgoep* transmits binary data, so it is not appropriate for string-based communication. And *btl2cap* requires more work from developers, setting and working around maximum message sizes. However, *btspp* hides these complexities from the developer, so this is the logical connection choice for a project aimed at simplifying our development process [9].

Therefore this makes Bluetooth transport layer connectivity to client applications without overburdening the client with code and asynchronous messaging considerations. This will let us achieve a straight forward factoring of function and leave code simple to develop hence low to maintain [9].

III. THE PROPOSED NETWORK ARCHITECTURE

As previously discussed there are different types of mobile networks, each one of them comes with different characteristics, properties, advantages and disadvantages for their users.

Therefore in order to identify which one of them is suitable and most likely to be adapted and can be used in different contexts, we had to briefly review their features before making our selection.

For the implementation of a network for mobile learning purposes at universities, we have to consider certain requirements from students such as price sensitivity, service and technology availability, ease of use and usefulness. A mobile learning environment should allow peer to peer communication as well as up-load and down-load to and from a central server. In this respect Bluetooth has the advantage of being widely used in most mobile phones [14], and it is available free of charge to all users, thus making transferring multimedia files like learning images and material cheap, easy [15] and convenient [10].

Furthermore, mobile phones with Bluetooth are normally problems free, establishing connection depends also on the operating system, supported specification and programming interface [15]. Bluetooth is also considered one of the possible wireless protocols to allow sharing and transferring of information in an ad-hoc learning environment. GPRS and Wi-Fi could be used as well. But GPRS requires students to subscribe for the service and pay for the download content, and Wi-Fi on the other hand requires higher power consumption, which also comes at a high cost for student. All this makes Bluetooth connectivity the most convenient and most suitable for mobile learning purposes in a university setting.

Via the wireless network, mobile phones can detect and locate the nearby devices with other students on campus, hence making sharing and uploading images and learning materials from the server to student's mobile phone fast and efficient, thus making learning more interactive, portable, collaborative and flexible. This will be explained further in the next section.

A. *Overview of the Network Infrastructure*

Our proposed architecture is implemented as a peer-to-peer mobile network, but also creates an "ad-hoc" mobile social network. Also these kinds of networks can be integrated together with the already existing client-server systems. They are called Ad-hoc Messaging Network (AMNET) [12][13].This network is different from the normal peer-to-peer (p2p) connection as it does offer a connection to the server. This connection is achieved via allocating one peer to act as an agent, and the communication between the P2P ad-hoc network and the client-server network is normally handled by the peer agent that acts as a facilitator of the communication network and offer other peers a service to interact with the connected server. The agent peer behaves like a normal peer, but it recognizes its specific role during the runtime. Because of this self managing there is no need to configure the agent peer, as it acts like a proxy for the main server, thus making the server an extra service in the P2P network. The illustration of the combined P2P and client-server network is displayed in fig. 2. and fig. 3

This would enable content exchanged between two different network types [11]

Therefore we do not have to do many changes to the already existing network at the university. And the transformation takes place easily and effectively in minimum time. In fact, the student will not notice any difference between the deliveries.

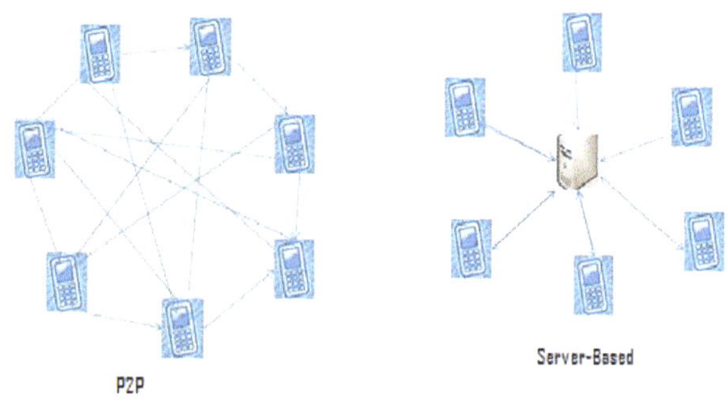

Figure 2. Peer-to-Peer Vs Server-based Network

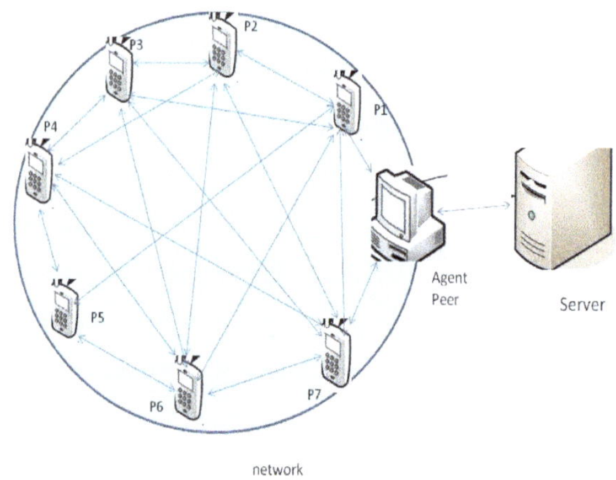

Figure 3. Agent peer to facilitate P2P and Client Server Network

For our system we pursue an approach which follows few good techniques within AMNet. Firstly by keeping routing as simple as possible, and secondly by considering the content as the point of interest and deny the ambition to bring all internet features to mobile networks, this is also achieved by considering both networks, client-server and ad-hoc peer-to-peer. Our approach is to separate the two different networks by a peer client, and allocate it as an agent peer to facilitate data transmission between the two networks, hence providing a best-fit solution mechanism for certain applications' requirements.

Therefore the proposed network follows the AMNet (Ad-hoc Messaging Network) approach [12], but the implementation of the peer agent to integrate to our existing client-server network is unique, effective and efficient, especially when all method of communications between the two networks must pass through the agent peer first, so any security threat will be handled by the agent peer before it actually reaches the vulnerable wireless network.

By deploying this method we are forming a strong secure network infrastructure that is capable of combing both networks effectively and efficiently. As a matter of fact this technique would fit seamlessly within the already existing network client-server structure at the university and can wrap the two networks together in an efficient manner, thus making communication and transmission of material among clients and server fast, reliable effective and secure.

B. Advantages of the Proposed Network

In general this type of network infrastructure can provide benefit to the university as well as to the students themselves, especially because the connection is achieved via an Ad-hoc Bluetooth network connectivity where any mobile phone embedded with Bluetooth can communicate and share resources free of charge. Fig. 3 shows that this type of network can help to support speedy access and efficient retrieval of the learning content, between peer-to-peer and server and Peers via the agent peer.

Therefore mobile network of devices using Bluetooth technology protocols is to allow one host device to interconnect with many active client devices (because a three-bit MAC address is used). Up to 255 further client devices can be deactivated, or parked, which the host device can bring into active status at any time. Also material can be transferred between the host and one client, but the host switches rapidly from server to agent peer and from client to client. Now to set up a connection, a client would perform an inquiry to find any available device to connect, and any host can be configured to respond to such inquiries. The host requires its owner to accept but the connection itself can be held until it goes out of range. Full illustration of this network and its protocol are beyond the scope of this paper and will be available in future publications.

To sum up our proposed architecture provides the following advantages for mobile learning in a university setting:

- By connecting both networks, (Ad-hoc peer-to-peer and client-server), the limitation of each particular network will be overridden by the alternative network.
- Each client on the network is free to share its own resources with other client and at the same time able to connect to the server via the peer agent.
- Since p2p network is not centralized, thus making administration difficult, therefore all administrations are handled on the client-server side.
- Cost is reduced as the client-server network already exists, and the integration of p2p network can be established at a low cost.
- It offers more security, since security related issues are handled by the client-server, and data must pass the peer client before it reaches each client on its own.
- As a method of proper mobile phone integration within the current learning environment.
- This architecture and concept guarantees portability, flexibility and availability of the learning content, when this content is saved on the mobile phone, and could be then transferred, shared and copied.

IV. CONCLUSION

Mobile phones and their networks are shaping our future; they are fusion technology that is capable of providing what once considered science fiction.

The proposed network architecture is not a replacement for any other delivery mechanisms, but rather an extension or an addition to the learning environment. The new mobile learning environment is to strengthen the other learning methods at university, by providing a mobile access to learning content and at the same time gives students the freedom of choice on selecting an alternative mobile method to access learning content.

This paper illustrated a valuable communication and network for a vivid ad hoc network that can be integrated with a server-based network. It also depicted an approach towards a technical solution that realizes all requirements for combining client-server and peer-to-peer ad-hoc network.

However, the limitation of this approach is that a lot of data have to be stored on the mobile device, and this sometimes can form a barrier due to limited storage space, limited capacity and less functionality in earlier versions of some mobile devices. But with the advancement of mobile devices and their rapid innovative development, for example the recent release of the iPad, iPhone 4, the blackberry mobile phone and many others, are already overcoming this limitation and will continue to do so in the in the near future.

REFERENCES

[1] Khaddage F., Lanham E. and Zhou W. (2009)"*A Mobile Learning Model for Universities, -Re-blending the Current Learning* Environment"*International Journal of Interactive Mobile Technologies.* ISSN: 1865-7923

[2] Rich Mogull, "*Peering Inside a Mobile Phone Network*" Networking and Communication website.

[3] Voda Phone Official Website.

[4] Cell phones with Wi-Fi coming before end of this year.

[5] Smith K. and Cap O. *"Impact on Social Change: Benefits and Barriers to School Culture and the Integration of M-Technology".* International Journal of Interactive Mobile Technologies (iJIM) – Volume 2, Issue 1, January 2008.

[6] Bettstetter C. ,Vogel H. J. ,Eberspacher J. ,(1999) VGSM Phase 2+ General Packet Radio Service GPRS: Architecture, Protocols, and Air Interface, Technische Universität München (TUM)

[7] EDGE, World record data speed on T-Mobile Hungary's GSM EDGE network supplied by Ericsson,(2004).

[8] IEEE 802.15, Standard Association. Wireless Personal Area Networks.

[9] Liscano R. (2006) *Introduction to Bluetooth Networking* Carleton University.

[10] Bygrave C. (2009) IBM Java Technology Center (JTC) (United Kingdom). *Developing IBM's Java 2 Platform, Standard Edition (J2SE) implementations.*

[11] Gradecki, J. D., editor (2002). *Mastering JXTA: Building Java Peer-to-Peer Applications.* Wiley Publishings, Indianapolis, USA.

[12] Khaddage F. , Lattemann C. (2009) *"Towards an Ad-hoc Mobile Social Learning Network Using Mobile Phones" The 12th International Conference on Interactive Computer Aided Learning,* ICL 2009 proceedings, September 2009, Villach, Austria ISBN: 978-3-89958-481-3

[13] Fuchß C. Stieglitz S. and Hillmann O. (2006) "*Ad-hoc Messaging Network in a Mobile Environment,*" presented at International Conference of Internet Technology and Secured Transactions, London, 2006.

[14] Kaasinen, E (2005) *"User acceptance of mobile services – value, ease of use, trust and ease of adoption"* VTT PUBLICATIONS Technical Research Centre of Finland 566

[15] Tourrilhes, J. (2003) On-Demand Bluetooth: Experience integrating Bluetooth in Connection Diversity, Research Paper Hewlett Packard Laboratories

AUTHORS

Dr Ferial, Khaddage is PhD. Lecturer of Computing and Information Technology at Deakin University, Faculty of Science and Technology, School of Information Technology, 221 Burwood Highway Burwood Victoria 3125 Australia (faye.khaddage@deakin.edu.au)

Prof. Dr Christoph, Lattemann is Professor of Business Administration and Information Management at Jacobs University, School of Humanities and Social Science business Administration Campus Ring 1 28759 Bremen Germany, (c.lattemann@jacobs-university.de).

17

Tablets (iPad) for M-Learning in the Context of Social Constructivism to Institute an Effective Learning Environment

P. Thinley, J. Reye and S. Geva
Queensland University of Technology, Brisbane, Australia

***Abstract*—With the proliferation of mobile devices, educational institutions have experimented with various mobile devices to implement mobile learning (M-Learning). Mobile devices have been used to facilitate, support, and enhance and extend the reach of teaching and learning. Although there are very few empirically evaluated studies on M-Learning projects, these studies reported that mobile devices brought a transformation to the educational process. To be able to view M-Learning as a rich, collaborative and conversational experience, whether in the classroom or outside we need good mobile applications. Studies have revealed that effective learning happens when teachers and learners are actively participating in the knowledge building process. Therefore, there is a need for applications that create effective learning environments which are learner-centred, knowledge-centred, assessment-centred and community-centred.**

***Index Terms*—M-Learning, Tablets, Social Constructive Learning.**

I. Introduction

With the proliferation of mobile devices, educational institutions have experimented with various mobile devices to implement mobile learning (M-Learning). Mobile devices have been used to facilitate, support, and enhance and extend the reach of teaching and learning. But, use of mobile devices has not moved further than when it was first adopted for M-Learning. It has been used to provide and access mere information, rather than knowledge.

Unlike the other learning activities, mobile learning started with the assumption that learners are always on the move from physical location to location, learning topics. Studies have reported that students learn effectively when they actively participate in learning. And in a mobile learning environment, learners are active as they initiate their learning activities and control their own educational progress and outcomes. Therefore, mobile learning has been predicted to bring great transformations to the traditional mode of learning and outcomes.

Although there are very few empirically evaluated studies on M-Learning projects, these studies reported that mobile devices brought a transformation to the educational process. As reported in EDUCAUSE 2011, M-Learning is still in need of distinct frameworks or models for using and developing mobile applications for learning. Without a proper framework, it would be challenging to evaluate outcomes from any M-Learning projects.

As M-Learning has been conceptualised with the assumption that learners are always on the move, M-Learning has been viewed as an isolated activity. To be able to view M-Learning as a rich, collaborative and conversational experience, whether in the classroom or outside we need good mobile applications. These mobile applications can act as a platform on which an effective learning environment can be created.

Tablets with more advanced features (iPad, Samsung Galaxy, etc.) have been widely embraced by universities. They have become popular educational devices because of the availability of significant number of educational applications. There are apps available ranging from study aids to collaborative and interactive learning apps. Most of these applications support traditional learning activities instead of enhancing them.

Applications that are available have demonstrated behaviourist, constructivist and collaborative perspectives of learning theory. Studies have revealed that effective learning happens when teachers and learners are actively participating in the knowledge building process. Therefore, there is a need for applications that create effective learning environments which are learner-centred, knowledge-centred, assessment-centred and community-centred.

II. M-Learning

M-Learning is defined as a form of learning accomplished with the use of mobile technology. Mobile devices that made their way into the hands of students and teachers include laptops,tablet-PCs,smart-phones and personal digital assistants (PDAs). Recently tablets havebeen added tothegrowing list of mobile devices used for teaching and learning purposes. Educational institutions have started adopting tablets because they are portable and the WiFi connectivity makes access to learning resources, communication, and collaboration, more convenient. Tablets are observed to be more suitable M-Learning devices because of their bigger display screens and advanced technological features.

Teaching and learning environments have used mobile devices to demonstrate technological feasibility and pedagogical opportunity [17]. The US National Research Council produced a synthesis of research into educational effectiveness and they concluded that students learn effectively when the learning environment is learner-centred, knowledge-centred, assessment-centred and community-centred [14]. These research findings match the social-constructivist approach of learning where students play an active role in a learning context and teachers and students collaborate to facilitate knowledge construction.

III. Social Constructivism and M-Learning

Social constructivism in education is when learners are encouraged to interact and participate in learning to construct, transfer and share knowledge effectively. Studies [5] and [18], have reported numerous advantages of increasing classroom engagement and participation. It is stated that discussions helped improve students' ability to test their ideas, collaborative skills, improve problem solving skills and also build deeper understanding of what they are learning. By adopting mobile technologies for teaching and learning, it has been expected to exploit the social-constructivist pedagogical approaches to learning.

Computer-supported collaborative learning (CSCL) is a pedagogical approach where learning is characterized by the sharing and construction of knowledge among learners using technology as primary means of communication and it is grounded in social constructivism.

To support teaching and learning process effectively we need to integrate the technological capabilities available in a way that enhances teaching and learning process instead of just replacing the traditional form teaching and learning. Laurillard's [8] "Conversational Framework" has been developed based on several learning theories (instructional, constructivism, social constructivism), in which communication in any direction between teachers and learners is seen as central to learning process. Numerous studies have reported that discussion, interaction and reflection during the learning process provide positive learning outcome [5], [18]. Therefore, "Conversational Framework" can be used to test the value of M-Learning technologies and also to ensure that these technologies enhance teaching and learning.

A. Expectations of Mobile Devices in Education

The use of mobile devices in educational institutions had great expectations from both teachers and students. It was expected to change the didactic traditional classroom learning to a more active and engaging learning environment, as reported in EDUCAUSE report 2011. According to the survey conducted by EDUCAUSE, students expected to use mobile devices in the following areas:

i. General Communication
ii. Instruction
iii. Administration
iv. Research

Institutions have reported their readiness and priority in deploying the following mobile services:

i. Web presence
ii. Learning/course management service
iii. Library catalogue and library services
iv. Student recruitment and admissions
v. Administrative services for student information

Most universities are ready with infrastructure as well as with human resources to adopt M-Learning and provide mobile learning applications. EDUCAUSE report 2011 also reported that the mobile devices will enable great opportunities in delivering course content and related teaching/learning activities. Students' responses to the EDUCAUSE survey also highlighted that they would value mobile applications that help in coordination of students and learning resources, and also applications to enhance pedagogical activities that integrate into the learning process.

But in reality, Litchfield, Dyson, Lawrence, and Zmijewska [9], reported that the majority of M-Learning projects have focused on improving interactivity in the classroom or on ubiquitous learning, i.e. increasing students' access to learning materials anytime anywhere. Therefore, mobile devices have been used to replicate the traditional teaching and learning techniques when it can be used for more than just information delivery and retrieval in education. Studies [5] and [18] have reported that effective learning happens when there is constant communication among the peers as well as between learner and teacher. These communications can take place in the form of asking question, receiving response, peer-discussion and feedback. Therefore, to make effective use of mobile technology for teaching and learning, its use can be evaluated using the "Conversational Framework".

B. Trends of Mobile Use in Higher Education

Zeng and Luyegu [22] reported that the combination of wireless technology and mobile computing has resulted in escalated transformations in the educational world. However, in 2010 some of the researchers were still predicting that mobile and wireless communication technology could play an important role in the revolution of education [10]. This clearly indicates that the use of mobile and wireless communication technology for learning and teaching has not been evaluated and its impact in higher education learning has not been formally reported either. It might not have been reported firstly, because frameworks for using and development of mobile applications are still not clearly defined. Secondly,mobile computing in higher education still remains at a nascent level for many institutions, as reported in EDUCAUSE's 2011 report.

Mobile devices are largely being used for delivering information and barely being used to actively engage students in the learning process. Uses of mobile device for learning as reported by various studies are described below. It is not clear whether these devices have been used in pedagogical appropriate ways.

Mobile phones with their restricted screen size have limited their use to sending SMS (Short Messaging Service) messages, with prompts for course requirements, assignment due dates, polling answers for quizzes, and sending URL links to additional learning resources.

Tablet PCs have been used in the classrooms as a digital whiteboard to present lecture slides. Additionally, with software products such as Microsoft OneNote, PowerPoint, and Journal, instructors are able to write on the digital whiteboard as they would do on a traditional blackboard [1], [19]. Such devices have also been used to take notes, read electronic books, and for in-class assessments, with the use of systems such as Lecturer's Assistant, Class Presenter System, Classroom Learning Partner, etc.

There are diverse reports regarding the impact of these mobile devices on student learning. For example, Cavus and Uzunboylu [3] reported that mobile devices promoted critical thinking, while Chao and Chen's [4] study revealed that there was no significant difference in knowledge retention when student used mobile devices. Nevertheless, mobile devices in general have made communication and learning materials accessible anytime and anywhere. Even more so, with technologically advanced mobile devices like tablets that are now available.

IV. Tablets for Education

Tablets such as iPads have been designed and developed primarily as an interactive content consumption device which allows multi-touch experience. It provides a platform for audio visual media including books, periodicals, movies, music, games, and web content. Though the iPad was never specifically developed as an educational tool, educational institutions have widely started adopting iPads in classrooms, as a teaching and learning tool to take advantage of its advanced attributes. Its' Wi-Fi and 3G connectivity allows the user to experience just-in-time learning opportunities, connection and convergence to other devices, networks and technologies. Thereby, studentsand teachers are not restricted to traditional limitations of time and space related to learning.

Other mobile devices have been used for learning but have limitations such as small screen sizes, lack of data inputcapability, short battery life, limited processing power, and low storage capacity. Tablets have been used as mobile learning devices and they can be promising M-Learning devices, if they are effectively integrated into teaching and learning environments, to support social-constructivist pedagogies.

A. iPad for Teaching and Learning

Many universities worldwide have discussed and reported on the general uses of the iPad but none of the universities have determined a significant pedagogical use of the device in higher education. Student and teachers have used iPad to take notes, listen to lecture recording and browse web for additional learning materials but they have not been used to improve engagement and interaction in a formal learning environment. Additionally, the universities have not yet formally measured the impact of the iPad for general use in teaching and learning contexts.

Universities have adopted the iPad as a collaborative tool, a standardized mobile device to integrate into curriculum. Stanford University of Medical Science, Seton Hill University, Universityof Adelaide and several other universities around the world have distributed iPads to incoming class of students as early on as 2010. The core goals of these projects were to improvethe student learning experience and also to replace traditional textbooks and other teaching materials with online study tools. On the contrary Macquarie University has been cautious in adopting the iPad as a standard tablet device. The university has kept its option open to differenttablet PCs expected to be released in 2011. These universities have adopted the iPad but they have not yet formally measured and reported the effectiveness of the iPad as a tool to enhance teaching and learning

B. Impact of iPad in Education

All the technologies that have been adopted for teaching and learning purposes have merely digitized what was available before rather than offering more unique, dynamic and engaging pedagogies (Rudd, 2011). Therefore, Tablets such as the iPad the capacity to be adopted as a teaching and learning tool that supports active learning pedagogical approaches.

Many universities adopted iPads in their curriculum with the expectation of improving the teaching and learning experience. It can be observed from iPad projects carried out in the universities, that the iPad has been effectively used as a content delivery device, complementing other forms of conventional content delivery methods. The University of Minnesota, Stanford University, the University of California-Irvine and the University of Central Florida are some of the medical schools that have handed iPads to students for electronic learning. These iPads were equipped with productivity, content delivery, and content consumption apps, such as HumanAnatomy, Epocrates, MedCalc, PDF Reader, and iAnnotate.

There are various reports released by the universities on iPad pilot projects or iPad study programs conducted in those universities. These reports focus primarily on the potential of the iPad as an eReader. Additionally, some universities such as the University of San Francisco have concluded that the iPad has limitations on its use as an effective instructional tool [2].

Furthermore, the medical school of Stanford University has conducted an iPad project and reported that the use of the iPad did not significantly contribute to increased student learning outcomes. But these universities have left it to the students to explore the use of the iPad for learning, instead of the faculty integrating the device in the design of their teaching and learning activities.

According to Perkins [17], learners do not just take in and store the information given; they interpret, elaborate, and test these elaborations, which lead to social constructive learning. Based on the number of universities eagerly adopting the iPad as an eReader, uses of the iPad can be further explored to support this social constructive nature of learning, by integrating it into the teaching andlearning process in the classroom.

V. Taxonomy of Educational Apps

Students and teachers have used numerous educational apps that are available on the Apple appstore (www.**app**le.com/au/**ipad**/from-the-**app**-store/) and some of the universities have even developed their own apps. The following taxonomy of educational applications has been developed to classify apps used for formal learning. In this context, formal learning is defined as learning where a university/department sets the goals and objectives of learning [6].

These categories developed by the authors have been chosen after analyzing applications that are (or have been) used by the students and teachers in higher education. These educational apps ranged from interactive study aids to productivity apps. Each category is further described below.

i. Content Consumption and Creation Applications
ii. Content Delivery Applications
iii. Collaborative and Interactive Learning Applications
iv. Course Management Applications

v. Teaching and Learning Enhancement Applications

A. Content Consumption and Creation Applications

Students and teachers have used these applications to read, take notes, concept map their ideas, create presentations, draw diagrams, and make use of spreadsheets to capture and analyse tabular data sets. These applications have helped learners and teachers to organize their learning materials but they have not significantly contributed to improvement of student learning outcomes.

B. Content Delivery Applications

This category of applications is used to enhance the delivery of lectures and distribute podcasts of the lectures and discussions. These applications are used to make the lectures more dynamic, by enabling writing on the slides as would occur on a traditional blackboard. This category of application made lecture presentation dynamic and exciting, therefore its use have been restricted to lecture delivery and presentation.

C. Collaborative and Interactive Learning Applications

Lecturers use this category of applications to encourage students to engage, participate in the lecture that isbeing delivered, and provide them with platforms to collaborate and learn. This category of applications includes applications that encourage any kind of communication among teachers and students, including classroom polling applications, as they help in initiating classroom discussion.

D. Course Management Applications

Applications that universities use for administration purposes such as unit enrolment, class allocations, unit tracking, grade management, and to manage lecture and course content. Some of the universities have developed custom applications for course management, while most universities use Blackboard Mobile™ application. Course Management Applications have made managing and organizing courses more efficient.

E. Teaching and Learning Enhancement Applications

These applications are used by teachers and students to support teaching and learning activities. These applications include file management applications that help people organise their files as the iPad does not have a file management system.

VI. Tablet Applications and Learning Theories

In Table I, the above categories are compared with the classification of mobile learning activities developed by Naismith et al. (2004).These authors have categorized activities around existing relevant learning theories.

Some iPad projects [3] and [9] have reported that the use of the iPad did alter academic workflows, such as making them paperless, making web resources more easily accessible to gather information, making it easier to handle notes, making marking more portable, and being able to listen to podcasts anytime [9], [13] and [14]. Though the iPad has received positive response as an educational technology tool, it has not yet accomplished the expectations of students that were highlighted in EDUCAUSE 2011 report.

TABLE I.
APPLICATION TAXONOMY WITH RELEVANT LEARNING THEORIES

Mobile Application Category	Example Mobile Applications	Learning Theory
Content Consumption and Creation	Consumption: inkling, GoodReader, iAnnotate Creation: Quick Graph, EverNote, Popplet , AutoCAD WS	Some of these apps have quizzes so the partially support Behaviorist learning theory
Content Delivery	KeyNote, SpeakerNote Lite, Podcasts, Syncpad	Constructivist learning theory
Collaborative and Interactive Learning	Edmodo, WordPress, iClicker, iResponseConvore, tweet and Post-it PopNotes	Behaviorist learning theory Collaborative Learning
Course Management	Blackboard Mobile™ and respective university applications	Not Applicable
Teaching and Learning Enhancement	Splashtop Remote Desktop, iDownload Pro, Air Sharing, DropBox	Not Applicable

As mentioned above, that report showed that students would value mobile applications that help in coordination of learners and learning resources, and applications to enhance pedagogical activities that integrate into the learning process. The educational applications that are available demonstrate that iPads have just been used to replace traditional modes of teaching and learning. They have indeed made learning and teaching efficient, but there are no formal reports on being able to create effective learning environments.

Therefore, universities need to develop applications that make the learning environment learner-centered, knowledge-centred, assessment-centred and community-centred. This learning environment would then match the social-constructivist approach to learning, where students play an active role in their learning, and teachers and students collaborate to facilitate knowledge construction.

VII. Conclusion

Educational institutions have experimented with almost every available mobile device, to enhance and to make learning more effective. The tablet is the latest technology that universities have started adopting. The availability of numerous educational applications for the tablet gives it an edge over all the other mobile devices.

Educational applications that are available demonstrate behaviorist, constructivist and collaborative perspectives of the learning theory. From the taxonomy presented above, it can be seen that most of these applications are used to support teaching and learning activities.

From the available educational applications, none of the applications demonstrate the social constructivist perspective of learning theory. There are few applications that provide interactive and collaborative pedagogy. Studies report that students learn when they are active participants in the learning process. Therefore, to make the most of the tablet, there is a need for educational applications that can capitalize on the benefits of effective learning pedagogies.

REFERENCES

[1] Anderson, R., Anderson, R., Davis, K. M., Linnell, N., Prince, C., and Razmov, V. (2007). "Supporting active learning and example based instruction with classroom technology." In Proceedings of the *38th ACM Technical Symposium on Computer Science Education*, 69–73, Covington, Kentucky, USA.

[2] Bansavich, J. C. (2011). iPad *study at usf* [Online]. Available: http://ipad.wiki.usfca.edu/file/view/iPad+Study+at+USF+Report.pdf.

[3] Belanger, Y. (2011). "The *Duke Digital Initiative – 2012 Update.*

[4] Cavus, N., & Uzunboylu, H. (2009).Improving critical thinking skills in mobile learning. Procedia Social and Behavioral Sciences, 1, 434438.

[5] Chao, P.Y., & Chen, G.D. (2009).Augmenting paper based learning with mobile phones. Interacting with Computers, 21, 173185.

[6] Cofer, D. (2000). "Informal Workplace Learning. Practice Application Brief." NO 10. U.S. Department of Education: Clearinghouse on Adult, Career, and Vocational Education.

[7] Corden, R.E. (2001). Group discussion and the importance of a shared perspective: Learning from collaborative research. Qualitative Research, 1(3), 347-367.

[8] Eichenlaub, N., Gabel, L., Jakubek, D., McCarthy, G., and Wang, W. "Project iPad: Investigating Tablet Integration in Learning and Libraries at Ryerson University." *Computer in Libraries.*

[9] Hardy, B. and Suter, T. (2011). "*iPad Pilot Summary and Course Summary.*

[10] Laurillard, D. (2002). Rethinking University Teaching. A conversational framework for the effective use of learning technologies.

[11] Litchfield, A., Dyson, L. E., Lawrence, E., and Zmijewska, A. (2007). "*Directions for m-learning research to enhance active learning.*" [Online]. Available: http://www.ascilite.org.au/ conferences/ singapore07/ procs/ litchfield.pdf.

[12] Liu, G. Z., and Hwang, G. J. (2010). "A key step to understanding paradigm shifts in e-learning: Towards context-aware ubiquitous learning." *British Journal of Educational Technology.*

[13] Marmarelli, T. (2011). "*The Reed College iPad Study.*

[14] Murray, O. T., and Olcese, N. R. (2011). "Teaching and Learning with iPads, Ready or Not?." *TechTrends.*

[15] Naismith, L., Lonsdale, P., Vavoula, G., and Sharples, M. (2004). "Literature review in mobile technologies and learning." Birmingham: NestaFuturelab.

[16] National Research Council. (1999). *How People Learn: Brain, Mind, Experience, and School.*Washington, DC: National Academy Press.

[17] Perkins, D.N. (1991). Technology Meets Constructivism: Do They Make a Marriage? *Educational Technology.*

[18] Sharples, M., Taylor, J., and Vavoula, G. (2005). "Towards a theory of mobile learning." Paper presented at the *mLearn 2005 4th World conference on mLearning*, Cape Town. Available:

[19] Taylor, J., and Evans, D. (2005). Pulling together: Keeping track of pedagogy, design and evaluation through the development of scenarios—A case study. *Learning, Media and Technology*, *30*(2), 131–145.

[20] Weber, K., Maher, C., Powell, A., & Lee, H.S. (2008). Learning opportunities from group discussions: Warrants become the objects of debate. Educational Studies in Mathematics, 68 (3), 247-261.

[21] Willis, C. L., and Miertschin, S. L. (2004). "Tablet pc's as instructional tools or the pen is mightierthan the 'board!" In Proceedings of the 5th conference on Information technology education,pages 153–159, Salt Lake City, UT, USA. ACM.

[22] Zeng, R., and Luyegu, E. (2012). "Mobile Learning in Higher Education." In A. Olofsson, & J. Lindberg (Eds.), *Informed Design of Educational Technologies in Higher Education: Enhanced Learning and Teaching* (pp. 292-306). Hershey, PA: Information Science Reference.

AUTHORS

P. Thinley is with the Electrical Engineering and Computer Science School, Queensland University of Technology, Brisbane, Australia (e-mail: palmo.thinley@ student.qut.edu.au).

J. Reye, is with the Electrical Engineering and Computer Science School, Queensland University of Technology, Brisbane, Australia (e-mail: j.reye@ qut.edu.au).

S. Geva is with the Electrical Engineering and Computer Science School, Queensland University of Technology, Brisbane, Australia (e-mail: s.geva@ qut.edu.au).

A C++ App for Demonstration of Sorting Algorithms on Mobile Platforms

R. Meolic and T. Dogša
University of Maribor, Maribor Slovenia

Abstract—**This paper presents a systematic approach for designing a C++ app for demonstrating sorting algorithms on smartphones and tablets. It is part of an on-going project on the usages of new technologies in education. The general properties of mobile platforms are discussed together with details specific to demonstrating sorting algorithms. The implementation of Insertion Sort is given as an example. The obtained results (e.g. design rules and algorithms) have been tested within a prototype application.**

Index Terms — **Usability, Qt, Engineering Education, Mobile learning**

I. Introduction

The American Dialect Society named "app" the word of the year for 2010. Though it is a short for "application" it is rarely used to denote just any computer program but only those intended to run on a mobile platform i.e. on smartphones, tablets, and similar devices. Apps are also special in their methods of distribution. Authors upload and users download them from special web-stores called markets. This is the most effective way of software distribution ever. Just anyone can publish or sell an app. From amongst millions of these apps, many of them are oriented towards mobile learning. It is expected that mobile platforms will influence the education process at least as much as personal computers and the internet. The benefits of mobile learning include learning from rich interactive content (in contrast to books), flexible learning locations (in contrast to PCs), and at flexible learning times (in contrast to classrooms).

The C++ programming language was born in the early 1980s and it was first standardised in 1998. The current standard was ratified in 2011. Although C++ itself is widespread and ubiquitous, C++ apps are not. Default system development kits (SDK) for today's more popular mobile platforms do not emphasise C++. Apps for Apple's iOS are written in Objective-C, Microsoft's Windows Phone apps are created with C#, and Google's Android apps are Java programs. On the other hand, all these platforms allow for the developing of C++ apps, iOS by simply integrating different programming technologies together (e.g. Objective-C++) and Windows Phone and Android by provided support for (ambiguously named) native code. In a standard model–view–controller (MVC) design, the "model" and "controller" parts which consist of data and functions for data manipulation are suitable targets for C++ implementation, whilst the "view" part which is responsible for data representation and user interface is usually bound to the default technology. When considering the broader spectrum of mobile platforms, the situation is more challenging. For many of them, C++ apps are first class citizens, i.e. C++ is used thoroughly.

Sorting is a relatively narrow and straightforward subject as long as we are not creating science from it [1]. In [2] the author states that sorting is the fundamental algorithmic problem in computer science and that learning the different sorting algorithms is like learning scales for a musician. Indeed, sorting is the first step in solving a host of other algorithmic problems. Today, the pseudocodes of different sorting algorithms and their implementations in different programming languages can easily be found on the internet (e.g. project Rosetta Code). However, the most respected reference is the 3rd volume of Knuth's encyclopaedia [3].

Engineers do not learn computer algorithms only for being able to write their straightforward implementation — efficient implementations are already present in libraries. The goal of knowing various algorithms is to be capable of adapting and extending existing ones and to help in inventing completely new algorithms. Indeed, one can benefit a lot from studying a high quality algorithm's implementation. Graphical debuggers allow for easy observation of every memory bit, and the tracking of any program flow. But these tools are really oriented towards testing and debugging, not teaching. A purpose-built demonstration of an algorithm may differ considerably from the debugger approach:

- Pseudocode can be used instead of a real programming language;
- Program flow can be reversed;
- Actions can be explained in advance;
- Explanations can go beyond simple comments;
- Additional functionality can be added such as an examination mode for testing the user's knowledge.

When overviewing the markets we found apps dedicated to sorting algorithms that already include some of these functionalities, e.g. the app described in [4] encourages learning by awarding students with points. However, existing apps are still pretty elementary, the ultimate app (will there ever be one?) should include a large set of functionalities already available in Java and JavaScript-based applications on the Web. Let us mention some of them that could serve as the origins of ideas.

The well-known web site called Sorting Algorithm Animations (*http://www.sorting-algorithms.com/*) gives a plain visual comparison between different sorting algorithms and their efficiencies. The main properties and very formal pseudocode are also given for each algorithm.

Project JHAVE includes demonstrations of different sorting algorithms (*http://jhave.org/*). It is a client-server project where each demonstration is implemented in a special scripting language. The demonstrations look different from each other because they are precisely created to explain each particular algorithm.

D. K. Nester set up an interesting web page, that shows pseudocode for various sorting algorithms and enables the tracing of their executions (*http://www.bluffton.edu/~nesterd/java/SortingDemo.html*).

This paper is an updated report on the on-going project as recently presented at the International Conference on Computer Supported Education [5]. It is further organised as follows. In Section II, the design of an universal framework is discussed. Section III provides specific details of tracking sorting algorithms. In Section IV, a prototype app is described.

II. An Universal Mobile Framework for Demonstration of Sorting Algorithms

In order to help developers create consistent apps, vendors provide default application frameworks for their platforms and may also offer a dedicated development environments (IDE). In contrast, there is a tendency to use third-party application frameworks which supports cross-platform development. Surely, multi-platform support can be the more easily addressed with web technologies but such solutions are not always suitable or flexible enough [6]. In the presented project we aim for portability but we also want to support complex data structures that cannot be efficiently implemented using scripting languages. We have chosen Qt (*http://qt.digia.com/*), a default application framework for BlackBerry 10, Sailfish OS, and Ubuntu Touch, which is actively being extended to also become also a third-party framework for Android and iOS. Qt supports all standard libraries and adds a special signals and slots mechanism for communication between objects. Effective user interfaces for Qt-based apps are designed for using the declarative language QML. Qt Creator (Fig. 1) is very helpful while developing Qt-based apps.

Mobile platforms share many common properties such as touchscreen display, virtual keyboard on demand, and various sensors. However, developing apps that can be used on different platforms is often extremely difficult [7]. For example, when taking into account the limited screen size and resolution of an average device it is important to create a minimalistic GUI. We found a helpful review of a mobile interface design in [8]. Based on the feedback from users, it is suggested that:

- All description texts that do not provide extra knowledge should be omitted;
- The appropriate colours and text sizes are very important for increasing readability and clarity;
- The usages of icons and images should be limited;
- Concepts from Web applications may help users to transfer known user experience to the mobile app, but input methods and menu organisation should resemble standard mobile functionalities.

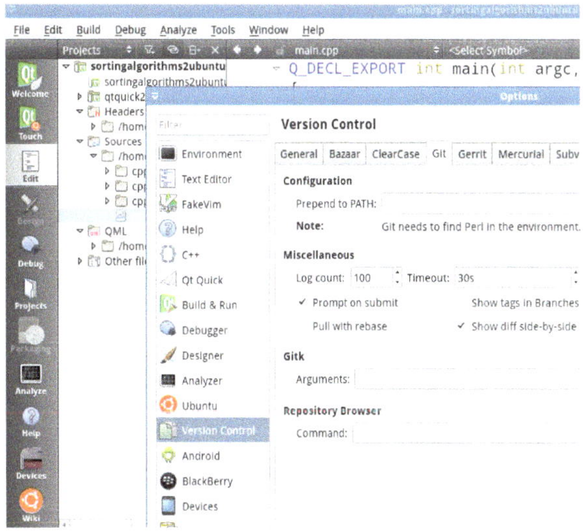

Figure 1. Qt Creator is a powerful IDE for cross-platform development

There is a lot of information, numerical and graphical, that we want to present to the user. However, we cannot show it all at once because of the limited space and because this would be too demanding for the user. Some possible solutions are:

- Let the device show different data in the portrait and landscape modes;
- Use pop-ups to show detailed data about items;
- Implement different screens that can be navigated by using tabs or swipe gestures;
- Use a hierarchical presentation where different groups and subgroups of data can be shown (expanded) or hidden (collapsed).

There is also a wide choice of mechanisms for controlling the app's run:

- Tapping and other gestures on the touchscreen;
- Rotating and other movements perceived by different sensors;
- Visual gestures observed by a camera;
- Voice commands, etc.

Let us discuss the advantage of orientation detection, only (Fig. 2). The portrait view is an obvious choice to show source code. Thus, the greater part of the screen may be reserved for that component. Thus the table of elements must be as small as possible but readable. Before an algorithm is chosen, the area reserved for source code can be used to show statistics. The landscape view may serve

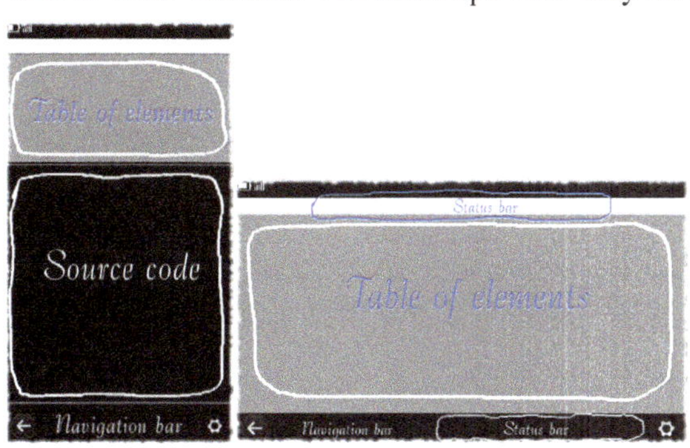

Figure 2. Portait and landscape view show different data

for presentations and interactions unrealised on the portrait view. Large elements make the user's interaction with them easier, thus almost all the area should be occupied by a table of elements. A status bar with one-line comment about the algorithm's current action could be added to allow for tracking the algorithm even without seeing it completely.

III. Tracking Sorting Algorithms on Mobile Platforms

Sorting algorithms can be demonstrated in many different ways. In order to compare their effectiveness it is best to run them concurrently and use only global view (e.g. each element is represented by a bar). To emphasise the differences, animations with fancy effects (e.g. colours or sounds) can be created. In order to explain the implementation details, pseudocode may be involved.

Detailed tracking of the algorithm's run seems to be the crucial functionality for deep understanding. Due to limited space let us discuss this aspect only. We are interested in the basic form of sorting algorithms where the elements to be sorted are stored in the table and no optimisations are used (i.e. each time the algorithm needs a value of an element it performs a memory read). Our goal is using the same framework for demonstrating the various sorting algorithms. In order to display details, differing algorithms may require different approaches. However, in our approach we sacrifice some flexibility for uniformity because we obtain more direct comparisons between different solutions. Also, shared platforms for demonstrating the problems and solutions are well accepted in computer science teaching books.

In order to study an algorithm one has to view it in one form or another. We went for the pseudocode in the form of a simplified source code because this is concise representation whilst also suitable for the usage within a basic computer science course. Hence, each algorithm is composed of sentences that are either assignments, decisions, or flow control statements. We preferred C-style syntax for assignments, keywords, and parenthesis but we have discarded the semi-colons at the end of sentences (to reduce the text width and also to make it more appealing for developers using scripting languages). An indent of 2 spaces is the most we could afford on a small screen. Syntax highlighting is used to make pseudocode comprehensible. We omitted variable declarations but assumed that all the elements and indices are integers. The following naming scheme is used:

- The number of elements is denoted by `n`;
- The value of the i-th element is denoted by `[i]`, i.e. the name of the table is discarded;
- The loops use variables `i` and `j`, whereas other variables are named as usual in the literature.

As well as the syntax, we paid careful attention to the presentation of the semantics. The inclusion of compound and otherwise complicated sentences is undesirable. Flow control by using the "for" loop is a typical example of a complex statement that is composed of an initial assignment, a condition to stop, and control actions applied at the end of every looping. Thus, we preferred "while" or "foreach" loops over the "for" loop within the pseudocode.

In order to avoid an inconsistent handling of different semantic parts, we identify a set of atomic operations (i.e functions) that reveal the concept of sorting algorithms. Atomic operations are those that will create actions on the screen, either by simply showing/changing some values or applying some graphical effects, e.g. highlighting a part of the screen, doing some animation etc. Moreover, the atomic operations correspond to an algorithm's steps during step-by-step tracing. To some extent, the atomic operations coincide with the pseudocode statements, but we would have lost all the flexibility by equating these two formalisms. The following atomic operations are necessary and sufficient:

- Reading the value of an element;
- Changing the value of a variable;
- Comparing the values;
- Starting/continuing/finishing a loop;
- Swapping and moving elements.

We consider swapping and moving elements to be atomic operations. They are autonomous principles usually taught along with the sorting algorithms.

Once atomic operations are being identified, we can determine the necessary highlighting effects. Two mark-up functions are quite enough for a simple app:

- Function `setMark` that denote which element is currently observing;
- Function `setSpecial` that denote elements with some special feature, e.g. they are already in order.

Functions `setMark` and `setSpecial` can be used for the same element simultaneously and therefore their visual effects must be compatible. All other informations will be presented through labels and comments. Swapping and moving elements are supposed to be animated and hence we do not need extra effects for these operations.

The user studying sorting algorithms could be interested in different statistics about the efficiency, e.g. the number of required data comparisons, memory reads, memory writes, etc. Static bounding of statistical data to the pseudocode statements is adequate for many situations, e.g. each swap of two elements consists of 2 write operations (plus 2 read operations if the values are not already stored in local variables). But once again, such fixation is inflexible and, for example, in the case of compound decisions or in the case of single-line loops it is also very impractical. Thus we propose explicit manipulation of statistical data via special functions.

In order to illustrate the presented approach, let us observe the source code for demonstration of Insertion Sort given in Fig. 3. The bold lines correspond to the lines of pseudocode that is shown to the user. Function `CC` is used to identify and describe single steps. Functions `setLabel` and `incrLabel` are used to update the values of labels and statistical counters. A log from running the presented implementation is illustrated in Fig. 4. There, elements marked with functions `setMark` and `setSpecial` are illustrated as framed or shaded, respectively. For each step, the corresponding pseudocode line number and comment are displayed. Please note, that it is up to controller to implement step-by-step and backward running.

```
insertionSort(table T, int n) {
  int i, j, key;
  setLabel("i",-1);
  setLabel("j",-1);
  setLabel("key",-1);
  setLabel("read",0);
  setLabel("write",0);
  setLabel("sourceLine",1);
  setSpecial(0);
  CC("Skip first element");
  i = 1;
  while (i < n) {
    setLabel("sourceLine",2);
    setLabel("i",i);
    setMark(i);
    CC("Outer loop now at i");
    key = T[i];
    incrLabel("read",1);
    setLabel("key",key);
    setLabel("sourceLine",4);
    CC("[i] stored into variable key");
    j = i - 1;
    if (j >= 0) {
      setLabel("sourceLine",5);
      setLabel("j",j);
      setMark(j);
      CC("Inner loop starts at j=i-1");
      while (j >= 0 && T[j] > key) {
        incrLabel("read",1);
        setLabel("sourceLine",6);
        CC("Decrement j ([j]>key)");
        clearMark(j);
        j--;
        setLabel("j",j);
        if (j >= 0) setMark(j);
      }
    }
    setSourceLine(7);
    if (j >= 0) {
      incrLabel("read",1);
      CC("Stop inner loop ([j]<=key)");
      clearMark(j);
    } else \{
      CC("Stop inner loop (j<0)");
    }
    if (j != i-1) {
      setLabel("sourceLine",8);
      CC("Move [i] to index j+1");
      clearMark(i);
      move(i,j+1);
      incrLabel("write",i-j);
    }
    if (j == i-1) clearMark(i);
    setSpecial(j+1);
    setLabel("sourceLine",9);
    if (j >= 0) setLabel("j",-1);
    CC("Increment i");
    i++;
    setLabel("key",-1);
  }
  setLabel("sourceLine",11);
  setLabel("i",i);
  CC("All elements sorted");
}
```

Figure 3. Implementation of Insertion Sort

3	2	5	1	4	01 – Skip first element
3	2	5	1	4	02 – Outer loop now at 1
3	2	5	1	4	04 – [i] stored into variable key
3	2	5	1	4	05 – Inner loop starts at 0
3	2	5	1	4	06 – Decrement j ([j]>key)
3	2	5	1	4	07 – Stop inner loop (j<0)
3	2	5	1	4	08 – Move [i] to index 0
2	3	5	1	4	09 – Increment i
2	3	5	1	4	02 – Outer loop now at 2
2	3	5	1	4	04 – [i] stored into variable key
2	3	5	1	4	05 – Inner loop starts at 1
2	3	5	1	4	07 – Stop inner loop ([j]<=key)
2	3	5	1	4	09 – Increment i
2	3	5	1	4	02 – Outer loop now at 3
2	3	5	1	4	04 – [i] stored into variable key
2	3	5	1	4	05 – Inner loop starts at 2
2	3	5	1	4	06 – Decrement j ([j]>key)
2	3	5	1	4	06 – Decrement j ([j]>key)
2	3	5	1	4	06 – Decrement j ([j]>key)
2	3	5	1	4	07 – Stop inner loop (j<0)
2	3	5	1	4	08 – Move [i] to index 0
1	2	3	5	4	09 – Increment i
1	2	3	5	4	02 – Outer loop now at 4
1	2	3	5	4	04 – [i] stored into variable key
1	2	3	5	4	05 – Inner loop starts at 3
1	2	3	5	4	06 – Decrement j ([j]>key)
1	2	3	5	4	07 – Stop inner loop ([j]<=key)
1	2	3	5	4	08 – Move [i] to index 3
1	2	3	4	5	09 – Increment i
1	2	3	4	5	11 – All elements sorted

Figure 4. The log from running the Insertion Sort

IV. THE PROTOTYPE APP

We have created a prototype app for Nokia N9. This smartphone uses Meego OS and allows native C++ apps (devices with more recent Sailfish OS or Ubuntu Touch are unavailable as yet and Qt for Android is still a kind of beta software). The prototype app is planned to have Teaching mode (demonstrating and teaching), Practice mode (checking the user's knowledge), and Game mode (learning by playing). In version 0.2, only a part of Teaching mode is implemented with about 2,300 lines of C++ code and about 1,800 lines of QML code. Gnome sort, Insertion Sort, and Quicksort are included. Fig. 5 shows that the structure of the C++ code is carefully designed to separate the "model" part (a pure C++ code) from the "controller" part (a Qt specific code). Fig. 6 and Fig. 7 give some screenshots from prototype app.

Furthermore, we briefly discuss the implemented algorithms. Please, compare the given pseudocodes with the implementations, as given in Fig. 3 and [5].

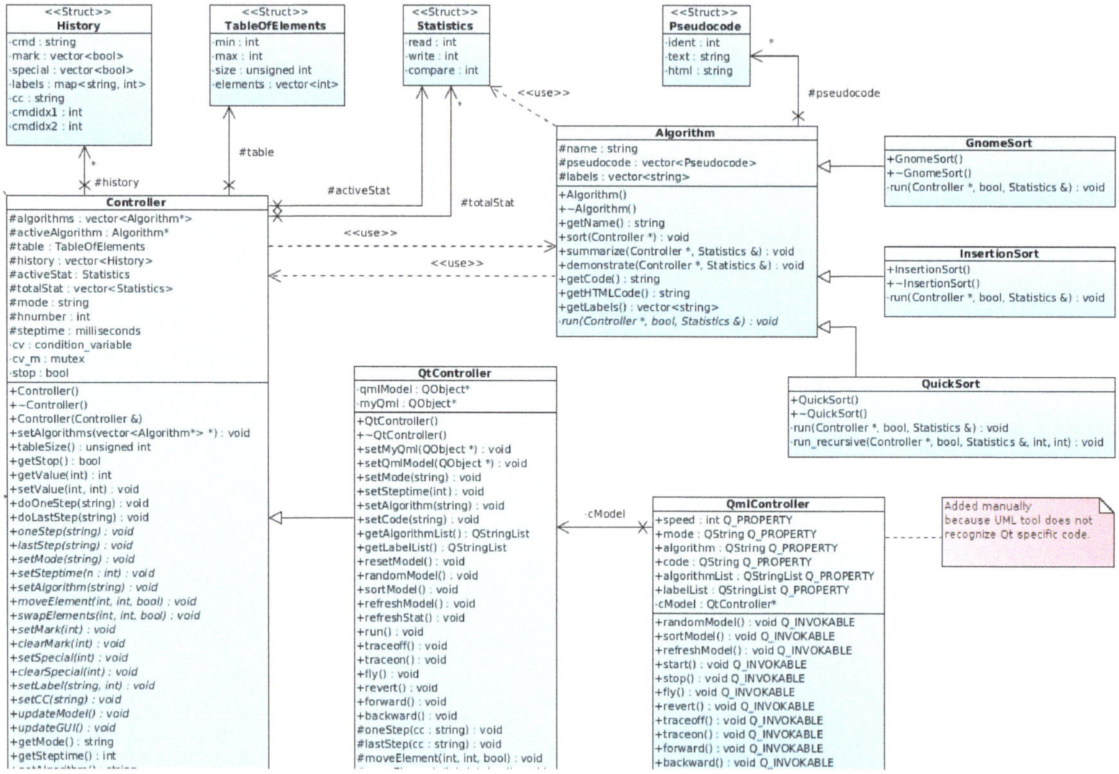

Figure 5. Class diagram showing a C++ part of prototype app

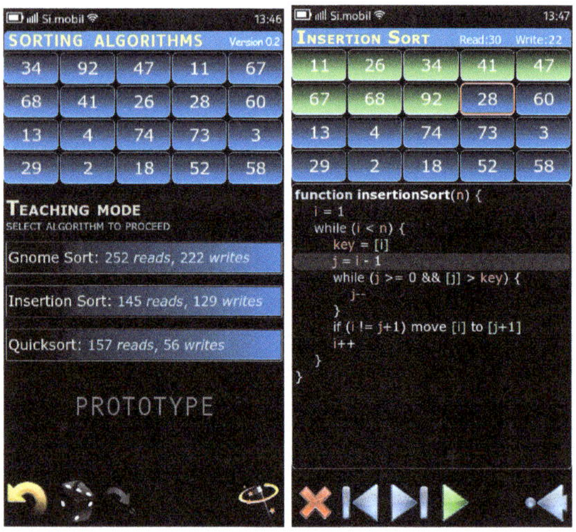

Figure 6. When algorithm is not selected, app shows the list of implemented algorithms together with short statistic data; when the algorithm is running, app shows the actual table of elements and pseudocode with marked current line

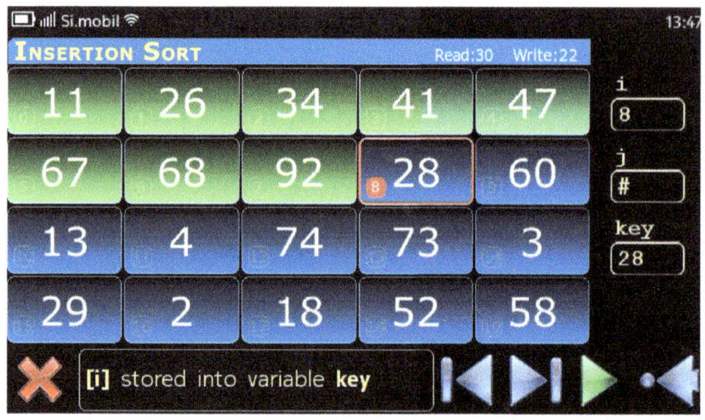

Figure 7. In landscape mode, app shows actual table of elements, the value of labels and the comment for current action

A. *Gnome Sort*

Gnome Sort is advertised as the technique used by the standard Dutch garden gnome to sort a line of flower pots. Gnome Sort is supposed to be the simplest sorting algorithm and it is indeed very easy to demonstrate it. The pseudocode of the Gnome Sort as shown in the prototype app is given in Fig. 9. The index used by the algorithm (yes, Gnome Sort uses only one index) is called "position" and is for brevity denoted by variable `i`. We use function `setMark` to mark the element on the index "position" and function `setSpecial` to denote the elements that are already in order.

B. *Insertion Sort*

Insertion Sort (Fig. 10) is a simple sorting algorithm that builds the final sorted array one item at a time. It is much less efficient on large arrays than more advanced algorithms, but for small arrays it can even outperform them. Moreover, Insertion sort handles nearly sorted arrays quite efficiently. When humans manually sort something (e.g. a deck of playing cards), most of them use a method similar to Insertion Sort. In our implementation of Insertion Sort we use function `setMark` to denote elements with indices `i` and `j`, that are currently being compared, and function `setSpecial` to mark the already sorted elements (see Fig. 3 and Fig. 4).

C. *Quicksort*

Quicksort is a divide-and-conquer algorithm that first divides the elements into two subgroups and then subsequently sorts these subgroups. Many versions exist of the original algorithm, which are either called a simple version or an in-place version. In our project we have chosen one of the in-place versions (Fig. 11), i.e. an algorithm that does not need an extra space for dividing elements into two subgroups. Moreover, we used a recursive version of the algorithm, which is also the more common. The presented variant is not the original version of the algorithm [9] but in our opinion it is the more appropriate one for the use in the class. Please, note that we omitted details of pivot calculation to make pseudocode shorter. In the prototype app we used the median of the first, middle, and last elements.

```
function gnomeSort(n) {
  i = 0
  while (i < n) {
    if (i == 0 || [i] >= [i-1]) {
      i++
    } else {
      swap [i] and [i-1]
      i--
    }
  }
}
```

Figure 8. Pseudocode of Gnome Sort

```
function insertionSort(n) {
  i = 1
  while (i < n) {
    key = [i]
    j = i-1
    while (j >= 0 && [j] > key) {
      j--
    }
    if (j != i-1) move [i] to [j+1]
    i++
  }
}
```

Figure 9. Pseudocode of Insertion Sort

```
function quickSort(begin, end) {
  left = begin
  right = end
  if (left < right) {
    calculate pivot
    while (left <= right) {
      while ([left] < pivot) left++
      while ([right] > pivot) right--
      if (left <= right) {
        swap [left] and [right]
        left++
        right--
      }
    }
    quickSort(begin,right)
    quickSort(left,end)
  }
}
```

Figure 10. Pseudocode of Quicksort

V. Conclusion

We have studied sorting algorithms with the goal of tracking their run by using a set of given visual effects whilst taking into account the limitations of the mobile platform. It is important to choose the proper presentation of the algorithm's flow and the proper type and amount of visual effects. We successfully managed three different algorithms, so we believe that our approach can be extended for most of the other sorting algorithms as well.

The paper has reported about a prototype app for a specific mobile platform. Nevertheless, we have researched and discussed all the problems in such a general way that the results are applicable on any mobile platform. Each statement in C++ code either belongs to the sorting algorithm, or collects statistical data, or serves as a part of the demonstration strategy. We were careful not to mix these roles. Hence, the obtained code is very portable. Furthermore, the statistical data can be obtained by only excluding all GUI statements.

Although not being the main goal of this project, the obtained platform itself was found usable when teaching software engineering. Today, students often underestimate the benefit of appropriate planning (e.g. preparation of graphical views) and appropriate testing (e.g. preparation of test cases). Developing an app similar to the presented one is quite a challenge due to the limited screen sizes on mobile devices and because the effective demonstration of sorting algorithms involves many navigation options.

References

[1] P. Vitanyi, "Analysis of Sorting Algorithms by Kolmogorov Complexity (a survey)". In *Entropy, Search, Complexity*, pp. 209–232, 2007.

[2] S. S. Skiena, *The Algorithm Design Manual*. Springer, 2nd ed., 2008.

[3] D. E. Knuth, *The Art of Computer Programming*, 2nd ed., vol. 3: Sorting and Searching, Addison-Wesley Professional, 1998.

[4] I. Boticki, A. Barisic, S. Martin, and N. Drljevic, "Teaching and Learning Computer Science Sorting Algorithms with Mobile Devices: A case study," *Computer Applications in Engineering Education*, vol. 21, pp. E41-E50, 2013.

[5] R. Meolic, "Demonstration of Sorting Algorithms on Mobile Platforms," In *The Fifth International Conference on Computer Supported Education*, pp. 136-141, 2013.

[6] W. Jobe, "Native Apps vs. Mobile Web Apps", *International Journal of Interactive Mobile Technologies*, vol. 7, no. 4, 2013.

[7] A. Holzinger, P. Treitler, and W. Slany, "Making Apps Useable on Multiple Different Mobile Platforms: On Interoperability for Business Application Development on Smartphones," In *Multidisciplinary Research and Practice for Information Systems*, volume 7465 of LNCS, pp. 176–189, 2012.

[8] J. Mirkovic, H. Bryhni, and C. M. Ruland, "Designing User Friendly Mobile Application to Assist Cancer Patients in Illness Management". In *The Third International Conference on eHealth, Telemedicine, and Social Medicine*, pp. 64–71, 2011.

[9] L. Khreisat, "Quicksort — a Historical Perspective and Empirical Study," *International Journal of Computer Science and Network Security*, 2007, vol. 7, no.12, pp. 54–65, 2007.

Authors

R. Meolic is with the Faculty of Electrical Engineering and Computer Science, University of Maribor, Smetanova ulica 17, SI-2000 Slovenia. (e-mail: meolic@uni-mb.si).

T. Dogša is with the Faculty of Electrical Engineering and Computer Science, University of Maribor, Smetanova ulica 17, SI-2000 Slovenia. (e-mail: tdogsa@uni-mb.si).

19

Lecturers and Students' Perception on Learning Dijkstra's Shortest Path Algorithm Through Mobile Devices

M. Seraj[1] and C.Y. Wong[2]
[1] Limkokwing University of Creative Technology, Cyberjaya, Selangor, Malaysia
[2] Multimedia University, Cyberjaya, Selangor, Malaysia

Abstract—**In recent years, many studies have been carried out on how to engage and support students in e-learning environments. Portable devices such as Personal Digital Assistants (PDAs), Tablet PCs, mobile phones and other mobile equipment have been used as parts of electronic learning environments to facilitate learning and teaching for both lecturers and students. However, there is still a dearth of study investigating the effects of small screen interfaces on mobile-based learning environments. This study aims to address two objectives: (i) investigate lecturer and student difficulties encountered in teaching-learning process in traditional face-to-face classroom settings, and (ii) to explore lecturer and student perceptions about learning the subject through mobile devices. This paper presents the results of a qualitative study using structured interviews to investigate lecturer and student experiences and perceptions on teaching and learning Dijkstra's shortest path algorithm via mobile devices. The interview insights were then used as inputs to define user requirements for a mobile learning prototype. The findings show that the lecturers and students raised many issues about interactivity and the flexibility of effective learning applications on small screen devices, especially for a technical subject.**

Index Terms—**Computer network, Dijkstra's shortest path algorithm, mobile devices, perception, small screen interface.**

I. Introduction

In the discipline of computer science, learning computer networks is a core subject for Computer Science undergraduate students. Students usually learn Network Routing Algorithms as a chapter in a computer networking course, in which they learn different types of routing algorithms such as Distance-Vector, Link State (LS) Routing Algorithm and so on (Tanenbaum, 2003; Forouzan & Fegan, 2007). Most computer science lecturers teach link state routing algorithm as part of network routing algorithms. Dijkstra's shortest path algorithm is used to identify and demonstrate the shortest paths between each of the nodes in the network when LS routing algorithm is to route in a network. Apart from learning, a number of Computer Science undergraduate students are also eager to delve further into the topic for doing research. Although electronic learning environments provide some facilities to learn complex concepts indirectly (Drigas, Koukianakis & Papagerasimou, 2006), students still encounter difficulties in coping with the complex concept of this routing algorithm. More importantly, understanding the technical terms requires hands-on practice rather than mere theoretical learning.

The use of electronic environments in educational technology started fairly recent to make the learning process easier and increase accessibility of learning and teaching (Sandrasegaran, Cheung, Hossein, Zia & Malwany, 1993; Zele, Hoecke, Lenaerts & Wieme, 2003). Learning via electronic environments, also called e-leaning, utilizes electronic and telecommunications technologies for information delivery, especially for educational purposes. With the advent of e-learning, most traditional learning limitations regarding learning times and places are now addressed in e-learning environments (Sandrasegaran, Cheung, Hossein, Zia & Malwany, 1993). Although e-learning technology offers different types of electronic learning environments such as web-based learning environments (WLE) and mobile-based learning environments (MLE), mobile-based learning has recently become an important modality used in e-learning environments (Tong, Yang, Liu & Liu, 2006; Khodabandeh, Afshari &Manian, 2010).

Today, mobile devices, especially touch-based models, have become increasingly used by students. Students presently use different types of different types of mobile devices such as tablet PCs. They often use hand phones to communicate with their friends and family, and also indulge in other tasks such as browsing the Internet, playing games and engaging in different types of mobile applications (Patric, Gao & Wu, 2008). In this regard, mobile technology has been introduced to students as a significant tool to help teach both technical and non-technical subjects (Kuo, Wu, Chang, Chang & Heh, 2007).

Mobile learning technology is commonly regarded as a new generation of communication technology that stimulates the user's interest in learning. On the one hand, educators and researchers are interested in exploring how mobile devices can be used to engage learners to learn and practice subjects more frequently anytime or anywhere, without any special technical requirements (Hentea, 2004; Jun & Zhi-yi, 2010). Although portable devices are viewed as learning tools which enable students to save time and eliminate distance learning limitations (Arreymbi, Agbor & Dastbaz, 2008; Cemal, Dastbaz & Bacon, 2008; Fetaji, 2008b; Ye, Lee & Geng, 2010), the challenge is how a small screen interface (with small and low resolution screen) can be used to learn a technical subject, such as Dijkstra's shortest path algorithm for computer science students, while they are on the move.

The motivation for this work is the first author's experiences during his studies in the Bachelor of Computer Software Engineering Program, where he observed most of his course mates having difficulties in grasping the concept of routing algorithms, in particular Dijkstra's shortest path algorithm. From this researcher's point of view, students can solve this problem through frequent practice on this subject in a more interesting way. From the perspective of interface design, interactive learning for a technical subject somehow suffers from usability issues. Instructional designing of Dijkstra's shortest path algorithm on a mobile platform can be a challenge because of difficulties related to adapting its content on small screen interfaces. Thus, this paper aims to investigate the difficulties computers science students and their lecturers encounter when learning and teaching this topic, and explore their perceptions about learning via small screen devices.

II. Literature Review and Related Work

Portable devices such as mobile phones, PDAs, notebooks and tablet PCs are used to assist the learning process, overcoming the limitations of time and location. Mobile-based learning is flexible, dynamic and easy to use in ubiquitous learning environments. Furthermore, mobile devices can offer ways of learning that are not possible in traditional classrooms or other types of electronic learning technologies (Tan & Liu, 2004; Patric, Gao & Wu, 2008).

Most researchers mention that they have faced some restrictions with regards to instructional design, development and implementations, in part, due to small screen interfaces (Ye, Lee & Geng, 2010). Small screen size, poor resolution, limited storage capacity and limited processing capacity are the main difficulties that researchers face when developing mobile applications (Fetaji, 2008c; Ye, Lee & Geng, 2010).

Despite the above mentioned challenges, researchers and experts on mobile learning technology have reported many positive impacts of learning via small screen devices. It is believed that mobile technology provides opportunities for learners to learn in an indirect and unconscious manner. Expanding the scope of electronic learning environments and environmental awareness are the two significant issues which researchers report are favored by mobile technology. They highlight that content delivery via mobile devices can motivate learners to grasp the concept of learned subjects (Fetaji, 2008c; Ye, Lee & Geng, 2010).

An example of a mobile-based interactive application was developed in Taiwan by Tan & Liu (2004). It consists of a mobile learning server and mobile learning tools to assist elementary school learning English. The application is able to maintain the learner's interest, enhance learning assessment and teach instruction for indoor and outdoor learning activities. In this system, students can connect to the network using their mobile devices, PDAs or notebooks, and download course content. The result of the project indicates that the second part of each lesson, which was presented in a mobile-based format, had a greater impact than the traditional format of content delivery. At the end of the project, the researchers used questionnaires to gauge levels of satisfaction among the learners of learning English via portable devices (Tan & Liu, 2004).

Fetaji (2008b) states that mobile learning can provide students more activities and practices in their fields of study. Fetaji considers how interfaces can be designed to be more usable and suitable for users working with learning applications. Fetaji used a questionnaire to identify the quality of user interface attributes. The study revealed the limitations of using mobile technology on mobile devices such as small screen size, processors, different brands of mobile devices, etc. Fetaji also discovered that some lecturers are hesitant to use mobile devices for instructional purposes. They think mobile devices can be used to cheat and allow other illegitimate activities (Fetaji, 200b; Fetaji, 2008c).

Although there is considerable existing research that investigates the capabilities of small screen devices such as mobile phones in learning and teaching environments, there is still a lack of research to examine the lecturers and learners' ideas about whether it is suitable to practice technical subjects (Fetaji, 2008a), there is still a need to design and develop instructional applications based on their feedback and opinions. According to Seong (2006), user perception is the most important attribute that needs to be evaluated by the developer before creating a learning application based on mobile devices. Thus, researchers and developers should examine learner characteristics in terms of age, gender, nationality and the level of education, as well as their familiarity with mobile devices. Understanding learner characteristics before development is required to help learners better use the application. Difficult mobile learning applications that are complicated will dissuade learners from using them (Seong, 2006). As a result, we are interested in investigating lecturer and student perceptions about learning a technical subject delivered on small screen devices.

As mentioned earlier, this study aims to identify the issues and challenges faced by computer science students and the lecturers for learning and teaching Dijkstra's shortest path algorithm. We investigate their perceptions on learning Dijkstra's shortest path algorithm based on small screen devices with a lower resolution and limited processing capability. This paper aims to explore the lecturers and student perceptions and experiences about the current methods of teaching and learning a specific technical subject, including difficulties they faced and their opinions of learning a technical subject via mobile devices. The objectives of the study are addressed below:

1) To explore the opinions and issues faced by the learners, in particular any difficulties for them to cope with the concept of Dijkstra's shortest path algorithm in the current face-to-face classroom scenario.
2) To measure the lecturers' experiences about teaching the subject in the current face-to-face classroom situation.
3) To investigate the lecturers' perceptions about learning Dijkstra's shortest path algorithm via mobile devices.
4) To elicit the learners' opinions about the feasibility of learning and practicing technical subject matter through small screen devices.

III. Research Methodology

This study employs a qualitative research method using structured interviews (Kumar, 2011; Creswell, 2009) with students and lecturers to identify their issues and difficulties in learning and teaching the subject (Dijkstra's shortest path algorithm). The data collection via interview is subjective data, mainly in text-based descriptions. For instance, we acquire user feedback and comments from the students about the problems they faced in understanding the concept of the subject, as well as lecturer problems related to teaching and delivering content in the classrooms. The study draws conclusions of student and lecturer perceptions on the topic, by providing suggestions about learning the subject via mobile devices. In terms of sample selection, we use a snowballing technique by approaching all the lecturers who are teaching subjects related to computer network. Only 8 out of 10 lecturers from Faculty of Information Technology (FIT) responded and were willing to participate in the interview session. For the student sampling, we approached the lecturers who agreed to participate in our interview to introduce us to their students in their computer network classes.

In short, the respondents of the interview consist of 8 lecturers (6 males and 2 female) from Faculty of Information Technology (FIT) and 21 Computer Science Undergraduate students (9 males and 12 females). The first group of respondents includes 4 senior lecturers with PhD degree and 4 lecturers with Master degree in the fields of Computer Science or Information Technology. The lecturers are experts in teaching Computer Networks subject to Computer Science undergraduate students. They generally have appropriate knowledge in the respective field of study. Apart from this, all lecturers taught the subject and have experience doing research in the field.

The second group of the respondents is Computer Science students, who have either passed the Computer Networks subject, or are still currently studying the subject. All the 21 students fall under 18-23 age groups. Among the 21 students, 12 students are in the first half of their period of study (6 students are currently studying in third semester and the other 6 students are in their first semester). 9 out of 21 students are in second half of their undergraduate studies (6 students are currently studying in fifth semester and 3 student is in sixth semester). Generally, they had learned knowledge about the Computer Network subject and other related technical subjects. Apart from this, all the 21 students have worked with some learning applications before. They are also familiar with learning technologies, which are used by the lecturers as one of their teaching methods for information delivery (Biello, 2005; Morrison, Ross and Kemp, 2004; Alessi and Trollip, 2001; Mahadevan, 2009).

IV. PROCEDURE

During the interview, some guided questions were prepared. We used simple and everyday language for the interview. The interview questions consist of open-ended and close-ended questions. Any ambiguous questions, which contain more than one meaning and leading questions to dictate an answer, or a specific answer to the respondents were avoided (Kumar, 2011; Creswell, 2009). In this regards, careful consideration was taken about the quality of collected data regarding the number of interviews, quality of interviewers and quality of interactions (Kumar, 2011). We acquired feedback and also comments from the Faculty of Information Technology (FIT) lecturers and computer science undergraduate students.

The interview questions in this study are targeted towards lecturers and undergraduate students. The first type of interview questions is designed for lecturers with the objective of better understanding their perceptions and experiences in teaching computer networks, especially routing algorithms. The second type of interview questions is designed to understand whether the computer science students having any difficulties in grasping the concepts of the subject, as well as any issues the students encounter in practicing and learning the subject.

We first interviewed 8 lecturers who have either taught or a presently teacher the course on computer networks to the computer science undergraduate students. Before each interview started, we explained the purpose of the interview and gave provided explanations about the study. A consent form was presented to obtain permission from the respondents agreeing to voluntary participate in the interview. We used audio-taped and note-taking to record the respondents' answers and recommendations. The answers were then analyzed and a conclusion was drawn to provide lecturers' recommendations, perceptions and expectations about the topic.

We interviewed 21 computer science undergraduate students, all of whom had previously studied the subject or were at that time were studying the subject of computer networks. The students' knowledge of computer networks had been pre-screened by the researcher before each interview was conducted. This was done through the FIT lecturers' suggestions for the interview. The researcher also referred to the semester they were currently studying in the computer networks course and the scores of the exercise, assignments and general exams. The researcher first briefed persons participating in the interview of the students to get their responses and points of view, according to the questions and recommendations. Their answers were then analyzed concerning the difficulties they faced in the face-to-face classrooms and their perceptions on learning the subject delivered on mobile devices.

From this interview, we intended to obtain student feedback and comments on learning in the computer network course and opinions on learning the technical subject via mobile devices with a lower resolution. To structure the interview, a set of pre-determined questions were prepared for the students. The questions included questions about their study major, their interests in about computer networks and any difficulties they encountered coping with the concept of algorithms. The questions also attempted to discover the technologies used by the FIT lecturers, whether they were effective in motivating them to practice and improve their knowledge of the subject itself. Furthermore, questions concerning the use of other types of technologies and the degree to which they effectively delivered information, proved to feasible for practice without constraints in time and location, and their opinions about learning via mobile devices.

The respondents were prompted to express any issues and difficulties they faced in current face-to-face classroom assignments and lab practices, the current e-learning technologies that were used, teaching methods employed by lecturers, and the interactive instructional systems that students were currently exposed to. They provided useful opinions and essential content, principles and practices for the mobile-based learning prototype. For instance, the 21 computer science undergraduate students consisted of juniors and senior students of the FIT (Faculty of Information Technology). They provided some feedback about their difficulties in coping with the concept of the subject in the classrooms along with their perceptions about learning and practicing the technical subject while using handheld or small screen devices.

V. Interview Findings

A. Interviews with Lecturers

During the interview, we prompted some guided questions to the FIT lecturers in order to gauge feedback in subjective comments. The questions were designed with different considerations, including their teaching experience, current technology use in their courses and their perception about using mobile devices for information delivery.

The first part of the interview questions covered the lecturer's experiences on the subject. For instance, *"Which chapter (chapters) of the course titled Computer Networks is the most difficult to teach?"* In the lecturers' opinions, teaching the topic of signals and signal transmissions is the most difficult chapter to teach because there is still lack of technologies and tools available for lab practice. Apart from this, network security and physical layers are the two most difficult topics to teach from the lecturers' points of views. From the FIT lecturers' responses, we derived that there are some existing tools to teach some complex topics, but the tools are generally expensive and pricey, which limit their potential use in the classroom. As a result, the difficulty of comprehending a particular topic is due to insufficient tools, software and applications to assist students to grasp the complex concept. In turn, this is complicated furthermore by inadequate lab practice for students. In addition, one of the lecturers believed each chapter of the subject can be considered difficult to teach students. The difficulty of teaching a subject or a chapter on the subject of computer networks depends on each individual student and the level of his/her knowledge on different topics of the subject. To the lecturers, routing algorithms can be difficult to teach some students, especially those who do not have knowledge about algorithms in general and, more specifically, routing algorithms.

The next question asked was *"Do you teach Dijkstra's shortest path algorithm in the Link State routing as part of the Computer Routing Algorithms chapter?"* This question is deemed significant because we wanted to find out whether the FIT lecturers teach the technical subject or just merely their research interests. All 8 lecturers expressed they teach the specific topic, and 6 out of 8 lecturers also expressed they do further research in the field of computer networking.

The second part of the interview questions was designed to identify the technologies used by the lecturers. The information gathered reveals that all of the 8 lecturers use the same teaching activities, which include assignments, quizzes, practice and exams. The technology used to teach is mainly Microsoft PowerPoint to deliver content. Additionally, Cisco software such as Cisco Packet Tracer and Cisco System were also used to show the students how a network works, including data transfer, packet analysis, routing transactions, and connections between nodes, routers, etc.

The lecturers were also asked the question, *"Which other technologies can be used to assist them to teach the Dijkstra's shortest path algorithm as part of the Link State routing algorithms or in Network Routing Algorithms?"* The purpose of this question was to obtain lecturers' opinions about other type of technologies they felt could be used to assist them teach the technical subject. The 8 lecturers disagreed as to how useful mobile devices are in helping students to learn the technical subject. In terms of increasing student learning and study performance, they generally do not consider using mobile phones to teach the subject. However, they mentioned that the small screen interface on a mobile platform is useful if the learning application on mobile devices is interactive, allowing instant communication with a server to connect with other users and share knowledge among the lecturers and their peers. In addition, they also mentioned that applications possessing graphics, animations and other types of multimedia elements are useful to help students' self-study, especially with regards to hands-on practice for the technical subject.

B. Interview with Students

In the interviews with 21 computer science undergraduate students, the researchers wished to discover:

1) Basic information about the respondents' academic status, their thoughts about the subject and, in particular, the topic of Network Routing Algorithms.
2) The technologies used by their lecturers to teach the subject and evaluate. Any other technology used to motivate them to practice and enhance their study performance.
3) Their difficulties (if any) that students encounter with the concept of the subject, especially routing algorithms.
4) Their opinions about any types of technologies that can be used to motivate them for hands-on practice.

During this interview, some guided questions prompted interviewees from a general topic (network routing algorithms) to the main point of interest, Dijkstra's shortest path algorithm. From the feedback gathered, all 21 students except one expressed *"I have learned Network Routing Algorithms as a chapter of the computer network course"*. 15 of the 21 students agree with the sentence, *"Network Routing Algorithms is a difficult chapter to learn in the computer networks course"*. 18 of the 21 students also mentioned they had learned Dijkstra's shortest path algorithm in Link State routing algorithms as as part of the chapter on network routing.

The students were then asked, *"What are the types of technologies used by your lecturer to deliver the information to you in learning network routing algorithms?"* The question was included to identify the technologies used by the lecturers at Multimedia University to deliver information for their exercises and lab practices. The responses given were consistent with the lecturers that Microsoft PowerPoint was used to deliver instruction, while Cisco Packet Tracer was used for assignments, quizzes and lab practices.

The question, *"Which technologies can be used to motivate you to better learn Dijkstra's shortest path algorithm in the link state routing or other types of network routing algorithms?"* was formulated to obtain student perceptions and expectations about any other technologies that they perceive as useful for their learning. The students generally answered that the technologies actually being used (Cisco software, for example) is useful in showing them how a network works, network topology, how data transfers in a network, and so on. 6 out of 21 students were generally satisfied about the Cisco network technology used by their lecturers in terms of helping them enhance their knowledge of routing algorithms via lab practice. Having said this, 9 students mentioned that although the Cisco software helped motivate them to practice the subject, the software is quite difficult to work with, especially when there is lack of access for frequent practice. From the students' points of view, hands-on practical and interactive learning applications can be helpful to improve their learning.

VI. Discussions

From the lecturers' points of view, the answers demonstrate they generally disagree about using mobile devices, especially mobile phones for learning Dijkstra's shortest path algorithm in the link state routing algorithms. Apart from learning and delivering the information to students, when asked, *"In your opinion, which teaching and learning strategy can be used to deliver usable and meaningful content on mobile platforms?"*, they suggested that tablet PCs with larger screen interfaces and higher storage and processing capacity are better devices for students to practice and communicate with each other or with their lecturers. All lecturers believed using Tablet PCs help improve hands-on practice and interactive training as compared to mobile phones. Apart from this, they also mentioned that using multimedia elements such as graphics and animations are useful in helping students grasp the complex concept of computer networks course and routing algorithms, in general. The characteristics of interactivity and practical software with immediate feedback are other suggestions recommended by the lecturers.

The question *"What is the main issue of you having difficulties teaching the concept of Dijkstra's shortest path algorithm in link state routing algorithms?"* was also included in the questionnaire. Several responses were given to this question, including the following: 'technologies which are used for teaching', 'students are having less practice', 'complex concept of the subject', 'time of teaching' and 'others (please explain)'. The lecturers were also given a chance to answer this question from their point of view. All 8 lecturers opined that insufficient practice is the main difficulty related to teaching the concept to their students. On the other hand, this issue can be solved if there are some practical software and learning applications to eliminate time and location constraints. The lecturers felt that this would save resources in terms of teaching, learning, practice and communication between lecturers and students.

From the students' perspective, 15 out of 21 students agreed with the sentence: *"Network Routing Algorithm is a difficult chapter of the Computer Networks course"*. The majority of the respondents (18 students) mentioned that they had learned Dijkstra's shortest path algorithm as part of Link State Routing Algorithms. 9 out of 21 students felt they did not have sufficient practice time to fully understand the concept of computer routing algorithms. On the other hand, 12 students said that they had had enough time to practice the subject. However, they expressed having a portable hands-on application at hand could be useful for them to practice anytime and anywhere.

In terms of difficulties related to learning the subject, the answers given by the students showed that the most important problems were the complexity of the concept and insufficient lab practice. 12 of the 21 students responded that the complex concepts of the subject. Of these, 3 students were in sixth, three students in fifth and three students in third semesters, respectively. In other words, senior and junior students with more knowledge about the subjects enrolled in the Computer Science Program are more concerned about having insufficient time to practice, more than other issue related to learning Dijkstra's shortest path algorithm. In addition, other challenges faced by the students include the types of training technologies used by the lecturers and the inadequate time of training make the subject difficult to learn. Having said this, the difficulties of learning can be somewhat mitigated if there are learning applications which can eliminate time and location limitations for learning and teaching, especially if they can provide hands-on practice.

Student perceptions and feedback generally shows that they prefer using interactive instruments such as mobile devices to practice Dijkstra's shortest path algorithm or other algorithms at a convenient time and location. Given the low resolution capacity and small screen interface of mobile devices, the students do not consider mobile phone very useful as a self-study tool in learning Dijkstra's shortest path algorithm. Although the lecturers have not considered using mobile devices and small screen devices to teach the subject, they mentioned that using mobile learning applications can increase their learning efficiency and help them to do better on evaluations. It is important to learn that any application should be interactive and provide immediate feedback so that students can interact and share knowledge among their peer groups.

VII. Conclusion

In conclusion, the lecturers and students in the computer science course have not thought extensively in terms of whether or not mobile devices are suitable for learning and teaching the Dijkstra's shortest path algorithm. They mentioned that although small screen devices can help save time and eliminate location restrictions for students and lecturers to communicate with each other, they still have many limitations such as limited processing capability, low resolutions and high costs to support learning a technical subject.

This study suggests that mobile devices should be considered to better integrate mobile technology into the teaching and learning of technical subjects. In general, it is recommended that *"the learning application should be interactive and provide immediate feedback. It should also allow students to communicate with each other and encourage teamwork. Communicating with a server and using multimedia elements to deliver content is considered more useful and applicable for students to practice more frequently and understand the concept of a subject in an indirect and unconscious manner, thus acquiring greater practice at their convenience"*. Based on these recommendations complied from lecturer and student opinions, a mobile device can be effectively used if given a larger small screen interface on smart phones or tablet PCs. However, tablet PCs are judged to be more useful mobile devices than mobile phones for learning this technical subject.

Applying user and experts opinions and suggestions are very important for providing a design structure for mobile-based learning applications. Observing user perceptions of an m-based learning application and implementing them in the design and development process can increase user performance and their grasp of concepts in technical subjects. A simple navigation and usable interaction design will definitely support learning and motivate learners to learn technical subjects.

Our future study will identify a suitable design structure to motivate and encourage learners to learn about Dijkstra's shortest path algorithm on mobile devices. It will be integrated after comparing its effectiveness in teaching and learning via tablet PC and mobile devices. Furthermore, usability testing will be conducted for the improved application. The application will be tested by computer science lecturers who are familiar with the subject and their students. The application will be compatible with most mobile platforms by using Java programming language.

REFERENCES

[1] S.M, Alessi, and S.R, Trollip, *Multimedia for Learning: Methods and Development,* 3rd ed., Boston: Allyn and Bacon, 2001.

[2] J. Arreymbi, E.A. Agbor, and M. Dastbaz, "Mobile-education: a paradigm shift with technology," *Proceedings of World Conference on Educational Multimedia, Hypermedia and Telecommunications*, vol. 2008, 2008, pp. 5114-5122.

[3] A.D. Biello, "A model for developing interactive instructional multimedia applications for electronic music instructors," *(Doctoral dissertation, Nova Southeastern University, 2005, Dissertation Abstracts International, UMI Microform 3206011, 2005*

[4] N.M. Cemal, M. Dastbaz, and L. Bacon, "Research and design challenges for developing personalized e-learning system," *Proceedings of World Conference on E-Learning in Corporate, Government, Healthcare, and Higher Education*, vol. 2008, pp. 2536-2542, 2008.

[5] J.M. Creswell, *Research design: qualitative, quantitative, and mixed methods approaches,* 3rd ed., London: SAGE, 2009.

[6] A. Drigas, L. Koukianakis, and Y. Papagerasimou, "A web based e-learning and e-psychology modular environment," *Proceedings of International Conference on Next Generation Web Services Practices*, pp. 168-174, 2006.

[7] M. Fetaji, "Designing usable M-learning application: mobile view case study," *Proceedings of World Conference on Educational Multimedia, Hypermedia and Telecommunications*, vol. 2008, pp. 1251-1256, 2008a.

[8] M. Fetaji, "Devising a strategy for usability testing of M-Learning applications," *Proceedings of World Conference on Educational Multimedia, Hypermedia and Telecommunications*, vol. 2008, pp. 1393-1398, 2008b.

[9] M. Fetaji, "Case Study: Analyses of factors that influence mobile applicative software," *Proceedings of World Conference on Educational Multimedia, Hypermedia and Telecommunications*, vol. 2008, pp. 1468-1475, 2008c.

[10] B.A. Forouzan, and S. C. Fegan, *Data communication and computer networks,* 4th ed., New York: McGraw-Hill, 2007.

[11] M. Hentea, "Research on the factors of affecting the mobile learning," *Information Technology: Research and Education*, pp. 91-95, 2004.

[12] Y. Jun, and H. Zhi-yi, "Learning anywhere anytime with mobile devices," *Artificial Intelligence and Education (ICAIE)*, pp. 768-771, 2010.

[13] A. Khodabandeh, H. Afshari, and A. Manian, "Using mobile communication technology in high school education: Motivation, Pressure, and Learning Performance," *World Conference on Educational Multimedia, Hypermedia and Telecommunications*, vol. 2010, pp. 5025-5047, 2010.

[14] R. Kumar, *Research methodology: a step-by-step guide for beginners* 3rd ed., London: SAGE, 2011.

[15] R. Kuo, M.C. Wu, A. Chang, M. Chang, and J.S. Heh, "Delivering context-aware learning guidance in the mobile learning environment based on information theory," *Seventh IEEE International Conference on Advanced Learning Technologies*, pp. 362-366, 2007.

[16] S. Mahadevan, "Visualization methods and user interface design guidelines for rapid decision making in complex multi-task time-critical environments," *Dissertation Abstracts International, UMI Microform (Doctoral dissertation). Wright State University*, 2009.

[17] G.R. Morrison, S.M. Ross, and J.E. Kemp, *Designing Effective Instruction,* 4th ed., Hoboken: John Wiley & Sons, 2004.

[18] R.P.I. Patric, Q. Gao, and L.M. Wu, "Using mobile communication technology in high school education: Motivation, Pressure, and Learning Performance," *Computer & Education*, vol. 50, 1-22, 2008.

[19] K. Sandrasegaran, W. Cheung, R. Hossein, V. Zia, and A. Malwany, "A computer-based learning Environment for Electronic fault diagnosis," *Electrical and Computer Engineering*, pp. 237-240, 1993.

[20] D. S. K. Seong, "Usability Guidelines for Designing Mobile Learning Portals. The University of Nottingham", Malaysia, *The 3rd International Conference on Mobile Technology, Application and Systems – Mobility 2006.*

[21] T.H. Tan, and T.Y. Liu, "The Mobile-based interactive learning environment (MOBILE) and a case study for assisting elementary school English learning," *Proceedings of the IEEE International Conference on Advanced Learning Technologies (ICALT'04),* pp. 530-534, 2004.

[22] G A.S. Tanenbaum, *Computer networks* 4th ed., United States of America: Pearson Education, 2003.

[23] M.W. Tong, Z.K. Yang, Q.T. Liu, and X.N. Liu, "A novel content adaptation model under e-learning environment," *36th ASEE/IEEE Frontiers in Education Conference*, pp. 1-5, 2010.

[24] H.Z. Ye, R. Li, and M.L. Geng, "Research on the factors of affecting the mobile learning," *3rd International Symposium on Knowledge Acquisition and Modeling*, pp. 313-316, 2010.

[25] E.V. Zele, T.V. Hoecke, J. Lenaerts, and W. Wieme, "An electronic environment for physics laboratory work: ELO4PFYS," *EUROCON 2003, Computer as a tool, The IEEE region 8*, pp. 271-275, 2003.

AUTHORS

M. Seraj is with the Faculty of Multimedia Creativity, Limkokwing University of Creative Technology, 63000 Cyberjaya, Selangor, Malaysia.

C.Y. Wong is with the Interface Design Department, Faculty of Creative Multimedia, Multimedia University, 63000 Cyberjaya, Selangor, Malaysia

Classroom Response Systems in the Wild: Technical and Non-Technical Observations

Jonas Vetterick, Martin Garbe, Andreas Dähn and Clemens H. Cap
University of Rostock, Rostock, Germany

Abstract—**Classroom Response Systems (CRS) provide lecturers a communication channel to get feedback from their students. In lessons with large audiences, CRS allow students to ask questions or state issues as the lesson continues. During the development and usage of our new CRS "Tweedback", we observed several technical and nontechnical problems, which are likely to be general CRS issues. Observed problems are caused by necessary devices, their connectivity and lecturers' and students' different ways to use CRS. In this paper, we describe our observations of technical and nontechnical problems and suggest solutions, which may be applied generically to interactive feedback systems.**

Index Terms—**Classroom Response Systems, Audience Response Systems, Live Feedback, Large Audiences**

I. Introduction

Smartphones, Tablets and Notebooks have become students' everyday companions. They are used for communication, entertainment, and assistance in various scenarios. Over the last years, researchers started to use these devices to create an additional communication channel between lecturers and students in lessons with large audiences. So called "Classroom Response Systems" (CRS) allow lecturers to get feedback from their students as well as it allows students to get feedback from the lecturer while the lessons proceeds [3].

Whereas there are several types of feedback that can be given during a lecture, we use the following three types [18]. First, students may submit questions to both the audience and the lecturer. To indicate important questions an up voting mechanism may be implemented. We call this a Chatwall. Second, lecturers can challenge their students with quizzes, i.e. multiple choice questions. Third, students can provide feedback to lecturers by rating specific parameters of their speech, for example talking speed or volume.

This publication provides an overview on approaches, existing projects and current solutions for live feedback in lessons with large audiences developed at German universities. Whereas a live feedback application has to deal with many technical problems, nontechnical problems are often invisible – until they grow up and become an obstacle for users. Thus, we documented technical and nontechnical problems we observed in CRS and suggest solutions to them. We hope that future and existing projects can use these results to improve their work.

In general, we observed three main technical difficulties. First, the wireless communication infrastructure has to be available permanently and must carry the huge amount of data, which may occur in large lessons with more than 100 students. Second, lecturer's screen stand and screen angle can become important issues to lecturers. Insufficient stands were very frustrating experiences for lecturers. Third, a CRS should be aware that all user input has to be reviewed prior to processing, as users may otherwise misuse the communication channel offered by the CRS.

In addition to technical problems, we observed several nontechnical problems with CRS: Students want to reply to other students questions and they want to access a lesson's content even after the actual lesson ended. Furthermore, young students seem to have a higher motivation to write comments, which are unrelated to the content of teaching. This behavior abates with the time of use, but lecturers should be aware of it.

This paper provides solutions regarding these observations and resulting issues. A fine-grained publication model provides students a communication channel which enables them to address only relevant recipients. Moreover, a CRS integrated into the existing presentation application reduces the lecturer's distraction.

The rest of the paper is organized as follows. Section II presents and compares existing projects. Section III describes the technical observations we made regarding our CRS' use, whereas section IV covers nontechnical observations. In section V we present our solutions to technical and non-technical problems. Section VI concludes the article with a summary and an outlook on future work.

II. Related Work

The following sections give an overview on the most advanced projects and approaches, which allow at least one of the three live feedback types as previously defined. We compare several attributes of these projects (Table I). To categorize CRS we start by distinguishing between native and web applications, as devices not supported by native applications reduce the number of people, who can give feedback. Whereas native applications benefit from their underlying operating system and are able to use device sensors or can be integrated into the system (and are often easier to maintain), web applications can be used on nearly every mobile device [7].

Furthermore, we take a look at different usage barriers. A question which reveals the asking person's identity may build up a barrier, as students may hesitate to ask questions perceived as embarrassing. Anonymity or pseudonymity may decrease students' inhibition to provide feedback [4], because anonymity and pseudonymity prevent students from identification by their lecturers [10].

TABLE I.
COMPILATION OF CRS

	ARSNova[1]	inVote[2]	myTU[3]	Pingo[4]	Tweedback[5]	Smile[6]
Native or Web App	Web app	Web app	Native	Web app	Web app	native for lectures & web app for students
Needs Authentication	No	Only Teachers	Yes	Only Teachers	No	Yes
Anonymity or Pseudonymity	Anonymity	Anonymity	Anonymity	Anonymity	Pseudonymity	Anonymity
Open Source	Yes (GPL)	No	No	No (Planned)	No (Planned)	No
Single App or Integrated	Both	Integrated	Both	Single app	Single app	Single app
Overlay Functionality	Yes	No	No	No	No	Yes
Chatwall	No	No	Yes	No	Yes	Yes
Quiz	Yes	Yes	No	Yes	Yes	Yes
Speech Parameter	Yes	Yes	Yes	No	yes	Yes

Another barrier of use is integration into the university's software system. While an application can benefit from integration into existing university software system, it is often a problem to use this application without this software system [7]. Deeply integrated applications may exclude users, who are not members of this system, e.g. visiting lecturers.Finally we inspected projects for their implementation of the live feedback types Chatwall, Quiz and Speech Parameter.Kundisch et al. implemented Pingo, which aims to support the peer instruction concept for large classes [8] [14] [17]. By using a public accessible web application, it can be viewed on every modern browser, regardless of the used device. At the moment, lecturers have to create an account in order to use it, whereas students can access lessons without creating an account. Pingo features only quizzes and is not integrated into any existing presentation software. A great advantage of Pingo is its capability to create questions in advance. Moreover it provides an overview on all created and used lessons. According to their statistics[7], Pingo is well accepted (more than 1700 created lessons between October 2012 and November 2013).

Feiten et al. developed Smile [1] [2], an application that consists of two different parts. First, Smile provides a Windows application for lecturers. Second, students can access Smile using a web application, an app on iOS, or an app on Android. The lecturer's application is used to create and control lessons, so lecturers are able to create their Quiz questions in advance. Furthermore, Smile allows students to ask and answer questions, which may be used as replacement for a Chatwall. Moreover, students can state their general understanding: They can send a signal whenever they do not comprehend the lecturer. Smile has a highly adjustable user interface, featuring many settings. The developers found an elegant way to combine the Quiz feature with the lecturer's slides presented in a third party application: By overlaying the slides with a small transparent window containing the current quiz' results, the lecturer is able to integrate the quiz into his presentation easily. Whereas this user interface allows lecturers to configure the application for their needs, Weber et al. state that the usability and responsibility of Smile does not yet fit students' needs [19].

Gommlich et al. extended their university's personal learning environment myTU with a functionality, which allows students to rate the speed of the lecturer's presentation [6]. It also features a generic "Stop" button, which may be used whenever students have an issue regarding the lecture or the lecturer's presentation. Moreover, students may ask short questions, which are immediately shown on the lecturer's device. The myTU-App can be considered more an assistant for student's life than a pure CRS: It is used to manage students class schedules, their lending from the library, the cafeteria plan and provides many other tools supporting students' everyday life. Because this application has been developed for the TU Freiberg, it is deeply integrated with the university's software system and every user needs a university account to use the app on Android or iOS.

ARSNova is a CRS implemented as a web application. It may be used anonymously or after authentication with a university, Facebook, or Google account. It is possible to integrate ARSNova into different learning management systems such as Moodle, Ilias, or StudIP. It provides functionality to create real-time quizzes and to rate a lecturers Speech Parameters. Furthermore, a presentation view similar to Smile allows lecturers to overlay the presentation slides with the current quiz' results [12].

Tweedback implements all three previously described features, and is provided as a public accessible web application. Therefore, no accounts are needed to create or participate in a lesson. Tweedback provides pseudonymity for students and lecturers and is not integrated in the universities software system [5]. Currently it is subject to a refactoring process. Subsequently to this process, Tweedback will be released under an open source license

inVote is a project developed by Netzmanufaktur GmbH for TU Dresden. It serves real-time quizzes and is implemented as a web application. Lectures have to create an account prior to starting a quiz, whereas students can answer anonymously and without an account to these quizzes [16].

[1] https://arsnova.eu
[2] http://invote.de
[3] http://mytu.tu-freiberg.de
[4] https://pingo.upb.de
[5] http://tweedback.de
[6] http://www.smile.informatik.uni-freiburg.de
[7] http://pingo.upb.de/stats

Table I compares the discussed projects. It provides all properties and features considered important. It seems remarkable, that all presented projects provide anonymity or pseudonymity. Only Tweedback and Smile provide all three kinds of live feedback, whereas ARSNova and Smile stand out with their overlay functionality, which provides information directly on a lecturers presenting screen.

III. Technical Observations

Due to the wide spread presence of smartphones, it seemed plausible to use a purely web-based architecture. Most of our students own a smartphone, with which they can use the Universities Wi-Fi network to connect to the Internet. Moreover, GSM and UMTS connections can also be used. In contrast to CRS hardware solutions, e.g. "Qwizdom" [13] and "Powervote" [11], which build up their own communication infrastructure, web-based architectures use existing Wi-Fi infrastructure. Therefore, they also share problems, such as connectivity problems caused e.g. by massive peer-to-peer file sharing. Our observations regarding communication infrastructure, input devices, and software requirements are described in the following section.

A. Communication Infrastructure Availability

Prior to using a web-based CRS, Wi-Fi has to be available in lecture halls. Not only the Wi-Fi availability but also the capacity has to be checked. It is important to know the exact location of installed Access Points (AP) and which area they cover. It is difficult to make estimations about upper bounds of the number of users in a Wi-Fi network. This number depends highly on the users' behavior, installed hardware, and the web application. Consulting the corresponding IT administration and using their experience is one way to estimate the maximum number of users which an AP can handle in practice.

B. Input Device

The process of introducing a web-based CRS to lecturers consists of the following steps:

- Verification of communication infrastructure
- Learn how to use the CRS
- If necessary: Introduce additional display/input device

Lowering the initial barrier is important for a wide acceptance. Therefore, offering support for lecturers new to a CRS can help accomplishing the tasks mentioned above. Depending on the CRS' capabilities, an additional display/input device is needed. Only a small number of CRS, namely ARSNova and Smile, have some sort of overlay window for presentation slides. Anyhow, this overlay shows only a small amount of information. For example, it is possible to show a smiling face if users are satisfied with the current talk or a sad smiley if it is not understandable. For features showing more information content, such as a Chatwall, a small overlay is not feasible. To keep an overview on new content, it is important to avoid interrupting the teaching process. Therefore switching between presentation and CRS window on one device cannot be recommended. One solution is using additional display devices, e.g. tablet computers.

When mounting the tablet, horizontal and vertical angle are important to assure an optimal view on the screen. Many stands for tablets are designed to place the tablet in a fixed place for watching films or reading a book. Consequently, many stands are insufficient for a CRS scenario, in which lecturers interact with devices. Lecturers scroll a webpage, activate features and highlight questions. Many stands are not designed for pushing devices at the upper corners. This physical problem has – regardless of how trivial it may seem – to be respected when choosing a stand, as this detail has significant influence on the lecturers' experience with a CRS.

C. Platform

During the design of Tweedback only very few assumptions regarding the user's platform were made. The only requirement was a browser application and integrated Wi-Fi. This requirement is fulfilled by almost all modern tablets. Therefore, building a web application with user interface frameworks, such as the Bootstrap framework, and the use of standard technologies (HTML, JS, and CSS), minimizes restrictions on end-user devices.

On the server side, we were able to verify the experience with users the Pingo project [17] made. As we are working with pseudonyms, every user has a nickname such as "user123" or "user211". In early versions of Tweedback this information was saved in a cookie. First tests with Computer Science students lead to massive manipulation of usernames. By changing the cookie's content, students changed their nicknames. Therefore it is highly recommended to treat all transmitted data send by users as possible manipulation attempts.

IV. Nontechnical Observations

Further observations deal with user behavior and new usage scenarios for lecturers.

A Chatwall feature can have a restricted functionality by design. One possibility is to allow only new questions and to deny responses. This way, users can only ask questions and lecturers have to answer them. The idea of providing a restricted usage is also supported by Cognitive Theory findings [9][15] which states that working memory of human cognitive system consists of channels which can be overloaded. Too many pictures or too much text overloads these channels. Too much unstructured information has the same effect.

As a consequence, we decided not to allow users to chat with each other or to comment on other posts.

Contrary to our intention, users began to use our system for communication with each other anyway. They replied to other posts by prefixing new posts with terms such as "@user123". This behavior shows that there is a need for a reply function which contradicts with recommendations of didactic researchers.

In lessons with high activity, the Chatwall has to serve several needs: On the one hand, users want to see the latest posts and vote on them. On the other hand, users want to see the "hot topics", i.e. those questions with most votes. Providing different sorting of questions turned out to be important for lecturers. In general, lecturers use a "most voted" view of the Chatwall, as this view emphasizes questions and problems addressed by most students and thus probably important. If new questions are the lecturer's focus, the "latest posts" view can be used. Anyhow, in case of a large audience with many students who ask many questions, it can be hard to overview all new messages in short time.

We were able to distinguish between usage patterns, which seem related to the duration of CRS usage. We identified an initial phase concerning students who used Tweedback for the very first time. They began to "play" with it, creating lots of small messages without lecture-related content. The initial phase is followed by a productive phase, in which the amount of such messages drastically reduces. The productive phase lasted for the rest of the term. From the lecturers' view, especially in the initial phase many messages can be classified as spam, as they do not address a problem related to the current lecture. This behavior was expected and after some lectures, we observed a normal behavior where students asked only serious questions and used the system as intended.

We also experienced a demand to access all results after the lesson ended. Lecturers wanted to evaluate and analyze the data that has been generated by the students to learn and understand their students' problems. Students on the other hand wanted to share these emerged problems in order to get a deeper knowledge in specific topics, slides, formulas, definitions, or simply to solve issues that are not directly related to the actual lesson's subject. For this reason, durable access to the lessons results turned out to be very important.

In addition to the ability to have durable access to a lesson's data, some lectures demand a functionality to ask questions before the lesson starts. Lecturers may use such a feature to check their students' knowledge of the previous lessons or to ask questions which may give an overview on the students' motivation ("Which semester are you?", "Which study path are you associated?", ...). Because students are not entering the lesson simultaneously, it may be useful to ask such questions in a loop before the lesson starts.

V. SOLUTIONS

Our observations showed that users want to comment on questions and find themselves a way to override the system's restrictions regarding replies on other comments. Keeping this experience in mind, we will implement a reply-feature for the Chatwall. In the first place, this seems to contradict with the didactic researchers' recommendation to not create another communication channel between students. But the problem is that there already exist other communication channels once the smartphone is up and running within a lesson. All apps on smartphones which allow communicating with others present a form of parallel communication channel. Therefore, our aim is to fulfill students' need and implement a reply function as it may keep them in the CRS. Because there are already other communication channels available, we believe that uses will not be more distracted than before. In case of many replies to a question,, the Chatwall page will become very long. Additionally, this will make it more difficult to keep track of current questions in short time. Therefore, we suggest different views on the Chatwall. In the following, we will differentiate between three motivations of a reply and explain their resulting views.

First, private replies (which are exchanged between two users) will reduce messages perceived as "noise" by other users. For a better overview, these messages are not shown to other users than the regarding recipient since others were not addressed explicitly.

Second, semi-private (audience only) replies reduce messages perceived as "spam" by lecturers. We observed a relevant amount of questions which are directed to the audience, e.g. organization of collective lunch. Lecturers perceive these messages as spam. Allowing for semi-private messages lets the audience communicate without disturbing teachers.

A possible third kind of replies are public comments. Such comments are visible to all users who have joined a lecture. Separating these three reply motivations and creating a unique view for each of them enables users to communicate without disturbing those not interested in the communication. Figure 1 summarizes all three motivations and their covering area in an auditorium.

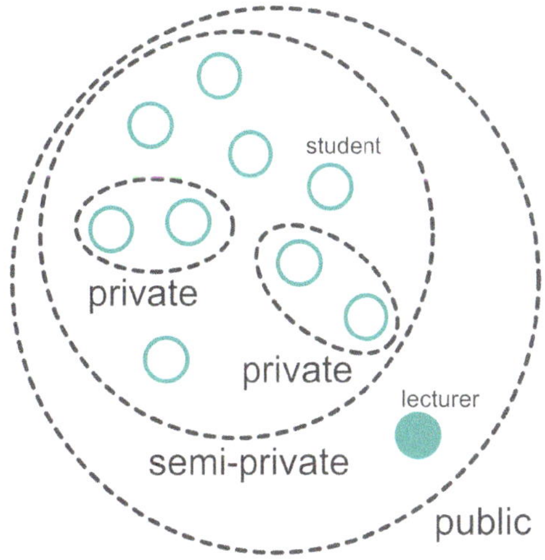

Figure 1. Shows a lecturer and the auditorium as well as the communication spaces of public, semi-private (audience) and private communications

Different needs for sorting, as stated above, can be implemented in a simple way. Depending on their needs, users can switch between two views of Question Wall. The view "voted most" will be primarily used by teachers. This way they keep an overview of questions many people voted for and they are probably important. The second view "latest questions" aims for students, as they can recognize new questions without wasting time to scroll down.

Up to now, we recommend using an additional tablet device when using a CRS in a lesson. This measure eases interaction with the Chatwall. This contradicts with the obvious interest of lecturers who want to carry as few devices as possible. To solve this problem, we plan to incorporate features into presentation software such as Microsoft PowerPoint. Once implemented, lecturers will not need to switch between devices or applications in case of Quiz and Speech Parameters.

There exist other solutions, e.g. to incorporate the Chatwall into presentation slides on audience devices. Our vision is that students can write their comments live directly into the slides. This would be a great advantage, as many questions refer to specific parts of a presentation. If students could leave comments directly in the slides, e.g. as annotation, the problem would be directly related to its origin. We imagine a web-based solution for this scenario, where presentation slides are available online. Users can view the slides simultaneously on the presenter's projection as well as on their own device. Furthermore, students may highlight or comment interesting parts visible to other students.

Concerning the feature of durable access to a lesson's user generated content, we implemented an interface showing all this information. Therefore we store and display all quizzes and all Chatwall posts. The Speech Parameter will be shown as a line graph, where each line represents the ratings of a specific parameter's development during the lesson.

VI. SUMMARY

Using students' smartphones to access a CRS has many advantages. To use a web based CRS, a communication infrastructure, such as Wi-Fi network, is necessary.

Getting an overview of the audience's feedback in short time is very important to lecturers. A Chatwall can contain many questions. Therefore, we recommend an external device (e.g. a tablet computer) to present this additional information to lecturers.

We observed that students use the communication channel in lessons to reply on questions – in groups or personally. Thus, we suggest providing private, semi-private (audience only) and public conversations. This allows users to comment on other users' posts and preserves a low spam level. It has to be evaluated whether this new possibility of commenting is distracting users significantly more than the existence of other communication channels on the smartphone.

Another observation were repeated requests for different sortings regarding the Chatwall. We identified two different sorting methods: 'by date' and 'by upvotes'.

Additionally, we observed the need for durable access to a lesson's results. This feature should be implemented in a read-only view, where all Chatwall posts and quizzes are listed. For Speech Parameters, a line graph showing ratings of each parameter over a lesson's duration seems a promising visualization.

Future research should focus on a wider application for this real-time communication channel between audience and lecturer. This may cover different types of live feedback, such as comments and annotations directly in the lecturer's slides. Furthermore, there is need for long-term didactical evaluations in CRS with large audiences.

REFERENCES

[1] L. Feiten, M. Buehrer, S. Sester and B. Becker, "SMILE-Smartphones in Lectures-Initiating a Smartphone-based Audience Response System as a Student Project," in *4th International Conference on Computer Supported Education (CSEDU2012)*, 2012, pp. 288-293.

[2] L. Feiten, K. Weber and B. Becker, "SMILE: Smartphones in der Lehre – ein Rück- und Überblick," in *GI2013, 43. GI-Jahrestagung, Workshop iLearn*, 2013.

[3] C. Fies and J. Marshall, "Classroom response systems: A review of the literature," in *Journal of Science Education and Technology*, vol. 15(1), Springer, 2006, pp. 101-109.

[4] M. Freeman and A. Bamford, "Student choice of anonymity for learner identity in online learning discussion forums," in *International Journal on E-learning* 3.3, 2004, pp. 45-53.

[5] M. Garbe, J. Vetterick and C. H. Cap, "Tweedback: Online Feedback System for Large Lectures," in *GI2013, 43. GI-Jahrestagung, Workshop iLearn*, 2013.

[6] F. Gommlich and G. Heyne, "Persönliche Lernumgebung - Architektur für Smartphones," in *GI2013, 43. GI-Jahrestagung, Workshop iLearn*, 2013.

[7] W. Jobe, "Native Apps Vs. Mobile Web Apps," *in International Journal of Interactive Mobile Technologies*, Vol. 7(4), 2013.

[8] D. Kundisch, M. Sievers, A. Zoyke, P. Herrmann, M. Whittaker, M. Beutner, G. Fels and J. Magenheim, "Designing a web-based application to support peer instruction for very large groups," in *International Conference on Information Systems (ICIS2012)*, 2012.

[9] S. L. Muthkumar, "Creating interactive multimedia-based educational courseware: cognition in learning," in *Cognition, Technology and Work*, Vol. 7, 2005 pp. 46-50.

[10] A. Pfitzmann and M. Köhntopp, "Anonymity, unobservability, and pseudonymity—a proposal for terminology," in *Designing privacy enhancing technologies*, Springer Berlin Heidelberg, 2001.

[11] Powervote, "Powervote: http://www.powervote.com [23.01.2013]," 2013.

[12] K. Quibeldey-Cirkel, C. Thelen, P.-C- Volkmer, D. Gerhardt, D. Knapp and J. Kammer, "ARSnova: ein Audience Response System für Inverted-Classroom-Szenarien mit Unterstützung von Just-in-Time Teaching und Peer Instruction," in *DeLFI 2013--Die 11. e-Learning Fachtagung Informatik*, Gesellschaft für Informatik, 2013.

[13] Qwizdom Ltd, "Qwizdom Q5: http://www.qwizdom.co.uk/q5.php [23.01.2013]," 2012.

[14] W. Reinhardt, M. Sievers, J. Magenheim, D. Kundisch, P. Herrmann, M. Beutner and A. Zoyke, "PINGO: peer instruction for very large groups," in *21st Century Learning for 21st Century Skill*, 2012, pp. 507-512.

[15] W. R. Robinson, "Cognitive theory and the design of multimedia instruction," in *Journal of Chemical Education*, Vol 81, 2004, pp. 10-13.

[16] L. Schlenker and S. Beyer, "Online In Der Vorlesung-Potentiale Digitaler Medien Für Aktives Lernen," in *11. Workshop on e-Learning (WeL2013)*, 2013.

[17] M. Sievers, W. Reinhardt, D. Kundisch and P. Herrmann, "Developing Electronic Classroom Response Apps for a Wide Variety of Mobile Devices: Lessons Learned from the PINGO project," in *11th World Conference on Mobile and Contextual Learning (mLearn2012)*, 2012, pp. 248-251.

[18] J. Vetterick, M. Garbe and C. H. Cap, "Tweedback: A Live Feedback System for Large Audiences," in *5th International Conference on Computer Supported Education (CSEDU2013)*, 2013.

[19] K. Weber and B. Becker, "Formative Evaluation des mobilen Classroom-Response-Systems SMILE," In *E-Learning zwischen Vision und Alltag (GMW2013 eLearning)*, 2013.

AUTHORS

Jonas Vetterick is with the Department of Computer Science, University of Rostock, Germany (e-mail: jonas.vetterick@uni-rostock.de).

Martin Garbe is with the Department of Computer Science, University of Rostock, Germany (e-mail: martin.garbe@uni-rostock.de).

Andreas Dähn is with the Department of Computer Science, University of Rostock, Germany (e-mail: andreas.daehn@uni-rostock.de).

Clemens H. Cap is with the Department of Computer Science, University of Rostock, Germany (e-mail: clemens.cap@uni-rostock.de).

UC1 Oscillator Remote Lab for Distant Electronics Education

Saida Latreche, Zehira Ziari and Smail Mouissat
Université Constantine 1, Constantine, Algeria.

***Abstract*—In this paper, we are interested in the transformation of a face-to-face active course into a web-based active course. Indeed, distance education has been rapidly become more important around the world due to the development of Information and Communication Technology.**

However, the roles of these later must evolve to accommodate these new models, with increased student responsibility and accountability.

First, we reports on the experience of Constantine1 University in the development of 'Electric oscillator circuit' on-line course via the Moodle platform. We describe the different stages of the online course and highlight its possibility, which is not available in the face-to-face courses.

Second, we present the remote lab of UC1 University, which is developed under eSience project. It aims to develop remote labs in the Maghreb country and to link them.

Therefore, we focus on the implementation of 'Wien Bridge Oscillator' experiment, which is deployed on the interactive version of iLab Shared Architecture (ISA).

***Index Terms*—eLearning, ISA, LabVIEW, oscillator experiment, remote laboratories.**

I. Introduction

The remote laboratory is an automated laboratory where students can experiment to find out the physical phenomena. The experiment in the remote lab is a real physical setup. In many engineering domain, where education includes laboratory work unit, the students must go to the site of the practical experiences. The concept of remote laboratory overcomes not only the problem of place inconvenience, but also the problem associated with practical work schedules [1, 2, and 3]. This new possibility to perform remote experiments laboratory allows a more flexible learning and teaching and also increases the means of learning presentation. Indeed, a teacher can achieve by this way real demonstrations, without moving the presented experience.

Elab is the result of a development, which began at the end of the 90th in the framework of remote laboratory.

From this date, many experiences have been emerged; we can cite especially the Ilab Shared Architecture (ISA) developed by MIT (Massachusetts Institute of Technology) team in August 2008. It uses the web services of the Microsoft windows platform [2, 4].

Later, in 2010 the Carinthia University of Applied Science team (Austria) has also developed an image processing online lab within the ISA too [5]. Then, in 2013, a java implementation of the batched ilab ISA has been developed by university of Queensland team [6].

Therefore, UC1 team adapted the ilabs Shared Architecture (ISA) for our remote experiments. The ISA is a web service infrastructure, which permits the access to a large variety of online laboratory [5]. It is produced by three software modules: the Labserver, which realizes a physical connection to the laboratory hardware, the User Interface (Client: students in our case) and the Service Broker (communication between the Client and the Labserver). The Service Broker generates multi services like authentication, user registration…

On another hand, many universities have been working for many years on the development of its courses to be delivered on the Internet known as the e-learning.

In this paper we illustrate our Remote Laboratory (UC1 eLab) deployed on the interactive version of iLab Shared Architecture (ISA) [2, 4] and the corresponding course implemented on a Moodle platform.

This remote lab is developed under eSience project. This later is a 'euro-maghrebin' project financed by the European community. It aims to develop and link remote lab in the Maghreb country (Algeria, Tunisian and Morocco Universities) and to link them.

II. The Corresponding Online Course

This eLab is the consequence of eSience Project which is a Tempus euro-Mediterranean project (2013-2015). Indeed, UC1 (University Constantine1) team has developed a remote lab which is the first remote laboratory at UC1 University and the first in Algeria.

To improve this eLab, we have also developed on-line corresponding courses. In this paper, we present the 'Electronic Principal Functions (FPE): The oscillators' course. The full module has a total weight of 7 ECTS credits and its aim is to permit to the Master1 student of electronics to develop knowledge of oscillator circuit. At the end of this UE ('Unité d'Enseignement' : study Unit), the student must be able to embed an oscillator and to optimize the amplifier and the feedback circuit for an optimal frequency.

We Use the Moodle (Modular Object-Oriented Dynamic Learning Environment) [7] platform which is very appropriate for distant education. The Fig. 1 presents the login page of the online course.

We give an account to all registered students and allow them to access to the areas courses and tutorials.

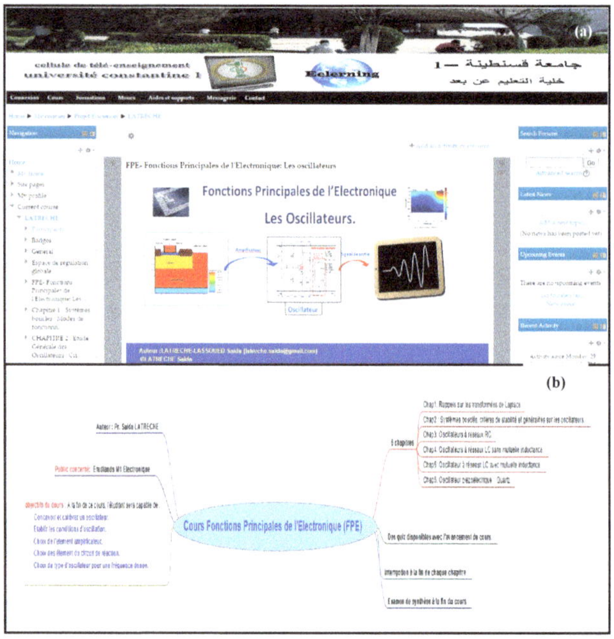

Figure 1. a-Login page of the FPE Course and b- the presentation of the course

Therefore, they consult in an interactive way the different resources. In addition, through forums and virtual meeting rooms, students are continuously in contact with the teacher. Constant assessment is carried out through:

- Quizzes,
- QCM, QCS,
- Homework.

Ongoing evaluations permit to detect possible weaknesses in understanding and to remedy immediately.

III. Design and Construction of the Remote Laboratory

The experimental setup (e-TP) used for the practical work of the 'Electronic Principal Functions (FPE): The oscillators' course are instrumented and controlled by computer, thanks to graphic interfaces. Therefore, it was relatively easy their adaptation to a remote manipulation by adding a communications module and image acquisition devices and sound.

Among the practical works of this course, one practice will be tested, which is the Wien bridge oscillator.

The Wien bridge oscillator is a device that produces low frequency quasi-sinusoidal oscillations without having a sinusoidal input source. This type of oscillator is widely used in commercial generator.

The Wien bridge oscillator is an oscillator, which uses a Wien network in the feedback chain. The operational amplifier used in this oscillator circuit is working as non-inverting amplifier mode. The aim of the remote practice is to study and calculate Wien bridge oscillations frequency. The produced sine wave has a frequency depending on R and C of the feedback circuit at operation DC power supply up to 12V. The oscillation frequency is given by:

$$f_o = \frac{1}{2\pi RC} \tag{1}$$

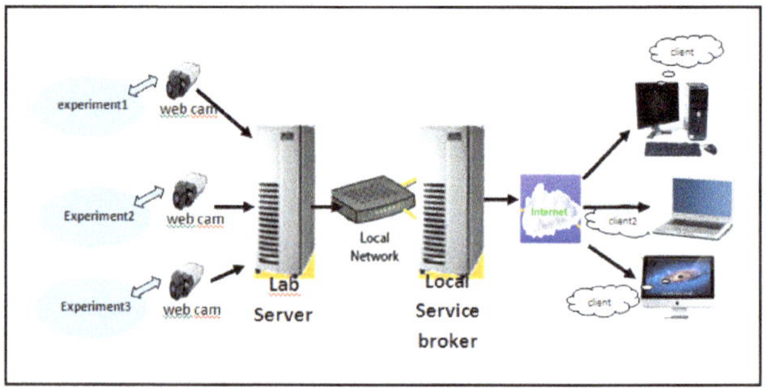

Figure 2. UC1 elab architecture.

A. *The eLab Architecture scheme*

Client server architecture was developed to enable collaborative work [5, 8]. The server is a computer equipped with necessary material to pilot the experience. This includes an acquisition system, a camera and a set of software (LabVIEW and drivers for control and signal acquisition). The server operates a list of customer requests (remote students) in order to use the system material optimally. It sends the results to the client. The client is a computer equipped with different functionalities, which enable the observation and the setting of the experiment remotely. The client application is a virtual instrument (VI) compiled for the chosen platform. After carrying out the experiment at the server, the remote client downloads the measurement results. Then, the student can analyze and discuss the experimental data in concordance with the theory.

Three modules are needed to build the client and server applications: the Labserver, the User Interface and the Service Broker.

In our case, the developed architecture is shown in Fig. 2. In addition to User interface, the system consists of:

- The Interactive Service Broker is the heart of the ISA. The Service Broker provides generic administrative services, such as authentication the user, authorization, reservation, data storage, user management, launching labs and for the preservation of the lab session with the considered laboratories [9, 10 and 11]. Therefore, the service broker has to communicate with the scheduling services, because they keep all data of the available time blocks in a laboratory and about user rights and reservations [12, 13].

- Web server, IP camera and several computers on the client side control a web Labserver that consists to realize a physical connection to the experimental module KL 200 (experiment1). The Lab server communicates with the service broker via the web server [13, 14].

B. *User Interface*

The students can access the remote lab through the laboratory's web interface, so they can manage their experiments. To get access to the eLab, the user must first authenticate himself. A login page will appear to permit to the student his identification by a username and a password (Fig. 3).

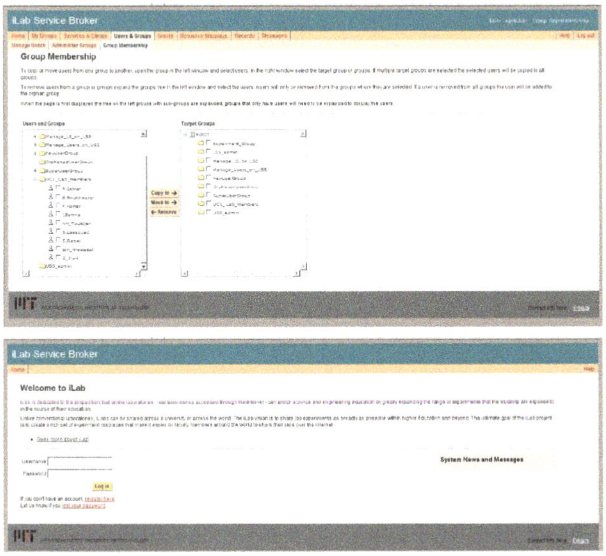

Figure 3. Service Broker, login page.

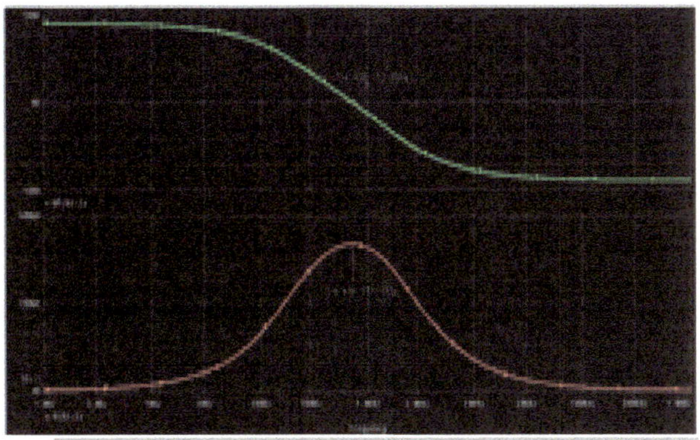

Figure 4. Response frequency simulation.

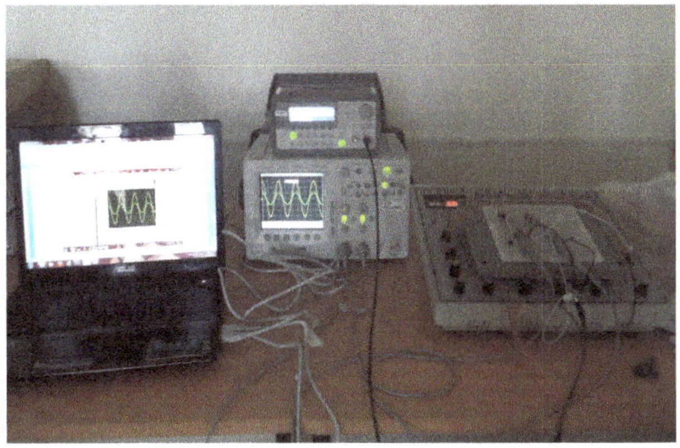

Figure 5. Wien bridge oscillator.

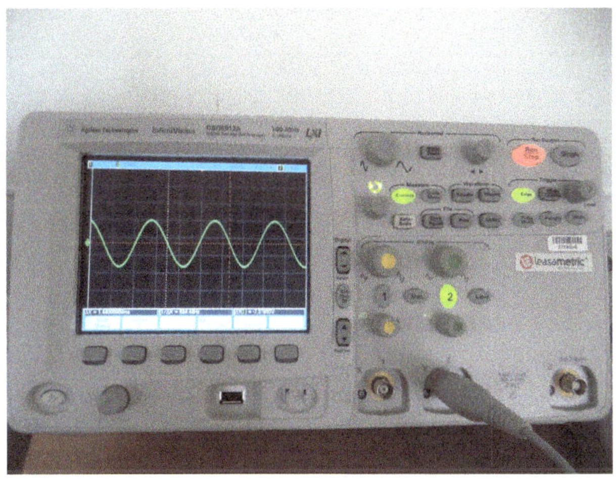

Figure 6. Output signal of the Wien Bridge oscillator.

After logged-in, the user can reserve for the chosen experiment. The date and time for this later can be picked from a schedule table.

C. PSpice Simulation

The simulation of the feedback circuit was performed by means of the Electronics Workbench Professional simulator, based on the PSpice software.

In Fig. 4 are shown both the gain in tension and the shift phase as a function of the feedback circuit frequency. The resonance frequency (f_0~ 700 Hz) is deduced from the maximum of the gain, which indicates a zero phase.

D. Practical Part (remote part)

Fig. 5 shows a typical Wien bridge oscillator design based on experimental module KL 200 [15].

There are different types of tools to implement the online laboratories. However, the LabVIEW 8.6 software (Laboratory Virtual Instrumentation Engineering Workbench) is used for remote control of equipment. It is a platform and development environment for a graphical programming language from National Instruments [16]. The LabVIEW software is installed in the OnlineLab Lab Server which controls the whole system. In our case, we use the LabVIEW platform for both the student control interface and the control of the hardware. The connection system and communication GPIB/BUS National Instrument connected to the computer USB ports is compatible with the equipment used in this practical experiment such as Agilent 33220 generator, equipped with a GPIB connection and digital oscilloscope Agilent DSO6012A, equipped with a GPIB connection. Once connected to server, students can control real experimental equipment using a visual interface. In this case, the student chose his manipulation through the student interface as shown in Figure 7 (e g, Wien bridge oscillator circuit) and all parameter settings (amplitude and frequency of the input signal). The virtual instrument (VI) front panel gives the students the ability to control the experiment. Students can follow a series of instructions to measure the amplitude of the output signal and the temporal shift between the input and output signals to estimate the phase shift introduced by the feedback circuit for different frequencies. Students can after visualize the output signal generated by Wien bridge oscillator. With the use of diodes, the non-ideal operational amplifier can produce steady oscillations. Given the measurement results (the gain and phase response versus frequency and the oscillator output signal), students analyzes and interprets experimental measurements, in link with the theory of sinusoidal oscillators and writes a report which will be evaluated by the teacher. He deduces the essential characteristics of the feedback circuit and the Wien bridge oscillator; such as its oscillation frequency, attenuation, phase shift and sine wave output signal Vs (t).

The obtained oscilloscope signal (Fig. 6) showed a sinusoidal form. Its frequency is about 700 Hz.

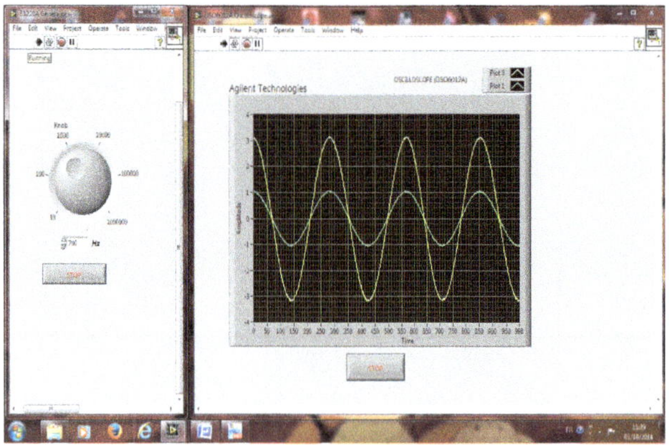

Figure7: Graphical user interface.

IV. Conclusion

Laboratory experiment is important for engineering and science students, to understand theoretical courses, as well as to prepare them for their professional career.

In this paper, we presented both a general architecture of UC1eLab for the remote implementation of practical laboratory experiment in the field of electronics and the corresponding on-line courses. Our remote lab is controlled by LabVIEW software and used acquisition interfaces as GPIB or DAQ mx 6008.

The construction of the remote laboratory on electronic oscillator circuits will enable students to experiment on specific characteristics of oscillators. A Wien Bridge Oscillator has been considered. The remote part of the experiment has demonstrated a good agreement with the PSpice simulation one.

In the future, we intend to further develop this remote lab in the aim to link the chain of experiments developed in the project Tempus eSience by Algerian, Tunisian and Morocco labs.

References

[1] T. Zimmer, M. Billaud, and D. Geoffroy, "A remote laboratory for electrical engineering education," IMCL2006, April 19-21, 2006 Amman, Jordan.

[2] V.J. Harward, J.A. del Alamo, S.R. Lerman, P.H. Bailey, "The iLab Shared Architecture: A Web Services Infrastructure to Build Communities of Internet Accessible Laboratories," *Proceedings of the IEEE* , vol. 96, no.6, pp.931-950, June 2008

[3] MIT iLab URL: https://wikis.mit.edu/confluence/display/ILAB2/Home - Last visited in 03/2010.

[4] *iLab Downloads – iLabs Dev– MIT Wiki Service.*

[5] Michael Niederstätter, Thomas Klinger and Danilo Garbi Zutin, "An Image Processing Online Laboratory within the iLab Shared Architecture", iJOE– vol. 6, Issue 2, May 2010,

[6] L.J. Payne, M.F. Schulz, "JAVA Implementation of the Batched iLab Shared Architecture", iJOE – vol. 9, Special Issue 3.

[7] Moodle Services. (2004). Welcome to the Moodle Service Network![Online]. Available: http://moodle.com/

[8] S.R. Williams, R. Blanchard, A. Mohammed, M. Bliss, R. Pancholi, M. Clowes,"The Development of a Remote Laboratory for Distance Learning at Loughborough University," *Proceedings of EDULEARN13 Conference*, 1st-3rd July 2013, Barcelona, Spain, pp. 4342-4349.

[9] J.L. Hardison, K. DeLong, P.H. Bailey and V.J. Harward, "Deploying Interactive Remote Labs Using the ILab Shared Architecture," 38th IASEE/IEEE *Frontiers in Education Conference*, October 22-25, 2008, Saratoga Springs, NY, S2A pp. 1-6.

[10] K. DeLong, V.J. Harward, P. Bailey, J. Hardison, G. Kohse, Y. Ostrocsky, "Three Online Neutron Beam Experiments Based on the iLab Shared Architecture", *IEEE Education Engineering* (EDUCON), Madrid , Spain pp . 145-150, 14 - 16 April 2010.

[11] A. Naddami, A. Fahli, M. Gourmaj, M. Moussetad, "Determination of Synchronous Machine Parameters using iLab Shared Architecture," iJOE, vol. 10, Issue 3, 2014, pp. 18-23.

[12] O. Akinwale, L. Kehinde, K. Ayodele, A. Jubril, O. Jonah et al. (2009). "A LabVIEW- based On - Line Robotic Arm for Students' Laboratory," ASEE, Austin, TX, June 14, 2009.

[13] M. Niederstätter, T. Klinger and D.G. Zutin, "An Image Processing Online Laboratory Within the ILab Shared Architecture," iJOE vol. 6. Issue 2, May 2010, pp. 37-40.

[14] J. Hardison, D. Zych, J.A. del Alamo, V.J. Ha rward, S.R. Lerman, S.M. Wang, K. Yehia, C. Varadharajan, " The Microelectronics Weblab 6.0 – An Implementation Using Web Services And The ILab Shared Architecture," *in Proceedings of the Conference on Exploring Innovation In Education And Research*, Tainan, Taiwan, 2005.

[15] KL-200 Linear circuit LAB teacher handbook.

[16] http://en.wikipedia.org/wiki/LabVIEW

Authors

Saida Latreche is with the 'Laboratoire Hyperfréquences et Semi-conducteurs' of Electronics Department of Constantine 1 University, Constantine 25000, Algeria. (e-mail: latreche.saida@gmail.com).

Zehira Ziari is with the 'Laboratoire de Microsystèmes et Instrumentation' of Electronics Department of Constantine 1 University, Constantine 25000, Algeria. (e-mail: zziari_zahira@yahoo.fr).

Smail Mouissat is with the 'Laboratoire de Microsystèmes et Instrumentation' of Electronics Department of Constantine 1 University, Constantine 25000, Algeria. (e-mail: sm_mouissat@hotmail.com).

A Mobile Based Tigrigna Language Learning Tool

Hailay Kidu
University of Gondar, Gondar, Ethiopia

***Abstract*—Mobile learning (ML) refers to the use of mobile and handheld IT devices such as Personal Digital Assistants (PDAs), mobile telephones, laptops and tablet PC technologies, in teaching and learning. Mobile learning is a new form of learning, using mobile network and tools, expanding digital learning channel, gaining educational information, educational resources and educational services anytime, anywhere .Mobile phone is superior to a computer in portability. This technology also facilitates the learning by yourself process. Learning without teacher is easy in this scenario. The technology encourages learn anytime and anywhere. Learning a language is different from any other subject as it combines explicit learning of vocabulary and language rules with unconscious skills development in the fluent application of both these things. This can only happen when language learners are exposed in authentic language use for many hours every week and this is something that mobile technologies can easily support. This article focuses on "Mobile Based Tigrigna Language Learning Tool" and implemented a tool for learn it.**

***Index Terms*—Mobile, Learning, Tigrigna and Language, Language learning**

I. INTRODUCTION

In Ethiopia, Tigrinya is the third most spoken language and the "Tigray" are the third largest ethnic group, after the Oromo and Amhara. In Eritrea, Tigrigna is by far the most spoken language, and they represent 55% of the population. Tigrinian, Tigrinyan,[15] is a Semitic language spoken in the Tigray Region of Ethiopia (its speakers there are called "Tigraway") by the Tigrinya people, where it has official status, and in central Eritrea, where it is one of the two main languages of Eritrea, and, and among groups of emigrants from these regions, including some of the Beta Israel now living in Israel. There is no generally agreed upon name for the people who speak Tigrinya. A native of Tigray is referred to in Tigrinya as tigrāwāy (male), tigrāweytī (female), tigrāwōt or tegaru (plural)[1]. In Eritrea, Tigrinya speakers are officially known as the Bihér-Tigrigna which means nation of Tigrinya speakers. Bihér roughly means nation in the ethnic sense of the word in Tigrinya Muslim native Tigrigna speakers are known as the Jeberti, an Arabic name which implies conversion to Islam among Horn Africans. Tigrinya is the third most spoken language in Ethiopia, after Amharic and Oromo, and by far the most spoken in Eritrea. It is also spoken by large immigrant communities around the world, in countries including Sudan, Saudi Arabia, Germany, Italy, Sweden, the United Kingdom, Canada and the United States.

Currently many people visit Tigray - or hope to do so one day - because of the remarkable manner in which ancient historical traditions have been preserved. Therefore anybody who wishes to visit or stay in Tigray and in Eritrea can learn the language for communication in formal class from schools, colleges or Universities. But there is no why that provides learning out of the formal class [2][3].

Tigrigna is the third most spoken language in Ethiopia, after Amharic and Oromo, and by far the most spoken in Eritrea. It is also spoken by large immigrant communities around the world, in countries including Sudan, Saudi Arabia, Germany, Italy, Sweden, the United Kingdom, Canada and the United States. Any interested Non- Tigrigna speaking citizens may also learn. Tigrigna language consists of around 300 alphabets and the language itself is highly inflectional, from one root word we can drive many words with different meaning. Therefore learning this language will take considerable amount of time. Even though the use of mobile phones is growing, there is no work done on Tigrigna mobile learning Tool.

II. RELATED WORK

Tigray forms the northernmost reaches of Ethiopia, and is located between 36 degrees and 40 degrees east longitude. Its north-south extent spans 12 and half degrees to 15 degrees north. It is bordered by Eritrea in the north, Sudan to the west, Amhara to the southwest and afar in the east. It is a very historic region noted for its custodianship of one of the powerful civilizations of the ancient world, which was centered here from at least 400 BC to the 10th century AD. Spreading far beyond modern Tigray, it molded the earliest culture of Ethiopia and left many historical treasures towering finely carved stele, the remains of extensive palaces, and the ancient places of worship are still vibrant with culture and pageantry [15].

The name of the language is Tigrinya. Tigrinya is descended from an ancient Semiticlanguage called Ge'ez. The Ethiopian and Eritrean Orthodox Churches officially use the Ge'ez as a liturgical language today, as in the past. The Tigrinya language is the direct descendant of Ge'ez, unlike Amharic (thought to be descended from a specific dialect or cluster of dialects of Ge'ez) and other southern Ethiopian Semitic languages[7][8][9]. In Ethiopia, Tigrinya is the third most spoken language and the "Tigray" are the third largest ethnic group, after the Oromo and Amhara. In Eritrea, Tigrigna is by far the most spoken language, and they represent 55% of the population. Tigrinian, Tigrinyan, is a Semitic language spoken in the Tigray Region of Ethiopia (its speakers there are called "Tigraway") by the Tigrinya people, where it has official status, and in central Eritrea, where it is one of the two main languages of Eritrea, and, and among groups of emigrants from these regions, including some of the Beta Israel now living in Israel. There is no generally agreed upon name for the people who speak Tigrinya.

A native of Tigray is referred to in Tigrinya as tigrāwāy (male), tigrāweytī (female), tigrāwōt or tegaru (plural). In Eritrea, Tigrinya speakers are officially known as the Bihér-Tigrigna which means nation of Tigrinya speakers. Bihér roughly means nation in the ethnic sense of the word in Tigrinya Muslim native Tigrigna speakers are known as the Jeberti, an Arabic name which implies conversion to Islam among Horn Africans. Tigrinya is the third most spoken language in Ethiopia, after Amharic and Oromo, and by far the most spoken in Eritrea. It is also spoken by large immigrant communities around the world, in countries including Sudan, Saudi Arabia, Germany, Italy, Sweden, the United Kingdom, Canada and the United States. Currently many people visit Tigray - or hope to do so one day - because of the remarkable manner in which ancient historical traditions have been preserved[10][11]. Therefore anybody who wishes to visit or stay in Tigray and in Eritrea can learn the language for communication in formal class from schools, colleges or Universities. But there is no why that provides learning out of the formal class.

Tigrigna is the third most spoken language in Ethiopia, after Amharic and Oromo, and by far the most spoken in Eritrea. It is also spoken by large immigrant communities around the world, in countries including Sudan, Saudi Arabia, Germany, Italy, Sweden, the United Kingdom, Canada and the United States[14][16][17]. Any interested Non- Tigrigna speaking citizens may also learn. Tigrigna language consists of around 300 alphabets and the language itself is highly inflectional, from one root word we can drive many words with different meaning. Therefore learning this language will take considerable amount of time. Even though the use of mobile phones is growing, there is no work done on Tigrigna mobile learning Tool [12][13].

III. Proposed Work

The proposed system has a client and the server application. A user with android mobile phone will be using this Tigrigna language Learning tool to practice basic skills of the language. He/she can learn the Alphabets of the language which helps for reading and writing, can also learn vocabulary words and different recorded audios will be used to help the user develop his/her listening skills. The tool can be used by a foreigner or anybody whose first language is not Tigrigna. To develop this application we use android 2.2(API) with SDK platforms as it provides the tools APIs necessarily to begin developing android applications on the android platform using android java programming language. It includes mobile device emulator- a virtual android operating system mobile device that runs on your computer. The emulator lets you develop and test android applications without using a physical android device. For editing and developing the user interface we use eclipse IDE since the tool is designed to give you a powerful and integrated environment in which to build android applications.

For the database we use the Built in SQL it that lets you store the user profiles in the database. The system shall be available for the learner as long as the learner mobile is working properly. The security of the system is related to the security of the mobile phone. Mobile phone security mechanisms such as pin code and password protection can also be used as security protection. The system is developed using object oriented approach and, therefore, adding new functionality or modifying some parts of the system will not affect the other parts of the system. The system is developed with easy to use and interactive user interface. Since the system runs on a mobile phone for better management of the small screen, the right type and amount of interactive user interfaces are used. In this future based on the learners selection the application will display all the lessons of Tigrigna alphabets. The Learner can view all the details of the alphabets in two row format user can select the second row of the alphabet then it will be displayed the sequence of the alphabets to the first row. When the learner clicks each alphabet button the learner can listen for the corresponding sound. In this future based on the learners selection the application will display all the lessons of Tigrigna words under the words tutorials are different sub category of tutorials. The Learner can view all the details of the words in three column table format the first column indicates the picture of the word the second column shows Tigrigna meaning of the word the third column shows equivalent meaning of English. When the learner clicks Tigrigna meaning of the word he/she listen text to audio synthesizer pronunciation. In this future based on the learners selection the application displays different options from the common Tigrigna conversations: like greetings, introduce one self, and leave taking under the sub categories there are different similar tutorials. The same to alphabet tutorials the corresponding sound will be displayed.

A. System Implementation

The Android application is developed by using Eclipse Helios Integrated Development Environment (IDE). Android SDK which includes a variety of custom tools that help us develop mobile applications on the Android platform is used. Android Emulator and ADT plug-in for Eclipse are the most important of these tools. ADT extends the capabilities of Eclipse to quickly set up new Android projects, create an application UI, add components based on the Android Framework API, debug applications using the Android SDK tools, and even export signed (or unsigned) .apk files in order to distribute application.

The GUI is designed using XML. Separating the user interface part from java code helps to easily modify the application without touching the java code. The sound files are stored on the sdcard of the emulator by pushing in to the emulator because the if the size of the application large it may be difficult to install on the device due to memory capacity. A total of 90 working hours spanning over 3 months have been spent on designing and developing this application and an approximate of additional 5 hours have been spent on testing the application for its correct functionality after the application has been fully developed.

Debugging of the application throughout the development is done using Dalvik Debug Monitor Server (DDMS). DDMS provides port-forwarding services, screen capture on the device, thread and heap information on the device, logcat, process, and radio state information, incoming call and SMS spoofing, location data spoofing etc. [22] Implementation of each feature in this application is explained in detail in the system description part. The basic reasons to choose Android platform, for this project, are the following:

- It uses free type, a free and open source bitmap and vector font engine, which supports Unicode text layout and glyph rendering. Ethiopic Unicode fonts, such as Microsoft's Nyala and GFZ's Geez Free Zemen, can be used to render text using free type on the Android platform. Other platforms such as the Palm OS lack the internationalization support required to fully meet the needs of Ethiopic software development.
- It is based on open source product, Linux kernel, which brings a big change from the current oligopoly of mobile OS makers that charge handset manufacturers various licensing fees.
- It comes with immediate availability to mobile versions of Google search, Google Maps, and Google Products that allow independent developers to create their own array of applications at a much less restrictive cost.

Eclipse version Helios is used as integrated development environment. In keeping with the Open Handset Alliance's theme of truly opening the mobile development market, Eclipse is one of the most fully featured, free, Java IDEs available. Eclipse is also very easy to use. The Open Handset Alliance has released an Android plug-in for Eclipse that allows Creating Android-specific applications, compiling them, and using the Android Emulator to run and debug them.

In this section several screen shots of android emulator have been inserted to show the working of different features in the application.

The Home screen of the system, i.e. the Tigrigna Language Learning Tool topics are illustrated in figure 2. This is the first screen that appears after the learner selects the Tigrigna icon from his/her mobile device. It contains the main Topics of the tutorials and exercises. The user can also view his/her profile by selecting view profile from the home screen of the application menu. Each of the menu option has a list of their own sub menus which includes different lessons.

IV. Conclusion and Future Enhancement

The Goal of this project is to build an android application which helps in learning Tigrigna language that allowed learners to learn the language anywhere at any time informally. The tool is developed using Eclipse IDE for Java, Version: Indigo Service Release 1, in mind about the design standards and maintainability of the code. Foreigners who want to visit Tigray or Eretria can use this application for communication. Mobile application promotes students to use foreign languages most frequently and also keep in touch with the culture. The intention of teaching outside the class room can be accomplished by making use of this application. Since the application can use in work areas or in home. The application has wide scope for other mobile operating system developers and client and server based application developers. As a future work, this learning tool has a vast enhancement opportunity: First, the language learning tool can be extended to other languages that are found in Tigray, Ethiopia, Eretria. Second, we can also design it in a client-server manner so that new lessons can be added to the system and the learner can download it whenever it is possible and use it next time. Third, since the tool uses a text to speech method to teach the words we can incorporate the best text-to –speech methods for Tigrigna language in the future. Fourth, this tool works only on android mobile phone therefore we can also extend it to other mobile phone platforms

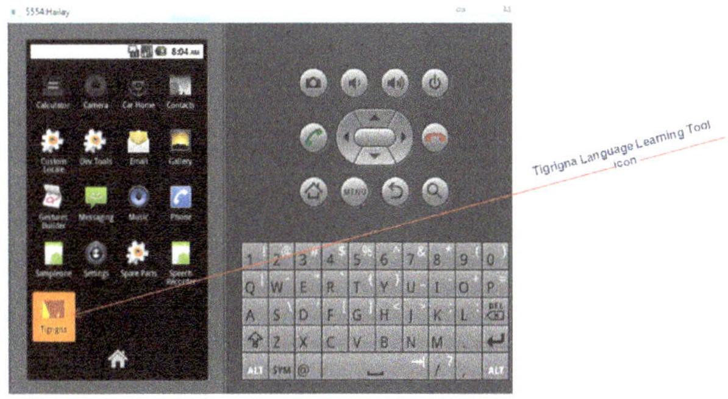

Figure 1. is the Android mobile home screen from which the Tigrigna Language Learning Tool is launched with the appropriate icon. I choose this icon because that icon is the people's flag so any one can easily identify it.

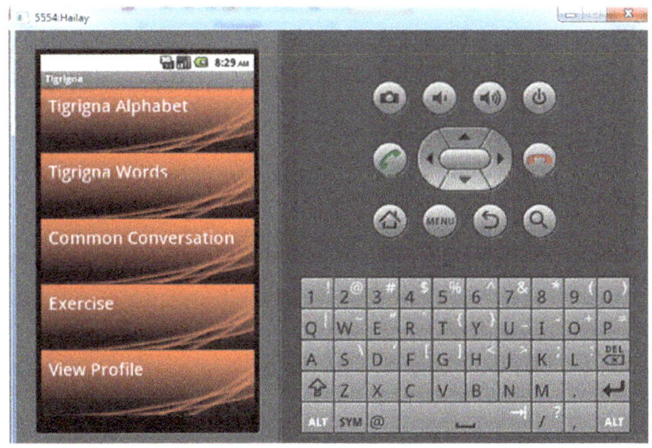

Figure 2. Home Screen of Tigrigna Language Learning Tool Application.

References

[1] Anastasios A. (2008) Requirements of Mobile Learning Applications. International Journal of Innovation and Learning, International Journal of Innovation and Learning. 5 (5): pp 457 – 479.

[2] Jill A. and Carol S. (2004) learning with mobile devices. Learning and Skills Development Agency.

[3] Xinghong L. (2009) Applied Research on the Mobile Learning in Foreign Language, Second International Conference on Education Technology and Training, pp 309-312.

[4] Morita M.(2003) The Mobile-based Learning (MBL) in Japan, proceedings of the First Conference on Creating, Connecting and Collaborating through computing. pp 203-270.

[5] Kong S., Ogata .H, Arnseth. H., Chan C., Hirashima T., and Klett F. (2009). Proceedings of the 17th International Conference on Computers in Education [CDROM]. Hong Kong: Asia-Pacific Society for Computers in Education. pp 325-525.

[6] Axumite T. (2011) Mobile Based Amharic Language Learning Tool. Addis Ababa University, Addis Ababa, Ethiopia, pp 1.

[7] Klaus R., Judith S., Norbert P., Graham A. (2011) Mobile learning: Crossing boundaries in convergent environments. ISSN 1753-3385.

[8] Vivian O. (2010) SMS-Based Mobile Learning System: A Veritable Tool for English Language Education in Rural Nigeria. Postgraduate strand at ICTD. National Open University of Nigeria. pp 1-2.

[9] Cui Y. and Bull S. (2005)Context and learner modeling for the mobile foreign language learner pp 353-367.

[10] Sara R., Shelley S.(2010) Design and Implementation of a mobile game based system to Support Chinese language speaking for International Students in a Chinese Environment.National Tsing Hua University. pp 428-430.

[11] Xinghong L. (2009) Applied Research on the Mobile Learning in Foreign Language. Second International Conference on Education Technology and Training. Pp 309-312.

[12] Milton J. (2006) Literature Review in Languages, Technology and Learning, Centre for Applied Language Studies. University of Wales Swansea pp 6.

[13] Maria L. and Vinci D. (2004) Possibilities of application of e-tools in education: mobile learning, Florence, Italy. Pp 200-2022.

[14] http://www.fotosearch.com/image-source/raw-foods-1/IGS653/ (last accessed on March 20,2012).

[15] http://www.absoluteastronomy.com/topics/Tigray-Tigrinya_people (last accessed April 8,2012).

[16] James F. (2008) A Programmer's Guide ,The McGraw-Hill Companies, pp 11.

AUTHOR

Hailay Kidu is a Lecturer, Researcher and software developer at the Department of Information Technology, University of Gondar. He received his first degree in 2007 by Information Technology, from Mekelle University, Masters Degree in 2012 by Information Technology from University of madras. He got special training in Leadership and Management Skills Development Training Program from Vision Ethiopian Congress for Democracy (VECOD), In ICT Business Incubation center from Mekelle Institute of Technology (MIT), In Solar Photovoltaic System (PV) from University of TUDelft. He was worked in the National software development Project called National Integrated Justice Information System (NIJIS) that was undertaken by cyber soft, Ethiopian software companies and chain's ZTE companies for the period of may 2012 up to December 2012. Mr. Hailay kidu is a member of different professional associations including the Information Communication Technology since 2012, and the Association of Ethiopian Teachers since 2001 and Tigray Academic and Scientific network since 2012.

Currently **Mr. Hailay kidu** is a senior Lecturer researcher and application developer at the University of Gondar, Gondar, ethiopia.He is teaching in the area of software engineering and theoretical computer science. Information Technology Department University of Gondar, p.o.box 196,Mobile, +251920253449,(e-mail: hailiyekidu@gmail.com)

Liquid Spheres on Smartphones: The Personal Information Policies

A. Serrano Tellería, M. Oliveira
Beira Interior University, Covilhã, Portugal

Abstract—**Data collected from the profiles and the digital identities has become a valuable currency for the mobile ecosystem, especially between users and providers. Services that required them are also described as the ground floor in direct linked with the infrastructures and as intermediate layers between networks, platforms and applications. The frontier debate between innovation and protection of privacy is shown off undefined and unstable. Therefore, a comparative analysis between 'Privacy Terms and Conditions' as well as the interrelation between operative systems (Apple iOS, Android, Blackberry and Windows Phone), social media platforms (Facebook, LinkedIn, Twitter and Google +) and applications (Instagram, WhatsApp, Line and Vine) were carried out focusing on Privacy issues. Two main tendencies were appreciated in relation with the two principal operative systems: Apple iOS closed environment and Google Android open source. They reconfigured the functional structure and design of platforms and applications in different ways. The liquid spheres observed varied from the first approach that tried to control every action and personal information from the binomial operative system-device and the second one that allowed the user actions and information to be more susceptible to interact with any kind of applications and platforms while the system was linked to information aggregation services to collect the data. Prominent aspects were the various stages of synchronization between the different levels of personal information (contacts, profile, digital identity and localization). Focusing on the case of Portugal, other complementary conclusions obtained from focus group and surveys showed a strong circumstantial pattern behaviour and a concern about privacy issues taking care of some actions while admitted checking if they had the terms and conditions involved - which were too ambiguous - but not reading them. Described also by other international previous researches, they showed lack of rationality in some attitudes and performances as well as limitations on the extension between knowledge and action.**

Index Terms—**Liquid Spheres, Operative Systems, Privacy, Terms and Conditions, Smartphones.**

I. An Undefined and Unstable Environment

The delimitation of the public and private spheres in the construction of the different profiles and the digital identities through smartphones on the Internet, especially with the expansion of the different type of applications and social media, has become a focus of attention both from providers and users perspective. It ought to be considered the ever-changing privacy policies of the platforms and the applications, the increasingly requested synchronization of different types of data between the operative systems and its environments, the applications and the platforms[1], the geo-localization, the emergence of companies to safeguard privacy, of platforms and apps that run the contents of others in order to improve the personal image –for example, the content curators-, to promote anonymity and to build on anonymous profiles. Therefore, the data flux in a continuum and diluted paths that reflects this undefined and unstable environment.

This uncertainty may be described from a technological perspective by the tension emerged between the willing to promote innovation and to protect user's data since mobile services –which stated to require this type of information to improve its services- are described as the intermediate layers between networks, operative systems, applications and platforms [1].

On the other hand and from a theoretical perspective, the tension arose between the transformation of people into merchandise [2] and the 'Right to be Forgotten', that European Commission is working on, specially focused on assignment an expiration date for personal data that should be applicable in the specific context of social networks. It occurs in a society where a "curious reversal" redefined the private sphere characterized by the right to confidentiality as a sphere that has become prey to the right to publicity [3]. Concerning policies, there exist two main tendencies: Europe's undefined one trying to control the market and USA and Asia letting it freewill and following the model of testing in progress.

Remembering Wellman's concept of portal[2] [4] applied to Human Being process of communication by smartphones, users exist in a timeless time [5] that is framed by the possibility of the perpetual contact [6] and in a kind of virtual configuration of the online space where the directionality and the distance are confused or undefined [7]. Moreover, in this undefined and unstable mobile ecosystem, users generate an ever-changing profile and digital identity, both conscious and unconsciously, as they have to deal with the liquid spheres constantly to be negotiated.

II. The Liquid Life and the Smartphones

Characteristics that described Digital Media were focused on the constant negotiation of rules where norms and values were not clear, on being a decentralized model with a multimedia and flexible format –constantly chang-

[1] Platforms referred to Facebook, Twitter, LinkedIn and Google +, while the environment (that may also be called platform) related to Apple iOS, Google Android, Blackberry and Windows Phone as operative systems with its relation to apps and mobile services.
[2] "It was I-alone that was reachable wherever I was: at a house, hotel, office, freeway or mail. Place did not matter, person did. The person has become the portal"[4].

ing, updating, correcting and being revised -, where content was insensible to distance and nonlinear, as well as on diverse resources fonts with fragmented audiences whose feedback were so valuable to bear in mind [8]. Those are intertwined with Bauman's metaphor of modern life, so liquid life: fluidity, transience, reticularity and dissolution of borders or boundaries defined [9].

Therefore, the characteristics of Digital Media are liquid ones as none of its frontiers are delimited and they are constantly being negotiated. Even more, the same delimitation might be considered as useless if bearing in mind that the content flows on diverse resources fonts and fragmented audiences –willing to expand-.

Concerning the process of communication by smartphones, the state of perpetual contact [6] enables people to recreate a network of protection similar to that of traditional societies [10], where people maintain a nomadic intimacy within a social system based less on location and more on themselves, so one can stay in touch on the go [11]. "This create a kind of nomadic intimacy in which the public space is no longer a full itinerary, lived in all its aspects, stimuli and prospects, but is kept in the background of an itinerant 'cellular intimacy'" [11].

More than any other media, stated Fidalgo [12] following Geser[3], is the mobile phone that restores the social relations typical of the small communities, a "throwback to pre-modern models of social life" [13].

In this trans-spatial communalism, individuals are losing the habit and confidence to think and decide for themselves due to the umbilical cord that keeps them connected - although physically far - to the original community. Nowadays, this permanent and ubiquitous connection is the cause of much tutored thought [12].

The mobile phone was described as the 7th Mass Media by Ahonen [14] because: It is the first personal mass media, it is permanently carried, it is always on, it has a built-in payment mechanism, it is available at the point of creative inspiration, it has the most accurate audience measurement, it captures the social context of media consumption and it offers a digital interface to the real world. But, as Katz summarized [15], mobile communication improves several dimensions of freedom and increases our choices in life, while it may also be turned against the user: invading personal privacy and causing emotional, political and technological distress.

The spheres liquidity favours the data recollection of personal information since perpetual contact and ubiquity may entail tutored though and lack of behaviour autonomy – for example, when users accept terms and conditions without reading them, just because other people did it before -. The nomadic and cellular intimacy 'on the go' is constantly delimiting the sphere boundaries in this diluted and unstable ecosystem. Too much effort is required from users to be properly updated and to know how to deal with technology, even more, when it is ever-changing. Furthermore, emotions are also liquid, as a result of managing permanent feelings and impulses as well as emerging motivations to be fulfilled.

Therefore, delimiting the different spheres is closely tied to the ability to manage wilfulness, taking into account the importance of the temporal priority as a relevant variable in the process. Also, the balance between the authenticity and the anonymity, the privacy and the functionality are considered key elements trying to distinguish what may be defined as public or private. Moreover, the game between the obscurity and the hyper-visibility that allows users to reach the spotlight of attention and the scope of the common space ought to be considered [16].

III. The Liquidity between the Public & the Private Spheres

If remembering Goffman (1959) description of the daily life performance, - where people move between the front-stage and the back-stage, between the public and the private spheres -, the integration of remote communications may be underestimating the importance of face-to-face interactions [17] and undermining the traditional rituals of separation in the different spheres of life [18].

In the emergence of the industrial capitalism and accelerated growth of metropolises of the nineteenth century, the representation of the individuals in public was no longer primarily a mechanism of social identification but also to be - and essentially – producer of personal meanings about each subject [19]. Thus, when and where the change between action and character as appearance occurred? The visual presentation stepped to be invested with meanings associated to personality. Under these conditions, the public system is expressed and transmuted into a system of personal representations. The personality in public considered - widespread belief - that appearance is an indicator of character, which results in the private individuals' anxiety [19].

A situation framed by the fact that electronic audio visual media were increasingly bringing elements of the individual "back-stage" for a facade region, favouring the expression of personal characteristics and exposing areas that before were private [20]. In the world of Goffman [21], people behaved but had no experience [19] where the media was converting the private space into merchandise [20]. Remembering McLuhan, the medium is the message and, it seems even more on the mobile devices, the (pro) consumer – (active) user has become the merchandise.

Bauman's metaphor of the liquid life - fluidity, transience, reticularity and dissolution of borders or boundaries defined -, consistent with the logic of the consumer society, provides a useful counterpart to address some of the characteristics of the mobility. Beyond the correlation between the impact of the digital technology and the digital features of the liquid society (which refers to the reflections on the acceleration, the dislocation, the consumption and the role of identity), the mobile medium particularly fits the parameters of fluidization of technological, institutional and cultural dimensions of the medium previously described by McQuail [22], [1].

IV. Characteristics and Structure of the Mobile Devices

The dissolution of the link between content and support, which had been the basis for the definition of genres and formats, reaches its peak with the expression of the distribution models based on storage services and cloud sync. The mobile environment is, in essence, a multi-device one, whose core lies in a conception of the mode of consumption and access to the content and services.

[3] Sociology of the Mobile Phone. University of Zürich. Online publications. Consult 18 February 2014. URL [http://socio.ch/mobile/]

The ubiquity, the diversification and the intertwining of the consumption scenarios, with a marked tendency towards transversal use of the media and access modes (multi-use), as well as its insertion into social dynamics where real identity games become objects of consumption; redefines the value perception of the contents by the users and converts them into a valuable source of the new resource of the digital economy: the personal information.

Focusing on smartphones, both Apple-iOS and Google-Android laid the basis for the environment – platform – (services, features and content) grouped around an app store, led either by an operative system – Google Android - or by the binomial terminal-operative system – iOS iPhone/iPad -. While Apple leads from its devices, Google does it based on an operative system linked to information aggregation services.

Apple relies on the tight control of their customer databases, which sells content and apps through iTunes and the App Store. Google also sells both through its Google Play Store, but its main bet does not need so much explicit information about the users, who become audiences for the commercial communication that handles Google and its technologies of contextual and behavioural targeting [23].

Both sell content, services and advertising to advertisers, but the formula for success is based on combining the three elements in different proportions and presentations, and this also leads to attract around nuanced public with different experiences. For example, studies tend to appreciate that iPhone users spend more money on their data plans and more time on Social Media that the owners of Android smartphones with similar performance [24] in [1].

Operative systems delimit programming environment, mobile platforms, user interface and experience as well as being the norm to which apps developers, providers and distributors ought to stick to. Amazon Android and Apple iOS prohibit access to rival shops. Meanwhile, platforms group the relation between the different actors through channel and services content distribution, configuring the app stores.

Thus, in addition to including specific relationships (usually external and often strained) with network operators; mobile platforms articulated hardware access (mobile and / or fixed), the operative systems and its user interface, a content management software / applications and a software development kit (SDK) with language and programming parameters specific to the environment – platform -. The content management software / applications also serves as a control of user activity, gathering information on the profile and preferences and limiting forms of income and execution in order to minimize the integration of foreign content to the platform [1].

V. The Personal Information

The use of personal information adds value to advertising (discriminated and targeted), is the basis of business models and a clear challenge to the ones of traditional content industries, offering information about users behaviours. Therefore, this raises important questions about privacy and transparency, as well as cookies tracking and data mining. In this environment, consumer protection laws are in a changing and challenging context, where it may be distinguished two main tendencies: USA and Europe.

OECD Privacy Guidelines[4] have described personal information as the information that relates to an individual identified or identifiable. There is some controversy about providers and users on the one hand and, on the other, administrations where are the limits of what is strictly personal information. A restricted conception covers the "who we are" personal data such as name, address, identification, financial records, sanitary ware, etc. A broader concept does the "what we do" behavioural data such as search, navigation, shopping, etc.; and "how, where and when we do", usage habits such as location, time, frequency, time spent, and so on [1].

A. *The Profile and the Digital Identity*

A profile is structured information about the users of digital services. Its concrete structure depends on the way the information has been gathered, the technology involved, the types of data or what the law allows. In the mobile ecosystem, this profile is usually richer than in the generic case of Internet and also is linked more clearly – mobile identifier device to the IP address - with a particular individual. Moreover, in the case of having ambient intelligence (context-awareness), the user profile may even include biophysical parameters, charge level of the battery and a rich set of variables on the environment; what may include, in addition to the location and the orientation, temperature, humidity, noise, etc. [1].

Especially mobile technologies show the biggest challenges regarding the use and the abuse of personal information. Main examples are location, and its access to Wi-Fi. A compilation of information that suppliers of the major mobile platforms, Apple-iOS and Google-Android argue that is done to accelerate the provision of different services based on the user's position [25]. According to the theory of multilateral markets, the use of personal information has become a bargaining chip between users and providers, where a strategic relationship occurs between agents involved in the platform: providers, users and advertisers.

The concept of Digital Identity (ID) arises from the combination of three main factors [26]: the concrete manifestation of the self-image of the person on the digital service in question (a professional social network, for example), the elements arising from the protection / limitation / modification or concealment of certain information that users consider according to their preferences (highlight certain employment) and the implementation of policies regarding provider data (creating a unique user ID and log all activities that occur when accessing the social network in question). Therefore, a certain individual may have multiple digital identities.

In structuring information about the users profile and combining the three factors to obtain the possible multiple digital identities, the liquid spheres are once again representative. The personal data flux in an environment supported by the norms of platforms, applications and other

[4] *OECD Guidelines on the Protection of Privacy and Transborder Flows of Personal Data.* Consult 28 February 2014. URL [http://www.oecd.org/internet/ieconomy/oecdguidelinesontheprotectionofprivacyandtransborderflowsofpersonaldata.htm].

services, each one with their own market interests and their programming code.

B. *Concerns about Privacy*

At this point, several researches showed the users concerns about privacy but they were not willing to pay for their protection in the case of UK [27] or even willing to sell it. A study indicated that 32% of Canadians would be willing to sell their digital data to the right company for the right price and 45 % would sell at least some of it[5]. Both focused on the fact that it was neither provided information nor tools to ensure contextual integrity, being the use of it out of context what most worries. Contextual integrity refers to the use of personal information in the specific context of the service employed and not outside the concrete relationship established between supplier and consumer [28].

In the USA, findings from January 2014 survey at Pew Research[6] manifested an increase of 7% in information stolen from 2013, with 18% of adults recognizing it as important ones and 21% of them with an email or social networking account being compromised. Half reported becoming more worried about the amount of personal information available online since 2009 (33%). About anonymity in September 2013[7], 86% took steps online to remove or mask their digital footprints and 55% to avoid observation by specific people, organizations, or the government.

Continuing with the case of Canada, many of those who participate in social networking used these spaces to create profiles expressly for public distribution. At the same time, and according to Dr. Burkell[8], the potential hazards of sharing personal information online – from concrete risks such as identity theft to more esoteric risks such as the erosion of personal autonomy as a result of surveillance – were too remote to influence their decisions; especially, when compared with the immediate and tangible benefits of that same sharing.

This is especially true when the collection, sharing and analysis of the personal information occurred invisibly, without our consent or even our knowledge. "Quite frankly, from an individual perspective, we still feel – many of us – as if we have privacy. And on a day-to-day basis ... the kind of surveillance we're under, the kind of oversight we're under is not evident to us," commented Dr. Burkell, "The risks we are running are not risks that are immediately evident to us".

Feijóo and Gomez-Barroso [1] explained two main tendencies to reduce privacy risk: by design and by law. Regarding technology, the PET (technologies enhanced privacy) may help to rebalance the relationship between users and providers, extracting general patterns of consumption and protecting the particular information. Other possibility will be that users may move their data between suppliers, portability of personal data, which would help to reduce the market power of these providers with respect to users, increase competition and, ideally, facilitate services that appeared more respectful with the use of personal information to prevent leakage of users to other suppliers.

These advances must be accompanied by a regulatory framework. There appear to be conclusive user's lack of rationality when confronted with digital services that are based on the personal information provided by them. To establish and implement the necessary regulation faces considerable challenges: the balance between the innovation and the consumer protection, the risks associated with invasion of privacy in the short, medium and long term and the fact that personal information is handled and transferred in a globalized world where geographic boundaries are irrelevant.

In Europe, the debate is still pretty theoretical without a stable regulatory framework, debating the 'Right to be Forgotten' by the European Union. Meanwhile, USA and Asia are leaving the market to dominate this debate. They follow a process of trial and error on innovations that are offered to users and proved somewhat uncertain to the benefits of the whole society. The European proposals on their territory are intended to be applicable regardless of the origin or the geographical location of the supplier, what requires a number of international initiatives and agreements that still seem distant.

C. *Perception of Privacy*

In the context of communicative functions, Jin Park [29] analysed three dimensions of the impact of digital literacy behaviours related to online privacy: a) familiarity with the technical aspects of the Internet, b) awareness of common and institutional aspects and c) understanding of the current privacy policy. The analysis showed a strong predictive capability of user's knowledge, but results were mixed when representing the interaction between the knowledge and experiences. There were limitations on the extensions of the knowledge and the action related to personalized information.

Moreover, these limitations were divided by socio-demographic characteristics such as age, gender, income and education. The study demonstrated the presence of a second-level digital divide in Internet privacy, apart from the level of access, both strongly influenced by temporal priority.

In this context, Fathi [30] distinguishes the following areas: perspective of security, authentication against impersonation, leakage resilient schemes, identity-based encryption for privacy, anonymity for privacy, private information retrieval for privacy and trust.

The lack of awareness about the immediate risks, the lack of rationality of users when providing personal data, the relevance of the temporal priority in literacy as well as the limitations between interaction of knowledge and action / experiences intimately related to memory, the liquid spheres from a technical perspective as from users behaviour one; lead to a worrying environment if bearing also in mind that Privacy Policies / Terms and Conditions

---

[5] "45% of Canadians willing to sell their digital data", in: *CBC News Business*. 30 January 2014. URL [http://www.cbc.ca/news/business/45-of-canadians-willing-to-sell-their-digital-data-1.2517427]. Consult 4 February 2014.

[6] "More online Americans say they've experienced a personal data breach". *Pew Research*. 14 April 2014. URL [http://www.pewresearch.org/fact-tank/2014/04/14/more-online-americans-say-theyve-experienced-a-personal-data-breach/]. Consult 14 April 2014.

[7] "Anonymity, Privacy, and Security Online". *Pew Research*. 5 September 2013. URL [http://www.pewinternet.org/2013/09/05/anonymity-privacy-and-security-online/]. Consult 14 April 2014.

[8] "Trading privacy for security in the online world", in: *GRAND NCE*. 4 February 2014. URL [http://grand-nce.ca/newsandmedia/news-container/2014/trading-privacy-for-security-in-the-online-world]. Consult 4 February 2014.

are described as ambiguous and confused for a long time – 2005 [31] -, and unfortunately, still does as our research and others concluded. Moreover, it ought to be strength the relevance of language in the cognition process.

Therefore, in this liquid mobile environment where the data fluxes without frontiers, users deal with a great difficulty in applying their knowledge to actions; partly because of their own lack of awareness, partly because the information about Privacy Policies / Terms and Conditions as well as about technologies is ambiguous and confused. Temporal priority as a key factor also aggravates their ability.

The Architecture of Intimacy [32], the Architecture of Disclosure [33] and the Interface Design of Exposure [16], representing Facebook a primary example; strengthens this liquid environment, setting a design structure that encourages exposure and visibility to the detriment of protection and privacy.

D. *Some Tendencies*

The concept of 'continuous partial attention' proposed by Stone[9] explains how "being –always- on" affects the quality that users deliver to each of their tasks, under less "mind share". Focused on the identity, it refers how people think about their lives and priorities, which may be also affected. In this regard, the "Self" may lose the sense of conscious communication choice, since the media are always hold on in the background [18]. Hypothesis supported also by Starner[10], whose research to date suggests that our ability to multitask is not as great as we think, "when we multitask we do less well on more tasks".

Experience related to the concept of "mobile identity", introduced by Stald [34] and focused on youth, identity and mobile communications. It is mainly characterized by the "fluidity of identity" – constantly to be negotiated – based on four axis: 1) availability; 2) experience of presence where the social presence in public space is being invaded by mobile communication in progress; 3) personal log for activities, networking and communication of experiences, a role which has implications both for the relationship between the individual and the group, as for the emotional experience; and 4) learning of social norms.

These tendencies are closely related to the three technology revolutions according to Pew Research Centre 2014: 'Broadband, Mobile and Social', where contacts, location and synchronization between them seemed to be an increasingly valuable resource in the mobile ecosystem, as showed by recent business models strategies like the purchase of Instagram or WhatsApp by Facebook. Another tendency underlined by MIT in its technology reviews was that some mobile apps were starting to add anonymity to social networking. Specialized ads for mobile devices, mainly local, were also a growing trend.

In this environment, less than 40% of web traffic came from Human[11] where a new generation theory of the user interface stated that there ought not to be a user interface; information ought just to be around [35]. Moreover, the ability to add new features to mobile search has just begun, with proposals for the future as MindMeld, Expect Labs (2013), a personal assistant that infers the future user behaviour from the analysis of their conversations [1].

VI. Objectives and Methodology

On the one hand, operative systems delimit programming environment, mobile platforms, user interface/experience and the norms to which apps developers, providers and distributors ought to stick to. On the other hand, environments – platforms - group the relation between the different actors through channel and services content distribution, configuring someway the apps stores. They articulate, apart from network operators, the operative system and the user interface/experience, the content management software/applications that serve as a control of user activity and the SDK, the software development kit.

Therefore, our research focused on the analyses of the interrelationships between the operative systems and its environment, the platforms and the applications choosing the privacy features configurations: Apple iOS (6.4.1), Android (2.3.5), Blackberry (5) and Windows Phone (7.5) with the platforms: Facebook, Twitter, LinkedIn, Google + and the applications Instagram, Vine, WhatsApp and Line.

Secondly, a comparison between the conditions and terms of privacy of the four platforms and the applications mentioned were carried out. Finally, the installation of all these platforms and applications were made both in Apple iOS and Android systems mainly as well as Blackberry and Windows Phone in order to appreciate similarities and differences in the process, in the results, in the interface design and in the action visibility.

Finally, some incipient conclusions from an exploratory focus group and a survey carried on in north Portugal will be added as a complement.

VII. Discussion

A. *Operative systems-environment, applications and platforms.*

Main differences were found between the Apple iOS, Blackberry and Windows Phone system and the Google Android one. If establishing a comparison with solid and liquid spheres, the first group was observed as solid and the second one as liquid, whereas iOS Jailbreak or Amazon OS, for example, were placed in the middle of an ongoing process. From the date our analyses started until nowadays, an approximation and even dissolution of boundaries have been appreciated as a tendency. Closed ones have initiating a process of openness in its system and vice versa.

Generally, Apple iOS, Blackberry and Windows Phone made efforts trying to create 'Closed Environments' while Android followed the 'Open Source' market conception.

---

[9] "What is Continuous Partial Attention?" In: *Lindastone.net*. URL [http://lindastone.net/qa/continuous-partial-attention/]. Consult 8 February 2014.

[10] "Multiplexing versus multitasking". In: *The Technium*. URL [http://kk.org/thetechnium/2011/03/multiplexing-vs/]. Consult 11 February 2014.

[11] "Report: Bot traffic is up to 61.5% of all website traffic" by Igal Zeifman. In: *Incapsula.com*. URL [http://www.incapsula.com/blog/bot-traffic-report-2013.html]. Consult 24 January 2014.

Cisco Visual Networking Index: Global Mobile Data Traffic Forecast Update, 2013–2018. URL [http://www.cisco.com/c/en/us/solutions/collateral/service-provider/visual-networking-index-vni/white_paper_c11-520862.html]. Consult 1 April 2014.

This division in operative systems related to most of other aspects analysed due to the main root characteristic observed: whether it controlled more or less all functions and relationships between its hardware and the applications and platforms installed, as well as other services. As explained in the theoretical framework, Apple iOS relies on the tight control of its user database through the binomial OS-device, while Google Android does on an operative system linked to information aggregation services and its technologies of contextual and behavioural targeting.

In this sense, Apple iOS asked the user for permission before downloading any kind of app; whereas, in the rest, users had to confirm or not the access to them after downloading. This aspect was well exemplified when the Apple iOS asked for the control of all its registers – to be able to share content after, for example - as well as all the access of them to the hardware and to the mobile data.

Another example was that the user did not need to be connected to the app or platform (with it opened) to carry on actions through it. For instance, it was observed that the user was able to share the contents through different apps and platforms without having them officially opened. User data were defined in the operative system so it was able to be used by any app and its content sharing menu in Apple iOS and in WhatsApp for Windows Phone.

On the contrary, Android and Blackberry needed to have all the applications and platforms, specifically in the case of Facebook and Twitter, installed and opened for their options of content sharing menu were able to appear in others.

Deepening into concrete questions made to all systems, analysis started measuring whether they let modify all 'Privacy Settings' or not and at what level. Remembering previous main difference explained, Apple iOS controlled all of them, including those specific ones from applications and platforms, through its operative system by itself. Any time users want to alter the configuration of 'Security' or the 'Privacy Settings' of them, they had to change it through the operative system.

On the contrary, Android established its 'Privacy Settings' in direct link with the applications and the platforms so they had to be modified from the same (apps and platforms). Due to the fact that they may come from unknown and uncontrolled authors, Android provided a specific option to let or not let install them - "installation of applications from unknown authors" -. As an example, users were not able to register the applications and the platforms that employed the GPS and, then, alter its configuration in the operative system 'Settings'.

Then, the main aspects concerning the interrelationships between OS, its environment, apps and platforms will be listed and employed as epigraphs. They were the features or applications that employ personal information for different kind of purposes.

A.1. Privacy Settings

Elements analysed were: 'Location', 'Contacts', 'Calendar', 'Reminders', 'Photos', 'Bluetooth', 'Twitter', 'Facebook', 'Phone ID', 'Safari', 'Chrome', 'Internet Explorer', 'BB Browser', 'Opera Mini', 'Backup', 'Feedback-Data Sense' (when you signed the contract), 'Transmission of data application usage' (feedback application) and 'Sharing files between applications'.

The applications listed are the ones that established a direct link between the device and the user. For example, 'contacts' were high susceptible of offering great quantity of user data. A characteristic observed in Android was the fact that contacts did not have specific privacy settings in the operative system and neither in the applications and platforms. Windows Phone followed same patterns as Android when asking about if they allow modifying options of 'Privacy Settings' in the operative system. Blackberry was also similar to Android regarding 'Privacy Settings', being the only one that offered specifically the possibility of defining 'Firewall' protection – others might have it but they did not specify it -. Another unique feature of Blackberry was that the installation of the application had to be made through the computer connection.

Then, three tendencies were able to be described from 'Location' to 'Phone ID' elements: the first one was Apple iOS with all categories included in 'Settings' and with a list of authorized applications; also allowing access to Phone ID. Second one was Android with 'Location' and 'Bluetooth' in 'Settings', specific 'Google Calendar', 'Twitter' and 'Facebook' also in 'Settings' only if you had previously installed the application and allowed access to Phone ID.

Third one was Blackberry and Windows Phone with much in common as, for example, the inability to find the 'Phone ID' in 'Settings' as well as 'Contact', 'Calendar', 'Reminder' and 'Photos' and without a list of authorized applications. Blackberry had its 'Bluetooth' defined in 'Settings' while Windows Phone had 'Location' and 'Bluetooth' in 'Settings'. From this part of the analysis, two main different models of operative systems and market models, as mentioned in the beginning of the chapter, were able to be distinguished: Apple iOS, Blackberry and Windows Phone with a control over all applications and apps stores installed through its operative system and Google Android that allowed multiple apps stores being controlled by others.

Each operative system had a default browser installed (Safari, Chrome, BB Browser, Internet Explorer). These browser settings disposed of 'Private Browsing' and their own privacy, similar to a PC browser. In the case of Windows Phone, the option was the similar one 'do no track'. All of them supported 'Opera Mini' browser. Regarding browsing settings, a liquid convergence in the privacy settings were observed. That is to say, that all browsers had already found a common balance point to deal with privacy control.

Apple iOS and Windows Phone did not allow applications to have access to the common mobile files' system (except to the camera roll) while Android did it to part of the files. To end up with, Apple iOS, Blackberry and Windows Phone were able to be named as 'Controlled Developers' or 'Closed Environments'; while Android was able to be called 'Open Source'. Here, it was observed a clear delimitation of models (liquid-solid). However, the tendency is to open the closed ones and vice versa in order to share items with more security. As an example, the upcoming iOS 8 will share information between apps like Android already allows.

A.2. Features

In the following group, 'Java support', 'Flash support', 'Security', 'Social Media Integration', 'Social Gamming',

'Movie Store', 'Music Store', 'Book Store', 'Default Browser', 'Cloud Support', 'Cloud Messaging', 'Wireless Cloud Support', 'Parental Control', 'Remove Clear Data', 'Internet Wi-Fi', 'Internet 3G/4G', 'SIM/telephone' and 'Notifications System-Messages' were categories studied.

Only Android supported 'Flash', and with Blackberry 'Java'. It was someway a bit dangerous because of the ligations arose next to the hardware. In this regard, this aspect seemed to continue fixed, without variations.

Regarding 'Security', Android and Blackberry encrypted personal folder and Windows Phone had multilayer, secure boot, sandboxing and an encrypted sync. All of them presented 'Social Media Integration' while Apple iOS (Game Centre) and Windows Phone (Microsoft XBOX Live) its own 'Social Gamming' as well as 'Wireless Cloud Support'. There, it would be interesting to know whether it is an encrypted backup and if the connection is secure. The convergence point seemed that it is going to be the encrypted personal data and the wireless transmissions.

About Stores, nothing especially interesting might be mentioned. Most of them had their own applications, being Apple iOS: iTunes and iBooks, Android: nothing specific for music and Google Books, Blackberry: Rovi and Music Store RIM and Windows Phone: Zune, whereas the latter two might have also Amazon-Kindle. All offered as well 'Cloud Support & Messaging' being for Apple iOS: iCloud and iMessage, Android: Google Sync and Google Cloud Messaging, Blackberry: Third Party and Windows Phone: Sky Drive and Windows Messaging.

Only Apple iOS and Android systems offered 'Parental Control' and all of them allowed 'Remote Clear Data' except Android by default, but it could be offered by Samsung services. The reason might lay on the existence of different developers involved in the market and without an agreement between them. Internet Wi-Fi, 3G/4G, SIM/telephone, notification system and messages were placed in applications. That is to say, all applications and platforms had freedom to choose / set / select those features or not, they can be controlled by the user in the same.

A.3. Official Applications

The following ones have been studied: 'Maps, Google Play and Search, Gmail, Youtube, Pandora Radio, Apple iTunes, Cooliris, purchases in applications, Twitter, Yahoo Messenger, eBay Mobile, Amazon Mobile, LinkedIn, Flikr, Instagram, WhatsApp, Skype, Line, Viber, Foursquare, Pinterest and Facebook'. Main aspect to underline was 'Purchases in Applications' where Apple iOS controlled it as well as Windows Phone through MS Market/XBox Live and while Android and Blackberry did it through Google Wallet, Paypal or the application itself as in the case of Kindle.

A logical difference was that Apple iTunes only existed in Apple iOS. Android was the only one that allowed 'Google Play' and while Apple iOS had 'Cooliris', Android and Windows Phone had 'LiveShare' –Blackberry anything -. Flikr was not been found for Blackberry, neither Instagram for Windows Phone.

Despite all the differences, the convergence point seemed to be that all environments are going to have all the applications available in the future.

B. Privacy Terms and Conditions

Generally, it seemed that 'Privacy Terms and Conditions' were written to protect the company rather than the user since they were sometimes very ambiguous, opened to different interpretations and likely to expand its content to define future situations, the benefit of whom? Many aspects were analysed but, due to a matter of space, only those more linked to the delimitation of privacy were selected. One characteristic to underline was its volatility and constant change, therefore, its liquidity.

On the one hand: Facebook[12], Twitter[13], LinkedIn[14] and Google[15] (as Google +, our aim of study, followed the same terms and conditions of Google). On the other hand: Instagram[16], Vine[17], WhatsApp[18] and Line[19]. The first four were considered web services (although they had applications for mobile phones). The last 4 were App services (App called) because they required a previous installation of the application on the mobile device (even in the case the user only employed the web service), which makes sense as they are specifically designed to be used by a 'mobile phone behaviour'. Another difference was that the terms and conditions of the first group were more completed, developed and offered in more languages.

Three of the web services required a valid email to conclude the registration (only Google not), where Facebook required also date of birth and sex. Applications linked with calling functions required obviously your phone number and in the case of 'Line' a password was required for 'phone-book' multi-device function.

Regarding minimum age required to register: Twitter specified 13, LinkedIn 18 while Google did not specify it (leaving the requirement opened to the case of some additional products) neither Facebook, although this last one requested it to do the registration (13) and for some applications (18). A different approach was observed in the applications: "Instagram does not knowingly collect or solicit any information from anyone under the age of 13", Vine did not discriminate but indicated the service was for older than 13, "You affirm that you are either more than 16 years of age, or an emancipated minor...." WhatsApp and Line did not stipulate it but they had a child protection policy.

Differences were observed about who was the owner of personal information: Facebook did not particularize it although indicated that were the ones that you decided to share, neither Twitter pointing out the user allowed the company to use it. LinkedIn and Google designated that

[12] <https://www.facebook.com/about/privacy/; date: 11th December 2012. Facebook, 1601 Willow Road, Menlo Park, CA 94025 USA

[13] <https://twitter.com/privacy; date: 3rd July 2013. Address not specify on it, in the web: Twitter, Inc.1355 Market St, Suite 900. San Francisco, CA 94103

[14] <http://pt.linkedin.com/legal/privacy-policy; date: 12th September 2012. LinkedIn Corporation, 2029 Sierlin Court, Mountain View, CA 94043 USA.

[15] < http://www.google.com/intl/pt-PT/policiestica /privacy/; date: 24th June 2013 -analyzed two documents, 'Terms of Use and Privacy Policy'-. Google Inc. 1600 Amphitheatre Parkway, Mountain View, CA 94043 USA.

[16] <http://instagram.com/about/legal/privacy; date: 19th January 2013.

[17] <https://vine.co/privacy/; date: 21st January 2013 –analyzed two documents, 'Terms of Service and Vine Privacy Policy'-. 1355 Market St., Suite 900 – San Francisco, CA 94103.

[18] <http://www.whatsapp.com/legal/; date: 7th July 2012. 3561 Homestead Road, #416, Santa Clara, CA 95010-5161

[19] <http://line.naver.jp/line_rules/en/; 1st April 2013.

the owner was the user but they also denoted that they were controlled by LinkedIn to protect users data and that the employment of Google did not confer the user any ownership over their services or accessed content. Instagram and Line did not itemize it, neither Vine that added "is a video sharing platform, so most of the information you provide us is information that you choose to be made public" and WhatsApp "...you retain your ownership rights in you 'Status Submissions', but you have to have the rights in the first place." This was a clear example of the ambiguity found in the analysed terms and conditions.

They all collected data from its use and from other sites or applications except WhatsApp and Line, it might be because they lacked of external partners. Also they all utilized cookies or similar technologies and gathered specific metadata to catch other information. Only Facebook and Twitter particularized the option of 'do not track' – tracing - although all indicate that they might prevent the browser use cookies.

They all offered the possibility to alter the information, disable, suspend or eliminate totally the account but, in the case of Google, it was the only one that did not refer specifically to the act of closing an account and it also did not ensure to be able to delete data account. Twitter and Line were the only ones that did not concrete where information was stored, what might be interpreted as this aspect will depend on each country law system. Concerning a specific data protection policy, once more, they were ambiguous or they did not describe a detailed one. They simply stated they will do the best they can and recommended the user to behave properly and help at this point.

One difference between the first group (called mainly sites, Facebook, Twitter, LinkedIn and Google) and the second one (applications Instagram, Vine, WhatsApp and Line) was that, regarding the possible circle of public to share the content with, they were more or less restricted to contacts saved on the mobile phone account. It meant that when answering to the question of allowing other sites, applications, platforms, services or users to access the account, the range was wider in the first group.

The majority of them detailed, or simply did not specify leaving this aspect opened, that users information was shared with other platforms, sites, applications, services or other users in the name of different purposes: personalized content, improve services, inform friends, make suggestions, etc. Here it had to be stress the fact that they would disclose information by law request or to protect their services, leaving this aspect again opened to various interpretations.

All of them had the concern to create targeted or custom ads, where LinkedIn had a specific part to describe how to deal with it. All first group and WhatsApp offered concretely the option of blocking ads. Facebook, LinkedIn and Google were adhered to some kind of regulatory authority. Also all of them, except Line, indicated the terms of sale of the service or the company as well as the updates of the 'Terms and Privacy Policy'. In practice, users had to be aware of them by their own as not always they would be informed through a direct contact from the company.

Another difference between groups was that none applications specify a policy targeted to children while only Twitter did it. Neither Google nor any applications detailed if they fulfilled 'Safe Harbour[20]' and 'TRUSTe[21]' rules or 'Shine the Light' Law of California[22] (direct marketing). Facebook and LinkedIn stated they did it and Twitter just mentioned the first one.

The interest for collection personal information and its sharing through social media is so widespread and relevant that all environments and platforms accepted being opened to others in order to obtain as much data as possible, generating a liquidity state for the data flux.

C. Exploratory Focus Group and Survey

C.1. Exploratory Focus Group[23]

Sociability, coordination and its maintenance were outstanding while issues related to personal and professional life were also considered very important, in some cases essential. There, it ought to be remembered the characteristics of fluidity of identity as well as the cellular and nomadic intimacy, constantly to be negotiated.

The issue of privacy appeared in two ways: in the level of content/data and in the level of interaction contexts. There was a concerned in the perception of privacy when producing a mobile phone appropriation by others. They stated reluctance to lend it beyond a momentary use to call and the ignorance – no control - over the audience. There was a notion of existence some risk of exposure/personal aspects to last in time online, but the practices seemed to be unconcerned. Participants tended to consider that the data was always private, but in some cases (images) did not contradict the possibility of share/publish them.

This observation matched previous studies mentioned: users concerns were too remote to influence their decisions - although they took some steps to prevent - compared with the immediate and tangible benefits of that same sharing – emotion involved and, even in the case of Canada, to sell personal data for a good prize -. Moreover, the risks were not so evident because they continued to act as if they had privacy. These attitudes remind some kind of lack of rationality described in the theoretical frame.

The group discussion highlighted the difference between two types of risk about the privacy of content/information: the data stored/fixed in a device (computer/phone's) and the data that became accessible by third parties at any time and place online – contextual integrity - . The concern of disseminating pictures of them had to do with the embarrassing situations and aesthetics of the images. Its publication in open or closed circles was produced by the "good common sense". They showed high tolerance to invasion/harassment commercial use of not allowed personal data.

At this point, they showed an initial awareness of the different sources to collect their personal data. However, the possibilities are wider as it was described in the theoretical frame concerning profile and digital identity. Therefore, the required knowledge ought to be transmitted in this area.

To end up, some general appreciations like the anxiety cause due to the fact of "having to be always available", to

[20] <http://export.gov/safeharbor/; consult 28th September 2013

[21] <http://www.truste.com/; consult 28th September 2013

[22] <http://en.wikipedia.org/wiki/California_Shine_the_Light_law; consult 28th September 2013

[23] Covilhã, Portugal, July 2013th. 3 males and 3 females, ages: 19, 2*20, 2*21, 23.

answer depending on who is calling, the reluctance to speak in public near unknown people, not to constrain sociability or interrupt conversations by mobile phone and the strategy to minimize risk of nuisance (silence mode/vibrator) are pointed.

C.2. Exploratory Survey[24]

Basic function, such as 'do calls' and 'send and receive text messages' were those that respondents privilege in their use, several times a day. In a second group, those respondents who used their device to, also several times a day, 'visit sites, browse the Internet, searching information' and still 'visit social networking sites'. 71% used the phone for taking photos or videos between one and three times per week. Closeness with people was crucial to lend or borrow mobiles. There exists a strong sense of property "it's my phone".

68% checked if the application offered "application permission" but 61% did not read those "permissions" before installing, being 32% who did. 55% did not allow apps to access the lists of contacts or information. Half took, therefore, some steps to prevent the synchronization of data.

Half might live without mobile phones, 32% would felt their lack and 18% considered that they did not have such a powerful role in their lives. Its use was strong circumstantial, very dependent on circumstances, since there were no general rules self-stipulated by the actors in relation to the different life situations, they dealt different behaviours according to the context. 63,5% agreed "I feel uncomfortable when I have to make a call and there are strangers around me", while "agree" and "totally agree", added 94,6%.

This strong circumstantial use highlighted relevant characteristics described in the theoretical frame: the fluidity of identity, the cellular and nomadic intimacy and the relevance of temporal priority in the digital literacy.

VIII. Conclusions

Regarding operative systems on smartphones, two main models were observed, iOS 'Closed Environment' controlling everything around the smartphone: Equipment, SDK, Apps, Market (App Store) as well as users and Android 'Open Source' allowing the freewill behaviour. Following this concept and concerning Privacy, iOS controlled the user data from applications and platforms while Android could or could not, depending of what allowed in each application and platform as well as their relationship with Google as an intermediate.

One important difference between them was that, after accepting both an installation, only through iOS user could alter the 'Privacy Configurations' related to its environment while in Android users had to reinstall if they did not agree with the whole package formerly accepted. Moreover, iOS offered the possibility to alter, for example GPS or contacts, in each application or platform whereas in Android the user could accept it or not for all of them.

In terms of security characteristics, the tendency seemed to be for all the environments to get closer, less liquid. That is to say, to adapt to the requirements of user privacy concerns, as it is going to be explained.

The lack of control about how and who accessed to the data and contents published, according to the literature review and the exploratory studies, was what users aware and worried about but they also seemed not to take many actions to prevent, except localization and the access to the list of contacts when installing or setting applications. To avoid synchronization and to use a new mail account to register could be partial solutions for these concerns.

As one and main example of the volatility and ambiguity of the 'Terms and Conditions' analysed, Google cannot guarantee full delete of users information account and none of them specify a data protection policy, leaving this question mainly to user's good sense, who had to be aware of the possible updates of the 'Terms and Conditions'. At this point, Google is right now working on a proposal for the 'Right to be Forgotten'.

The liquidity observed in the society and in the mobile environment as well as between the public and private spheres, can also be applied to the operative systems. iOS, Blackberry and Windows Phone started being a solid sphere (lack of external versions in its market) that are experiencing a process of liquidity (allowing access to others despite keeping a tight control). Meanwhile, Android begun totally liquid (open) and is increasingly becoming more solid (due to its concerns about Privacy, for example, Amazon getting closed). This tendency could lead to reach a convergence meeting point where both brands / developers and users would have the control of Privacy. Both spheres / operative systems and its development are dealing with a process of liquidity.

Bearing in mind such defining features of smartphones like instantaneity and ubiquity; concerning literacy, the continuous partial attention and its relation with memory; the lack of rationality in some attitudes and performances; the limitations on the extension between knowledge and action; the strong circumstantial pattern behaviour; the volatility and ambiguity of the 'Terms and Conditions', adding the liquidity and mobility of our society and technology itself; the users deal with liquid spheres where the constant data flux escapes from a clear awareness of it and a notion of the risks involved in deep.

Therefore, the users manage a constant negotiation of circumstances based on the evaluation of each scenario framed by the ambiguity and the immediacy, which also determines the reflectivity (required time to analyse) as well as the perception of the risk involved in every action. Moreover, the possibility of receiving stimuli of all types constantly influences how to establish the priority level properly as well as how to protect their privacy at the different layers and stages according to their possibilities.

References

[1] J. M. Aguado, C. Feijóo, I.J. Martínez (cords.), *La comunicación móvil. Hacia un Nuevo ecosistema digital,* Barcelona: Gedisa, 2013, p. 18, p. 30, p. 31, p. 308, p. 309, pp. 318-320, p. 41.

[2] Z. Bauman, *In Search of Politics,* Stanford University Press, 1999 / *Em busca da política.* Rio de Janeiro: Zahar, 2008.

[3] Z. Bauman, *Vida para consumo. A transformação das pessoas em mercadoria,* Rio de Janeiro: Zahar, 2000, p. 71.

[4] B. Wellman, "Physical Place and Cyberplace: The Rise of Personalized Networking". *International Journal of Urban and Regional*

[24] Portugal: from 15th July until 21st August 2013, online. It had 74 answers and largest number of respondents was in the age group between 30 and 45 years old – nobody less than 18 and more than 65 -; 56,8% women and 43,2% men. 73% employed, being 85,7% women and 81,2% men with higher education. 67,7% had a smartphone and 31 % not.

Research, 2001, 25, pp. 227–52.

[5] J. Katz, M. Castells (eds), *Handbook of Mobile Communication Studies,* Cambridge: MIT Press, 2008, p. 49.

[6] J.E. Katz, M. Aakhus (eds), *Perpetual Contact. Mobile Communication, Private Talk, Public Performance*, Cambridge: Cambridge University Press, 2002.

[7] A. Fidalgo, A. Serrano Tellería, J.R. Carvalheiro, J. Canavilhãs, J.C. Correia JC, "Human Being as a Communication Portal: The construction of the Profile on Mobile Phones", *Revista Latina de Comunicación Social*, 2013, 68.

[8] K. Kawamoto, *Media and Society In the Digital Age*, New York: University of Washington, 2003, p. 33.

[9] Z. Bauman Z, *Liquid Life*, Cambridge: Polity, 2005.

[10] S. Isabella, "Mobile Phone: Users Practices and Innovations Between Public and Private Sphere". In *COST 298 Conference. Participation in the Broadband Society*, 2009, p. 7.

[11] L. Fortunati, "The Mobile Phone: Towards New Categories and Social Relations". In *Information, Communication, and Society*, 2002, pp. 514-528.

[12] A. Fidalgo, "Conectados e tutelados. Uma revisitação tecnológica da esfera pública", in *Public Sphere Reconsidered: Theories and Practices*, Beira Interior University: LabCom, 2011, p. 68, pp.68-70.

[13] H. Geser, "Is the cell phone undermining the social order?", in *Thumb Culture*, Bielefeld: Transcript Verlag, 2005, p. 25.

[14] T. Ahonen, "Mobile as 7th of the Mass Media. Cellphone, Cameraphone, IPhone, Smartphone", in *Futuretext*, 2008.

[15] J. E. Katz, "Mainstreamed Mobiles in Daily Life: Perspectives and Prospects", in Katz, Castells (eds.), *Handbook of Mobile Communication Studies*, Cambridge: MIT Press, 2008.

[16] A. Serrano Tellería, "The Construction of Profiles on Mobile Phones. Public and Private Spheres", in *DESIGNA 2013: INTERFACE International Congress*, 21 & 22 November, Beira Interior University: LabCom, 2014.

[17] M. Mazmanian, *Some thoughts on blackberries*, Memo, 2005.

[18] S. Turkle, "Always-On / Always-on-You": The Tethered Self", in Katz, Castells (eds), *Handbook of Mobile Communication Studies*, Cambridge: MIT Press, 2008, p. 128, p. 129.

[19] R. Sennett, *The Fall of Public Man*, New York: Penguin Books, 2002, p.36.

[20] J. Meyrowitz, *No Sense of Place: The Impact of Electronic Media on Social Behavior*, New York: Oxford University, 1985.

[21] E. Goffman, *The presentation of Self in everyday life*, New York: Garden City, 1959 / *A Apresentação do Eu na Vida de Todos os Dias*, Lisboa: Relógio d'Água, 1993.

[22] D. McQuail, *Mass Communication Theory*, London: Sage, 2006.

[23] A. Castellet, *El ecosistema del contenido móvil: actores, líneas de evolución y factores de disrupción*, Ph.D dissertation, Spain: Murcia University, 2012.

[24] J. McDermott, "IPhone owners consume more entertainment than Android", in *Advertising Age, adage.com*, 25 March 2013.

[25] S. B. Wicker, "The Loss of Location Privacy in the Cellular Age". Communications of the ACM, 2012, 55(8), pp. 60-68.

[26] A. Cerra, C. James, *Identity Shift*, Indianapolis: John Wiley & Sons, 2012.

[27] D. Potoglou, D. Patil, C. Gijón-Tascón, J. Palacios, C. Feijóo, "The value of personal information online: results from three stated preference discrete choice experiments in the UK". *European Conference on Information Systems, ECIS 2013*, Utrecht, Netherlands, 2013, pp. 1-12.

[28] H. Nissembaum, *Privacy in context. Technology, policy and the integrity of social life*, Stanford: Stanford University Press, 2010.

[29] J. Jin Park, "Digital Literacy and Privacy Behaviors Online", in *Communication Research*: Sage, 23 August 2011.

[30] H. Fathi, "Security and Privacy Challenges in Globalized Wireless Communications", in: Prasad et al. (eds.) *Globalization of Mobile and Wireless Communications: Today and in 2020, Signals and Communication Technology*, Springer, 2011.

[31] J. Fernback, Z. Papacharissi, "Online privacy as legal safeguard: the relationship among consumer, online portal and privacy policies", in *New Media and Society*, 2007, 9, 715.

[32] S. Turkle, *Alone Together: Why we expext more from technology and less from each other*, New York: Basic Books, 2011.

[33] J. Marichal, *Facebook Democracy. The Architecture of Disclosure and the Threat to Public Life*, United Kingdom: Ashgate, 2012.

[34] G. Stald, "Mobile Identity: Youth, Identity, and Mobile Communication Media." In: Buckingham D. *Youth, Identity, and Digital Media.* The John D. and Catherine T. MacArthur Foundation Series on Digital Media and Learning, Cambridge, MA: The MIT Press, 2008, pp. 143–164.

[35] J. Manyika, "The impact of disruptive technology. A conversation with Eric Schmift", in: *McKinsey & Company, mckinsey.com*, 2013.

AUTHORS

A. Serrano Tellería. Ph.D Assistant Professor. Posdoc, Ph.D candidates (5) coordinator of the European FEDER project 'Public and Private in Mobile Communications' in LabCom, Beira Interior University, Portugal (April 2013 - 2015). Lecturer of Cyberculture (2013-14). Ph.D. (2010) and Bachelor in Journalism (2002) by the University of The Basque Country. Extraordinary Ph.D Award (2012). First cycle of English Translation and Interpretation (2000). Management of European projects and International Cooperation ones in Documenta (2012-13). Responsible of the Communication Department in the Territorial Development Agency "Campoo Los Valles" (2011). Master in Innovation Management (Full grant, 2010-11). Visiting researcher at the School of Communication in Federal University of Salvador de Bahia, Brazil (Grant of the Spanish Ministry of Science and Innovation, 2009). Cultural Management (2005). Master in Theatre and Performing Arts (2004-05). Theatre teacher (2003-04). Online journalist in *Elcorreo.com*, Vocento (2002-03). Email: anaserranotelleria@gmail.com.

M. Oliveira. Ms.C Computer Scientist. Master's in Computer Engineering from the University of Beira Interior in 2009. He has been performing duties as senior technician in the areas of web development and administration of systems in the Online Communication Laboratory (LabCom) at Beira Interior University since 2005. Email: marco.oliveira@labcom.ubi.pt.

ANNEX. FEEDBACK FROM ITS INSTALLATION

Aspects related to 'Privacy' and tools concerning definition of personal image and 'Public and Private' Spheres were analyzed. Tests had been carried on four mobile phones with different operative systems. It should be underlined that each application is continuously being updated so the different aspects analyzed may be altered after this observation and comparative study (April 2014).

A. Facebook

Focusing in 'Privacy Settings', Android asked users to allow who could search your 'chronology' by name. Blackberry and Windows phone sent you with a direct link to Facebook webpage to read there 'Privacy Terms and Conditions'; in this last case, there were no options to 'privacy settings' in the profile, the user could only set 'localization'. There were more options about 'push notifications' in iOS than in the rest of the operative systems.

B. Twitter (4.3.0.8 for Blackberry, 5.11 for iOS)

In its webpage, there were headlands for 'Security' and 'Privacy' that were not found in the mobile application.

iOS asked to refresh 'user' settings', 'localization' and questions about different types of 'permissions'. Only 'login verification' was placed in the same, iOS controlled the rest through its operative system. Android and Blackberry requested access to: Location, contacts, accounts, access to the memory card, internet access, location, system tools, network, synchronization, informal personal contact data, account authenticator, manage lists of accounts, control the vibrator, discover accounts known and Google services.

Blackberry asked about content access, Twitter license contracts and in 'settings': refresh, Blackberry messenger, possibility to add GPS to the tweet, profile edition and assistance. Interface design of Windows Phone was similar to web one but it only offered the option of allowing GPS, new followers, instant messages or if the user is mentioned in a tweet. To 'log out' the user had to remove data account.

C. LinkedIn (6.2 for iOS)

Installing the application, iOS system asked if the user would like to receive 'push notifications" to alter sound, icons and contacts while Android one 'count and configuration', 'privacy settings' and 'application permissions': ID Phone, system tools, personal information, manage the accounts list, USB storage, full access to Internet, hide/discover accounts, control of vibrator by hardware, system tools-sync, Internet communication.

Windows Phone only offered 'settings' and 'notifications' whereas Android 'settings: Sync, notifications, push and about 'Privacy Policy: Access to website and Copyright Information'. iOS app shared similar configuration, presenting a difference about 'feedback' when the user receives messages about the application.

D. Google (+) (Android, 4.5.1 for iOS, 2.11.166 for Blackberry)

Official applications neither for Windows Phone nor for Blackberry were found. iOS asked permission to access photos and create a security copy of them while Android did to: control hardware (audio, photo and video), and exact location, communication, system tools, telephone calls, personal data, accounts and storage. Other optional ones were: automatically update when connecting to Wi-Fi, create security copies of the photos and "allow Google to provide suggestions based on the people you communicate with most often on this phone".

E. WhatsApp (2.11.4 for iOS)

Android asked access for 'Personal Information, Hardware Control, User Location, total access to Internet, services that costs money, calls, read status and Phone ID and to change system definitions'. Permission to access your contacts was also requested in Blackberry.

F. Instagram (4.1.4 for iOS, 4.1.2 for Android)

Questions about the installation in Android were: 'Hardware control, recording audio, taking pictures and recording videos, system, personal information, location, storage, network'. iOS version just required access concerning photos. No versions found for Windows Phone and Blackberry.

G. Vine

iOS version asked the user if older than 17 years old, to access Twitter accounts, localization, push notifications and contacts while redirecting to the web about 'Privacy Policies'. No findings for Android, Windows Mobile and Blackberry.

H. Line (3.9.0 for iOS, 3.9.1 for Android, 1.9.15 for Blackberry)

iOS asked about: push notifications, email, telephone, Facebook account and authorized apps while Android requested more permission: 'Hardware control, localization, services implying payment, communication network, system utilities, telephone calls, personal data, messages and storage'. Blackberry just asked about contacts.

24

Learning Physics Down a Slide: A Set of Experiments to Measure Reality Through Smartphone Sensors

L. Martinez and P. Garaizar
University of Deusto, Bilbao, Spain.

Abstract—**During the last decade, the use of smartphones among teenagers in their daily life has grown significantly. Despite the efforts to use tablets in learning processes, these teenagers are often prompted to switch off their personal devices before entering a classroom. Moreover, most mobile learning applications do not take advantage of the device sensors (e.g., touchscreen, accelerometer, or gyroscope). In order to overcome this situation, we have developed Serious Physics, a free mobile app that allows using smartphones as measuring tools to conduct experiments on Physics.**

Index Terms—**educational programs; educational technology; physics education; sensors;**

I. Introduction

There are literally thousands of mobile educational applications available in the app markets [1]. These apps offer new learning experiences, ranging from simple interactive books to complex Augmented Reality simulations [2]. In regard to learning Science, there are content-based apps such as interactive periodic tables of elements (e.g., Merck PTE HD) or exploration tools to navigate through all parts of a 3D model of the human body (e.g., 3D Anatomy), and problem-based apps, often designed with a challenge-solution approach (e.g. Mathway, Algebra Touch). Something similar happens with applications for learning Physics (e.g., Constant Table, Learn Physics). Moreover, most mobile learning apps do not take advantage of the device sensors to enable an experimental approach. For this reason, we decided to develop Serious Physics, an Android-based free mobile app to conduct experiments on Physics using the smartphone as a measuring tool.

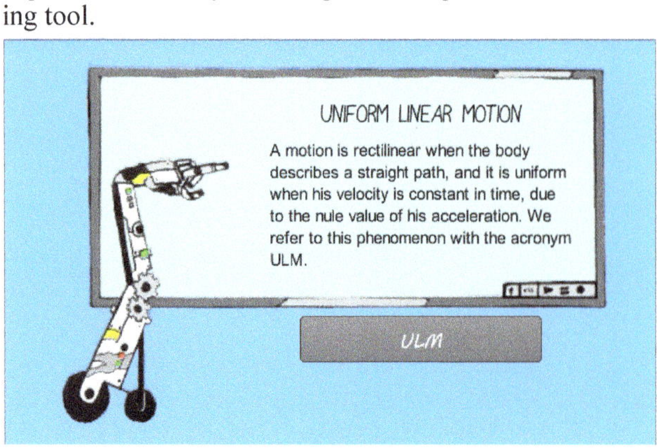

Figure 1. Serious Physics: Uniform Linear Motion explanation.

II. Aims and Scope

Our proposal addresses the improvement in the learning of Physics by 12 to 18 year-old students in the Basque Country (Spain). Therefore, Serious Physics provides theoretical and practical contents based on the curriculum designed by Dept. of Education of the Basque Government, and also a set of experiments that take advantage from the smartphone sensors to gather real measurements. Considering its scope, Serious Physics is freely available in English, Spanish and Basque, and relies on an extensible architecture that allows to design and build new experiments to cover other topics of the curriculum.

III. Related Work

In addition to the commercial apps mentioned before, there are some previous research works that address the issue of adopting an experimental approach in the learning of Physics. Kuhn and Vogt noticed that the use of mobile phones as experimental tools has been a neglected topic in the field of educational research [3]. They focused their research on the mobile device as a mean of documentation, due to the possibilities that offer the microphone and the camera of the device. Surprisingly, Kuhn and Vogt did not cover the use of other sensors included in mobile devices. In that sense, it is particularly interesting how they propose an experiment to estimate the gravity using the microphone instead of the accelerometers of the smartphone as Peters proposed three years before [4].

Chevrier et al. developed a set of experiments within the iMecaProof project to help in the teaching of classical mechanics [5]. In this project, users are provided with an application with features divided in different levels of expertise (from 0 to 3) which include data gathered by the sensors of the mobile device to support theoretical explanations. SPARKvue is another project that allows to gather sensor data from a mobile device (and from external sensors too) [6].

Using a completely different approach, Gabriel and Backhaus combined the GPS sensors of the mobile devices with Google Maps to develop an application intended to support the teaching of Kinematics [7].

IV. Serious Physics

In addition to its theoretical and practical contents (see Fig. 1), Serious Physics enables the use of a mobile device to conduct several experiments on Kinematics. As can be seen in Table 1, there are many new scenarios where Serious Physics users can learn about Kinematics in an experimental way. The modular architecture of the software (based on free software libraries, such as LibGDX, OpenGL, or Box2D) allows covering other topics and scenarios on the top of it.

TABLE I.
EXPERIMENTS AND SCENARIOS

Concept	Sensors	Scenario
Uniform Linear Motion	Touch screen	Indoors, self-study
Uniformly Accelerated LinearMotion	Accelerometer Gyroscope Magnetometer	Indoors, free fall, outdoors, park with swings and slides
Circular Motion	Accelerometer Gyroscope	Outdoors, park with tire-swings and roundabouts

How students and teachers might use these features to improve their understanding of physics? Imagine a school scenario where: (a) students have mobile devices (e.g., *smartphones, tablets, phablets*) able to run Serious Physics; and (b) teachers are willing to go to a playground with them to perform a set of experiments. If that were the case, teachers might begin the explanation of Kinematics with a discussion session for groups or couples of the theoretical content offered by the application. Once the concepts involved have been assimilated, students could define a series of experiments to test them. In these experiments, students define the initial conditions and the expected results according to the underlying theoretical models, as well as the boundary cases that may lead to complications in the procedure. Not all of these experiments require visiting a playground, some of them can be conducted indoors without any problems. However, collating scientific experimentation with experiences as fun as going down a slide can be an incentive even for students not interested in Physics.

In order to get as close as possible to the scenario presented before, we conducted three experiments in which we try to prove empirically that they meet the underlying theoretical models.

A. Experiment 1: Uniform Linear Motion

Description. Through this experiment we want the student to understand the foundations of Newton's First Law of motion in a practical way. This law is often stated as *"An object at rest stays at rest and an object in motion stays in motion with the same speed and in the same direction unless acted upon by an unbalanced force"*. To this end, the application captures the motion of a finger along a rectilinear path on the screen and then displays the speed of the finger. Students can then check whether the previous calculations correspond to those measured by the application or not.

Sensors. Current phones and tablets feature touchscreens that are able to detect multi-finger movements. We gather the speed of the finger from the touchscreen in pixels/s and then convert it to m/s.

Procedure. We conducted a series of eight measurements of the time and space covered by a finger sliding on the screen of a mobile device in order to have an estimate of the accuracy and precision of the measurement. Considering that the sliding finger travels 5 cm along the screen (i.e., three quarters of a regular screen) without any acceleration in approximately 0.1 s, students can use the well-known equation of Uniform Linear Motion (1) to predict the results of the tests:

$$v = \frac{s-s_0}{t}; v = \frac{0.05\,m - 0\,m}{0.1\,s}; v = 0{,}5\,m/s \qquad (1)$$

Results. The speed values gathered during the eight attempts are shown in Table 2.

TABLE II.
UNIFORM LINEAR MOTION

Attempt	Speed (m/s)
1	0.43726072
2	0.7753074
3	0.48552343
4	0.5373729
5	0.48787132
6	0.43785048
7	0.56286395
8	0.6531293

As we can see, the results are compliant with the predicted value (M: 0.522, SD: 0.075).

Given the simplicity of this experiment, it might be autonomously conducted by students during class, at home as homework, or even during the trip to the playground. After that, students can compare the results with their predictions and with the results of their classmates. Another interesting approach is to conduct the experiment first and then try to figure out which factors are involved in the Uniform Linear Motion and which are the interactions between them. To facilitate this task, teachers can provide students with a list of possible factors and ask them to prove or discard their implication in Uniform Linear Motion.

B. Experiment 2: Uniformly Accelerated Linear Motion

Description. In the experiment of Uniformly Accelerated Linear Motion of Serious Physics, students can analyze both the movement of the mobile device in free fall (on a soft surface such as a pillow or a bed), as the downward movement of the mobile device along an inclined plane. We recommend starting with the first case because it is easier to understand and calculate since it does not require estimating the value of the acceleration. Instead, acceleration corresponds with the value of g (i.e., 9.8 m/s^2) in this case. In the second case, the value of the acceleration depends on the angle of the inclined plane and the friction coefficient of the surface on which slides the mobile device. Equation (2) is used to calculate the distance traveled. In the first case students can estimate it by simply replacing a with g, whereas in the second case it is necessary to calculate the acceleration using equation (3).

$$d = \frac{1}{2}at^2 \qquad (2)$$

$$a = g(\sin\alpha - \mu\cos\alpha) \qquad (3)$$

Sensors. Two different mobile sensors are used in this experiment. First, we estimate the angle of the plane using the magnetometer. Second, we log the acceleration changes in each axis via the accelerometer.

Procedure. We conducted a series of eight tests throwing the mobile device along a slide. In all these tests, Serious Physics recorded the time (detecting the beginning and end of the movement) and the acceleration in order to estimate the inclined plane angle (i.e., the slide), the distance traveled and the height from which it has been thrown. All tests have been performed on a slide with a height of 2 m and an inclination of 40 º.

TABLE III. UNIFORMLY ACCELERATED LINEAR MOTION

Attempt	Length (m)	Height (m)
1	3.22	2.06
2	2.91	1.86
3	3.65	2.336
4	3.34	2.13
5	3.50	2.24
6	2.98	1.90
7	3.25	2.08
8	3.09	1.97

Results. The results of these tests are shown in Table 3. As can be seen, there is some variability in them due to small differences in the execution of each test (Heigth, M: 2.072, SD: 0.163; Length, M: 3.243, SD: 0.252). Fig. 2 shows a screenshot of Serious Physics after performing one of these tests, providing estimates of the angle, distance and height.

C. Experiment 3: Circular Motion

Description. Students find the circular movement more difficult to understand, not only due to the trigonometry involved, but also because it is much more common to have first-hand experience of Uniform Linear Motion (e.g., using a conveyor belt at a constant speed on a airport) or Uniformly accelerated Linear Motion (e.g., traveling in a car that accelerates slowly on a straight stretch of road), rather that first-hand circular movements. Through this experiment provided by Serious Physics we intend to address this problem, providing first-hand experiences of circular movements that students can easily control.

Sensors. Gyroscopes are used to estimate the number of spins made by the mobile device.

Procedure. Analogously to previous experiments, we carried out a series of eight tests using the circular motion experiment of Serious Physics. The circular motion is defined by equation (4).

$$\omega = \frac{n2\pi}{t} \quad (4)$$

As long as it is necessary to know the value of the radius in this equation, it will be provided by the user of the application. As it happened in the experiment of Uniformly Accelerated Linear Motion, the user has to indicate the start of the experiment and then the application will start counting the time from the moment it detects movement (this time measuring procedure ends automatically when the mobile device stops completely). After completing a test, the application provides the angular velocity (ω) in

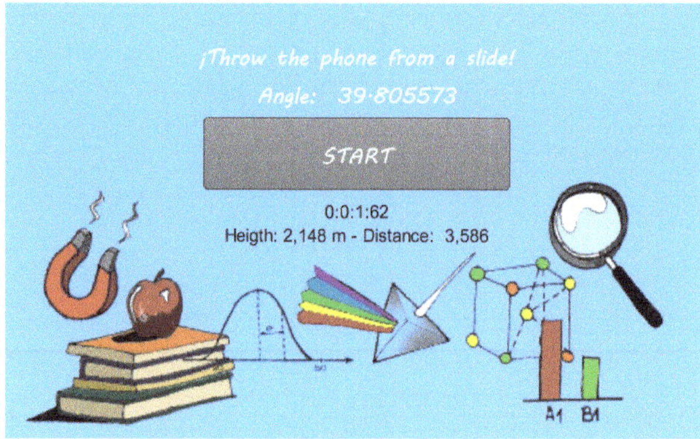

Figure 2. Serious Physics: Uniformly Accelerated Linear Motion experiment

rad/s, the speed ($v = \omega r$) m/s and the frequency of rotation ($f = \omega / 2\pi$) in rpm (see Fig. 3).

Results. The results of the tests are shown in Table 4. In contrast to previous experiments, it makes no sense to analyze the stability of the measurements taken, because each test was conducted with a different number of revolutions. This approach, which can also be used in other experiments already mentioned, is interesting in order to improve the comprehension of circular motion by the students. It allows checking compliance of gathered results with the underlying theoretical model using different parameters and understanding the relations between them.

TABLE IV. CIRCULAR MOTION

Attempt	ω (rad/s)	v (m/s)	f (rpm)
1	9.08	2.724	86.70
2	11.15	3.345	106.52
3	12.74	3.822	121.71
4	8.13	2.439	77.94
5	8.98	2.694	85.79
6	9.65	2.895	92.19
7	10.88	3.264	103.94
8	9.2	2.760	87.89

V. DISCUSSION

Along the experiments described above, we have seen how mobile devices can be used for learning physics from a practical approach. Considering the popularization of mobile devices (more than 100 million smartphone owners in the U.S. in 2012 [8]), especially among students, it is not hard to imagine a near future in which science learning is facilitated by the use of sensors available on mobile devices.

Not only traditional pedagogies could take advantage from this approach. Newest proposals like the "flipped classroom" [10] in face-to-face education or Massive Open Online Courses (MOOCs [11]) and Small Private Online Courses (SPOCs [12]) could also benefit greatly from it. When a teacher decides to use a flipped classroom methodology, theoretical contents are individually attended by students as their homework, while class time is devoted to tutored practice. In this scope, Serious Physics can be used for both tasks. At home, to learn about the

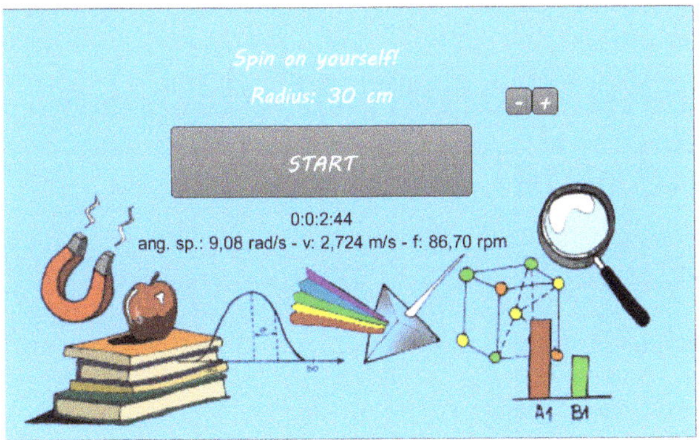

Figure 3. Serious Physics: Circular Motion experiment

theoretical concepts like other conventional mobile applications about Physics, but also in class during practical experimentation with examples like those discussed in the previous section. In regard to online learning, some of the recurring criticisms of this type of learning is its excessive emphasis on content distribution and the limited time devoted to practice. The approach proposed by Serious Physics could help to balance this inequity between theory and practice in these courses.

However, what we have shown in this article is only the beginning of a long road. Thanks to the modular architecture of Serious Physics, it is relatively easy to add new experiments to the application. These are some of the ideas for new experiments in which we are working:

Energy consumption experiment: In this experiment students will record a journey on foot or by any other means of transportation (e.g., bicycle, bus, car, train) using the mobile device's GPS and then the application will provide information about the speed average, accelerations, or the energy consumed in easily understandable terms (e.g., on this trip has been spent the energy equivalent to that provided by 6 kg. of french-fries eaten by a person).

Pressure waves experiment: In this experiment students will emit a sound wave via the speakers of the mobile device and subsequently capture how it resonates on a wall using the microphone. Serious Physics will then estimate the distance between the mobile device and the wall.

Electromagnetic waves experiment: In this experiment the students will be able to transmit information using the frequency variations of an electromagnetic wave, either visually (using the screen of the transmitter and the camera of the receiver) or via infrared waves.

Magnetic fields experiment: In this experiment students will understand the basics of magnetic fields using the magnetometer present in most mobile devices to provide the compass feature.

Moreover, we are open to suggestions, comments or collaborative projects that seek to benefit from the diverse and powerful set of sensors that incorporate today's mobile devices. Our ultimate goal is that teachers and students realize that they have an advanced laboratory in their pockets waiting to unleash its potential to facilitate science learning.

VI. CONCLUSIONS

Serious Physics represents another step towards the adoption of the smartphone as a learning tool. Not only as a mere content player, but also as a sensor-based mobile laboratory.

Although its functionality is still limited and there is a long way to go in improving and extending its features, we consider that Serious Physics faces a problem mentioned repeatedly by students once they complete their education process: What use are all the equations learned? When I shall have occasion to use them in real life? Memorizing an equation is used to solve an exercise correctly and, therefore, to pass exams. Discovering through experimentation is a completely different thing (i.e., infer what are the factors that affect a particular physical phenomenon and come to the underlying theoretical model with the help of a tutor). This results in a much more insightful, applicable and hard to forget learning.

REFERENCES

[1] Dickens, H., & Churches, A. (2011). Apps for Learning: 40 Best iPad/iPod Touch/iPhone Apps for High School Classrooms. The 21st Century Fluency Series. Corwin, A SAGE Publications Company. 2455 Teller Road, Thousand Oaks, CA 91320.

[2] Cobcroft, R. S., Towers, S. J., Smith, J. E., & Bruns, A. (2006). Mobile learning in review: Opportunities and challenges for learners, teachers, and institutions..

[3] Kuhn, J., Vogt, P. (2013). Applications and Examples of Experiments with Mobile Phones and Smartphones in Physics Lessons. Frontiers in Sensors (FS), Volume 1 Issue 4.

[4] Peters, R.D. (2010). Physics Education using a Smartphones Accelerometer.

[5] Chevrier, J., Madani, L., Ledenmat, S., Bsiesy, A. (2012). Teaching Classical Mechanics using Smartphones.

[6] SPARKvue. https://itunes.apple.com/de/app/sparkvue/id361907181

[7] Downhill and Projectile Launcher. http://smartphonephysics.com/freephysicsappsandroid.html

[8] Gabriel, P., Backhaus, U. (2013). Kinematics with the assistance of smartphones: Measuring data via GPS – Visualizing data with Google Maps.

[9] comScore, (2012). 2012 Mobile. Future in Focus. comScore, February 2012.

[10] Bergmann, J. (2012). Flip Your Classroom: Talk To Every Student In Every Class Every Day Author: Jonathan Bergmann, Aaron Sams, Publisher: Inte.

[11] Siemens, G., and Downes, S. (2008, 2009). Connectivism & connected knowledge: CCK08, CCK09. http://ltc.umanitoba.ca/connectivism

[12] Fox, A., & Patterson, D. (2013). What We've Learned from Teaching MOOCs. https://www.edx.org/blog, May 8, 2013.

AUTHORS

L. Martinez and **P. Garaizar** are with Deusto Institute of Technology (DeustoTech), University of Deusto, Bilbao, Spain.

25

Interactive Design and Gamification of eBooks for Mobile and Contextual Learning

José Bidarra [1], Mauro Figueiredo [2], Carlos Natálio[1]

[1]Open University of Portugal, Lisbon, Portugal
[2]University of Algarve, Faro, Portugal

***Abstract*—Given the global spread of mobile platforms like iOS and Android, it is relevant to consider the potential of these devices to teach and learn. Our exploratory research was designed to evaluate the possibility of using common eBook formats to create *gamebooks* (gamified books) that are effective in teaching and learning. After analyzing the features available in many free or open tools, we decided to use iBooks Author to build a model of a dynamic book that may function as an educational game for "Environmental Studies", aimed at children in the 4th grade, mostly 9 to 10-year olds enrolled in Portuguese schools. This paper presents the design and features of this interactive book titled "Adventures in the Guadiana River", developed using the Apple iBook format, and especially set for the iPad retina. Because this is a proprietary format, we also built a prototype following the EPUB3 standard and developed new ways to interact with content within this universal format. Preliminary tests with our prototypes revealed very good usability and a promising pedagogical potential for the proposed models.**

***Index Terms*—mobile learning; e-learning; multimedia; eBook; gamebook; iBooks Author; EPUB3**

I. INTRODUCTION

We are presently witnessing a new era in which the classic "passive receiver" of the communication is apparently assuming a role as an active interlocutor, able to intervene and to express a critical opinion. This perspective takes shape if we look at various blogs and social websites, like YouTube or Facebook. We are actually talking about innovations that result in the development of a networked society, in constant interaction, where individuals take quick decisions, are globally informed, and integrate various digital media. In this context, it is wise to consider the integration of digital media and mobile devices (iPad, iPod, tablets, smartphones), allowing students to set personal goals, to manage educational content and to communicate with others in the right context [1].

It is also important to note that this is a generation of digital natives; these students are the generation of digital games and social networks. We cannot ignore that they are no longer the same students for which the education system was designed traditionally. These students grew in a new technological environment, with its own techno culture, and they will live in a more demanding, competitive and complex world. See, for example, the prospect of Heide and Stilborne [2], for whom "the technological revolution has produced a generation of students who grew up with multidimensional and interactive media sources; a generation whose expectations and world views are different from those that preceded it" (p. 27). Moreover, this is a generation that lives in a very complex world and will have to make very difficult decisions (economy, environment, social welfare, etc.). Unfortunately, school education often does not establish the bridges which allow students to develop skills to better understand this complexity [3].

With the emergence of (cloud) distributed learning technologies, students can learn in a more informal setting using mobile devices, without being confined to a room full of computers. Furthermore, teachers and educators emphasize the importance and need for "authentic learning activities", where students can work with real world problems [4]. Therefore, the development of educational activities for students that combine learning resources from the real world with those from the digital world has become an important and challenging research topic. This may be accomplished, for example, by using mobile communication and wireless technologies, which can be moved to any place, allowing for scientific experimentation, augmented reality, image collection, resource sharing, and communication with colleagues.

In the current context of mobility, we wanted to explore the possibility of integrating the typical structure of a game in a new type of electronic book (eBook) to be used in tablets. The concept is, in essence, an interactive eBook, which enables the integration of text with images, audio, video and animations. We understand that there are clear pedagogical benefits in the development of an innovation that combines a book with a game – let's call it a *gamebook* - with potential in various learning contexts (formal, informal or non-formal). It is this new possibility that we pursue in our investigation, evaluating the potential of the technology in terms of tools and interfaces.

The mobile learning concept is part of a societal model that assumes digital skills as vital in this second decade of the XXI century. We refer to, in particular, the ability to (1) analyze and produce digital information where and when the user wants, (2) make decisions in the context of an information society, (3) apply creative skills and innovation (technological and methodological), (4) engage in collaborative work and (5) master an operational knowledge of digital media and global communications.

This article reports on our exploratory investigation on eBook formats (EPUB and iBooks) and the creation of gamebooks that may be effective in various learning situations. Ongoing research involves the evaluation of three major components that are clearly distinguishable: (1)

assess the potential of the current eBook technology, (2) study the role of narratives in games, and (3) determine the potential of gamebooks for certain educational applications.

II. Designing Mobile Learning

While reviewing the literature we found several definitions of mobile learning. It should be noted, for instance, the view of Sharples [5], who defines mobile learning as the learning process that is not limited to a fixed place and takes advantage of mobile computing technology. Considering the popularity of mobile devices (e.g, smartphones, tablets and laptops), we realize that most researchers in this field have adopted this definition, which we think remains valid in the second decade of this century.

Traxel [6] also defines mobile learning as learning that is supported by a portable or mobile device. This mobile device encourages learning through its ease of access to information, the ability to easily move and its ability to manage very diverse media (text, image, audio, video, animation, etc.). The integration of portable devices in the instructional model is assumed, in the opinion of this author, to be a possible way to increase the efficiency of learning. Portable devices have been used for a significant period in curriculum areas such as languages [7], [8], [9], [10], mathematics [11], social studies [12], [13], [14], and science [15], [16].

On the other hand, compared with traditional education based on textbooks, learning with eBooks, coupled with mobile devices, seems to be an increasingly attractive option; one which can trigger the interest and motivation of students. Considering that interactive eBooks may eventually have teaching potential, they can also be distracting and produce some dispersion [17], [18]. On the positive side, Glasgow [19] showed that student motivation is higher when children interact with multimedia materials, especially those children with reading difficulties. Supporting this view, Bearne [20] argues that most children today are immersed in vast multimedia experiences and are used to combining diverse resources to interact with. In a recent study, Larson [21] suggests that using digital reading devices promotes new literacy practices and extends the connections between text and other relevant learning resources. Despite these encouraging results, however, the development of multimedia tools and methodologies to improve outcomes in student learning remains an important and challenging question for authors and institutions [22].

Mobile devices such as tablets may actually promote a new way of "being in school", anytime and anywhere, so students do not have sit in front of a computer located in a room or lab [23]. As we shall see, mobile devices have unique features that interest us, namely, in terms of interaction and exploration modes in instructional processes. However, the iPad and other tablets are still relatively recent in the marketplace, so research of their use in education is relatively inconclusive [24], [25], [26], [27].

During a study conducted in 2009, at the University of Wisconsin in Madison, we verified the existence of well-defined patterns in the use of mobile devices by students [28], explicitly that:

- mobile learning evolves around the social environment of the student (sharing podcasts, images, texts and notes), and is not linked to the classroom;
- learning activities are mainly based on online resources and on interaction with other people (often through academic work);
- collaboration networks and group work are an important aspect of mobile interaction anytime, anywhere;
- the ease of instantly publishing content online stimulates students to become researchers and authors;
- the possibility to easily capture, record and publish multimedia transforms students into producers and critics (for example, when interacting in Facebook, Twitter, Flicker or YouTube).

III. Gamification in Mobile Learning

Research in educational technology needs to go far beyond learning with multimedia to recognizing the role of new learning experiences; for example, one that games and simulations can reveal. The learning model which we call "ludic" - based on games - can be used in formal or informal education by well-defined age groups, and can be introduced in various scientific fields. But how do we broadly define "game"? For our educational purposes a working definition may be the one proposed by Klopfer [29], for whom a game is a goal-oriented activity, based on specific rules, which players perceive as enjoyable. In an educational context, considering that "ludic" is not usually a priority in most activities, a game may be the motivating factor that is needed in many learning resources.

Unfortunately, the production and deployment of educational games presents some difficulties, particularly:

- The high development costs and an uncertain market make investment in educational games and innovations too risky for producers;
- Institutions resist adopting innovations and do not want to make unnecessary changes and investments, including the use of new technologies for learning;
- Instructors, institutions and publishers do not (traditionally) want to replace textbooks with educational games;
- The value of specific educational technologies (games, simulations, etc..) have not been proven in many cases;
- Parents and teachers still have very negative attitudes about the use of games in the classroom;
- Games are especially suited to teach higher order skills that are not typically assessed through examinations (multitasking, decision-making, strategic vision, etc.);
- Easy access to computers and the Internet cannot be taken for granted in educational institutions (the case of many developing countries).

Gamification is a relatively new concept that has acquired considerable momentum over the last years [30], [31], [32]. It is a concept that integrates the mechanics of gaming in non-game activities to make these more effective and enjoyable. When used in the educational field, *gamification* seeks to integrate game dynamics and game mechanics into learning activities, for example, using tests, quizzes, exercises, edu-games, badges, etc., in order to increase intrinsic motivation and foster participation of students.

In this context, we can define game mechanics as the set of rules and rewards that make up game play, a satisfying and highly motivational activity, in other words, making it more challenging and engaging. The most common game mechanics [31] include:

- *Points*: studies at the University of Chicago show that points are fantastic motivators and can be used to reward users/students across multiple levels or dimensions of a gamified activity. In general people love to be rewarded and, when interacting with a point system, they feel like they have gained something.
- *Levels*: these are often defined as point thresholds, so the students (or users) can use them to indicate a higher status and have access to bonus content.
- *Challenges, badges, achievements, and trophies*: the introduction of goals in an activity make students (users) feel like they are working toward a goal. Normally, challenges should be configured based on specific actions and should include user/student rewards when they accomplish certain milestones with badges, achievements or trophies.
- *Leader boards or "high-score table"*: in the context of gamification, high-score tables are used to track and display desired actions, using completion to drive valued behavior. In intrinsic motivation terms, they are one of the most important features of a game, bringing the aspiration factor to the process.

In a way, educational processes have always used *gamification* in learning activities by applying scores on assignments that can be considered points. Also, according to this perspective, the academic graduation may be considered a level achieved and a diploma earned is a form of granting a badge of confidence [31]. However, this game-based system does not seem very engaging for the students; we think that perhaps education processes could be improve by adding the gaming factor through a technology that involves eBooks, as these are tools that students use and accept in their learning.

In this context, to overcome many difficulties, we wanted to explore the possibility of integrating the structure of a simple game in the last generation of eBooks. The aim is to create an electronic book that enables dynamic integration of text with images, audio, video and animations, but which may be used as a game. It is reasonable to admit that the educational system can more easily adopt an innovation like a *gamebook* - an interactive and ludic book - than conventional electronic games. It is this new possibility that looks interesting to study, based on the emergence of powerful mobile platforms (iPad, tablets, smartphones, etc.), but also relying on previous studies about games-based learning. In brief, we wanted to align eBooks with two learning scenarios: mobile learning and game-based learning.

Furthermore, it makes sense to consider the integration of mobile and game learning in systems managed by students, allowing them to set personal goals, manage content and communicate with each other. In practice, these Personal Learning Environments (PLEs) are made up of several components, which may include social networks, virtual worlds and authoring software, interconnecting various learning resources suitable to the pedagogical contexts and skills to be acquired by each learner. We argue that eBooks (or *gamebooks*), due to their characteristics and potential for interactivity, can be a much valued part of these PLEs.

IV. EBOOK AUTHORING

At the most basic level, the eBook is an electronic book that can have a simple format, such as a text in PDF. However, there are currently more advanced standards, such as EPUB3 (universal) or iBooks 2 (Apple), that can integrate multimedia components. Ideally, an eBook should have sufficient quality for current devices with their high-resolution displays, be compatible with a wide variety of reader apps, and, if necessary, allow conversion to other formats. In actuality, however, there are more than a dozen formats that are not compatible with each other, and sometimes they do not adapt to higher resolution screens.

Over the years, teachers have often used multimedia encyclopedias and textbooks on CD, DVD, or online. Recently, however, a new generation of electronic books has emerged that can offer more interactive and dynamic learning experiences. Students of the "Playstation generation" tend to respond better to this dynamic and interactive content, due not only to its ability to display text, but also other media with hypertext links, search capacities and connections to online databases. One benefit of these eBooks for students is the possibility to select any word that they do not understand and find its definition immediately.

Another potential use refers to the capability to have several students simultaneously accessing the same book and thus sharing the same learning experience. More specifically, they can communicate and share files in the context of group work, all the while participating from different locations. The eBook can also contain tests, allowing students to self-assess their progress.

An economic advantage refers to financial aspects; currently, the cost of textbooks represents a high expense for students and their families. The introduction of electronic books contributes to reducing these costs effectively. Furthermore, the emergence of less expensive mobile devices is expected, which is currently already occurring in countries such as China and India. Altogether, the widespread use of eBooks will also benefit the environment by reducing paper consumption, bearing in mind that a device may store thousands of eBooks.

On a less positive note, despite the huge advances provided by this technology, many applications used to read eBooks currently do not possess really useful features such as the ability to highlight or otherwise mark text, bookmark or write notes. These are apparently trivial matters, but they are important for a student. In a paper book students usually highlight the most important parts and add annotations in the margins. In this way, when students read the book again, they can more readily focus on these parts.

Although there are several standards for eBooks, not all reading applications and mobile devices use the same. Thus, to reach a wide audience, it is necessary to adopt some of the most commonly used formats for reading electronic books: Adobe PDF in computers, Mobi for the

Amazon Kindle, EPUB for the Barnes & Noble Nook, and iBook for Apple's iPad. The Portable Document Format (PDF) is the most popular format for creating digital books and can be read by the free Adobe Reader installed on most computers. It is an open standard that enables the creation of eBooks with support to use sound, images, video, notes and bookmarks. Its main advantage is to keep the layout of the printed book within original quality on any device. However, this benefit is a major disadvantage on mobile devices, where text is sometimes difficult to handle in a PDF file because of its visualization, forcing users to increase or decrease text size in order to view the contents of the page. In other eBook formats, such as EPUB and iBook, there are no physical constraints to a fixed style and everything scales.

EPUB is a format based on open specifications, primarily written in XML and XHTML. The EPUB format is supported by a wide range of devices and platforms, including Android devices, Nook, iPhone, iPad, iPod, MobiPocket, Adobe Digital Editions, FBReader, Stanza, Sony Reader, and many other readers and applications. Of the most popular devices, the only one that does not support EPUB is the first Amazon Kindle. The EPUB specification is an open standard, allowing the creation of eBooks with sound, images and video (in its newest specification - EPUB3). This specification introduces innovative features to address structure limitations, such as: precise layouts specialized for comic books, support for MathML, support for multimedia, and introduction of notes.

The Amazon Kindle is an electronic book reader very popular in the US. Early versions used the Kindle proprietary format AZW. This is basically the Mobipocket format based on the Open eBook standard using XHTML. This specification supports images, notes and bookmarks. In late 2011, Amazon started selling the Kindle Fire along with the new file format Kindle Format 8 (KF8), which supports a subset of the features of HTML5 and CSS3. This is also a proprietary specification and expands the functionality of the earlier versions of the Kindle, in order for the Kindle Fire to support sound and video. Thus, eBooks can be created in these formats using a set of applications that convert PDF files in multiple formats such as EPUB, or MOBI for the Kindle, among others. For example, the widespread Calibre, which is a free application running on Windows, Mac OS X and Linux, offers a simple way to create digital books for multiple mobile platforms, starting with PDF or RTF files.

There are other free or open applications that support the creation of electronic books. For example, the application eCub allows the creation of simple books in EPUB or MobiPocket, from text files or XHTML. However, eCub is very limited, with no WYSIWYG capabilities. It is suitable for the production of simple eBooks with front and back pages with an image, index, a title page and it can convert content to a sound file (WAV or MP3). The eCub is free and is available for Windows, Mac OS X, Linux, FreeBSD and Solaris.

Booktype is an open platform, available since 2012, which allows editing and writing of eBooks for different platforms by exporting in PDF, EPUB, MOBI, ODT and HTML. This application also exports the eBook directly to Amazon, Barnes & Noble and iBookstore online stores, as well as to online printing sites. Digital books written with this application are immediately available in any of these platforms. When writing the eBook, the author does not need to worry about formatting, since it will automatically be formatted to work in these different platforms. Booktype also provides a set of collaborative tools for reviewers, editors, translators, designers, and authors, enabling the different participants to work collaborative in the production of an eBook. Some of the features offered by this platform include: intuitive drag-and-drop tools, chat, messages, adding images and text formatting. Booktype also maintains a history of all changes, which allows the author to compare different editions and return to a previous edition. It is even possible to use snippets (pieces of computer code). One of the disadvantages of this platform is the need for installing on a server and accessing via web browser, which requires some additional expertise.

The Firedocs eLML editor [33] can also be used to create eBooks. The eLML framework (eLesson Markup Language) is an XML platform for creating online classes using XML. It exports produced materials in SCORM, HTML, PDF and EPUB format. The main objective is to ensure that classes are modeled according to the ECLASS reference, which defines five distinct sections: Entry, Clarify, Look, Act, Self-assess, Summary. Current implementation only supports JPG, PNG, GIF and SVG images. It does not allow Java scripts and forms, so some of the functionalities as the glossary, references to labels and self-assessment tests are not available in the EPUB format.

SIGIL is an open WYSIWYG editor used by Google to create eBooks following the EPUB2 specification for Windows, OS X and Linux. This application imports, creates and edits XHTML documents and exports them into EPUB2 documents. EBooks created with SIGIL may contain text, pictures and links, but this standard does not support video or sound. It also provides multiple views of the work: book, code and a split mode. In the book view, it allows content edition in WYSIWYG mode.

Finally, we should mention the electronic books created in the iBook format from Apple. These eBooks are created with a free application - the iBooks Author tool. The format is proprietary, although based on the EPUB standard specification, with some differences in the CSS3 tags. This tool makes the process of creating eBooks very easy, by presenting a very complete set of integrated features, including: sound, image, video, dictionary, text underline, annotations, text-to-speech conversion, navigation and many widgets to enhance the interactive experience. The introduction of widgets in the iBook is an enriching experience for readers of an electronic book. iBook Author offers seven types of pre-defined widgets: (1) photo gallery, (2) video or audio media file, (3) review questions, (4) slide show; (5) interactive tagged images, to give detailed information on specific parts of an image or graph, (6) 3D models, and (7) objects created in HTML. The ease of creating widgets allows users to add any object to an interactive iBook, and there are many possibilities, from calculators, puzzles, maps, YouTube videos, among many others. Table I summarizes the characteristics of the authoring applications surveyed.

V. A PROTOTYPE FOR THE IPAD

In the process of creating a prototype of an electronic book, we started out with the idea of a *gamebook* – a book with a story that can be read sequentially or not. The main difference refers to the ability to choose different paths to the main characters or the unfolding of the history, as happens in games. The reader/player makes choices that affect how the story unfolds and his decisions have a significant impact on events and the final outcome. This will be the next step of our research.

Some choices may be as simple as turning right or left at the end of a road. Others may be much more difficult, requiring decisions about facts or occurrences, such as, problems arising in natural disasters, environmental pollution and climate change. In a particular story the student may be the main character - the hero of the story - or simply manipulate variables with consequences at the strategic level. Thus, not only decisions change the story, but also the sequence of choices can change the ending.

A *gamebook* may follow typical genres and formats of electronic games, such as puzzle, RPG, adventure, strategy, among others, where the reader will face threats, adversaries, and discover the truth behind an intricate story. Ultimately, in this project we strive to create a didactic narrative built on situations capable of providing expectation, suspense, challenge and other positive emotions. Based on the latter, it is theoretically possible to engage students in the study of a particular subject matter and obtain improved results in terms of attention, retention and understanding. In fact, the possibility and value of integrating game play in learning practices is indicated clearly by several researchers, who recognize its potential in making learning more meaningful and in assimilating new subject matter [34], [35], [36], [37], [38].

The development of a *gambook* prototype for the iPad is still in the early research phase. To build the model, we chose the Apple iBook format, essentially due to the vast potential for multimedia authoring with the iBooks Author tool, when compared with the current tools for the EPUB3 format (the most universal but the authoring tools are still in the early stages of development).

Within the initial exploratory stage of this research, and considering all the multimedia features and widgets available, we built a model of a dynamic book that can function also as an educational game for the curriculum of "Environmental Studies", suitable for children in the 4th grade program, with the title "Adventures in the Guadiana River" (fig. 1).

The ongoing research involves three strategic components, clearly distinguishable in our workflow:

1. Establish the potential of the technology: the most common formats and authoring tools that allow the creation of *gamebooks* that are effective in learning;
2. Identify the role of narratives and games: what kinds of narrative and game genres can be recreated as *gamebooks* for learning, and what learning activities can be prepared based on this typology.
3. Demonstrate real educational applications: according to a specific level of education, what pedagogical models can better integrate *gamebooks*, and in which subject areas are most effective.

In addition to a story that can be followed by reading the text consisting of a trip along the Guadiana River, there are several devices that make the development of the narrative more memorable. First of all, a basic text adapted to the target audience was illustrated with appealing images and galleries on various aspects of the river, emphasizing its environmental context. To ensure the en-

TABLE I.
FREE EBOOK AUTHORING TOOLS

	eCub	Book-type	Fire-docs	Sigil	iBooks Author
Photos	√	√	√	√	√
Graphics					√
Sound					√
Videos					√
Hyperlinks	√	√	√	√	√
Animation					√
Search					√
Dictionary					√
Underline					√
Markers					√
Notes					√
Digital Speech					√
Feedback		Msgs			
Tests and self-assessment					√
Widgets/programs/ apps					√

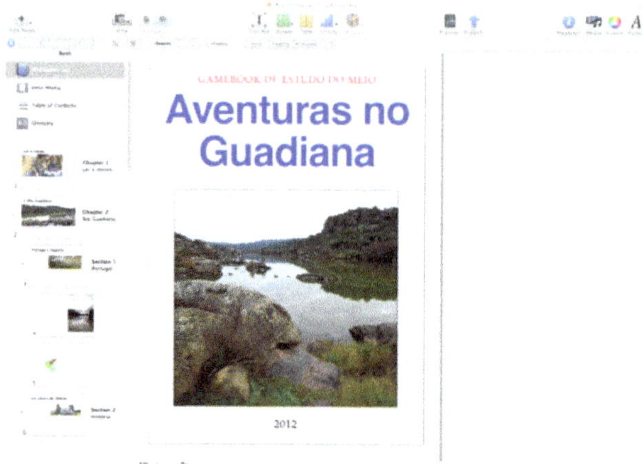

Figure 1. iBook prototype

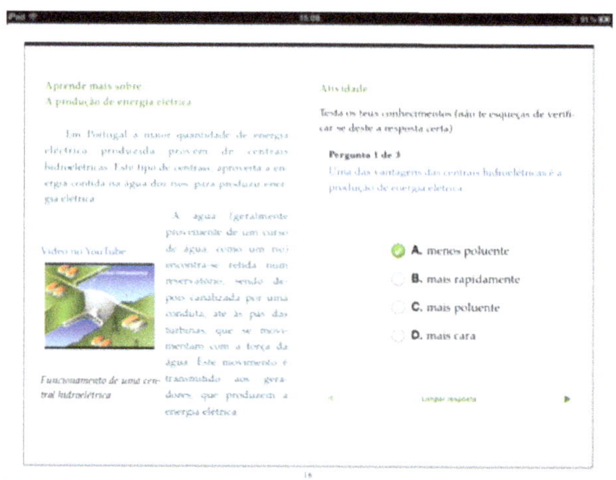

Figure 2. Video with multiple-choice questions.

gagement of readers, the narrative requires the completion of certain tasks that are supported by interactive devices based on iBooks' widgets, in particular, the inclusion of thematic videos and their exploitation through multiple-choice questions (fig. 2). To allow for geographic contextualization of themes, we opted to include Google Earth maps, accessible from a specific widget, so that the journey could be followed on the map.

We are still investigating ways to program HTML5 and new widgets so that user decisions have a significant impact on events and in final outcomes. Our ultimate goal is to allow students the possibility of selecting different paths affecting the unfolding of the story. The prototype already has a narrative based on a journey of discovery, with various unexpected situations to maintain student interest. On the other hand, while providing a suitable thematic framework, we also tested the playability of puzzle games with moving parts (fig. 3), which have the potential to motivate young users to interact more intensely with the iBook.

Initial usability tests of the model were held with a panel of a few 9 to 10-year- old children. The first results provided very positive reactions in terms of interface usability and user motivation. However, these also show that some aspects need revision, particularly in what concerns grasping the interaction mechanisms. For example, the use of images, videos, colors and graphics seem to work well on the iPad interface, but some aspects of the widgets have to be reformulated. In the case of Google Earth, there are maps that are difficult to read or puzzles too complicated to play; technical aspects which may eventually lead to some frustration.

It was observed that students interact with a *gamebook* as if they were playing, when in fact they are using a tablet to learn specific content. To this end, the study continues with the development and testing of this prototype in a broader educational context, where we will use a more expanded sample, based on a class of students in the 4th year of the 1st Cycle (Basic Education). In a second phase, more robust validation tests will be made with this and other target groups, particularly covering subject areas that are traditionally unattractive to students (Math, Physics, etc.).

VI. EXPERIMENTING WITH EPUB3

Experiments with the EPUB3 standard started with the creation of an eBook about a beautiful lagoon with a diversified fauna and flora in the middle of Portugal (Óbidos). In this project, we tried many different widgets and scripts that somehow emulated the effects achieved with the Apple format iBooks. Because the EPUB3 norm does not specify what kinds of JavaScript should be used, we had to try proven solutions, based on the jQuery library that may be applied in many platforms. For instance, some scripts used in touch platforms to interact with a panoramic photo:

```
<script src='scripts/jquery-1.8.3.js'
type='text/javascript'></script>
  <script src='scripts/jquery.ui.touch-punch.min.js'
type='text/javascript'></script>
  <script src='scripts/jquery.reel-1.2-bundle.js'
type='text/javascript'> </script>
```

Another device tried was Lightbox, a technique that uses a small JavaScript library that permits users to view and increase the size of images and other content overlapping to the main page, instead of causing a new page to load (fig. 4). When the overlay is displayed, the rest of the content of the page is dimmed for the user to focus only on the interesting part. Although the pages can be greatly expanded in tablets, one of the advantages of this plug-in is to include informational text. Depending on the settings, the overlay can be positioned, adapted to the size of the browser window on the fly. The plug-in determines which links are activated via the “href” attribute. Lokesh Dhakar wrote the plugin used:

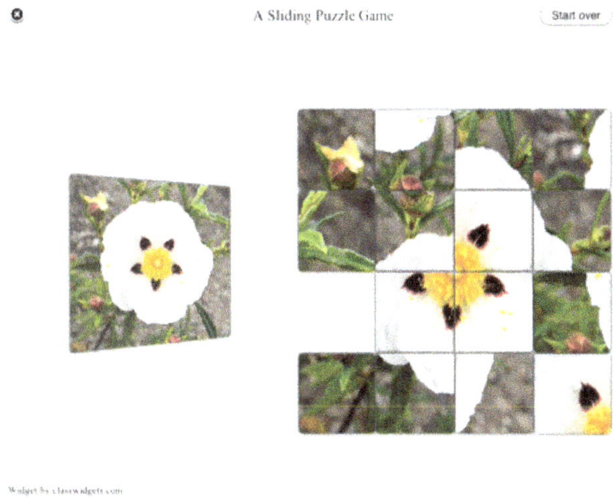

Figure 3. Puzzle game with pieces of flower (Esteva)

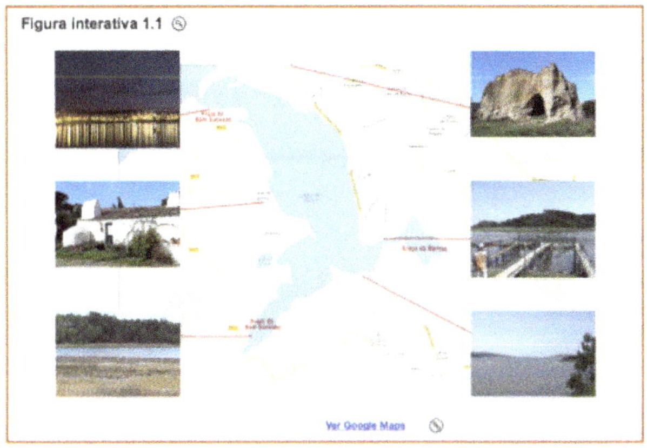

Figure 4. Page with images that expand (Lightbox).

```
  <script src="js/lightbox.js"
type="text/javascript"></script>
   <script type="text/javascript">
   jQuery(document).ready(function($) {
        $('a').smoothScroll({
                speed: 1000,
                easing: 'easeInOutCubic'
        });
        $('.showOlderChanges').on('click', function(e){
                        $('.changelog
.old').slideDown('slow');
                $(this).fadeOut();
                 e.preventDefault();
        })
(http://lokeshdhakar.com/projects/lightbox/)
```

Another script used was flexslider.js, a device that creates a customized slideshow with points (fig. 5):

```
<script type="text/javascript"
src="js/jquery.flexslider.js"></script>
<script type="text/javascript">
$(window).load(function(){
 $('.flexslider').flexslider({
   animation: "slide",
   start: function(slider){
   $('body').removeClass('loading');
   }
 });
     });
</script>
```

Because simple card games are appealing and easy to implement, a script called "card.js" was used because it contains a specific CSS style in the eBook for the identification of the birds in the lagoon (fig. 6):

```
<link type="text/css" rel="stylesheet"
href="css/card.css" />
<script type="text/javascript" src="scripts/jquery-
1.8.3.js"></script>
<script type="text/javascript"
src="scripts/card.js"></script>

<div id="boxbutton">
<span class="link">
<span id="count">0</span>tentativas.</span>

</div>

<div id="boxcard">
<div id="card1"><img src="img/01.jpg"
alt="card"/></div>
<div id="card.....

<div id="card13"><img src="img/01.jpg"
alt="card"/></div>                </div>
```

Many other scripts were tested, for example, to create a "time slider" with a historical map, to insert a "scrollbar" for a text box, to create different kinds of quizzes, and to present video and audio. One of the problems until recently, with the use of video and audio files, is that external plug-ins were needed to play videos (Flash Player, QuickTime, RealPlayer; Windows Media Player). With the advent of HTML5 and its adoption by the EPUB3 standard, this problem has been overcome. However, there are still issues relating to licenses and patents, which restrict the use of the formats. There are two main types for video coding in eBooks: MP4 for h.264 (iPad) and webm for VP8 (open format). In audio, the most common is the .MP3 format.

VII. Conclusion

After studying the current state of the art in eBook technology (free and open), we chose the Apple software iBooks Author as the most suitable for the construction of our first model, mainly due to its ability to integrate sound, image, video, dictionary, text underline, annotations, text-to-speech, ease of navigation, and widgets that can enhance the interactive experience. Nevertheless,

Figure 5. Page with interactive slideshow.

Figure 6. Card game with wild birds.

given the high cost of equipment and the limited distribution of the iPad in schools at present, the development and testing of a prototype that uses the open standard EPUB3 which may be used in any portable device (Android or iOS) was also tried. With this in mind, doctoral students are already developing new tools with EPUB3.

Preliminary tests with our prototypes suggest that the iBook technology and the EPUB3 standard have great usability and pedagogical potential for a learning eBook model. The interface and interaction design for contextual learning was the main focus. At first glance, the concept of a "gamebook" appears to have the potential to challenge students to become actively involved in the educational process, as it allows students to try different routes and distinguish what is important from what is secondary, enabling them to create and annotate material from various sources, while also encouraging the exploration of new issues. These are, however, aspects that need to be investigated and validated by future research in specific educational contexts.

References

[1] J. Bidarra, A. M. Sousa, F. Grazina, P. Simões, P. Azevedo, "Personal Learning Environments no contexto virtual de um mestrado em Comunicação Educacional Multimédia", in M. C. Rodriguez, R. A. Silveira, P. Escudeiro (Eds), *TICAI 2010*, IEEE, Sociedad de Educación, Capítulos Español, Portugués y Colombiano, 2011.

[2] A. Heide & L. Stilborne, *Guia do Professor para a Internet - Completo e fácil*. Porto Alegre – Brasil, Artmed Editora, 2000.

[3] J. C. Lima & B. Silva, "TIC e processos de autoregulação da aprendizagem", in L. Almeida, B. Silva, & S. Caires (orgs.), *Atas do I Seminário Internacional "Contributos da Psicologia em Contextos Educativos"*, pp. 747-759. Braga, CIEd, Universidade do Minho, 2010.

[4] J. S. Brown, A. Collins & P. Duguid, "Situated cognition and the culture of learning", *Educational Researcher*, vol. 18, no.1, pp. 32–42, 1989.

[5] M. Sharples, "The design of personal mobile technologies for lifelong learning", *Computers & Education*, vol. 34, pp. 177–193, 2000.

[6] J. Traxel, "Learning in a mobile age", *International Journal of Mobile & Blended Learning*, vol. 1, no. 1, pp. 1-12, 2009.

[7] L. Bomar, "iPods as reading tools", *Principal*, vol. 85, no. 5, pp. 52–53, 2006.

[8] K. B. Patten & D. V. Craig, "iPods and English-language learners: A great combination", *Teacher Librarian*, vol. 34, no. 5, pp. 40–44, 2007.

[9] L. P. Shoemaker, "Handhelds for reading and note taking", *Learning and Leading with Technology*, vol. 35, no. 2, p. 36, 2007.

[10] A. Moura, *Apropriação do telemóvel como ferramenta de mediação em mobile learning: estudos de caso em contexto educativo*, Doutoramento em Ciências da Educação, na especialidade de Tecnologia Educativa, Instituto de Educação, Universidade do Minho, 2010.

[11] L. M. Lary, "A baker's dozen: 13 Palm applications for mathematics (and math related!) instruction", *Learning and Leading with Technology*, vol. 39, no. 9, pp. 22–27, 2004.

[12] A. Dixon, "Finding your way: GPS and geocaching", *Learning and Leading with Technology*, vol. 34, no. 8, pp. 29–31, 2007.

[13] R. Royer & J. Royer, "What a concept! Using concept mapping on handheld computers", *Learning and Leading with Technology*, vol. 31, no. 5, pp. 12–16, 2004.

[14] D. L. Vess, "History to go: Why iTeach with iPods", *The History Teacher*, vol. 39, no. 4, pp. 479–492, 2006.

[15] J. Roschelle, W. R. Penuel, L. Yarnall, N. Shechtman, & D. Tatar, "Handheld tools that "informate" assessment of student learning in science: A requirements analysis", *Journal of Computer Assisted Learning*, vol. 21, no. 3, pp. 190–203, 2005.

[16] R. Tinker, P. Horwitz, S. Bannasch, C. Staudt, & T. Vincent, "Teacher uses of highly mobile technologies: Probes and podcasts", *Educational Technology*, vol. 47, no. 3, pp. 16–21, 2007.

[17] C. Burrell, & J. Trushell, "Eye-candy" in "interactive books"—A wholesome diet?, *Reading*, vol. 31, no. 2, pp. 3–6, 1997.

[18] K. I. Matthew, "The impact of CD-ROM storybooks on children's reading comprehension and reading attitude", *Journal of Educational Multimedia and Hypermedia*, vol. 5, nos. 3–4, pp. 379–394, 1996.

[19] J. N. Glasgow, "It's my turn! Part II: Motivating young readers using CD-ROM storybooks", *Learning and Leading With Technology*, vol. 24, no. 4, pp. 18–22, 1996.

[20] E. Bearne, "Multimodal texts: What they are and how children use them", J. Evans (Ed.), *Literacy moves on: Popular culture, new technologies, and critical literacy in the elementary classroom*, pp. 13–29. Portsmouth, NH: Heinemann, 2005.

[21] L. Larson, "Digital readers: The next chapter in e-book reading and response", *The Reading Teacher*, vol. 64, no. 1, pp. 15–22, 2010.

[22] J. Bidarra & O. Martins, "Exploratory learning with Geodromo: An interactive cross-media experience", *Journal of Research on Technology in Education* (JRTE), vol. 43, no. 2, pp. 171–183, 2010.

[23] J. Brand & S. Kinash, "Pad-agogy: A quasi-experimental and ethnographic pilot test of the iPad in a blended mobile learning environment", *Proceedings 27th Annual Conference of the Australian Society for Computers in learning in Tertiary Education* (ASCILITE), Sydney, Australia, http://works.bepress.com/jeff_brand/18/, 2010.

[24] O. Murray & N. Olcese, "Teaching and Learning with iPads, Ready or Not?", *TechTrends*, vol. 55, no. 6, pp. 42-48, 2011.

[25] M. Manuguerra & P. Petocz, "Promoting student engagement by integrating new technology into tertiary education: The role of the iPad", *Asian Social Science*, vol. 7, no. 11, pp. 61-65, 2011.

[26] M. Weisberg, "Student attitudes and behaviors towards digital textbooks", *Publishing Research Quarterly*, vol. 27, no. 2, pp. 188-196, 2011.

[27] B. McClanahan, K. Williams, E. Kennedy, & S. Tate, "How use of an iPad facilitated reading improvement", *TechTrends*, vol. 56, no. 3, pp. 20-28, 2012.

[28] J. Bidarra, "Emerging Digital Media, Games and Simulations: A Challenge for Open and Distance Learning", *Revista de Ciências da Computação, 4*, Universidade Aberta, Lisbon, 2010.

[29] E. Klopfer, *Augmented Learning*, The MIT Press, Cambridge, Massaschusetts, 2008.

[30] S. Deterding, K. O'Hara, M. Sicart, D. Dixon, & L. Nacke, *Gamification: Using Game Desing Elements in Non-Gaming Contexts*. CHI 2011, pp. 7-12, Vancouver, BC, Canada, ACM, 2011.

[31] J. J. Lee & J. Hammer. *Gamification in Education: What, How, Why Bother?* Adademic Exchange Quarterly, p. 146-151, 2011.

[32] K. M. Kapp, *The Gamification of Learning and Instruction - Game-based Methods and Strategies for Training and Education*. San Francisco, Pfeiffer, 2012.

[33] R. Weibel, S. Bleisch, S. Nebiker, J. Fisler, T. Grossmann, M. Niederhuber, C. Collet & L. Hurni, "Achieving more sustainable e-learning programs for GIScience", *Geomatica*, vol. 63, pp. 109-118, 2009.

[34] M. Prensky, *Digital game-based learning*, McGraw Hill, New York, 2001.

[35] J. P. Gee, What video games have to teach us about learning and literacy, Palgrave Macmillan, New York, 2003.

[36] K. Squire & H. Jenkins, "Harnessing the power of games in education", *InSight*, vol. 3, no. 1, pp. 7–33, 2003.

[37] J. Kirriemur & A. McFarlane, "Literature review in games and learning", *NESTA Futurelab Series*, NESTA Futurelab, Bristol, 2004.

[38] S. Johnson, Everything bad is good for you: How today's popular culture is actually making us smarter, Riverhead Books, New York, 2005.

[39] J. Bidarra, M. Figueiredo & C. Natálio, Designing eBook Interaction for Mobile and Contextual Learning, *Proceedings 8th International Conference on Interactive Mobile Communication Technologies and Learning* (IMCL 2014), Thessaloniky, Greece, 2014.

AUTHORS

José Bidarra has a Ph.D. in Educational Communications from Universidade Aberta (the Portuguese Open University), where he is currently Assistant Professor in the Department of Science and Technology. He is head of the Informatics, Physics and Technology Section (SIFT), and coordinator of a master degree in multimedia. His current research interests focus mainly on the application of multimedia and digital media in distance education, including eBooks, games and simulations. Many of his master and doctorate students are developing new methodologies to engage learners in effective experiences with digital media. Most of the research is conducted at Universidade Aberta and at CIAC (Center for the Arts and Communication Research, University of Algarve); other research includes a fellowship at the University of Wisconsin – Madison, USA (e-mail: jose.bidarra@uab.pt).

Mauro Figueiredo has a Ph.D. in computer science at the University of Salford, Manchester, UK. He is an adjunct professor at the University of Algarve in Portugal. He is member of the CIMA (Marine and Environmental

Research Center) and of the CIAC (Center for the Arts and Communication Research). His research interests are about the use of Information and Communication Technologies for education and, in particular, about eBooks, games and augmented reality (e-mail: mfiguei@ualg.pt).

Carlos Natálio was a research student at the Portuguese Open University and is now a researcher of eBooks, multimedia and interactive systems. He is also involved in digital authoring and corporate training on eBook design (e-mail: carlosraminhos@gmail.com).

Developing a Remote Laboratory for Heat Transfer Studies

R. Ennetta and I. Nasri
Higher Institute of Industrial Systems of Gabes, University of Gabes,
Salaheddine El Ayoubi Street, 6011, Gabes,Tunisia.

Abstract—**Remote labs that enable students to conduct real-world experiments at a distance using a computer and Web-based tools can be the only realistic method of performing practical experiments in the context of distance learning.**

The e-Sience Tempus project, "Maghreb Network of remote labs," offers an innovative pedagogical approach that integrates eLearning and is complementary to classroom learning. The main objective of this project is to create an efficient remote labs network in the Maghreb region for the modernization of higher education in technological sciences.

The current paper summarizes the work carried out on a heat exchanger bench to be fully accessed and controlled remotely. Through this new device a student of Mechanical Engineering is introduced to many fundamental aspects of heat transfer.

Index Terms—e-Sience project; heat exchanger; Maghreb; remote lab;

I. Introduction

The Web has spurred our imagination as to how education can be radically transformed and enhanced through the adoption of ICT. While the use of Web-based education to date has seen as a significant innovation, it has not had the revolutionary impact originally envisioned and has certainly not displaced the traditional settings of classrooms, schools, colleges and universities [1], especially in educational laboratories. Those practical educational aspects need more and more development.

As we know, the use of laboratory experiments is a critically important aspect of education. Experience in teaching has shown that a complementary approach combining theoretical and practical exercises is vital for effective learning [2]. According to Hansen (1990) [3], students retain 25% of what they listen to, 45% of what they listen to and see, and 70% when they manipulate, control and modify experiments, putting into practice what they are learning.

In the context of distance learning, remote access to distance laboratories can be the only realistic method of performing practical experiments [2]. These remote labs enable students to conduct real-world experiments at a distance using a computer and Web-based tools [4].

Increasingly, teaching institutions are offering remote access to distant laboratories as part of an overall eLearning strategy. Remote experimentation provided as part of a Web-based learning approach affords a number of critical benefits, allowing flexible access to on-campus resources free of time or geographical constraints. However, adapting and redeveloping existing experimental resources for this purpose is a very hard task.

In the Maghreb region, the strong demand for top technicians and engineers needs an increased training capacity. Enrollment growth and quality of courses are often incompatible.

To avoid typical constraints of traditional laboratories such as scheduling as well as cost of equipment and location, remote operation of real plants can be incorporated into engineering courses [5].

The e-Sience Tempus project, "Maghreb Network of remote labs," [6] offers an innovative pedagogical approach that integrates eLearning and is complementary to classroom learning. The main objective of this project is to create an efficient remote labs network in the Maghreb region for the modernization of higher education in technological sciences.

As a partner of this project, we had to adapt and develop distant access solutions for two remote labs to allow students to perform practical experiments on heat transfer and mechanical vibration. These two remote labs will be a part of the Maghreb Network of remote labs.

In this paper, a heat exchanger experiment remote lab is presented, through which a student of Mechanical Engineering is introduced, in both theoretical and practical ways, to many fundamental aspects of heat transfer. We present the work carried out to adapt and redesign this device to be fully accessed from the Internet.

II. The e-Sience Project

The main objectives of this project were

- to build a network of remote labs in the Maghreb region
- to implement of a state-of-the-art solution
- to establish a best practice for the modernization of higher education in technological sciences.

This remote labs network will enable students to conduct real-world experiments at a distance and to assess the educational potential of such a system.

The specific objectives of this project are

- Adaptation of content to the technological changes in sciences while taking into account the expectations of the market's needs,
- Setting of three remote measurements platforms in the Maghreb region and their networking,
- Implementation of remote practical activities,
- Development of eLearning modules and courses,

- Evaluation of the developed course,
- Accreditation of the learning units in accordance with the Bologna process,
- Use of resources,
- Dissemination of the project at the national level in partner countries and at international level,
- Sustainability through enrollment in university courses in partner countries.

At this time, the e-Sience project is well advanced and the majority of remote labs are being tested and evaluated.

At the Higher Institute of Industrial Systems of Gabes (ISSIG), as a partner of the e-Sience Tempus project, we adapted and developed distant access solutions for two remote labs to allow students to perform practical experiments on heat transfer and mechanical vibration. These two remote labs are directly accessible through the ISSIG e-Lab server.

The current paper summarizes the work carried out on a heat exchanger bench to be fully accessed and controlled remotely. Through this new device a student of Mechanical Engineering is introduced, in both theoretical and practical ways, to many fundamental aspects of heat transfer.

III. The Heat Exchanger bench

Engineers learning about thermodynamics and heat transfer need to know how well different heat exchangers work. They can use this information to decide the correct heat exchanger for their own designs. TD360 Heat Exchanger bench [7] shows students how different small-scale heat exchangers work. They mimic the most common heat exchangers used in industry and compare how well they work for different flow rates and temperatures.

The TD360 module (Fig.1) is a compact frame with two water circuits (hot and cold) and instruments to measure and display water flow and temperature. This module can work with various types of heat exchangers (Concentric Tube Heat Exchanger, Plate Heat Exchanger, Shell and Tube Heat Exchanger and Jacketed Vessel with Coil and Stirrer). Students test each of the optional heat exchangers and record the flow and temperature changes to see how well the heat exchangers work. If they have one or more of the heat exchangers, students can compare them to see which is best for any application.

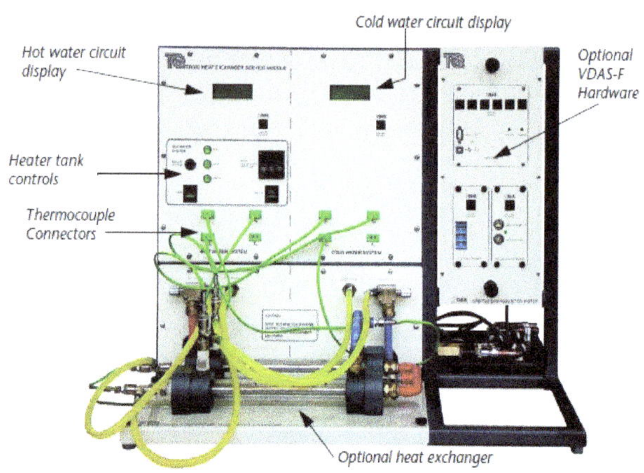

Figure 1. TD360 Heat Exchanger bench

In the present experiment we used only the Concentric Tube Heat Exchanger (Fig. 2).This heat exchanger is a simple shell and tube heat exchanger. It has two tubes, one inside the other. The outer tube is the shell. The inner tube carries the water from the hot circuit; the other tube carries the water from the cold circuit. Heat transfers between the two tubes. You can connect the water circuits to give contra-flow (counter-flow) or parallel flow experiments. This heat exchanger is in two equal parts with extra thermocouples at the mid-point.

These experiments help students understand more clearly how the temperature changes along the heat exchanger.

Fig.3 shows an example of the results of the effect of varying the cold flow rate on the performance of the heat exchanger in both parallel flow and counter flow connections. In this case the hot flow rate and heater temperature are fixed.

IV. The remote controlled Heat Exchanger Bench

The architecture of the ISSIG e-Lab is presented in Fig.4. This e-Lab is composed of two remote labs: the mechanical vibration remote lab and the heat exchanger remote lab.

To provide access to these remote labs, we use the MIT iLab Shared Architecture (ISA) [8]. In ISA, the Service Broker is the core of the architecture and provides user authentication, authorization, experiment data storage and access to scheduling services [9].

The heat exchanger bench was adapted for remote operation. The adaptations included control of the hot water supply pump, the heater and the cold and hot water supply flows (Fig. 5). To provide students with an overview of the whole heat exchanger bench, an IP camera was located in the laboratory.

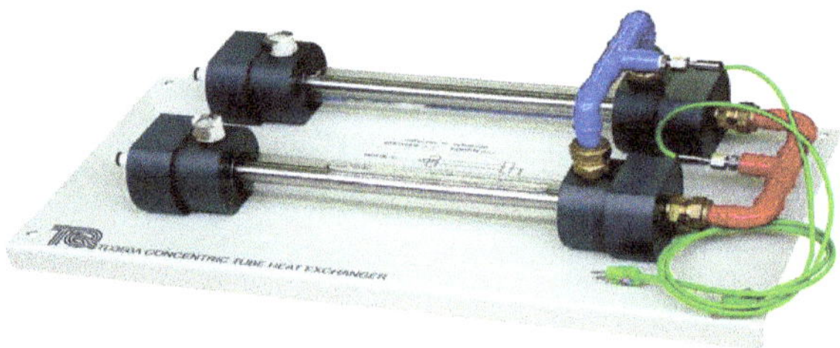

Figure 2. Concentric Tube Heat Exchanger (TD360-a)

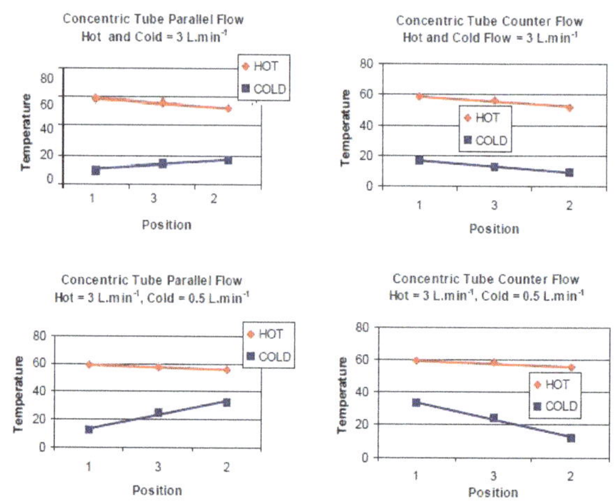

Figure 3. Example of heat exchanger experiment results

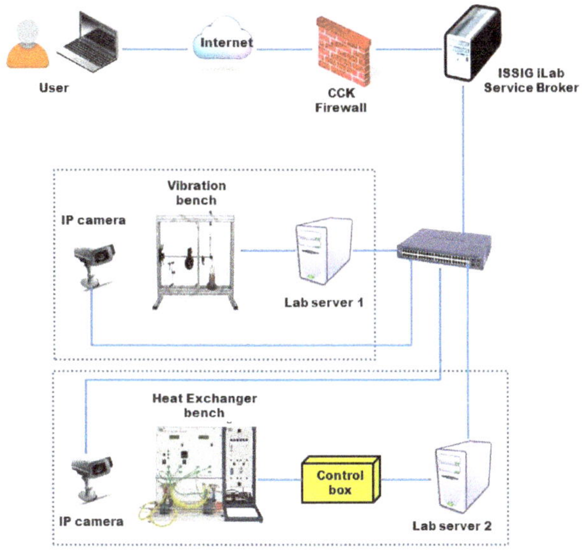

Figure 4. The ISSIG e-Lab architecture.

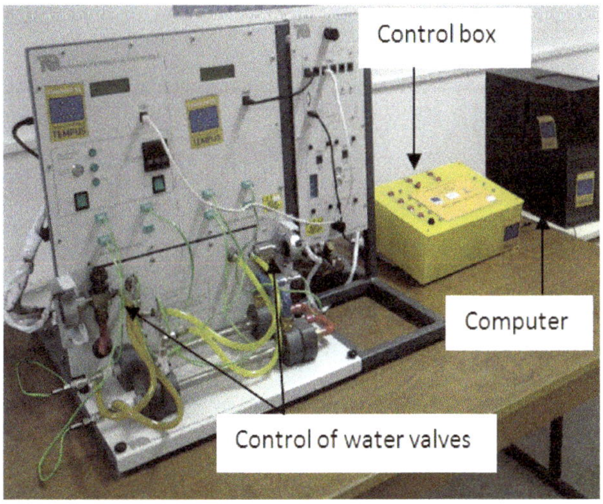

Figure 5. Heat exchanger remote Lab.

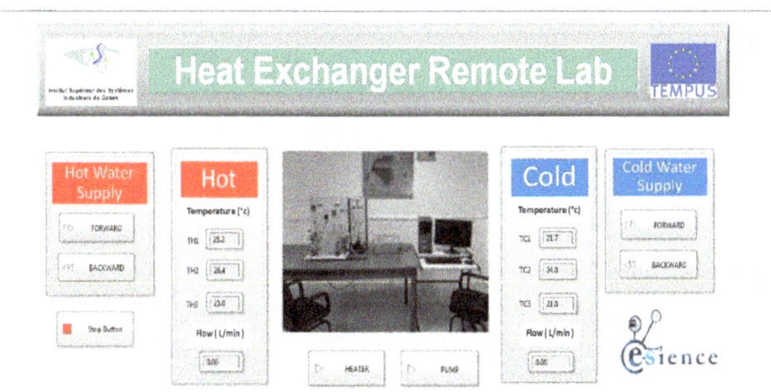

Figure 6. Screenshot of the user interface developed at ISSIG to control the heat exchanger remote lab.

The user interface (UI) application created using Labview software is shown in Fig. 6. This UI was developed to enable students to control the heat exchanger remotely.

The UI included buttons for controlling the different parts of the heat exchanger bench and displayed the feeds from water supply flowmeters, thermocouples and the IP video camera.

V. Pedagogical Evaluation

The evaluation of our remote labs will be in the framework of the whole evaluation strategy adopted by all e-Sience project partners [10].

The evaluation strategy of the project will focus on five different but interrelated directions: (a) usability of remote labs; (b) learners' attitude toward remote labs; (c) technical evaluation of remote lab operation; (d) evaluation of the eLearning content, namely the teaching units previously described; and (e) the learning outcome.

The evaluation will be conducted in two sequential phases:

- Phase I : pilot evaluation
- Phase II : large scale evaluation

The first phase will be the pilot evaluation of all remote labs in small scale usage. During this phase we plan to assess the remote labs' usability and proper functioning as well as the learners' attitude toward remote labs.

The first evaluation phase of the ISSIG remote labs will be conducted with a population of 30 students who will perform their experiments on every remote lab.

According to the results of the first stage evaluation of the remote labs, the operation and user interface will be improved and they will be deployed for large scale usage. After that period of usage, the large scale evaluation will be conducted. The large scale evaluation will focus on the usability of the remote labs; the learners' attitude toward remote labs; evaluation of the eLearning content; and assessment of the learning outcome [10].

VI. Conclusion

The main objective of e-Sience Tempus project was to create an efficient remote labs network in the Maghreb region. As a partner of this project, ISSIG had to adapt and develop distant access solutions for two remote labs to allow students to perform practical experiments on heat transfer and mechanical vibration.

The current paper presented the work carried out on a heat exchanger bench to be fully accessed and controlled remotely. This includs control of the water supply pump, the heater, and the cold and hot water supply flows. Also, an IP camera was located in the laboratory to provide students with an overview of the whole device.

In addition, a user interface application was developed to enable students to control the heat exchanger remotely. It incorporated buttons for controlling the different parts of the heat exchanger bench and displayed feedback from the sensors (flowmeters and thermocouples) and the IP camera.

To test and validate this remote lab an evaluation phase will be conducted. It will be in the framework of the whole evaluation strategy adopted by all e-Sience project partners.

Acknowledgment

This project has been funded with support from the European Commission. This publication reflects the views of only the author, and the Commission cannot be held responsible for any use which may be made of the information contained therein.

REFERENCES

[1] C. Allison, A. Miller, I. Oliver, R. Michaelson and T. Tiropanis, "The Web in education," *Computer Networks*, vol. 56, pp. 3811–3824, 2012.

[2] M.J. Callaghan, J. Harkin, T. McGinnity and L.P. Maguire, "An Internet-based methodology for remotely accessed embedded systems," *2002 IEEE International Conference on Systems, Man and Cybernetics*, vol. 6, pp. 6-9 October 2002.

[3] E. Hansen, "The role of interactive video technology in higher education: case study and proposed framework", *Education Technology*, pp. 13–21, 1990.

[4] E. Scanlon, C. Colwell, M. Cooper and T. Di Paolo, "Remote experiments, re-versioning and re-thinking science learning", *Computers & Education*, vol. 43, Iss. 1–2, pp. 153–163, August–September 2004.

[5] E. Fabregas, G. Farias, S. Dormido-Canto, S. Dormido, F. Esquembre, "Developing a remote laboratory for engineering education," *Computers & Education,* vol. 57, pp. 1686–1697, 2011.

[6] T. Zimmer, D. Geoffroy, A. Pester, R. Oros, Th. Tsiatsos, S. Douka. "eSience: Setting up a network of remote labs in the magrebian countries," *International Conference on Engineering Education and Research (iCEER 2013)*, Marrakesh, Morocco, pp. 1-5 July, 2013. http://www.esience.org

[7] http://www.tecquipment.com/Thermodynamics/Heat_Transfer/TD360.aspx

[8] http://ilab.mit.edu/wiki, The iLab Project.

[9] D. G. Zutin, M. E. Auer and I. Gustavsson, "A VISIR lab server for the iLab shared architecture," *2011 IEEE Global Engineering Education Conference (EDUCON): Learning Environments and Ecosystems in Engineering Education,* pp. 30-33, Amman, Jordan, 4-6 April 2010.

[10] T. Tsiatsos, S. Douka, A. Mavridis, S. Tegos, A. Naddami, T. Zimmer and D. Geoffroy, "Evaluation plan and preliminary evaluation of a network of remote labs in the Maghrebian countries", *iJOE International Journal of Online Engineering*, Vol. 10, Iss. 5, pp. 15–20, September 2014.

AUTHORS

R. Ennetta and I. Nasri are with the Mechanical Engineering Department of the Higher Institute of Industrial Systems of Gabes (ISSIG), Gabes University, Salaheddine El Ayoubi Street, 6011 Gabes, Tunisia (e-mail: ridha.ennetta@issig.rnu.tn, ibrahimnasri2013@gmail.com).

In-service and Pre-service Middle School Mathematics Teachers' Attitudes and Decisions Regarding Teaching Mathematics Using Mobile Phones

W. Daher[1, 2] and N. Baya'a[2]
[1] An-Najah National University, Nablus, Palestine
[2] Al-Qasemi Academic College of Education, Baqa, Israel

Abstract—**Several researches examined students' mathematics learning using mobile phones, while very few researches examined mathematics teaching using this new educational tool. This research attempts to analyze in-service and pre-service teachers' attitudes and decisions regarding mathematics teaching with mobile phones using activity theory. More specifically, three case studies are analyzed in this research: One concerns an in-service teacher who used mobile phones in her class, the second case study involves a pre-service teacher who collaborated with the in-service teacher to teach mathematics lessons using mobile phones, and the third case study is about 15 pre-service teachers who were observers of the experiment but did not use mobile phones in their teaching. We held one-hour semi-structured interviews with the in-service teacher, with the leading pre-service teacher and with the group of other observer pre-service teachers. This happened three times during the academic year: at the beginning of the experiment, after three months and after it ended.**

The research findings indicate that different factors have an impact on the attitudes and decisions of the teacher: history of the teacher using technologies in teaching, perceptions of the teacher using technologies in teaching, community's teacher, rules regulating the use of technologies in teaching mathematics, and division of labor. For example, questions as to who decides which technologies to use in the classroom and who should prepare the learning material for the students. Contradictions were identified in mathematics teachers' activity when utilizing mobile phones in their teaching. These contradictions hinder or slow such utilization. Community, especially its leading members, mediated and helped overcome the activity contradictions that arose throughout the experiment regarding the teaching processes in and/or out of the mathematics classroom.

Index Terms—**middle school, mobile phones, mathematics teachers, teachers' attitudes, teachers' decisions.**

I. INTRODUCTION

Mobile and wireless devices, especially mobile phones, have become increasingly common among young students. This provides new possibilities, opportunities, and challenges for education [1]. This is especially true for mathematics education, where students can use technologies to assist them to conceive independently and/or with the guide of the teacher mathematical concepts and relations. The possibilities and opportunities regarding the use of mobile phones in mathematics education have been reported in the past few years in the literature, specifically regarding students' learning (see for example [2], [3]). We have also conducted our own research on different aspects of students' learning of mathematics using mobile phones. Specifically, we have studied different educational aspects of mathematics learning related to the mobile phone environment, where these aspects involved the behavioral aspect of students' learning ([4], [5]), the affective aspect of students' learning ([6], [7]; [8]), the social aspect ([9], [10]), the cognitive aspect ([8], [5]) and the meta-cognitive aspect of this learning [8].

In contrast to the substantial research conducted on students' learning with mobile phones (see for example the above examples on such research), little research has been done on issues related to teachers' instruction in a mobile phone environment. The current research attempts to address this issue by using activity theory as a conceptual framework to examine the attitude of teachers who use mobile phones in mathematics teaching. Three professional groups participated in this study: an in-service mathematics teacher; a leading pre-service teacher and a group of pre-service teachers. Using the framework of activity theory we try to understand what affects mathematics teachers' attitudes and decisions regarding the use of mobile phone in his/her instruction.

Though little research has been conducted on teaching using mobile phones, some research was done on using other mobile technologies in teaching. We describe some of this research below.

II. LITERATURE REVIEW

We will here present a review of the literature that deals with the educational aspects that concerns this research: M-learning and activity theory.

A. M-learning

Different aspects of using mobile technologies in education drew the attention of researchers who studied the impact of these technologies on education. Some of these issues are: the functions and advantages of using mobile technologies in education, factors that influence the use of mobile phones in education, constraints on using mobile technologies in education, and teachers' acceptance of these technologies.

Researchers pointed out various advantages of using mobile technologies in education: enabling mobility, activity and interactivity, communicability, working meaningful content, motivating students, etc. Ref. [11] said that mobile technologies can improve the quality of education for they allow location independence, time independence and meaningful content. Furthermore, there is evidence that learners from diverse backgrounds may benefit from mobile technologies because of their emphasis on activity and interactivity as well as communication potentialities [12]. In addition, these technologies advance high-quality experiential learning [13]. Specifically for teachers, the communicational and the mobility facilities of the mobile technologies help monitor and guide students' learning, as well as facilitate group communication [14].

Ref. [14] described some of the mobile technologies' advantages in the case of integrating wireless LAN, wireless mobile learning devices, an electronic whiteboard, an interactive classroom server, and a resource and class management server to build a wireless technology enhanced classroom (WITEC) that supports everyday activities unobtrusively and seamlessly in classroom contexts. The actual use of WITEC revealed the advantages of the electronic system that includes mobile components. For example, in contrast with traditional classrooms, WITEC enabled the teacher and students to concentrate on teaching and learning itself rather than spend too much time and effort on tedious tasks. Furthermore, it helped the teacher to monitor each student's learning easily, which facilitated further guidance, as well as enabled engaging students in learning activities and facilitating students' group collaboration. We see then that wireless and mobile technology helped maintain the classical functions of computer technologies (for example facilitating computing tedious tasks) as well as providing more advanced functions in communication and collaboration.

Advantages of using mobile technologies in education encourage teachers' use of these technologies, where various reasons encourage teachers to use mobile technologies in their teaching [15]. Ref. [15] examined the reasons that encourage teachers' involvement with mobile technologies. They found four main reasons that the participating teachers mentioned as encouraging them to become involved in the mobile learning program: (1) keeping up to date with new technology and developing skills with technology; (2) using technology to motivate and engage students to learn better; (3) using technology to improve their teaching; and (4) having to be part of the program.

The use of mobile technologies in education is influenced by constraints that may slow the success of mobile technologies use in the classrooms [16], for example, the screen small size. Below we describe in more details some studies that treat teachers' use of mobile technologies.

Constraints on using mobile technologies in education influence the acceptance of these technologies by teachers. Ref. [11] conducted a study on the acceptance of mobile phones for teaching and learning with a group of pre-service teachers in Hong Kong. The study showed that the majority of the respondents generally accepted the use of mobile phones for teaching and learning. Specifically, the participants somewhat agreed that mobile phones can improve education quality. In addition, their curiosity to use mobile phones for teaching and learning was high. Moreover, the respondents were interested in seeing mobile learning become part of their curriculum. Furthermore, inhibiting factors perceived by the participants were small screens, text input, slow connectivity, high expenses, and the difficulty of mobile phone use.

The findings of the previous researches regarding the functions and advantages of using mobile technologies in education encouraged us to suggest mobile phones as a teaching tool to our mathematics pre-service teachers, as well as to one training teacher in the school where our pre-service teachers were trained. Doing so, we were interested in verifying their attitudes and decisions regarding teaching mathematics with mobile phones. To analyze these attitudes and decisions we will use the activity theory framework.

B. *Activity theory as conceptual framework*

Activity theory is a cultural-historical theory of activity which was initiated by a group of Russian psychologists in the 1920s and 1930s. This approach was led by Lev Vygotsky (1896-1934) and his colleagues A.N. Leont'ev and A.R. Luria [17]. Ref. [18] says that activity theory, as initiated by Vygotsky, is centered around the idea of mediation which is represented in a triangular model in which the conditioned direct connection between stimulus (S) and response (R) was transcended by a complex mediated act, where Vygotsky's idea of cultural mediation of actions is commonly expressed as the triad of subject, object, and mediating artifact (Figure 1)

Ref. [19] proposed a model (Figure 2) that conceptualizes all purposeful human activity as the interaction of the elements: subject, object, tools, community, rules, and division of labor.

In what follows, we present a description of the elements that constitute the activity system (Ref. [20]).

In the model, the subject refers to the individual or group whose agency is chosen as the point of view in the analysis. The object refers to the 'raw material' or 'problem space' at which the activity is directed and which is intended to be transformed into an outcome with helping tools, where the tools can be physical and/or symbolic, external and/or internal. The community comprises multiple individuals and/or groups who share the same object and who construct themselves as distinct from other communities. The division of labor refers to both the horizontal division of tasks between the members of the com-

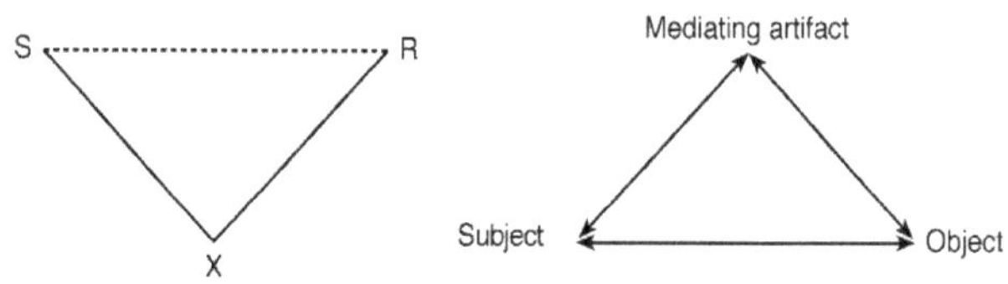

Figure 1. Vygotsky model of activity

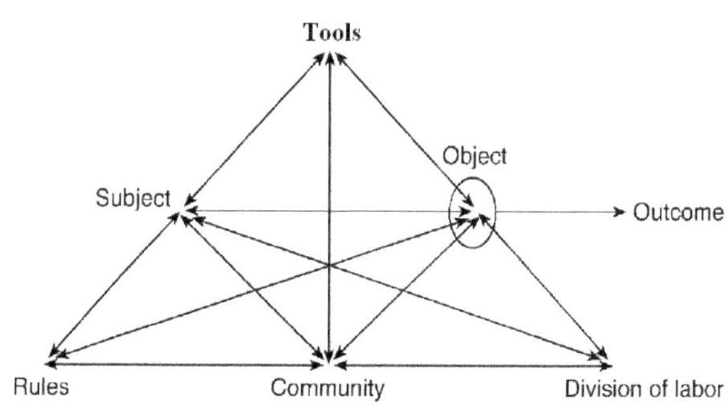

Figure 2. The structure of a human activity system

munity and to the vertical division of power and status. Finally, the rules refer to the explicit and implicit regulations, norms and conventions that restrict actions and interactions within the activity system. The outcome is the result of the subject's involvement with the object of the activity, and which is influenced by the activity different elements. These elements of the activity theory fit the goal of analyzing teachers' attitudes and decisions due to the fact that they represent the conditions which influence teachers' work, especially when introducing a new technological tool to the classroom. We describe below these elements as related to our research.

The subject refers to the in-service teacher, the leading pre-service teacher or the rest of the pre-service teachers who were observers of teaching mathematics with mobile phones. The object refers to the mathematical educational environment (mathematics students, mathematics topic, mathematical activities, inter alia) which were to be developed into a mobile mathematical one, i.e. an environment that suits using mobile phones to teach and learn mathematics. The tool refers to the mobile phone which is suggested as a learning instrument in and/or out of the mathematics classroom. The community comprises the school in which the in-service /pre-service teacher teaches/practices, as well as the college in which the pre-service teacher learns, including the in-service /pre-service teacher herself. The division of labor refers to who should teach, who should prepare the learning activities needed to be used for learning mathematics with the mobile phone, who decides which tools to use in the mathematics classroom, who decides which mathematical material to cover in an academic year, etc. The rules refer to the norms and conventions regarding teaching/learning mathematics in the middle school/college or in the specific school/college, for example the question as to whether the computer or mobile phones should be used in the classroom or whether the in-service /pre-service teacher should prepare additional learning material for his/her students, inter alia. The outcome is the participants' decisions and attitudes towards mobile phone use in mathematics teaching.

C. *Contradictions in activity theory*

According to Ref. [21], as reported in Ref. [22], activity theory uses the term 'contradiction' to indicate a misfit within an element, between them, between different activities, or between different developmental phases of a single activity. For Ref. [19], if these contradictions are handled constructively they will lead to development of the activity, as well as to the learning outcome.

It is our intention to use activity theory to examine the in-service and pre-service teachers' instructional attitudes and decisions concerning the use of mobile phones in mathematics teaching (accepting these tools for mathematics teaching, when to use them, and how to use them). Doing so, we agree with the claim of Ref. [23] that when considering individual and institutional perspectives, as well as the evolution of technology use over time, activity theory and its contradictions provide a useful tool for examining various aspects of educational technology use. This claim is especially true when we consider teachers' attitudes and decisions regarding the use of mobile phones in the mathematics classroom because these attitudes and decisions are influenced by various factors which activity theory involves.

D. *Activity theory as a tool for analyzing educational processes*

Activity theory was used by researchers to identify the influences of the various aspects of an educational phenomenon. For example, Ref. [22] used activity theory to examine pedagogical, social, and technological issues in ICT integration process in Turkish classrooms. They concluded that the following aspects should be taken into consideration when carrying out ICT integration processes: organizational culture, changing roles of teachers and students with regards to ICT, inflexible time table curriculum, support of the school administration, mediation of the ICT coordinator, and collaboration among teachers.

Another research which examined a whole educational phenomenon using activity theory framework is that of Ref. [24] who used activity theory to study the potential and actual interrelationships between the constituent elements that comprise the complex process of music education. Welch concluded that there is much to be gained by exploring the principles of activity theory and applying them to the relationships and contributions between top-down and bottom-up perspectives in an educational process related to music.

Other researchers used one aspect of activity theory to analyze an educational phenomenon. For example, Ref. [25] attempted to explain how students' learning outcome evolved from their online learning experiences. The researcher found that there were three levels of contradictions that the students faced during their online learning: conflicting-object contradictions, inter-activity contradictions, and technology-related contradictions. He also found that contradictions may be invisible to some members who are part of the subject of the activity, but some members may have the capacity to identify the contradiction, yet they may not have the desire or motivation to make necessary changes to further learning and development within the activity. The researcher concluded that contradictions might not be resolved although they are visible.

Ref. [26] designed a qualitative collective case study to investigate how students revised after writing center conferences. To do the investigation, he used the activity theory framework. The study suggested that the object of the activity system of a writing conference should include the situation definitions that students hold about how to revise and how to participate in a writing conference.

Ref. [27], following Radford ([28], [29], [30]), analyzed mathematical learning at pre-apprenticeship training class for pipe-trades conducted at trade union run school in British Columbia, Canada. More specifically, an apprentice's thought processes in relation to reading a measuring tape was analyzed, together with his interactions with the researcher. This analysis, using Radford's approach to activity theory, considered activity theory as a lens to analyze individuals' activity and not only system's activity as suggested by Ref. ([18], [19], [31]). We will follow Ref. ([28], [29], [30])) and Ref. [27] in analyzing the attitudes and decisions of middle school in-service and pre-service teachers regarding the integration of mobile phones into their mathematics teaching, i.e. we consider the current research participants (the in-service teacher, the leading pre-service teacher and each of the other pre-service teachers) the individuals whose attitudes and decisions will be studied concerning mobile phone use in mathematics teaching.

III. Research Rationale and Goals

Using mobile tools in mathematics education is increasing, where this use changes the educational scene and enables out-of-class mathematics learning and teaching [32]. Researches who have studied the use of mobile tools in mathematics education have focused on students' learning rather than on teacher's instruction. This research attempts to shed light on this issue. To do so, we utilize activity theory as teacher's instruction is influenced by various factors that activity theory involves, such as school community and rules. In additions, activity theory can explain teachers' decisions in the light of the contradictions which they confront in their work.

The current research findings can help us establish insight into teachers' attitudes towards the use of new and mobile technologies in their instruction, as well as their decisions regarding the integration of these technologies into mathematics teaching.

IV. Research Question

For the purpose of this research the following research question was formulated:

How do the educational activity elements influence in-service and pre-service teachers' instructional attitudes and decisions concerning the use of mobile phones in mathematics teaching?

V. Research Design and Methodology

A. Research setting *and* participants

This research formed part of a larger research project in which we carried out an experiment that involved using mobile phones in teaching and learning mathematics. We carried out the project as an initiative in Al-Qasemi Academic College of Education, and obtained the approval of the research center in the college to conduct research, including interviewing the college pre-service teachers. Participation in the research was voluntary, and the respondents signed an informed consent.

The experiment took place in one middle school in Nazareth in the framework of the practical training of middle school pre-service teachers who majored in mathematics and computers in a teachers' college. The pre-service teachers were in their third year of study and were required to use various technological tools in their teaching including the mobile phone. One pre-service teacher was responsible at the training school of the other pre-service teachers' use of technology in their instruction. This responsibility meant three things: (1) the pre-service teachers could consult him regarding their use of technology in their instruction; (2) the leading pre-service teacher collaborated with one in-service teacher in the training school to prepare and teach lessons which integrated mobile phones in mathematics lessons; and (3) the leading pre-service teacher participated in the meetings held by the researchers regarding the use of technology of tools, namely, mobile phones, in the pre-service teachers' instruction. In the current research we call this pre-service teacher the 'leading pre-service teacher'.

This article describes three case studies related to the participants' instructional attitudes and decisions associated with using mobile phones in mathematics teaching. The first case study is about one in-service teacher who collaborated with the researchers and the leading pre-service teacher to use mobile phones in her class. The second case study concerns the leading pre-service teacher who collaborated with the in-service teacher and the researchers to teach mathematics lessons using mobile phones, while the third case study involves 15 pre-service teachers who were experiment observers but did not use mobile phones in their teaching. Those pre-service teachers could have used mobile phones in their teaching but they were not obliged to do so, because neither all the mobile phones of those pre-service teachers nor those of the school students had java which is needed to make midlets work.

B. Mathematical software and learning materials

The middle school students worked with mobile phone software (midlets) that supports algebra and geometry learning. The midlets can be downloaded from Math4Mobile site of the Institute for Alternatives in Education that operates within the Faculty of Education at the University of Haifa [33]. To perform the activities assigned to them, the students used the algebraic midlets and various tools and technologies embedded in their mobile phones. The middle school students used mobile phones in tasks related to the linear and quadratic function.

C. Data collecting tools

Interviews with the teachers: We held one-hour semi-structured interviews with the in-service teacher, the leading pre-service teacher and with the group of the other observer pre-service teachers. This happened three times during the academic year: at the beginning of the experiment, after three months and after it ended. The issues discussed in the interviews were related to the activity theory components and sub-issues associated with them, for example personal history and perceptions (related to the subject), the subject's perceptions of the conditions for the success of teaching mathematics with mobile phones (related to tool, rules, community and division of labor), and decisions and attitudes towards use of mobile phones in mathematics teaching (related to the object and outcome of experimenting teaching mathematics with mobile phones).

D. Data analysis

We used the constant comparison method [34] to arrive at categories related to the activity theory components. The constant comparison method is a process of inspection consisting of concept construction, through a constant comparison of emergent codes and the development of relationships among the concepts ([35], [36]). Here, it was applied keeping in mind the activity theory components and themes related to each of them. For example, we kept comparing each participant's sentences (given in the interview) which implied how division of labor and community rules were represented and influenced the participants' attitudes and decisions regarding the use of mobile phones in mathematics education. Doing so, we arrived at the themes related to the division of labor in our specific context: who should teach, who should prepare learning activities needed to be used for learning mathematics using mobile phones, who decides which tools to use in the mathematics classroom, who decides which mathematical material to cover in an academic year, etc. Furthermore, we looked for the values of the different themes in each one of the three case studies.

E. Relevance of the data analysis to the setting and theme of the study

The goal of the current study was to describe an activity theory model for teaching mathematics using mobile phones. One way to do so was to use the constant comparison method to arrive at themes and categories related to activity theory components when teaching mathematics with mobile phones. This analysis process has been followed in previous researches for similar goals. For example, Ref. [26] used it to analyze how students revised after writing center conferences in activity theory terms. Furthermore, Ref. [25] used it to explain how students' learning outcomes evolved from their online learning experiences, using activity theory contradictions. Moreover, Ref. [36] analyzed two individual teachers' experience of mathematics and explored, using activity theory, how these experiences impacted their teaching. In the three previous researches, as in the current one, the emphasis is on the individual interacting with an educational situation.

VI. FINDINGS

At the beginning, we describe the activity system's elements regarding each of the three case studies in which this research involves: the in-service teacher's case, the leading pre-service teacher's case and the other observer pre-service teachers' case.

A. The subject

Two factors distinguish the subject: his/her history and his/her perceptions regarding the use of technology in teaching mathematics. Below we describe these factors regarding each subject in the current research.

The in-service teacher as subject

History: The in-service teacher has a history of using mathematical software in teaching mathematics. This history was not successful, for the students did have conceptual knowledge of the mathematical topics they learned, but they did not have appropriate procedural knowledge of these topics; therefore, they had great problems when they advanced to high school where teachers depended greatly on procedural knowledge.

Perceptions: The in-service teacher thinks that procedural knowledge of mathematical topics is essential for the students; thus, although she assumes that students enjoy learning using computers she thinks that this should not be done in every lesson or in the whole lesson. She also thinks that it is difficult to integrate real life into mathematics teaching and learning in middle or high school.

Leading pre-service teacher as subject

History: The leading pre-service teacher took two courses focusing on the use of technology in the classroom, where one of them (didactics of teaching mathematics) focused on technology use in mathematics teaching. More specifically, he performed different tasks on this use, so he was aware of the advantages of using technology in mathematics education. In addition, the leading pre-service teacher participated in the meetings held between the two researchers and in which they discussed the experiment advancement, the activities that should be carried out and students' learning using mobile phones.

Perceptions: The leading pre-service teacher thinks that technology makes teachers' instruction and students' learning easier. Furthermore, he thinks that working with mathematical software adds a visual aspect to students' mathematical investigations, which assists the students in arriving at mathematical relations, so technology should be used in mathematics teaching. In addition, he argues that the mobile phone is already in students' hands, and they like working with it, so it would better be utilized as a learning tool.

Observer pre-service teachers as the subject

History: The pre-service teachers' history is similar to that of the leading pre-service teacher, but they did not participate in the meetings with the researchers.

Perceptions: The observer pre-service teachers' perceptions regarding the use of technology in mathematics education were similar to those of the leading pre-service teacher, but they did not mention the idea of utilizing mobile phones for its availability in students' hands.

B. The object

The difference between the object as the in-service teacher and the object as the pre-service teachers (including the leading one) lies in the fact that the object of the in-service teacher involved her own class, where she was responsible of covering all the material in the mathematics book, as well as of having a good level of students' achievement in mathematics. On the other hand, the object of the pre-service teachers was a classroom in which they were being trained, where they did not have the responsibilities that the in-service teacher had.

C. The tool

All the subjects emphasized the need that at least half of the learners in a class should be equipped with mobile phones. In addition, all the subjects pointed out that the school administration should be flexible and allow students to use mobile phones in mathematics classes.

D. The community

The community in the case of the in-service teacher as the subject comprised the school in which the teacher taught. It included the teacher herself, the principal of the involved school, as well as the whole community of teachers. The community in the case of the leading pre-service teacher, as well as in the case of the observer pre-service teachers consisted of the community of the pre-service teachers in the college where they studied, the instructor of the mathematics didactics course (the first author of this article) and the supervisor of the pre-service teachers in the college and the training school (the second author of this article). In addition, this community consisted of the in-service teacher who participated in the experiment and the students' school where the experiment was carried out.

E. Division of labor and community rules

In the case of the in-service teacher, the principal expected her to cover most of the mathematics book for grade 9. More specifically she was expected to prepare her students for the secondary school mathematics, as well as to make her students have high achievement in mathematics.

It was not expected from the in-service teacher to prepare additional learning materials for her students. It also was not expected from her to use computers or mobile phones in her mathematics teaching. At the same time, the school principal expected her to cooperate with the supervisor and his pre-service teachers and start using mobile phones in her teaching on Mondays, when they are present in her lessons.

In the case of the leading pre-service teacher, the researchers advised him to cooperate with the in-service teacher to prepare learning materials that suit the mobile phone environment. The researchers also advised him to teach those learning materials to grade 9 students alone or with the in-service teacher.

In the case of the observer pre-service teachers, it was expected that they prepare additional learning materials for the students. They also were required by their supervisor to use electronic tools (for example videos, PowerPoint presentations, electronic spreadsheets, applets) in their teaching, but not necessarily mobile phones.

The in-service teacher and the leading pre-service teacher emphasized the need for appropriate mathematical activities, so the students will be seriously involved in learning mathematics and not distracted by mobile phones.

F. Outcome

Tables 1, 2 and 3 describe the participants' attitudes towards and decisions regarding the use the mobile phone in mathematics teaching in the three cases. After having analyzed the collected data using the constant comparison method, we arrived at four themes that comprise the participants' attitudes and decisions: Place of use, frequency of use, target of use and role of the tool. The tables display the values of the previous themes in the three research cases. We chose this display format to present in-service and pre-service teachers' attitudes and decisions more clearly and neatly.

TABLE I.
ATTITUDES AND DECISIONS REGARDING MOBILE PHONE USE IN MATHEMATICS TEACHING: THE CASE OF THE IN-SERVICE TEACHER

Aspect	Attitudes and decisions
Place of use	Inside the classroom
Frequency of use	Not in every lesson and not in the whole lesson
Target of use	To help students discover mathematical relations
Role of the tool	It should be secondary to the role of teacher

TABLE II.
ATTITUDES AND DECISIONS REGARDING MOBILE PHONE USE IN THE MATHEMATICS TEACHING: THE CASE OF LEADING PRE-SERVICE TEACHER

Aspect	Attitudes and decisions
Place of use	Inside as well as outside the classroom
Frequency of use	In every lesson
Target of use	Introducing a new mathematics topic, investigating and discovering mathematics relations or summarizing with the students the conclusions of their investigation
Role of the tool	It has a main role like the teacher

TABLE III.
ATTITUDES AND DECISIONS REGARDING MOBILE PHONE USE IN THE MATHEMATICS TEACHING: THE CASE OF OBSERVER PRE-SERVICE TEACHERS.

Aspect	Attitudes and decisions
Place of use	Mostly inside the classroom, but sometimes outside the classroom
Frequency of use	In every lesson
Target of use	To introduce a new mathematical topic or to summarize it
Role of the tool	it complements the teacher's role

VII. DISCUSSION

Figure 3 shows the components of each element of the activity system as related to teaching mathematics using mobile phones.

Figure 3 indicates that two characteristics distinguish the subject: his/her history and his/her perceptions. These characteristics influence the outcome of subject's activity, in our case the attitudes and decisions regarding the use of mobile phones in mathematics teaching. This influence of the history and perceptions of teachers is acknowledged in the literature. For example Ref. [38] identified experience (one facet of one's history) and sense of efficacy to implement the innovation as influencing teachers' attitude. Moreover, the educational history of the subject has been identified as an important factor influencing the student resistance to the use of new technological tools for learning (Ref. [39], as reported in Ref. [23]), but in our case it influenced the teacher's attitudes and decisions regarding mobile phone use in mathematics teaching. Furthermore, the above findings agree with other studies' findings about the factors that influence teachers' attitudes. For example, Ref [40] identified teachers' perceptions of the importance and difficulty of implementing innovations as influencing teachers' attitude towards this implementation.

The community also influenced the participants' decisions and attitudes towards integrating technology into the mathematics classroom. Community is reported to influence the subject's activity in other researchers, where the subject can be the students or the teachers. Ref. [32] found that the involvement of the school principal influenced students' learning with mobile phones. Ref. [41] found that the need of leadership involvement could hinder the integration of interactive boards into the classroom. In our current experiment, the pre-service teachers were influenced by their instructors' activity, so they had more favorable attitude towards mobile phone integration into their teaching than that of the in-service teacher who was not influenced as much as the pre-service teachers by the researchers.

Rules and division of labor also are important factors influencing the subjects' decisions and attitudes towards mobile phone integration into their mathematics teaching. Up to the time of our experiment, there were no declared national rules in the Israeli schools that impose using technology in the mathematics classroom, so teachers did not find themselves obliged to perform such integration. Furthermore, mathematics teachers were not expected to prepare technology based learning materials for their classrooms; especially mobile phone-based learning materials. This made them reluctant to integrate new technology into their classrooms; an act that requires them to prepare appropriate learning material.

Division of labor influence on the subject attitude, as described above, is also mentioned by Ref. [42]. Moreo-

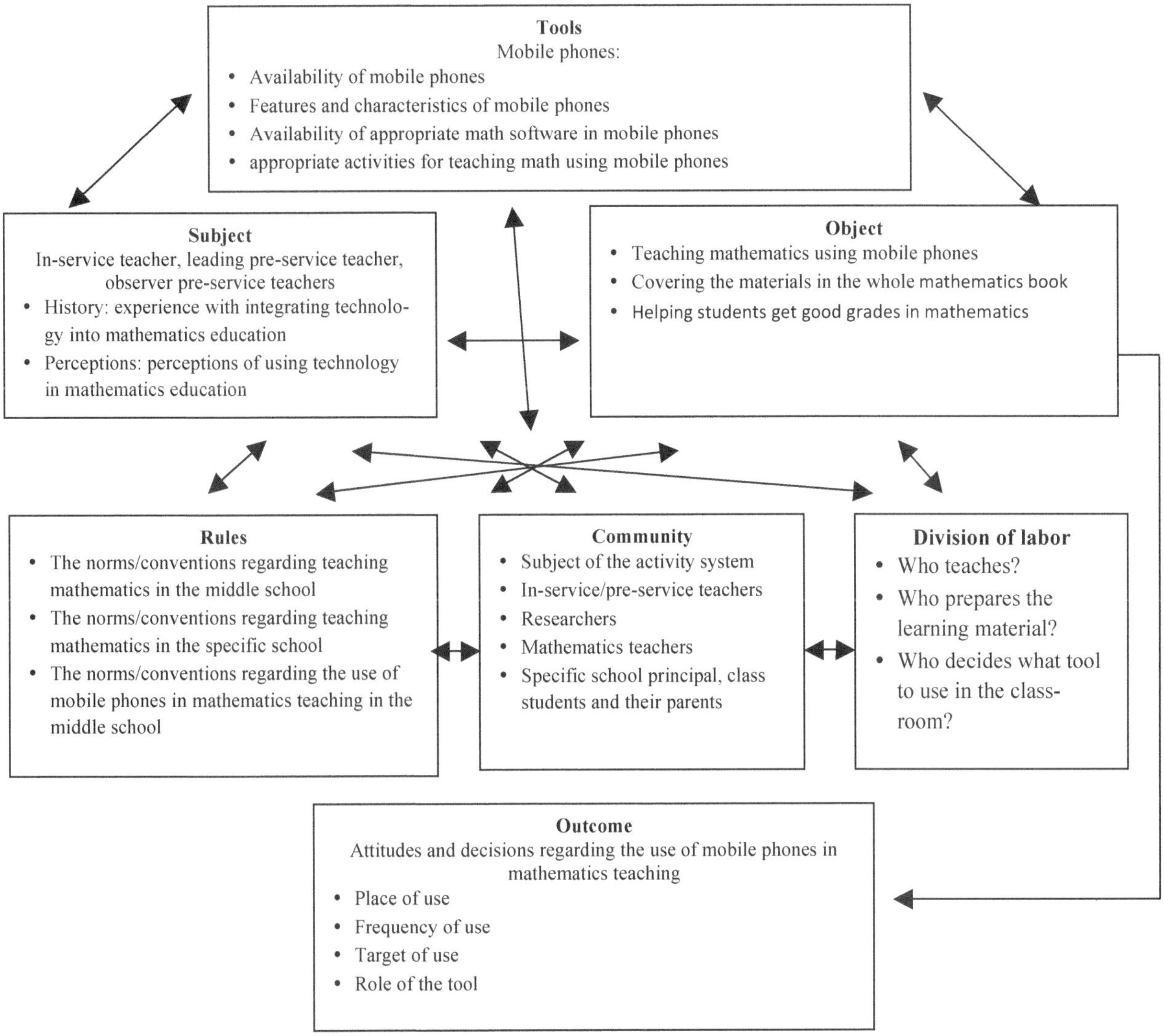

Figure 3. Components of the activity system's elements as related to teaching mathematics using mobile phones.

ver, Ref. [43] argued that changing the division of labor between the teacher and students would change their attitudes towards their roles in the classroom. Here, the in-service teacher changed her attitude towards mobile phone use in mathematics teaching in spite of the traditional rules and division of labor prevailing in her class. This change happened due to the new experience of the in-service teacher as a participant in an experiment which considers mobile phones as teaching and learning tools in the mathematics classroom. The change led to new division of labor that the mobile phone helped attain because it encouraged a more active role of the students (see for example Ref. [44]).

As for the attitudes and teaching decisions of the subjects at the end of the experiment, they differed among the subjects in the four aspects: place of the educational use of mobile phones, frequency of the mobile phone use in mathematics teaching, target of mobile phone use and tool role. Regarding the place of use, the leading pre-service teacher thought that mobile phones should be used inside the classroom as well as outside the classroom. This attitude could have been developed as a consequence of his several meetings with the researchers, where they emphasized the great potential of mobile phone in mathematics learning, especially outside the classroom, where the ideas of the outside activities developed. The observer pre-service teachers thought that mobile phones should be used mostly inside the classroom, but sometimes outside the classroom. This attitude could have been developed because they witnessed the activities carried out by the students outside the classroom, and saw their enthusiasm to carry out such activities. At the same time, they were not actually involved in developing these activities. This lack of involvement affected their attitude towards mobile phone use in education [45]. With regard to the in-service teacher, she thought that mobile phones would better be used inside the classroom. Probably, her responsibility for covering the learning material discouraged her from considering the option of the outside use of mobile phones. This attitude of the in-service teacher goes side by side with the expectations of her community, specifically the school principal and the parents of the students.

We find differences in the subjects' attitudes regarding the frequency of use too. All the pre-service teachers, including the leading one, thought that mobile phones should be used in every mathematics lesson, while the in-service teacher thought that such use should not be in every lesson or in the whole lesson. Here too, the community affected this difference in attitudes, where the pre-service teachers were influenced by their instructors (the researchers) whose instruction emphasized the role of technology in mathematics education, while the in-service teacher was influenced by her community, where this community, as mentioned above, did not expect mobile phones to be used in mathematics teaching. The subjects' history also caused the difference in their attitudes towards the frequency of mobile phone use in mathematics teaching, which was described above.

As for the target of use, the in-service teacher was impressed by the features and characteristics of mobile phones that were made apparent during its use in the mathematics teaching during the experiment. The features included the illustration abilities enabled by the mobile phone midlets, while the characteristics included the mobile phone availability and its ease of use. Therefore, she adopted the target of discovering mathematical relations using mobile phones. On the other hand, the observer pre-service teachers, who did not participate practically in carrying out mobile phone activities, thought that mobile phones could be used in the math lesson for just introductory and summary purposes. As for the leading pre-service teacher, who carried out the actual activities with the students, he could see the diverse advantages of mobile phones, which convinced him of the varied teaching targets using it; i.e. introducing a new mathematics topic, investigating and discovering mathematical relations or summarizing with the students the results of their investigation. This picture of the subjects' attitude towards the target of using mobile phones in mathematics teaching was also influenced by the subjects' history, rules, division of labor, and community, as described above.

If we look at the subjects' attitudes towards the role of mobile phones in mathematics teaching, we find different attitudes, where the leading pre-service teacher had the most favorable attitude regarding this role, saying that mobile phones have a main role like the teacher, while the in-service teacher had the least favorable attitude, saying that they should be secondary to the role of teacher. The observer pre-service teachers had an attitude between the previous two attitudes, saying that mobile phones complement the teacher's role. The previous three different attitudes consider the role of teacher as a main one, which agrees with previous studies. For example Ref. [46] described the role of the teacher as critical when working with technology in the classroom, while Ref. [47] emphasized the complementary nature of the role of technologies in subject teaching.

The above influences indicate some contradictions within the activity which hinder or slow the process of meeting the objective of the activity; i.e. using mobile phones in mathematics teaching. One contradiction was between the subject and the tool in the case of the in-service teacher. This contradiction occurred as a consequence of the teacher's unsuccessful history in using technology in mathematics teaching. Another contradiction occurred between division of labor and the subject, in the case of the in-service teacher, because she was not expected by the school community to prepare learning materials for her classroom. Both of the previous two kinds of contradictions were reported by Ref. [22] to slow the integration of ICT into a Turkish school. Here too, they could have slowed to the minimum the use of mobile phones in mathematics teaching if the in-service teacher did not receive help from the researchers and the leading pre-service teacher. In other words, these contradictions were lessened by the collaboration of the in-service teacher with the researchers at the beginning, and afterwards with the leading pre-service teacher. This collaboration made the teacher aware of the possibilities of mobile phone use in mathematics teaching for students' learning. Thus, she came to realize the worthiness of preparing new appropriate learning materials to make the students utilize the potentialities of mobile phones in their mathematics learning. Moreover, the new experience lessened the negative impact of the previous teaching experience of the in-service teacher with technology.

Other contradictions can be found in the activity system described above. One such contradiction is between the community and the tool, where some members of the community, in our case the principal, prohibit mobile phone use in the school area including the classrooms. This contradiction was solved from the beginning when the researchers met the principal and discussed mobile phone use as a learning tool in the 9th grade mathematics classroom. Another contradiction is between the tool and the object, where the in-service teacher and pre-service mathematics teachers thought that using tools in the mathematics classrooms could slow the teaching process of the mathematical topics, which made them reluctant to use tools in their classroom. The contradiction in which the tool is a component is reported in other researches but for different reasons. For example, Ref. [48], reported a case where the teacher considered the computer as suitable for lower level cognitive skills, whereas he wanted to advance higher levels of these skills. The contradiction in our research in which the tool was involved was lessened when the teacher and the pre-service teachers watched the researchers utilize mobile phones successfully to introduce complicated mathematical topics in relatively short time.

All the contradictions are shown in Figure 4, represented in the dashed lines, but it is important to note that the two contradictions in the case of the subject (the in-service teacher) were present more than in the other two cases.

These contradictions were solved primarily by the mediation of the community, especially by leading members in the community (the researchers and the school principal), which indicates the importance of the role of the leading members in the community in maintaining the momentum throughout the experiment [49]. In our case, the leading members mediated and overcame the educational contradictions that arose as a consequence of introducing new technological tool in teaching. In other words, the attitudes and decisions of the leading members of the community concerning mobile phone use in teaching mathematic made them develop an initiative that influenced in different ways, according to the specific condi-

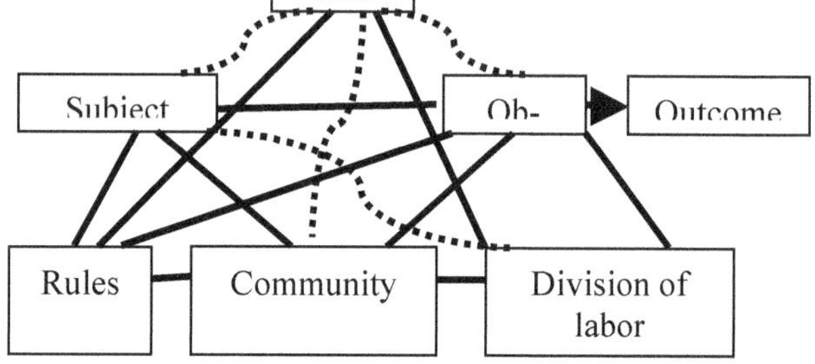

Figure 4. Contradictions in the case of in-service teacher.

tions of the community, the participating in-service/pre-service teachers' attitudes and decisions regarding mobile phone use in teaching mathematics.

VIII. CONCLUSIONS

The research goal was to examined decisions and toward teaching mathematics when using mobile phones. It was found that the teacher's history related to using technologies in teaching had an impact on his/her attitude and decisions regarding utilizing new technologies in teaching. This makes it necessary to consider teachers' history in implementing technology in teaching before asking them to do so. It also means preparing these teachers for technology integration, so their attitudes and decisions regarding teaching mathematics with the mobile phone are considered and treated. This could be done especially through reflection on the participating teachers' experiences [50]. Mathematics teachers will not integrate successfully technology into their classes unless they change their historical negative attitudes towards this integration, which could also be done through the guidance of educational technology experts [51].

Other factors which influenced the participants' attitudes and decisions were: teachers' perceptions of using technologies in teaching, teacher's community, rules of using technologies in teaching mathematics and division of labor, for example who decides which technologies to use in teaching and who should prepare the learning material for students. These factors were also described elsewhere as influencing students' learning of mathematics with mobile phones [32], where another theory of analyzing data was used, namely, the grounded theory. This emphasizes the prevalence of these factors as influencing the integration of mobile phones (and we suppose technology in general) into the learning and teaching of mathematics.

As for the pre-service teachers who participated in the research, it can be concluded that their history, together with the characteristics of their school community made them more apt to accept mobile phones as teaching tools. They have been exposed in their preparation as future teachers to new educational technologies and at the same time were part of a community in which the leading members initiated projects that utilized such technologies. They also were required to utilize such technologies in the classroom. So, history, community and rules made the difference in the activity outcome related to the different subjects. The community, especially its leading members, also mediated and helped overcome the activity contradictions that arose throughout the experiment regarding the teaching processes in and/or out of the mathematics classroom. This emphasizes that school conditions and school leadership, manifested by the mobile phone project leadership represented by the researchers, are key issues in efforts to change instruction [52] and enrich it with new technologies in our case.

REFERENCES

[1] Cobcroft, R., Towers, S., Smith, J., & Bruns, A. (2006). Mobile learning in review: Opportunities and challenges for learners, teachers, and institutions. In *Proceedings of Online Learning and Teaching (OLT) Conference 2006* (pp. 21-30). Queensland University of Technology, Brisbane. Available from http://eprints.qut.edu.au/5399/1/5399.pdf

[2] Botzer, G. & Yerushalmy, M. (2007) Mobile Applications for Mobile learning. In *Proceedings of the Cognition and Exploratory Learning in the Digital Age* (CELDA). December. Algarve, Portugal.

[3] Kizito, R. N. (2012). Pretesting Mathematical Concepts with the Mobile Phone: Implications for Curriculum Design. *The international review of research in open and distance learning, 13* (1).

[4] Daher, W. (2010a). Building mathematical knowledge in an authentic mobile phone environment. *Australasian Journal of Educational Technology, 26*(1), 85-104.

[5] Daher, W. & Baya'a, N. (2012). Characteristics of middle school students learning actions in outdoor mathematical activities with the cellular phone. *Teaching Mathematics and its Applications: An International Journal of the IMA, 31* (3), 133-152.

[6] Baya'a, N. & Daher, W. (2009). Learning mathematics in an authentic mobile environment: the Perceptions of Students. International *Journal of Interactive Mobile Technologies, 3 (Special Issue, IMCL 2009),* 6-14.

[7] Daher, W. (2009). Students' perceptions of learning mathematics with cellular phones and applets. *International Journal of Emerging Technologies in Learning, 4*(1), 23-28.

[8] Daher, W. (2011a). Learning mathematics in the mobile phone environment: Students' emotions. *Journal of Interactive Learning Research, 22(3)*, 357-378.

[9] Daher, W. (2010b). Mathematics learning community flourishes in the cellular phone environment. *International Journal of Mobile and Blended Learning (IJMBL),* 2(2), 1-17.

[10] Daher, W. (2011b). Building mathematics cellular phone learning communities. International *Journal of Interactive Mobile Technologies (iJIM), 5*(2), 9-16.

[11] So, S. (2008). A Study on the Acceptance of Mobile Phones for Teaching and Learning with a Group of Pre-service Teachers in Hong Kong. *Journal of Educational Technology Development and Exchange, 1*(1), 82-92.

[12] Litchfield, A.J., Dyson, L. E., Lawrence, E. & Zmijewska, A. (2007). *Directions for m-learning research to enhance active learning.* In ICT: Providing choices for learners and learning. Proceedings ascilite Singapore 2007.

[13] Dyson, L.E., Litchfield, A., Lawrence, E., Raban, R., & Leijdekkers, P. (2009). Advancing the m-learning research agenda for active, experiential learning: Four case studies. *Australasian Journal of Educational Technology, 25*(2), 250-267.

[14] Liu, T.C., Wang, H.Y., Liang, J.K., Chan, T.W., Ko, H.W. & Yang, J.C. (2003). Wireless and mobile technologies to enhance teaching and learning. *Journal of Computer Assisted Learning, 19*(3), 371-382.

[15] Ng, W., & Nicholas, H. (2012). A framework for sustainable mobile learning in schools. *British Journal of Educational Technology.* 44 (5), 695–715.

[16] Armatas, C., Holt, D.M., & Rice, M. (2005). Balancing the possibilities for mobile technologies in higher education. In *Proceedings of the 22nd Annual Conference of the Australasian Society for Computers in Learning in Tertiary Education (ASCILITE): Balance, Fidelity, Mobility: Maintaining the Momentum?* Brisbane (pp.27-35).

[17] Engestrom, Y. (2003). *What is Activity Theory?*

[18] Engeström, Y. (2001). Expansive learning at work: toward an activity theoretical reconceptualization. *Journal of Education and Work, 14*(1), 133-156.

[19] Engeström, Y. (1987). *Learning by expanding: An activity-theoretical approach to developmental research.* Helsinki: Orienta-Konsultit.

[20] Centre for Activity Theory and Developmental Work Research. (2003-2004). *The Activity System.*

[21] Sujan, M. A., Rizzo, A. & Pasquini, A. (2002). Contradictions and critical issues during system evolution. In *Proceedings of the 2002 ACM Symposium on Applied Computing*, Madrid (pp.711-715). New York: ACM.

[22] Demiraslan, Y. & Usluel, Y. K. (2008). ICT integration processes in Turkish schools: Using activity theory to study issues and contradictions, *Australasian Journal of Educational Technology, 24*(4), 458-474.

[23] Murphy, E. & Rodriguez-Manzanares, M.A. (2008).Using activity theory and its principle of contradictions to guide research in educational technology. *Australasian Journal of Educational Technology, 24*(4), 442-457.

[24] Welch, G. (2006). Addressing the multifaceted nature of music education: an activity theory research perspective. In: *Proceedings of the 21st Seminar on Research in Music Education* (pp. 221-229). Bali, Indonesia. Hong Kong Baptist University.

[25] Terantino, J.M. (2009). *Transformational processes and learner outcomes for online learning: An activity theory case study of Spanish students*. Unpublished doctoral dissertation. University of South Florida.

[26] Van Horne, S.A. (2011). *An activity-theory analysis of how college students revise after writing center conferences.* Unpublished doctoral dissertation. University of Iowa.

[27] LaCroix, L.N. (2011). Mathematics learning through the lenses of cultural historical activity theory and the theory of knowledge objectification. In M. Pytlak, T. Rowland, & E. Swoboda (Ed.), *Proceedings of the Seventh Congress of the European Society for Research in Mathematics Education (CERME7)*, Rzeszów, Poland, 2462-2471.

[28] [28] Radford, L. (2006). Elements of a cultural theory of objectification. *Revista Latinoamericana de Investigación en Matemática Educativa, Special issue on semiotics, culture and mathematical thinking*, pp. 103-129.

[29] [29] Radford, L. (2007). Towards a cultural theory of learning. In Pitta-Pantazi, D., & Philippou, G. (Eds.), *Proceedings of the Fifth Congress of the European Society for Research in Mathematics Education (CERME – 5)* (pp. 1782-1797). Larnaca, Cyprus, CD-ROM, ISBN - 978-9963-671-25-0.

[30] Radford, L. (2008). The ethics of being and knowing: Towards a cultural theory of learning. In L. Radford, G Schubring, & F. Seeger (Eds.), *Semiotics in mathematics education: Epistemology, history, classroom and culture* (215-234). Rotterdam: Sense Publishers.

[31] Engeström, Y. (1993). Developmental studies of work as a testbench of activity theory: The case of primary care medical practice. In S. Chaiklin, & J. Lave (Eds.), *Understanding practice: Perspectives on activity and context* (pp. 64–103). Cambridge, England: Cambridge University Press.

[32] Baya'a, N., & Daher, W. (2010). Middle school students' learning of mathematics using mobile phones: Conditions and consequences. *Journal of Interactive Learning Research, 21*(1), 1-25.

[33] Yerushalmy, M., & Weizman, A. (2007). *Math4Mobile mobile environment.* The University of Haifa.

[34] Glaser, B.G., & Strauss, A. (1967). *Discovery of grounded theory: Strategies for qualitative research.* Chicago, IL: Aldine.

[35] Athens, L. (2010). Naturalistic inquiry in theory and practice. *Journal of Contemporary Ethnography, 39* (1), 87-125.

[36] Fran, S. M. (2013). The Constant Comparative Analysis Method Outside of Grounded Theory. *The Qualitative Report, 18*, 1-25.

[37] Page, S., & Clark, J.A. (2010). Experiences of learning and teaching mathematics: Using activity theory to understand tensions in practice. In L. Sparrow, B. Kissane, & C. Hurst (Eds.). *Proceedings of the 33rd annual conference of the Mathematics Education Research Group of Australasia: Shaping the future of mathematics education, 2*, 462-468. Fremantle, WA: MERGA.

[38] Guskey, T.R., & Gates, S.L. (1986). Synthesis of research on the effects of mastery learning in elementary and secondary classrooms. *Educational Leadership, 43*(8), 73-80.

[39] Blin, F. (2005). *CALL and the development of learner autonomy: An activity theoretical study*. Unpublished doctoral dissertation, Institute of Educational Technology, The Open University, UK.

[40] Sparks, G.N. (1983).Synthesis of research on staff development for effective teaching. *Educational Leadership, 41*(3), 65-72.

[41] Navarrete, C. (2011). Identifying affordances and barriers to student-centered, collaborative learning in the integration of interactive Whiteboard Technology. In M. Koehler, & P. Mishra (Eds.), *Proceedings of Society for Information Technology & Teacher Education International Conference 2011* (pp. 2996-3001). Chesapeake, VA: AACE.

[42] Martinez, R., Cheng, A.S., Smith, K., Smith, M., & Yoon, S. (2002). *Indiana University Instructional Systems Technology (IST) graduate student attitudes towards group work*. Y520: Strategies for educational inquiry Section 5982, Fall 2002 Group Project Reports.

[43] Kollias, V., Mamalougos, N., Vamvakoussi, X., Lakkala, M., & Vosniadou, S. (2005). Teachers' attitudes to andbeliefs about web-based collaborative learning environments in the context of an international implementation. *Computers and Education, 45*(3), 295-315.

[44] Batista, S.C.F, Behar, P.A. & Passerino, L. (2012). Pedagogical use of cell phones in Calculus I: Advantages and difficulties. *International Journal on New Trends in Education and their implications (IJONTE), 3*, 1-14.

[45] Stokmans, M.J.W. (2001). The effect of involvement on the reading attitude-behavior relationship. In A.J. Scott (Ed.), XXVI *Annual Colloquium on Research in Economic Psychology: Environment and Wellbeing* (pp. 285-290). Bath, UK: University of Bath.

[46] Clements, D.H., & Nastasi, B.K. (1993). Electronic media and early childhood education. In B. Spodek (Ed.), *Handbook of research on the education of young children* (pp. 251–275). New York, NY: Macmillan

[47] Hennessy, S., Ruthven, K. & Brindley, S. (2005). Teacher perspectives on integrating ICT into subject teaching: commitment, constraints, caution and change. *Journal of Curriculum Studies, 37* (2), 155-192.

[48] Hardman, J. (2005). An exploratory case study of computer use in a primary school mathematics classroom: New technology, new pedagogy? *Perspectives in Education, 23*(4), 99-111.

[49] Herrington, J., Mantei, J., Herrington, A., Olney, I. W. & Ferry, B. (2008). New technologies, new pedagogies: mobile technologies and new ways of teaching and learning, In R. Atkinson & C. McBeath (Eds), *Annual Conference of the Australasian Society for Computers in Learning in Tertiary Education, 2008* (pp. 419-427). Melbourne, Australia: Deakin University.

[50] Carlson, S. & Gadio, C.T. (2002). Teacher Professional Development in the Use of Technology. In W. Haddad, and A. Drexler, (Eds), *Technologies for Education*, pp. 118-132. Paris: UNESCO.

[51] [Baya'a, N., Daher, W. & Khalil, R. (2009). The role of the ICT coordinator in school. *Jamaea'a, 13*, 257-285.

[52] Spillane, J., Halverson, R., Diamond, J. (2004). Towards a Theory of School Leadership Practice: Implications of a Distributed Perspective. *Journal of Curriculum Studies, 36* (1), 3-34.

AUTHORS

W. Daher is affiliated with An-Najah National University, Nablus, Palestine and with Al-Qasemi Academic College of Education, Baqa, Israel (e-mail: wajeehdaher@gmail.com).

N. Baya'a is affiliated with Al-Qasemi Academic College of Education, Baqa, Israel (bayaan@qsm.ac.il)

Security of m-learning System: A Collective Responsibility

S.A. Shonola and M.S. Joy
University of Warwick, Coventry, United Kingdom

***Abstract*—Innovation in learning technologies and services is driven by demands from Higher Educational Institutions (HEIs) in order to meet students' needs and make knowledge delivery easier. The technology could play a pivotal role in extending the possibilities for teaching, learning, and research in educational institutions. M-learning emerged from this innovation as a result of an unprecedented explosion in the number of mobile devices due to availability and affordability of mobile phones, smartphones and tablets among students. Competition in the mobile device industry is also encouraging developers to be innovative, constantly striving to introduce new features. Consequently, newer sources of risks are being introduced in the mobile computing paradigm at manufacturing level. Similarly, many m-learning promoters and developers focus on developing and delivering learning content and infrastructure for m-learning system without adequate consideration for security of stakeholders' data, whereas the use of these mobile technologies for learning poses a serious threat to confidentiality, integrity and privacy of those involved in teaching and learning, yet traditional security threats are also evolving. Against this backdrop, the stakeholders in education sector (i.e. education providers, educators, m-learning promoters and developers) should begin to consider the security implications of these devices in modern teaching and learning environments. The purpose of this paper is to identify the m-learning security issues that stakeholders may face, how they are being affected by these security threats, who among the stakeholders are affected or most affected by security issues in m-learning using three Nigeria universities as case studies and what are the responsibilities of the stakeholders in ensuring risk free m-learning.**

***Index Terms*—mobile learning security, m-learning security, security threats**

I. Introduction

M-learning is creating a new environment for teaching, learning and education delivery. It is a mobile-based learning process using internet technology to design, implement, select, manage, support and extend knowledge acquisition. As m-learning has many advantages such as flexibility and diversity, it has become a way of learning that augment classroom and e-learning. It is a growing trend that extends learning beyond lecture theatres and can be exploited to respond to the challenges of particular educational contexts, complement and enhance formal schooling, improve and assist learning for people of different ages and augment learning opportunities in communities where educational opportunities are limited. The development and adoption rate of mobile technologies are increasing rapidly on a global scale, many education institutions have aligned their curricula to accommodate m-learning and have invested in hardware and software resources to take full advantage of the new invention. There are numerous applications for mobile technologies in education, from the ability to wirelessly transmit learning modules and administrative data, to enabling learners to communicate with instructors and peers "on-the-go". Lecturers can give out lecture notes and instructions while on the move and students can listen to recorded lectures either online or offline anytime anywhere [1].

However, along with opportunities in m-learning come several challenges that need to be addressed. For instance, in an m-learning environment, educational materials and instructions can be passed to students through their mobile devices, but this adds another layer to the personal computer-based model of teaching and e-learning. Similarly, the pervasive use of m-learning may entail, among others, loss of privacy and attacks on users and institutional security. One important concern about the use of mobile devices in learning is the security risk and vulnerability issue relating to educational content and private information of stakeholders [2]. Higher Education Institution management, educators, and individual learners are profoundly worried about the growing security threats in m-learning. Concern about data security and privacy issues seems to be quite high among educators, education providers, m-learning developers and other stakeholders. Some perceived risks include unauthorised interference with the learning content and instructions by the learners. More importantly, as students are allowed to use their portable devices to access learning content and materials anywhere, this increases the security risks. Therefore, one security challenge is to ensure that students' have access only to their required learning material and instruction regardless of whether they learn in the lecture room or outside the classroom [3]. This paper specifically examines the collective responsibilities of stakeholders in ensuring risk-free m-learning environments.

This paper is an extension of work on Mobile Learning Security Concerns from University Students' Perspectives, reported in proceedings of the International Conference on Interactive Mobile Communication Technologies and Learning [4]. The original paper identified the security threats that students may face when using mobile devices for educational purposes and examined the perceived damaging effects of mobile learning on students in case there is a security breach. We extend our previous work by investigating the security threats which other stakeholders such as educational institution management, academic and support staff may face and discuss the collective responsibilities of the stakeholders in ensuring threat and risk free m-learning environment.

Our paper is organised as follows. The second section of this article is a review of related research on m-learning security. It summarises an existing study on m-learning security and evaluates the recommendations made in the literature. The third part discusses the research carried out on how security issues affect the use of m-learning (using three universities in Nigeria as a case study) from stakeholders' perspective, and who are responsible for ensuring risk free m-learning. It details the purpose of the research, the methodology and research questions. A brief overview of the analysis of the results of the research is presented in section four while section five gives a detailed discussion of the results gathered and statistical tests. The last part of this article highlights and discusses recommendations given to the security issues mentioned in the previous sections. The article concludes with problems encountered during the research and direction for future work in ensuring risk free m-learning environments.

II. Literature Review

In recent years there are many studies on mobile learning in academics institutions, although most of the studies are related to the development of mobile learning materials. Studies on the perception of stakeholders towards mobile learning security are few in number, and are examined in this section.

Guo et al. [5] observed that more academic institutions plan to integrate mobile security into their computing curriculum and found out that there are at least two challenges. The first challenge is the unique characteristics of mobile security, which is a new and evolving paradigm. Traditional security threats, such as malware or social engineering, are growing in this new environment using new attack vectors or adapting to the new platform. Similarly new components such as Global Positioning System (GPS) and services in mobile platforms introduce new sources of risks and few security courses cover the full spectrum of mobile security, especially those new and unique threats. The second challenge is the shortage of effective learning materials, compared to the rich learning materials available for general computer security or other special security areas such as web security or network security. Therefore, there is a need to explore a learning approach to mobile security from the perspectives of mobile devices (smartphones and tablets) and mobile applications (apps).

Charlesworth [6] stated that the use of mobile technologies poses a risk to the confidentiality, integrity and privacy of the data involved in the educational process for both the users and the services. The challenge, therefore, has to do with securing the systems and deploying proper security policies and procedures so as to be able to deter and repel attacks. It also requires insuring the integrity, privacy and confidentiality of the data stored and transferred for the needs of the educational process. Therefore, the provision of robust mechanisms to support user authentication, authorisation and non-repudiation, management of data, content copying, editing and downloading, safeguarding learner examination and assessment processes from attackers and impostors are some prerequisites for secure online learning arena.

Kamba [7] found out that some universities in Nigeria have websites and engage in m-learning for advertisement and information dissemination only and not for leaning purposes for various reasons, one of which is the security aspect. There are perceived risks, as such students interfering with the learning contents and instructions given to them by their lecturers and instructors. The security need in an m-learning platform is to protect the content, services and the personal data of the stakeholders and guarantee their confidentiality at all times [8]. In some cases, especially if the learning institution supplies learners with special mobile equipment, there exist worries about making them attractive to thieves. A key concern for educationalists and policy makers is ensuring that learners can use mobile technologies, and the online services and communication accessed through these technologies safely.

While the use of mobile technologies for academic purposes does not increase the risk that may already exist as a result of learners' personal ownership of mobiles, the designers and practitioners of education are, however, responsible for producing coherent and reliable accounts of the likely consequences of the proliferation of mobile devices in the higher education landscape [9]. The focus of this paper is on the challenges associated with mobile security specifically related to education sector stakeholders' concerns in the areas of safeguarding and protecting the privacy and security of individuals, data, systems and equipment. The outcomes of this research will contribute to the field of study in determining how various stakeholders in academic institutions are affected by the security issues in m-learning and in making recommendations for ensuring risk free m-learning environments.

A. *Research Questions*

This study attempts to answer the following questions:

1. What are the m-learning security issues that stakeholders may face?
2. How are the stakeholders in academic institutions being affected by the security threats in m-learning?
3. Who among the stakeholders are most affected by the security issues?
4. What are the responsibilities of the stakeholders in ensuring risk free m-learning?

III. Methodology

In identifying the challenges of m-learning security, two separate research activities were carried out as primary instruments to gather data. The first was an online and hard-copy questionnaire for collecting quantitative data from stakeholders, mostly the lecturers and learners who are the main users of m-learning. The second instrument was a series of interview with a sample of academic and support staff to obtain qualitative data on their observations and experiences on m-learning as well as their understanding of factors contributing to security threats in m-learning. The research study was conducted at three universities from South West Nigeria using qualifying demographics. Technical terms and concepts were explained in brief in the questionnaire and the researcher was available during the study to assist the respondents in understanding any part of the questionnaire. A total of 150 responses were gathered comprising of 120 questionnaire data and 30 interview responses, all of which were analysed. Secondary instruments for data collection are documented evidence, articles obtained from the academics and administrative staff as well as information obtained through research papers from relevant journals which were used to support the primary data gathered from the research study.

IV. FINDINGS

The findings of this work are organised into three sections in order to provide answers to the research questions as shown below:

Research Question 1: What are the m-learning security issues that stakeholders may face?

The research question is a multi-choice question to find out various security issues stakeholders might have encountered when using mobile devices to complement teaching and learning activities. Fig 1.shows the various security issues, topping the list is unauthorised access to learning content and materials (75.6%), followed by virus/malware attack on m-learning system (68%). Denial of service is considered the least threat to m-learning system (30.5%).

Research Question 2: How are the stakeholders in academic institutions being affected by the security threats in m-learning?

This section addresses how security issues in m-learning affect the stakeholders i.e. the education institutions, lecturers/faculty members and the learners in case of security breaches.

Educational Institution Management/Promoters: This part of the survey investigated the adverse security effects of m-learning on the HEIs who are interesting in incorporating m-learning in their curriculum. Educational institutions and their management are stakeholders in knowledge development and delivery and they risk suffering loss of confidential information, reliability, and goodwill as well as working hours of the developers and support staff to security issues as shown from the survey data. Around nine out of ten participants (87.78%) indicated loss of confidential information as the most adverse effect of security threat to m-learning for HEIs. Eight out of ten participants indicated loss of goodwill and integrity as the risk of m-learning threats to the HEIs. In case of successful hacking, there is a loss of reliability on the part of the institution which 77.8% of the respondents pointed out. 43.3% of respondents also perceived loss of working hours in correcting the anomalies, restoring students' confidence and rebuilding integrity of the institution after a security breach.

Academic/support staff: The educators may risk loss of confidential information, unauthorised change in learning content, loss of control during e-examinations, loss of privacy as well as psychological effect. From Fig. 3 below, around nine out of ten participants (93.3%) indicated that loss of confidential information is a major consequence of the m-learning security threat to educators while more than eight out of ten participants (86.67%) were of the view that loss of control mainly during e-assessment and online examinations is a threat to the academic staff. Also more than eight out of ten (83.3%) agreed that loss of content quality of learning materials is a concern while exactly three out of ten indicated psychological effects as a consequential result of m-learning security threats. Almost half of the participants (48%) are of the view that loss of privacy is a security threat to m-learning for the academics.

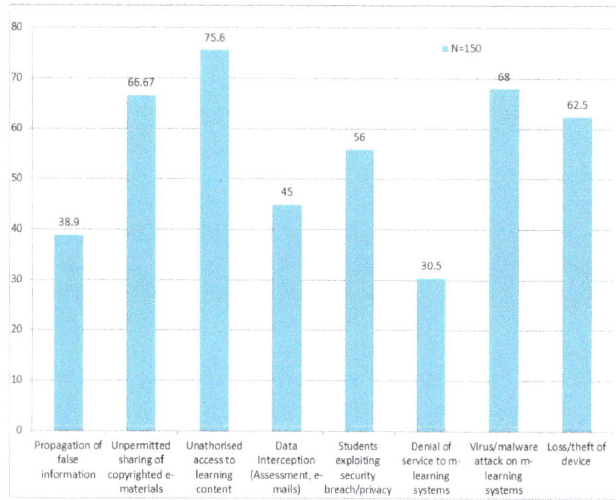

Figure 1. What are the m-learning security issues that stakeholders may face?

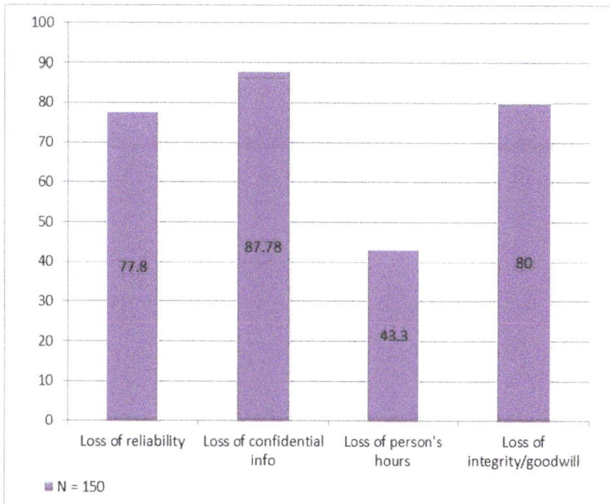

Figure 2. How the stakeholders (HEIs Management/ Developers) are affected by m-learning security threats

Students: Being one of the stakeholders in m-learning are exposed to the following risks as gathered from participants in the research study: loss of performance (65%), loss of study hours (71.67%) and loss of their confidential information (80%) as well as psychological effects (46.67%) as shown in Fig. 4 below.

Research Question 3: Who among the stakeholders are most affected by the security issues?

This section of the study reveals that among the different stakeholders in a university community, the university management are most affected by any security threats. The faculty lecturers and students as users are also affected next, after the management. Fig. 4 indicates that the vast majority of respondents (66.67%) indicated that the university management and promoters are most affected by m-learning security threats among all the stakeholders. This is highly likely to happen as they are the policy makers in charge of the smooth running of the university. The lecturers and support staff are next as perceived by 53.33% of the participants. Students are affected as indicated by four out of ten participants. The developers are least affected if the m-learning systems are being developed by consultancy outfit or software company as found out in the research study.

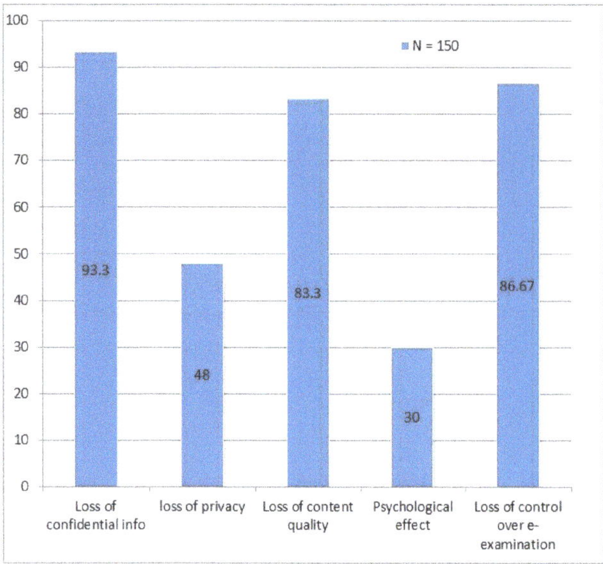

Figure 3. How the stakeholders (Academic/support staff) are affected by m-learning security threats

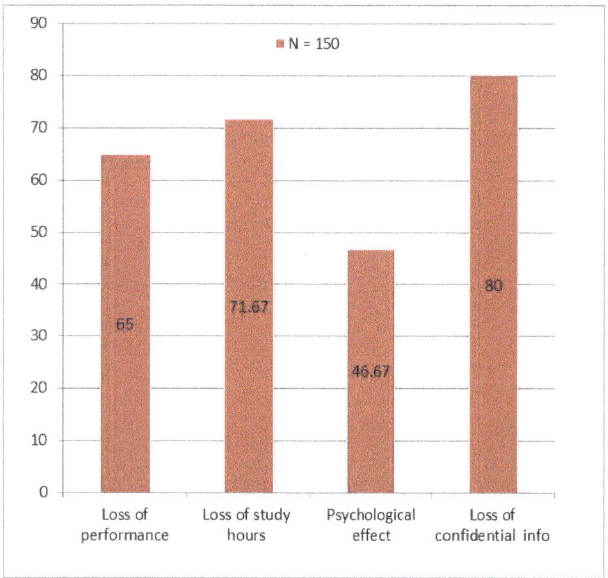

Figure 4. How the stakeholders (Students) are affected by m-learning security threats

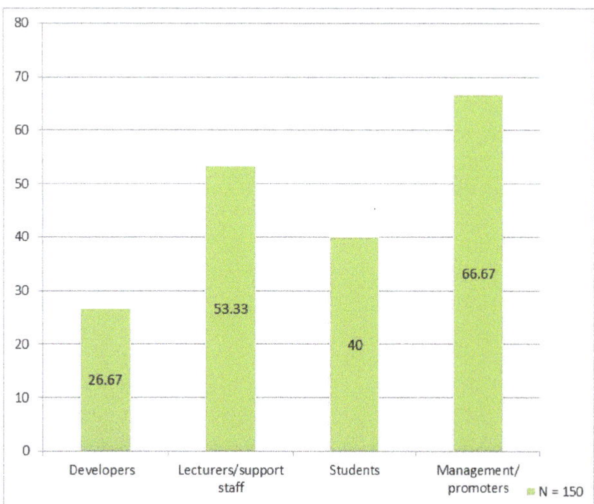

Figure 5. The affected stakeholders in m-learning

V. Statistical Analyses

The Research Question One on m-learning security issues that stakeholders may face was analysed using Mann-Whitney U test for dependency based on the responses of the main stakeholders as shown in Table 1 and Table 2 below.

Table 1: Ranking of main stakeholders in m-learning security

Stakeholders	Participants	mean of ranks	sum of ranks
Students	120	10	50
Academics/support	30	4	28
Total	150	6.5	78

Table 2: Mann- Whitney test

Test	m-learning security threats
U Mann - Whitney	0 (critical value of u = 5)
Z	-2.7608
p-value (2-tailed)	0.05

According to Table 2, since the value of Z is less than -1.96 or greater than 1.96, **there is significant difference between students' and educators' perspectives in relation to the security issues that each set of stakeholders face in the use of m-learning.** This implies that the educators and students have different views on the m-learning security issues and they are being exposed to different risks which may be due their different use of mobile devices in education. While academic and support staff use their m-learning for teaching and passing knowledge, the students use their m-learning devices for learning. These different viewpoints are further explained in the discussion section.

Additionally, the Mann- Whitney U test was also performed on the data obtained on **Research Question Two regarding how the stakeholders in HEIs are affected by the security threats in m-learning**. The results of the statistical tests are shown 1n Table 3 and Table 4 below.

Table 3: Ranking of main stakeholders in m-learning

Stakeholders	Participants	mean of ranks	sum of ranks
Students	120	7	35
Academics/support	30	4	20
Total	150	5.5	55

Table 4: Mann- Whitney test

Test	m-learning security effects
U Mann - Whitney	5 (critical value of u = 2)
Z	-1.4623
p-value (2-tailed)	0.05

According to Table 4, **there is no significant difference between students and academic/ support staff viewpoints in relation to the security effects on stakeholders in m-learning**. This implies that both stakeholders are exposed to the same or similar security risks when using m-learning. This assertion is true because, being stakeholders, they both suffer loss of confidential information and privacy as well as psychological effects. In addition, academics and support staff may suffer loss of control over e-exams and content quality while the effect on students may include loss of study hours and performance.

VI. Discussion

Fig. 1 illustrates security issues being encountered when using mobile devices for learning. Noticeable in the list is loss or theft of mobile devices. This is common in many developing countries since mobile devices are still regarded as precious possessions and in some cases where the HEIs supplies learners with mobile devices, there are concerns about making learners attractive to thieves. This result is in line with the work of Obodoeze et al. [10] which identifies that the second most challenging security concern affecting mobile users in Nigeria is the frequent or widespread losses of mobile devices by their owners to thieves or by negligence. It is also consistent with a survey conducted by Jupiter [11] on mobile device users that showed that one out of every three mobile device users has lost their device at some point in time. Unauthorised access to learning content and interception of confidential information and personal details such as grades, feedback and emails by students and outsiders, either for fun or malicious acts, is a security threat which accounted for 75.6% and 45% respectively. Similarly, unpermitted sharing of copyrighted e-materials by the students among themselves is a security issue common among learners in HEIs in Nigeria. This is made possible due to inadequate copyright laws and software piracy in Nigeria [12, 13]. Virus and malware attacks are also threats to the use of handheld devices for learning purposes. This study also supports the work of Obodoeze et al. [10] who identify the various forms of threats including virus and malware attacks and hacking as the biggest security challenges being faced by mobile device users in Nigeria. Virus and malware threats are normally encountered when downloading educational materials from an unknown source. Propagation of false or misleading information using mobile devices among learners is a threat to m-learning. This is quite common as some students spread incorrect information through social media [14]. Denial of service is also a threat that usually affects the availability of m-learning system, mainly due to irregular power supply to mobile learning servers, a problem that is common in developing countries. This study is also consistent with the results of Osang et al. [15] in which 64% of the respondents identify that the poor power supply situation in Nigeria is a threat to m-learning.

How the educational institutions, educators and students are affected by m-learning security threats are shown in Figs. 2, 3 and 4 respectively. Common effect among all stakeholders is loss of confidential information, leading to loss of privacy. The study is consistent with the work of Kambourakis [2], who states that loss of confidential information is one of the worries for lecturers and their confidentiality should be guaranteed at all times. It is also consistent with the work of Zamzuri et al. [16] which states that one of the reasons why students reject online systems is due to security reasons because they are worried about the loss of their private and confidential information. Fig 3 indicates that the loss of control mainly during e-assessment and e-examination is a worry for the academic staff as this can lead to examination malpractices and unpermitted collaboration during assessments. This result is consistent with the study conducted by Osang et al. [15], in which most of the educators believed that m-learning will ease examination malpractices. Again, the findings agree with the work of Kambourakis [2], which revealed that e-examination procedures carried out in an unsupervised or semi-supervised way is one of the difficult challenges within the m-learning context. Therefore educators, who are interesting in using any technology for educational purposes, will want to take ownership and control of such projects [15]. Fig. 3 also shows that loss of content quality of learning materials is a likely side effect of introducing m-learning systems that can make it possible for learners to tamper with learning materials if the security is weak. Altering learning content and grades without authorisation from lecturers and amending confidential documents can be feasible if there is a security breach in the m-learning system known to the students.

Fig. 4 reveals that at least 60% of the students feared loss of study hours and performance as consequences of a security breach in m-learning due to denial of service. This is possible when learners have viewed m-learning as a complement to classroom and relied on it as a learning portal. Thus the non-availability of service for a long period of time will have adverse effects on learners' study hours, revision time and consequently their performance. This finding is in line with the work of Kukulska-Hulme et al. [17] which shows that good m-learning improves learners' study retention and performances in their study. Therefore, learners need a reliable, highly available and dependable m-learning system to avoid being frustrated when using the system, which can indirectly affect their study performance.

Research Question 4: What are the responsibilities of the stakeholders in ensuring risk free m-learning?

Having identified and discussed the m-learning security issues that stakeholders may face and how they are being affected by the security threats, it is important to note that the responsibility of ensuring a risk free m-learning environment lies mainly with the stakeholders themselves as they are the people involved in managing and using the system. Their collective responsibilities in ensuring risk free m-learning are highlighted in the recommendation section below.

VII. Recommendations

Based on the findings and results presented above and the cited literature, the following recommendations are the collective responsibilities of each specific stakeholder in ensuring risk a free m-learning environment.

Educational Institution/Management: The education providers have the main responsibilities of running risk free m-learning systems. The research findings indicate that 81.8% of the respondents agreed that the university management is responsible for ensuring risk free m-learning. This is expected as the university authority is the policy maker and also accountable for smooth running of the university facilities including m-learning infrastructures. herefore, the educational institutions/ promoters can ensure risk free m-learning by performing the following steps.

- Create security awareness among other stakeholders and encourage them to be security conscious when using their mobile devices. Educators and students should be made aware of the potential risk of connecting to bogus free Wi-Fi which criminals may have set up in public places in order to collect personal data. Creating security awareness is vital as our study revealed that some users do not take the security of their mobile devices very serious, so there is a need to promote mobile security education among users [19]. With adequate knowledge, students will be more security conscious about the safety of their handheld devices, thereby reducing the rate at which small electronic gadgets are lost or stolen which is mainly due to their negligence.

- Implement separate wireless networks for academic users and visitors, to access the internet from their mobile devices whilst allowing restricted access to m-learning systems. This is to ensure that m-learning users have access only to their required or legitimate activities on the system and reduce unnecessary traffic to the servers.
- Implement mobile device management (MDM) systems for administering the m-learning devices in real-time, such as locate, track and gather information on the movement of devices, with the aim of remotely diagnosing and fixing software security problems, install and update software on devices and erase data on lost or stolen devices.
- In tackling the sharing of copyrighted e-materials, the university administrators can implement Digital Rights Management (DRM) solution. DRM is a technology that can be used for content protection in m-learning environment. It is a class of access control measures that are used to limit the use of digital content and devices. A DRM based m-learning system can focus on learning content protection and other basic procedures of m-learning facilities that can be secured.
- The university administrators can also operate a blacklisting method whereby websites or categories of websites deemed to be inappropriate or insecure are blocked from the university's network.

Academic staff: Around four out of ten (41%) of the respondents indicated that faculty lecturers are also responsible for ensuring risk free m-learning as they are in charge of running the academic programmes including the m-learning curriculum. The academic staff can ensure risk free m-learning through the following.

- Align the existing curriculum for m-learning with proper consideration for security as well as integrate new technology into their modules in a secure manner.
- Participate in the design and development of m-learning systems and mobile apps for their taught modules and research activities.
- Providing experts' opinions and contributions for the overall implementation and application of m-learning systems, in particular to their field of knowledge and research.
- Computer Science academic staff can develop a highly secured m-learning system following a standard security framework for implementation and participate in training the support staff on security concepts such as data encryption.

Developers/ Support staff: The responsibilities of support staff in ensuring risk-free m-learning include:

- Ensuring regular data backups are taken, installing firewalls on m-learning servers and having up to date anti-malware and anti-virus software installed on m-learning systems as well as installing all security patches.
- Putting in place proper security procedures and policies that will prevent hacking activities which may deny legitimate acts, a scheduled maintenance policy for m-learning servers and network infrastructure, as well as an uninterruptible power supply.

Students: In view of the students' concerns on m-learning, the following recommendations are offered for secure and effective m-learning.

- They should avoid connecting to unsecured public Wi-Fi as many of them connect to educational resources while on the move using any free available Wi-Fi. They should be aware of the credibility of the organisation providing the connection regarding the security and safety of free network facilitates before using it. For example, connecting to an unsecured and unverified wireless infrastructure increases the chances of putting personal data at risk.
- Security apps such as phone finders should be installed on mobile devices to enable locating them in case of lost or theft. Remote wipe apps should be installed to prevent unauthorised access to confidential and private information as well as learning materials stored on the devices if a lost device cannot be traced.

VIII. Conclusion

Until recently, stakeholders in education were unconcerned about security, mainly because users in academic fields tended to be non-malicious. Nowadays, security in learning environment is increasingly important because online learning applications and infrastructure have become 'business-critical' applications in HEIs [19]. Therefore, stakeholders, most especially the University management and academic staff are now seeing security issues in learning technologies as a part of the overall educational and business strategy of their institutions. Proper design of the technologies being used for online learning and adequate security management for m-learning system will lead to greater effectiveness, acceptability and usage of mobile learning [9]. The stakeholders of m-learning should focus not only on developing and using m-learning content and infrastructure but also make effort at securing the system because the users need a reliable, highly available and dependable m-learning system to avoid being frustrated when using the system.

References

[1] S.A. Shonola and M.S. Joy (2014), 'Mobile learning security issues from lecturers' perspectives (Nigerian Universities Case Study)'. 6th International Conference on Education and New Learning Technologies, 7-9 July, 2014, Barcelona, Spain. pp. 7081-7088

[2] G. Kambourakis (2013). Security and Privacy in m-Learning and Beyond: Challenges and State of-the-art. International Journal of u- and e- Service, Science and Technology. Vol.6, no. 3, pp.67-84

[3] F. Graf (2002) 'Providing security for elearning'. Computers & Graphics. vol.26, no.2, pp.355–365

[4] S.A Shonola and M.S Joy (2014), 'Mobile Learning Security Concerns from University Students' Perspectives'. 8th International Conference on Interactive Mobile Communication Technologies and Learning, November 13-14, 2014, Thessaloniki, Greece. pp. 165-172

[5] M. Guo, P. Bhattacharya, M. Yang, K. Qian, and L. Yang (2013). Learning mobile security with android security labware. In Proceeding of the 44th ACM technical symposium on Computer science education (pp. 675-680). ACM.

[6] A. Charlesworth (2009) "Code of Practice for the Further and Higher Education Sectors on the Data Protection Act 1998", Tech.

rep., JISC Legal, (2009),

[7] M. Kamba (2009). Problems, challenges and benefits of implementing e-learning in Nigerian universities: An empirical study. International Journal of Emerging Technologies in Learning (IJET), vol. 4, no.1

[8] C.D.C Luminita and C.I.N Magdalena (2012). 'E-learning Security Vulnerabilities'. Procedia - Social and Behavioral Sciences 46 (2012) pp. 2297 – 2301

[9] M. O. M El-Hussein and J.C Cronje (2010). 'Defining Mobile Learning in the Higher Education Landscape'. Educational Technology & Society, vol.13, no 3, pp. 12–2

[10] F.C. Obodoeze, F.A Okoye, C.N Mba, S.C. Asogwa and F.E. Ozioko (2013). A Holistic Mobile Security Framework for Nigeria. International Journal of Innovative Technology an Exploring Engineering (IJITEE), vol.2 , no.3, pp.1-11

[11] Juniper Networks (2012) '2011 Mobile threats report' Juniper Networks Inc, February, 2012

[12] K. M. Waziri (2011). Intellectual Property Piracy and Counterfeiting in Nigeria: The Impending Economic and Social Conundrum. Journal of Politics & Law vol. 4, no.2, pp.196-202

[13] B. A. Fabunmi (2009). The Roles of Librarians in Copyright Protection in Nigeria. International Journal of African & African-American Studies vol. 6, no.1, pp84-93

[14] P. O. Jegede (2009). Age and ICT-related behaviours of higher education teachers in Nigeria. Issues in Informing Science and Information Technology, vol.6, pp.770–777.

[15] B. F. Osang, J. Ngole and C. Tsuma (2013). Prospects and Challenges of Mobile Learning Implementation in Nigeria: Case Study National Open University of Nigeria (noun)'. A paper presented at International Conference on ICT for Africa 2013, February 20 -23, Harare, Zimbabwe

[16] Z. F. Zamzuri, M. Manaf, Y. Yunus, and A. Ahmad (2013). Student Perception on Security Requirement of e-Learning Services. Procedia-Social and Behavioral Sciences. 90(2013), pp.923-930.

[17] A. Kukulska-Hulme, M. Sharples, M. Milrad, I. Arnedillo-Sánchez and G. Vavoula (2009) 'Innovation in Mobile Learning: A European Perspective'. International Journal of Mobile and Blended Learning vol. 1, no. 1, pp. 13-35.

[18] S. Diaz (2012) 'Mobile security needs more than just software, needs education.

[Accessed on Jan 20, 2012].

[19] E. Weippl and M. Ebner, (2008). Security & Privacy Challenges in E-Learning 2.0. Proceedings of E-Learn 2008, Las Vegas, pp. 4001-4007

AUTHORS

S.A. Shonola is a PhD student and researcher at the Computer Science Department, University of Warwick, UK. (e-mail: s.a. shonola@ warwick.ac.uk).

M.S. Joy is an associate professor at the Computer Science Department, University of Warwick, UK. (e-mail: m.s. joy@ warwick.ac.uk).

The Readiness of Applying M-Learning among Saudi Arabian Students at Higher Education

Sulaiman Almutairy, Trevor Davies, Yota Dimitriadi
University of Reading, United Kingdom

Abstract—**This paper presents study investigating the extent to which Saudi students understand and are familiar with mobile learning. It investigates students' use of handheld devices and considers the daily activities for which they could be used while also evaluating m-learning in Saudi Arabia - a topic that has not yet received adequate attention from researchers. Initial data was acquired through a questionnaire carried out among 131 Saudi students at UK universities in July 2013. The study confirmed that students' are confident using mobile devices in their daily lives and that they would welcome more opportunities of mobile learning. The findings reinforce the fact that higher education policy makers in Saudi Arabia need to consider the possibility of creating mobile learning environments at academic institutions.**

Index Terms—**M-learning, mobile technology**

I. Introduction

The process of learning and accessing information has undergone rapid changes in the past decade. A global technology revolution is taking place, resulting in a transition from desktop computing to widespread use of mobile technology. Technology is changing the way people work, learn, conduct business, interact with one another and access information [1]. The learning process is no longer reliant on traditional teaching methods, but has expanded to include new technologies and new forms of learning, such as mobile learning (m-learning). Mobile technologies offer opportunities to deliver new and interesting methods of learning, both outside and inside the classroom.

M-learning involves the ability to acquire information anywhere, at any time, using handheld devices such as mobile phones and Personal Digital Assistants (PDAs). M-learning is defined as:

Any sort of learning that happens when the learner is not at a fixed, predetermined location, or learning that happens when the learner takes advantage of the learning opportunities offered by mobile technologies (p. 6) [2].

New figures released in 2014 by the International Telecommunication Union (ITU) predict that the number of mobile-cellular subscriptions will reach 7 billion by the end of the year. Thus, by the end of 2014 there will be almost 3 billion Internet users, two-thirds of whom will be located in the developing world. The ICT Commission states that, by the end of 2013, in Saudi Arabia, which is a developing country, the number of mobile phone subscribers had almost reached 51 million. The majority of these connect to the Internet using a wireless connection, or via a 3G or 4G mobile phone network [3], [4].

II. The Saudi Arabian Context

Saudi Arabia is currently experiencing a comprehensive developmental renaissance in various areas of life. The government has established an infrastructure that will ensure optimal use of information and communications technology in education. The average age of Saudi Arabians is now 25.3 years and the total population is 29,994,272. The estimated number of graduates from public education institutions is 450,000 students, all of whom subsequently attempt to secure a place in one of the country's 25 public universities; thereby, creating a demand for places which cannot be met.

If an m-learning environment were adopted in higher education it could address the issue of large numbers of students applying to universities, particularly the new generation of students born after 1980 and who grew up in the digital age, so-called 'digital natives' [5].

In Saudi Arabia, higher education policy makers should also consider the possibility of applying an m-learning environment in academic institutions.

III. Review Of M-Learning In Saudi Arabia

It is necessary to establish a clear definition of m-learning before undertaking a review of the literature regarding m-learning in the context of Saudi Arabia. As mentioned previously, m-learning can take place anytime, anywhere. M-learning is:

'the processes of coming to know through conversations across multiple contexts amongst people and personal interactive technologies' p.225, [6].

Many universities and institutions in Saudi Arabia have begun to implement facilities to support m-learning. For example, King Saud University has launched a new service offering users the opportunity to send text messages to the mobile phones of individuals or groups of students, directly from their PCs. These messages allow faculty managers, schools and administrative divisions to deliver timely and relevant information to their students [7].

In addition, a study conducted by Al-Fahad [8] concluded that wireless networks effectively facilitate access to resources at anytime from anywhere using handheld devices. Additionally, he asserted that SMS can be used effectively to support and improve student relationships.

However, m-learning in Saudi Arabia is still in the developmental stages [9]. The researchers found that over 75% of Saudi Arabian students surveyed expressed positive attitudes to m-learning, due to the flexibility of m-learning methods, and their scope to improve communication among learners.

IV. METHODOLOGY

The research method adopted for a study refers to the means through which a researcher intends to gather data. A survey based methodology is commonly used to collect results obtained from interviews and questionnaires. This reflects a specific approach which seeks to establish the relationship between variables and influential factors and trends.

This study acquired initial data during two weeks in July 2013, using an online survey comprising 30 questions. This data collection phase took place at a UK university with 131 Saudi Arabian students. The questionnaire collected information to provide general information regarding Saudi students employment of handheld devices in different settings, and to explore Saudi Arabian students' readiness to engage in, and familiarise themselves with, m-learning. It also examined and measured the extent of Saudi Arabian students' acceptance of the notion of an m-learning environment being formally introduced by higher education institutions. The research took the form of a descriptive study. The questionnaire was divided into four sections namely: (1) demographic data; (2) personal attitudes; (3) definition of m-learning; and (4) opportunities. A five point Likert Scale was used to measure the responses to the main questions, ranging from: strongly disagree; disagree; agree; strongly agree and not sure.

As mentioned previously, there are currently 51 million mobile phone subscribers in Saudi Arabia and between 110,000 and 125,000 male and female Saudi students are studying abroad [10]. The majority of these are able to connect to the Internet. Thus, a survey is therefore an appropriate research tool for such a study.

V. RESEARCH SAMPLE

The participants in this research are Saudi Arabian higher education students studying in UK universities. The sample comprises higher education students of both genders, aged 18 to 45 years.

The quality of a piece of research not only stands or falls by the appropriateness of methodology and instrumentation, but also by the suitability of the sampling strategy that has been adopted. (p. 92) [11].

The sample was selected randomly, from different institutions and levels of study, in order to involve a wide range of Saudi Arabian higher education students, who use handheld devices for learning purposes.

Table 1 presents the distribution of the study sample according to demographic variables. It reveals that out of a sample of 131 students, 42 were female and 89 were male. It also reveals the age distribution of the sample with the majority being in the age group 18-33 years. In addition, the majority are educated to either Masters' or PhD level.

VI. FINDINGS

This study initially aimed to investigate the extent of Saudi students' understanding of and their readiness to use the m-learning environment formally. It investigated the students' habits of use, preferred knowledge of handheld devices and the daily activities to which handheld devices can contribute.

Figure 1 reveals that Engineering, Education and Health Science are the three disciplines, which are most popular among Saudi Arabian students at universities around the UK.

All the Saudi students selected for this study had mobile phones. Of the 131 Saudi students, 130 had access to a 3G mobile phone network. Figure 2 shows that the majority of the Saudi students (119) 'always' and 'usually' carry their mobile phones with them and only 12 students only carry their phones with them 'sometimes'. Based on the answers to question 8, 130 students have access to resources using 3G mobile phone networks, which means that they are able to use their mobile phone as a tool to gain knowledge and information online.

TABLE I.
THE DESCRIPTION OF DEMOGRAPHIC INFORMATION OF THE STUDY SAMPLE

	Gender	Age	Educational level
Female	42		
Male	89		
18-23		3	
24-28		25	
29-33		49	
34-38		37	
39-44		15	
45+		2	
Bachelor			14
Masters			57
PhD			58
Vocational Training			1
Professional Training			1
Total	**131**	**131**	**131**

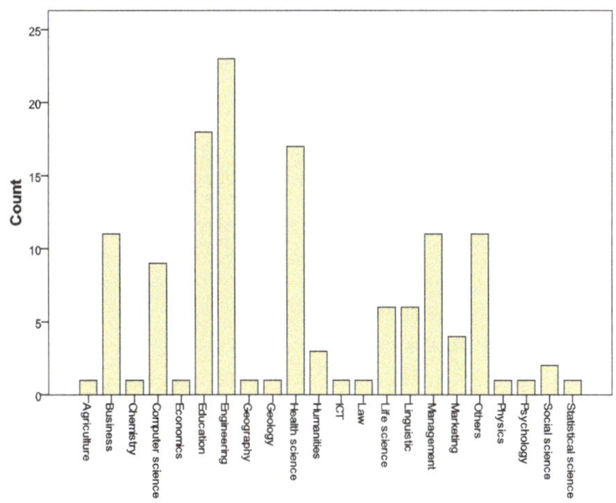

Figure 1. Subject distribution

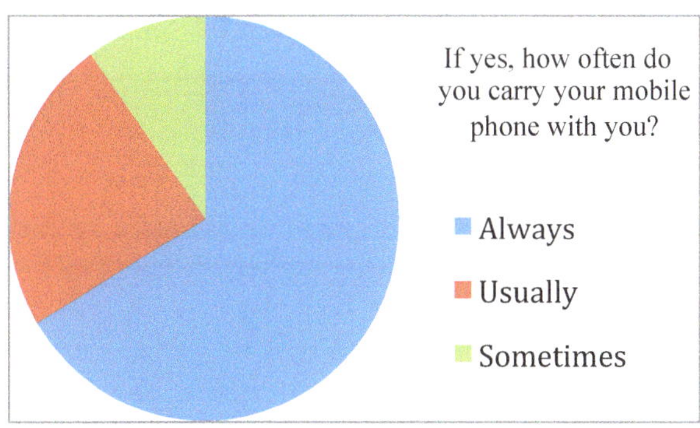

Figure 2. How often do you carry your mobile phone?

Table II shows that the majority of participants 'always' or 'usually' uses mobile phones at home, at the workplace, at university and during leisure time. It is clear that Saudi students are used to using their mobile phones at university.

The table above illustrates how familiar the Saudi students are with using different applications on their mobile phones. It shows that students 'very often' use different mobile applications (such as SMS, email, social networks, etc.). This means that the students are accustomed to mobile phones and their applications. This fact is highly conducive to m-learning.

Table IV reveals that more than 87% of the students are using mobile phones for Internet browsing 'very often' or almost 'always'. This is of note as it shows that students are using mobile phones in the same way that they use a laptop computer and therefore they have the ability to use mobiles instead of a computer to support learning processes.

The information given in Table V indicates that Saudi students are willing to accept m-learning as a formal method for future study. The majority of the students 'strongly agree' or 'agree' that they like to use mobiles phones for learning and are confident about such use. 78% of the participants believe that m-learning can provide new methods to acquire knowledge.

The level of understanding of m-learning from the Saudi students' perspective can be deduced from the results presented in Table VI. The majority of students agreed, or strongly agreed, that mobiles can be used to (1) read lectures from the university website using mobiles; (2) learning without a teacher; (3) listening to audio lectures; (4) reading e-books; (5) just-in-time learning; (6) knowledge/thought sharing; (7) learning and communicating at any time and from anywhere; and (8) an extension of e-learning.

The majority of the students who were surveyed are interested in owning a new mobile device with advanced features if this could improve their learning and performance at university. This is an important point, as m-learning requires the use of advanced mobile devices which can assist the learning process.

Table VII shows that the majority (80%) of students believes that m-learning will increase the opportunities for learning and can save time. It shows that students think that m-learning is easy, as it is possible to learn, meet needs and interests and gain feedback from lecturers more quickly than through traditional methods. A unique feature of m-learning is that learning can occur when the learner is not in a fixed location, as can be seen in Table 7 with rates of use of handheld devices equalling 86%: hence m-learning can be used anytime and anywhere.

TABLE II.
WHERE DO YOU USE YOUR MOBILE PHONE MOST OFTEN?

	Always	Usually	Sometimes	Almost Never	Never
Home	87 (66%)	31 (24%)	12 (9.2 %)	1 (0.8 %)	0
Work	46 (35%)	38 (29%)	42 (32%)	5 (4%)	0
University	49 (37%)	38 (29%)	34 (26 %)	10 (8%)	0
Leisure	56 (43 %)	31 (23%)	36 (28%)	8 (6%)	0

TABLE III.
WHERE DO YOU USE YOUR MOBILE PHONE MOST OFTEN?

	Very Often	Sometimes	Almost Never	Never
Call	77.1 %	22.1 %	0.8 %	0
SMS/MMS	37 %	50 %	13 %	0
Email	63 %	31 %	6 %	0
Camera	50 %	46 %	4 %	0
Entertainment	29 %	39 %	32 %	0
Podcast	38 %	38 %	14 %	0
Social Network	72 %	21 %	7 %	0
Weather	40 %	41 %	19 %	0
Reminder	49 %	35 %	16 %	0
Translate	45 %	39 %	16 %	0
Calendar	57%	36%	7%	0

TABLE IV.
MOBILE PHONE USE AS AN INTERNET BROWSER

	Very often	Always	Sometimes	Almost never	never
I use my mobile phone for Internet browsing.	42 (32%)	70 (54%)	15 (12%)	2 (1%)	2 (1%)

TABLE V.
ACCEPTANCE OF M-LEARNING

Statement	Positive responses
I like to use my own mobile phone for my own learning.	97 (74%)
I am confident when using my mobile phone for learning.	90 (69%)
M-learning will increase the quality of e-learning.	98 (75%)
M-learning provides me with new methods to learn.	102 (78%)

TABLE VI.
M-LEARNING FROM THE SAUDI STUDENTS' PERSPECTIVE

	strongly agree	agree	Not sure	disagree	Strongly disagree
Reading lectures from the university website	17 (13%)	66 (51%)	28 (21%)	17 (12%)	4 (2%)
Learning without a teacher	17 (13%)	62 (47%)	41 (32%)	9 (7%)	2 (1%)
Listening to audio lectures	32 (24%)	76 (58%)	20 (16%)	3 (2%)	0
Reading e-books	33 (26%)	60 (46%)	29 (22%)	8 (6%)	0
Just-in-time learning	19 (15%)	60 (46%)	45 (3%)	7 (5%)	0
Knowledge /thought sharing	36 (28%)	70 (54%)	19 (14%)	6 (4%)	0
Learning and communicating at any time and from anywhere	66 (50%)	53 (41%)	10 (8%)	2 (1%)	0
An extension of e-learning	43 (33%)	67 (51%)	17 (13%)	4 (3%)	0

TABLE VII.
M-LEARNING OPPORTUNITIES

Statement	Positive responses
M-learning will bring new opportunities for learning.	105 (80%)
M-learning can save my time.	103 (79%)
I find m-learning easy, as it is possible to learn what I want.	96 (73%)
M-learning meets my needs and interests.	82 (63%)
M-learning enables me to get feedback from lecturers more quickly than before.	78 (60%)
M-learning is more flexible than traditional learning; it can be anytime and anywhere.	113 (86%)
It is possible to achieve personal educational aims through m-learning.	82 (63%)

TABLE VIII.
APPLYING M-LEARNING ENVIRONMENT INSIDE THE CLASSROOM

	Very often	Often	Occasionally	Seldom	Never
Accessing resources	50 (38%)	46 (35%)	23 (18%)	4 (3%)	8 (6%)
Taking notes	26 (20%)	42 (32%)	38 (29%)	14 (11%)	11 (8%)
Translating	72 (56%)	41 (32%)	11 (9%)	3 (3%)	0
Texting	26 (20%)	35 (27%)	42 (32%)	15 (11%)	13 (10%)
Asking and answering questions	34 (26%)	29 (22%)	41 (31%)	11 (9%)	16 (12%)
Exchanging ideas and information with instructors and students	33 (25%)	37 (28%)	35 (26%)	10 (8%)	16 (13%)

Table VIII shows that the notion of applying the m-learning environment inside the classroom is welcomed by Saudi students in higher education. It shows that a majority of students will 'very often' or 'often' use their mobiles to access resources, take notes, translate, text, ask and answer questions and exchange ideas and information with instructors and other students. As mentioned previously, the sample selected for this study comprises Saudi Arabian higher education students whose first language is Arabic and who are studying at UK universities. It is interesting that the percentage using mobile phones for translation in the classroom is 88%, which demonstrates that Saudi students are already applying m-learning in their learning activities.

VII. CONCLUSION

This paper has presented the findings of a pilot study exploring the possibility of integrating m-learning into Saudi Arabian higher education institutions. It has also presented m-learning and the unique opportunities it provides from the perspective of Saudi students. The analysis of their responses revealed that the students are ready to use mobile phones for their studies. They believe that it is possible to achieve personal educational aims through m-learning and that using mobile phones in the classroom would help to expedite their learning.

The participants pointed out that the use of mobile phones inside the classroom has positive outcomes in terms of increasing study skills and knowledge acquisition. Therefore, higher education policy makers should consider the possibility of creating an m-learning environment at Saudi Arabian academic institutions.

VIII. FUTURE WORK

Because of cultural and religious norms there is gender segregation in the higher education system in Saudi Arabia. In terms of access to educational systems unequal opportunities and challenges are expected to exist between male and female students. M-learning may offer opportunities to tackle this problem, providing equal opportunities and bridging the gaps between Saudi male and female students.

The future work for this research will focus on women's learning environment in Saudi universities. M-learning can empower Saudi female students to overcome social and cultural obstacles and to claim their rights.

REFERENCES

[1] Ally, M. "State of Mobile Learning in Canada and Future

[2] Directions." Athabasca University, 2011.

[3] O'Malley, C., Vavoula, G., Glew, J.P., Taylor, J., Sharples, M., & Lefrere, P. "WP4 – Guidelines for learning/teaching/tutoring in a mobile environment" MOBIlearn deliverable. 2004, Available:

[4] http://www.mobilearn.org/download/results/guidelines.pdf

[5] International Telecommunication Union. Retrieved from:

[6] http://www.itu.int/net/pressoffice/press_releases/2014/23.aspx#.U5W8fvmwKkE

[7] Communication and Information. Retrieved from:

[8] http://www.citc.gov.sa/English/Reportsandstudies/Indicators/Pages/CITCICTIndicators.aspx

[9] Gasser, U. "What Companies Should Know About Digital Natives" 2009. Retrieved from:

[10] http://www.web-strategist.com/blog/2009/05/14/what-companies-shouldknowabout-digital-natives/

[11] Sharples, M., Taylor, J., & Vavoula, G. "A theory of learning for the mobile age." The Sage Handbook of Elearning Research. London: Sage, 2007.

[12] Altameem, T. "Contextual Mobile Learning System for Saudi Arabian Universities," International Journal of Computer Applications, vol. 21, no. 4, pp. 21-26, May 2011.

[13] Al-Fahad, F. "Students' attitudes and perceptions towards the effectiveness of mobile learning in King Saud University, Saudi Arabia," The Turkish Online Journal of Educational Technology – TOJET, vol. 8, no. 2, pp. 111-119, April 2009.

[14] F. Chanchary & S. Islam. "Mobile learning in Saudi Arabia – Prospects and challenges." The International Arab Conference on information Technology, Zarga University, Jordan, 2011.

[15] Pavan, A. "A new perspective on the quest for education: The Saudi Arabian way to knowledge society." Higher Education Studies, vol. 3, pp. 25-34, 2013.

[16] Cohen, L., Manion, L. & Morrison, K. "Research methods in education." 5 ed, Abingdon: Routledge, 2000.

AUTHORS

Sulaiman Almutairy, Trevor Davies, and Yota Dimitriadi are with the Institute of Education, University of Reading, United Kingdom (e-mail: s.m.s.almutairy@pgr.reading.ac.uk, t.c.davies@reading.ac.uk, y.dimitriadi@reading.ac.uk).

30

A Software Tool to Visualize Verbal Protocols to Enhance Strategic and Metacognitive Abilities in Basic Programming

Carlos Argelio Arévalo Mercado, Estela Lizbeth Muñoz Andrade and Juan Manuel Gómez Reynoso
Universidad Autónoma de Aguascalientes, Aguascalientes, México

Abstract—**Learning to program is difficult for many first year undergraduate students. Instructional strategies of traditional programming courses tend to focus on syntactic issues and assigning practice exercises using the presentation-examples-practice formula and by showing the verbal and visual explanation of a teacher during the "step by step" process of writing a computer program. Cognitive literature regarding the mental processes involved in programming suggests that the explicit teaching of certain aspects such as mental models, strategic knowledge and metacognitive abilities, are critical issues of how to write and assemble the pieces of a computer program. Verbal protocols are often used in software engineering as a technique to record the short term cognitive process of a user or expert in evaluation or problem solving scenarios. We argue that verbal protocols can be used as a mechanism to explicitly show the strategic and metacognitive process of an instructor when writing a program. In this paper we present an Information System Prototype developed to store and visualize worked examples derived from transcribed verbal protocols during the process of writing introductory level programs. Empirical data comparing the grades obtained by two groups of novice programming students, using ANOVA, indicates a statistically positive difference in performance in the group using the tool, even though these results still cannot be extrapolated to general population, given the reported limitations of this study.**

Index Terms—**Basic programming, Verbal Protocols, Dual Coding, e-learning.**

I. Introduction

A. The programming context

The difficulties many undergraduate students face when learning programming are still a common topic in cognitive, educational and technological research literature. The problem has been approached from many angles, such as the study of the cognitive behavior of novices and experts [1-3] some creative pedagogical strategies [4-6], the cultural environment of the student [7],[8], and of course the use of software tools [9-12].

In developing countries such as Mexico, programming skills are relevant for undergraduate students, given the increasing trends of first world economies to outsource programming, information technology and software related jobs [13-15].

But the most visible aspect of the problem is the almost universal pattern of high failure rates among first year computer science programming students. Depending on the source, it can be found that this failure rates range from 30% to 60% [16-19]

B. The traditional teaching model

Teaching programming is often based on the pedagogical pattern of: 1) presenting the topic, 2) showing a few examples, and 3) assigning practice exercises; that is, the *presentation-examples-practice formula* [20]. And so, a traditional programming course is mainly based on theoretical lecture sessions and practical work on computer laboratories, where most of the content is focused on the characteristics of the computer language being taught [21]. Refs. [22-24] agree that most introductory programming courses are reasonably good to emphasize syntax comprehension of programs, but that they do not reinforce the strategic kind of knowledge required to write programs.

A common perception of computer programming educators is the assumption that this strategic knowledge will develop itself as a byproduct of curricular design [25], while literature suggests that a more effective approach is that this knowledge has to be explicitly taught [21],[26].

C. Software tools

During the last four decades, researchers and designers have been trying to make programming more appealing to students and to the public. They have developed a wide variety of software applications, to make programming skills easier to acquire. From Logo [27], to Alice [11] a diversity of learning goals have been pursued: to develop problem resolution abilities, to develop logical thinking through games, or to facilitate the transition to general purpose programming languages by way of alternative and easier to use interfaces, among other goals. Ref. [28] did a survey of approximately 80 software tools designed to teach programming or to foster the interest in programming by way of games, animations or puzzles.

Other recent types of software tools designed to be an aid in teaching programming are: a) *program visualization*; by using graphics that enable the student to visualize the behavior of algorithms and data structures [29], b) *learning objects*, small instructional components that can be reused in several contexts [10], c) *concept maps,* that work like big knowledge "scaffolds" to represent the main concepts of programming, and to combine them with other teaching strategies and tools [30],[31], and d) *cognitive tutors*, that use declarative and procedural knowledge in the form of rules, to give guided feedback to the student [32],[33].

All these types of tools have reported positive results, and have had various degrees of success in the goal of teaching programming, but we argue that while some of them have been adopted by programming educators, most of them have focused only on a limited subset of the cognitive aspects (e.g. the transfer of mental models through graphics or interactive feedback) that cognitive literature report as critical.

D. Critical cognitive aspects

The act of programming is essentially a cognitive process of problem resolution that involves writing abstract structures of an algorithmical process. In other words, programming is a way to mentally create a solution to a problem, simultaneously combining a limited and predefined set of syntactic structures and statements, by way of a computer language.

Cognitive literature regarding the acquisition of programming skills is vast and complex. The subject was of special interest in the 1980s [34-36], and in recent years it has still been explored, so that other facets of the problem have been identified and more alternative solutions explored [25],[37],[38]. The most recurring topics found in relevant cognitive literature are: a) comparative studies of mental models of novices and experts, b) the development of programming strategies (also called, plans, schemas or clichés) for common types of problems, and more recently, c) the cognitive process called Metacognition.

1) Mental models

Ref. [3] defines a mental model as an internal representation of a system or complex task, whose construction enables the learner to comprehend and predict the behavior of that system or task.

Ref. [39] also tells that a mental model develops and refines through time, as a result of interaction between the subject and the target system, and that this mental model does not have to be very precise, as long as it is "functional".

In programming, a mental model refers to the image the programmer has about the invisible processing that occurs inside the computer, in the interval between an input and an output [35]. Ref. [40] clarifies that to write a program a person has to have many and very diverse mental models, referring, for example, as to how a loop, a data structure or decision structure behaves. Ref. [41] notes that the existence of a wide range of valid mental models is critical for the novice to acquire the ability to write programs, and if these mental models are not explicitly taught, the student will anyway create its own, of dubious quality and effectiveness.

2) Strategies

It has been found that, even though a student is in fact able to acquire valid mental models, and knows the correct syntax of a programming language, a key cognitive element is still necessary for him to write effective programs. This component is called "strategy" [42] (also known as schema, plan or cliché). Strategies are predefined solutions to stereotyped kinds of problems. The lack of a minimum amount of these strategies restricts the student ability to recognize certain types of problems, and therefore their solution. Ref. [43] indicates that an important aspect of strategies is that *they cannot be deducted from the final form of the program*. This means that a novice can study the final shape of a program, but unless explicitly taught by a teacher, he or she cannot see the process and strategies involved in its writing. The final form of a program can give the student information about the concepts and syntactic structures used, but not about the strategies and decisions applied during the writing process. These strategies are a lot more difficult to teach in the classroom and laboratories, but Ref. [44] notes that "in programming, there is considerable empirical evidence that suggests that strategies are the main basic cognitive component used in design and program comprehension."

Finally, Ref. [45] argues that the process of writing a program does not have to be understood as a "literal transcription" of a previously stored and typified solution, but rather as an iterative, exploratory, and incremental process determined by minor episodes of problem solving and constant re-evaluation of the effectiveness of applied strategies. That is, the effectiveness of a set of strategies is constantly monitored and evaluated by the programmer, in the process of writing the program.

This finding leads to another important aspect of programming (and of problem solving in general) called Metacognition.

3) Metacognition

Ref [46] described Metacognition as "awareness of a person's own cognitive process". While strategies allow a programmer to solve problems, Metacognition allows him to monitor its progress, apply his knowledge to new situations, and identify its own limitations. Ref. [47] indicates that through Metacognition a student can define the nature of a problem or task, select a useful mental representation, use the most pertinent strategy to implement it and put attention to feedback as to how he or she is making progress towards the solution.

In this context, favorable results have been reported through the use of instructional strategies such as "pair programming" [48] (that is, a pair of novice programmers monitor each other's progress, with constant feedback), and with the use of "think-alouds" [49], (instructing the students to verbally reproduce their thought process when writing a program, thus explicitly making such students aware of the decisions, and problem solving strategies they are applying). These are clear examples of Metacognition in the programming context.

Given that empirical evidence in cognitive literature suggests that these three cognitive components (valid Mental Models, Strategies and Metacognition) are critical to acquire the ability to program, we argue that a software tool designed to help students to learn to program, has to include some form of these elements.

II. Development of a Verbal Protocol Visualizer Tool.

A. Verbal protocols

Verbal protocols, as a method of representation and analysis of a person's thought processes, have a solid tradition in the context of cognitive psychology [50-52]

As a technique, verbal protocols were initially developed to study a person's short term memory processes (to what things he or she pays attention to, and in what order, when given a certain task?), but in time, they have been extensively used in other disciplines such as software engineering (e.g. usability studies [53],[54], software task analysis [55]), and even in programming teaching [49].

To develop our tool, we selected the method of verbal protocols, as a way to elicit (and later explicitly show to students) an expert programmer's series of decisions when given a certain basic programming type of problem: that is, a verbal protocol can show a student what elements of a problem the expert is paying attention to, how and in what form the programmer applies the basic programming structures (loops, decisions, data structures), and how does the programmer identifies when he or she made a mistake and has to backtrack and correct it.

To analyze a verbal protocol, a researcher has to rely on some kind of recording device (in the old days, a tape recorder) to be able to transcribe, apply a coding scheme and compare the verbalizations of a given set of subjects. At this point we opted for a different kind of recording method, using *video capture software*, to be able not only to record the verbal data, but also *the visual behavior* of the expert programmer.

For example, in our study, a recording session would consist of asking an expert programmer to write a program to solve a simple programming problem, such as the following, while verbalizing his or her thought process:

Write a program in C Language that, when given a quantity N of integers, gives the sum of all pair numbers, and the average of all uneven numbers.

Then, we used the video capture software to record audio and video activity taking place in the computer. The resulting product was a video file with visual and audio information that was later transcribed and edited to a database (Figure 1)

It has to be noted that, in all cases, video editing was needed to re-record the video segments of the protocols, because correspondence between the verbalizations of the programmer, and the visual information (the actual writing of the code) where very rarely synchronized (Figure 2)

For our experimental test, four edited protocols where produced, representing four types of problems with different levels of difficulty (Figure 3).

Once the tool was in its final form, students could access and visualize the protocols using a web based interface, typing a keyword, an specific phrase or an author's (programmer) name (Figure 4).

B. Dual coding

Dual coding theory (DCT) describes that to process sensorial stimuli from the environment, the human mind has two independent but connected memory subsystems: one for visual and one for verbal information. The visual subsystem handles concrete images and sounds. The verbal subsystem records language and abstract information. According to the theory, both systems function independently but are intimately connected: when a verbal representation is created as a response to a visual image, or when an image is created as a result of seeing or hearing a word, it is said that a *referential connection* has been made, and thus, *dually coded.* [56-58].

Empirical data of DCT studies [56],[57],[59] shows that the brain can retrieve information better when it is dually coded.

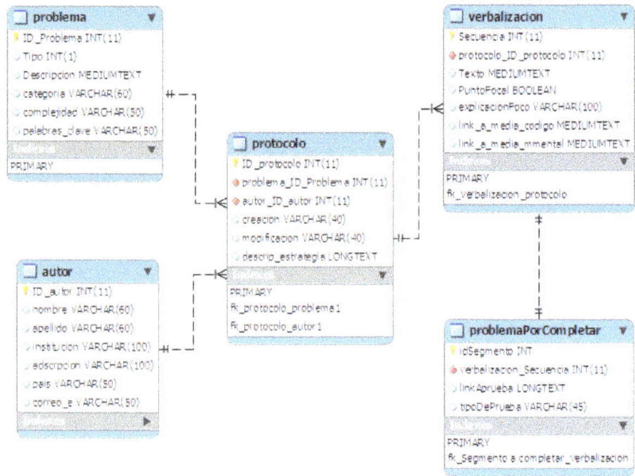

Figure 1. Data Model used to store and retrieve verbal protocols

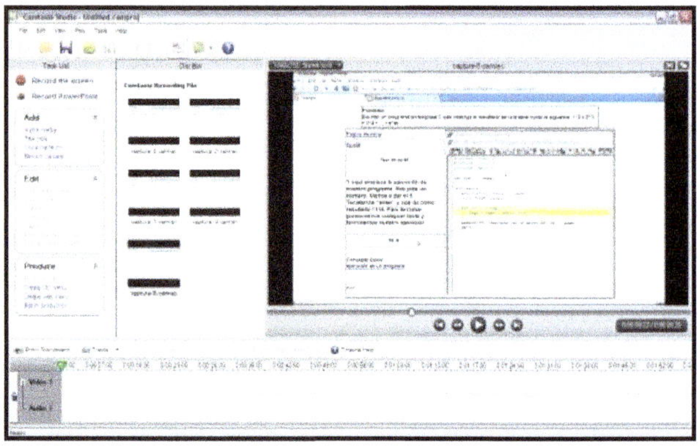

Figure 2. Video editing of verbalization segments of protocols

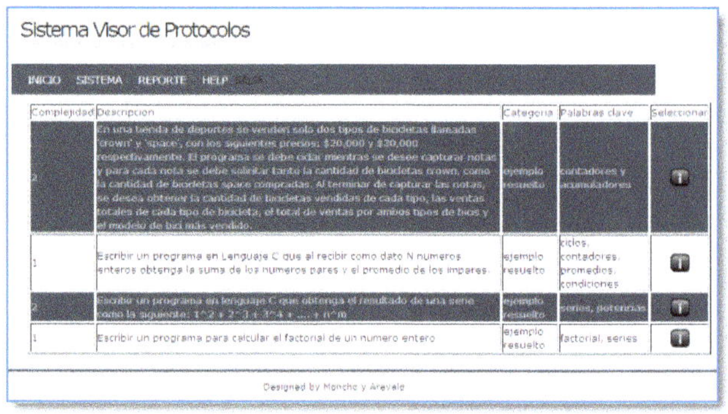

Figure 3. List of protocols available in the tool

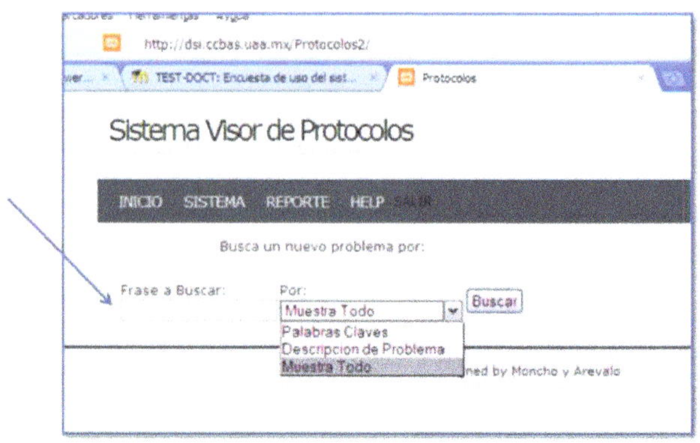

Figure 4. Searching interface

In our study, we tried to apply this principle to the design of the user interface of the tool, by allowing the student to browse the verbal and visual information of the protocols (Figure 5). The protocols were divided in segments that students could study and analyze by reading the verbalizations, and watching the video segment corresponding to the writing of the code.

The interface was subject to several tests to further refine its usability. For example a feature was added to allow a student to "jump" directly to a specific step of the protocol (Figure 6).

C. *Experimental conditions*

To measure the tool's capability to transfer strategic knowledge to novice undergraduate programmers, we designed a standard test, consisting of three basic programming problems, to assess the student's skills. The test was written to evaluate the following specific abilities:

a) Recognize types of problems that involved combined structures of repetition (loops) and selection (if).
b) Effectively write repetition and decision structures.
c) Recognize and effectively apply problems that involve counters.
d) Make calculations involving exponents.

It has to be noticed that the test was designed using previously applied questions and problems, taken from our internal programming academy quiz repository. This repository of tests dates back to the year 2008. The specific sample of the three test questions was randomly selected to design the measurement instrument, but taking into account their similarity to the desired specific abilities to be measured. 15 historical undergraduate student's results of both computer science and electronic engineering were selected.

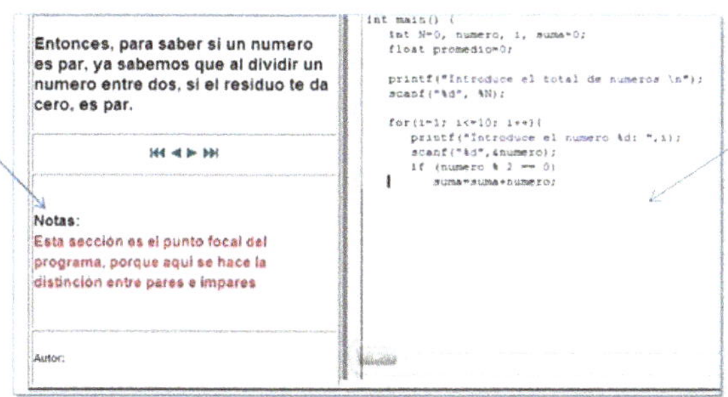

Figure 5. Dual coding user interface.

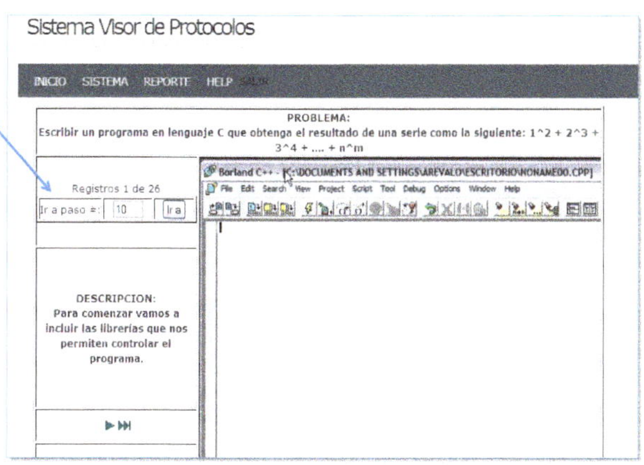

Figure 6. Usability modifications to user interface

These previous results were analyzed to verify if the instrument's behavior was normal and without significant bias. We used a grading scale of 0 to 10.

Descriptive statistics (Table I), normality tests (Table II**Fehler! Verweisquelle konnte nicht gefunden werden.**), the corresponding histogram of the instrument data (Figure 7), and a Q-Q Plot (Figure 8) are shown.

Given the small size of the sample we look at the Shapiro-Wilk test for normality asumption. In this case, the Sig. value is greater than 0.05, wich indicates that the data is normal.

Also, in Figure 8, we can see that the data obtained with the instrument (that is, the grades obtained) also behave normally, except for one observed value.

TABLE I.
MEASUREMENT INSTRUMENT DESCRIPTIVE STATISTICS

		Grade
N	Valid	15
	Mean	5.7340
	Median	5.8300
	Mode	8.83
	Std. Deviation	2.27753
	Variance	5.187
	Skewness	-0.679
	Std. error of Skewness	0.580
	Kurtosis	0.879
	Std. error of kurtosis	1.121
	Range	8.50
	Minimum	0.33
	Maximum	8.83

TABLE II.
NORMALITY TESTS

Instrument	Kolmogorov-Smirnov[a]			Shapiro-Wilk		
	Statistic	*df*	*Sig.*	*Statistic*	*df*	*Sig.*
Grade	.129	15	.200*	.939	15	.370

*. This is a lowerbound of true significance.
[a] Lilliefors Significance Correction.

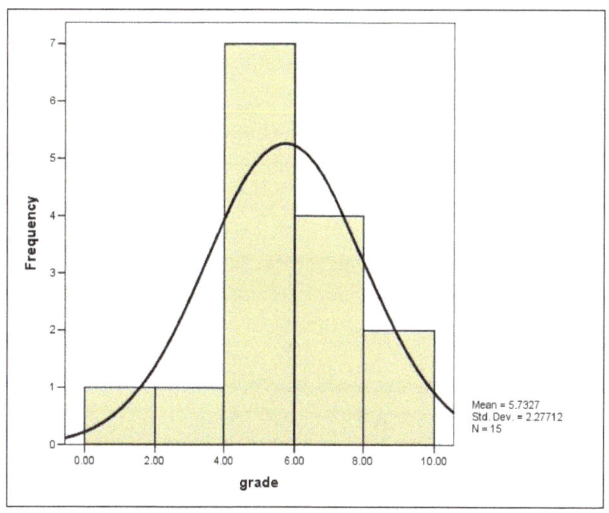

Figure 7. Measurement instrument histogram

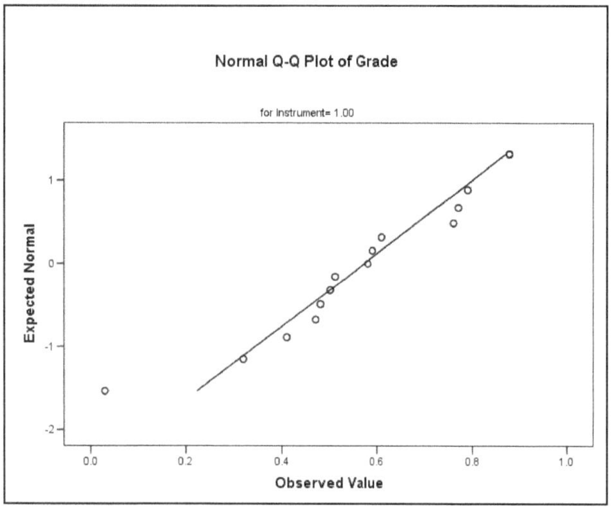

Figure 8. Q-Q Plot of grades obtained with instrument

To test the effectiveness of our tool, a semi-experimental setting was designed, using two groups of programming students: one from 2nd semester Computer Science students and another from 2nd semester Electronic Engineering students, both from Autonomous University of Aguascalientes (UAA), México. Selection of participants was not random. Complete groups were invited, given that the experiment was conducted during a period of normal classes.

The Electronic Engineering group (n=20) was selected as the control group (TRAD). The computer science group (n=18) was to serve as the experimental group (EXP). 55% of the (TRAD) group had previous programming experience from highschool, while 41% of the (EXP) group had previous programming experience.

Two days before the application of the test, four excersices were given to both groups for them to study and practice. This excersices were the same ones loaded in the Protocol Visualizer Tool.

Our research model had as independent variable the "teaching method", so that the control group (TRAD) had a teacher giving a traditional lecture using blackboard and laboratory computers to explain the solving procedure of the given practice excersices, and the experimental group (EXP) used the Protocol Visualizer Tool to study the solving procedure of those same excersices. Our dependent variable was "performance": that is, the grade obtained through the instrument.

Controlled conditions for both groups were:

- Explanatory lecture sessions had a one hour duration for both groups.
- The given time to answer the instrument was limited to an hour for both groups.
- A "motivation" factor was introduced for both groups in form of "extra points".
- Characteristics of the lab computers used from both groups were the same.
- Previous programming experience was similar in both groups (55% for TRAD and 41% for EXP).
- At the time of the experiment, the instructional content given (during normal classes) to both groups was the same. Both groups where studying basic data structures in C Language.

Uncontrolled conditions were:

- Teachers from both groups were different, but they came from the same programming academy, taught the same content, and had the same experience in teaching.
- It was not possible to record the individual answering time of the test participants.
- Selection of participants was not random. In both cases, complete groups where invited to participate in the study.

III. RESULTS

A. Descriptive statistics

The test was graded by professors of our internal programming academy staff.

Table III shows the results of the descriptive statistics of both groups. Mean results for EXP group was 6.33 and 3.79 for TRAD group. The Median value for EXP group was 7.5, and 3.7 for the TRAD group. Mode for the EXP group was 3.30 and 2.30 for the TRAD group. Mean values indicate that the experimental had performed 25% better than the control group; but Modes results suggests that both groups had bad performers. Standard deviation of the EXP group was 2.99, which indicates bigger dispersion of data than the TRAD group (2.35). This result indicates that the TRAD group performed "uniformly bad", and that the EXP group had more "better than average" performers.

Comparative histograms of both groups' results are shown in Figures 8 and 9 . It can be seen that EXP group is negatively skewed to the left, meaning that the majority of its frequencies are grouped towards the upper values of the scale. TRAD group frequencies shown in Figure 9 show the opposite behavior.

B. ANOVA

Given the descriptive statistics results, it can be inferred that the EXP group had better performance than the TRAD group. To see if this performance was statistically significant, we ran an ANOVA test with the results shown in Table IV.

These results indicate that there was a statistically significant different performance ($p<=.006$) between the two groups.

TABLE III.
COMPARATIVE DESCRIPTIVE STATISTICS OF GROUP PERFORMANCE

		Grade (EXP group)	Grade (TRAD group)
N	Valid	18	20
	Mean	6.3389	3.7900
	Median	7.5000	3.7000
	Mode	3.30(a)	2.30
	Std. Deviation	2.99257	2.35683
	Variance	8.955	5.555
	Skewness	-.499	0.689
	Std. error of skewness	0.536	0.512
	Kurtosis	-1.412	0.443
	Std. error of kurtosis	1.038	0.992

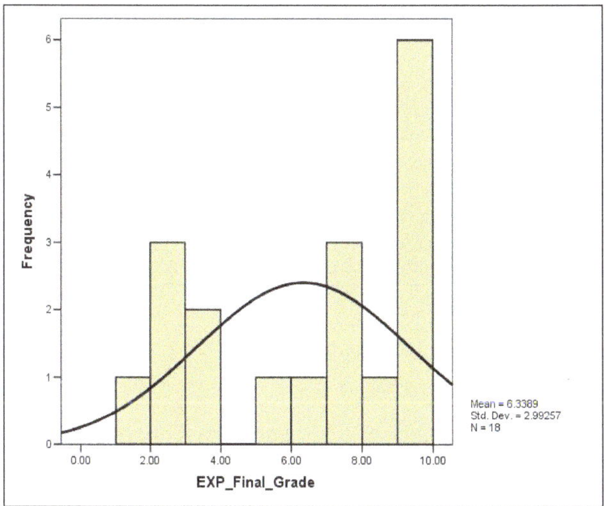

Figure 9. Test results of EXP group

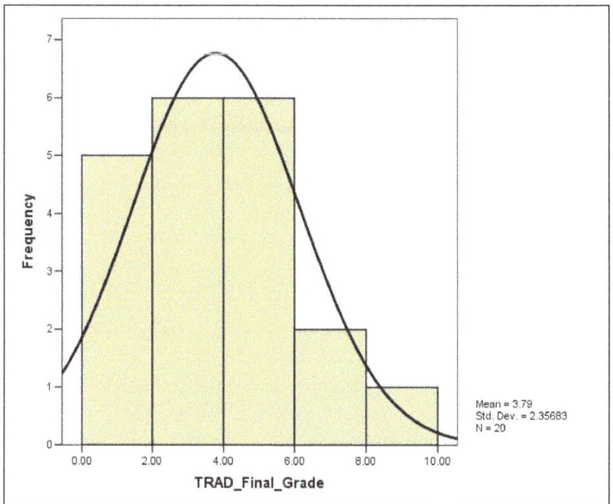

Figure 10. Test results of TRAD group.

TABLE IV.
ANOVA TEST OF "PERFORMANCE" VARIABLE FOR TRAD AND EXP GROUP

	Sum of squares	df	Mean square	F	Sig
Between groups	61.549	1	61.549	8.596	0.006
Within groups	257.781	36	7.161		
Total	319.330	37			

C. Correlations

Programming literature suggests that there is a positive correlation between previous experience and performance in the first undergraduate programming course [37],[60],[61]. We ran a correlation test to see if a correlation could exist between previous experience of these participants and the results obtained in the test. As mentioned earlier 55% of TRAD group participants have had previous experience in programming, and 41% of EXP participants had studied programming in high school. Pearson correlation results are shown in V and Table VI.

TABLE V.
EXP GROUP PEARSON CORRELATION.

		Previous experience	Grade
Previous experience	Pearson correlation	1.000	0.334
	Sig. (2-tailed)		0.175
	N	18	18
Grade	Pearson correlation	0.334	1.000
	Sig. (2-tailed)	0.175	
	N	18	18

TABLE VI.
TRAD GROUP PEARSON CORRELATION.

		Previous experience	Grade
Previous experience	Pearson correlation	1.000	-0.139
	Sig. (2-tailed)		0.558
	N	20	20
Grade	Pearson correlation	-0.139	1.000
	Sig. (2-tailed)	0.558	
	N	20	20

Pearson results show that, for both groups, there is no significant correlation between their previous programming experience, and their performance in the study.

IV. CONCLUTIONS

A. Explicit strategy learning

These positive results are promising in the sense that, under the conditions of the study, a significant improvement (25%) in performance was obtained.

It can be interpreted that when students where using the tool and explicitly studied (by reading and seeing) the problem solving procedure, they learned a small set of strategies that where effective to those kinds of problems.

Also, the verbalization feature of the protocols, allowed the students to understand why the expert programmer was using a specific kind of programming structure.

In the protocols, a metacognitive element was implicitly present during some "backtracking" episodes. For example: in one case, a programmer identified that she had omitted the declaration of a variable, and that was causing a syntax error in the program. In other case, (in the final steps of the protocol) the programmer noted that she needed a counter variable to obtain the average of the uneven numbers. These incidents showed the students that the process of writing a program is not linear, but incremental and constantly monitored.

Our results also seem to support previous studies related to some instructional strategies in the context of cognitive load theory [62]. That is, that showing a "worked example" [63-65] can be an effective instructional strategy, given the reduced "memory load" that the student is submitted to.

B. Benefits

We argue that the tool can be helpful in decreasing high failure rates, if the amount of protocols loaded in the tool is sufficiently big to cover a significant range of problem categories. That is, if students are given a wider range of problem solving strategies, the transit between the initial states of learning [66] and towards an automation of strategies can be made more efficiently.

Also, programming teachers can share their knowledge with a wider range of students, whom, in turn, can constructively compare different kinds of solutions for one same type of problem.

It is important to note that the tool was designed as an aid and complement to programming teachers, and not as a substitute to them.

C. Limitations.

Given the uncontrolled conditions reported in previous sections of this paper, the positive results obtained by the EXP group using the Protocol Visualizer Tool, cannot be generalized to be valid for all the population of first year undergraduate programming students.

An uncontrolled variable was teaching style: that is, even though both teachers were part of the same academy and had similar background and experience, the two groups having different teachers could have had an unmeasured effect on the final results.

Also, duration of the actual study was limited to three days, given the limited availability of the students who voluntarily participated.

D. Future studies

Future studies need to be longitudinal in nature, so that the effect of a longer exposure of the students to the tool can be measured.

A randomly selected sample of participants is desirable, but this kind of scenario is not always possible (or practical) given the nature of every day lectures.

It is also possible to load the tool with protocols that involve problems related to other programming languages or paradigms such as Java, C#.

Also, the graphical user interface can still be improved through usability test, in order to use within other Web-enabled platforms (such as mobile devices).

Lastly, it is planned to extend the functionality of the tool, by adding pedagogical features such as *completion problems*, in selected segments of the protocols, to be in accordance to suggestions given by Refs [67],[68].

REFERENCES

[1] V. Fixx, S. Wiedenbeck, y J. Scholtz, "Mental Representations of Programs by Novices and Experts," Amsterdam, The Netherlands: ACM Press New York, NY, USA, 1993, pp. 74-79.

[2] L. Ma, J. Ferguson, M. Roper, y M. Wood, "Investigating the viability of mental models held by novice programmers," *ACM SIGCSE Bulletin*, vol. 39, 2007, pp. 499-503.

[3] C.E. George, "Experiences with Novices: The Importance of Graphical Representations in Supporting Mental Models," A.F. Blackwell & E. Bilotta, 2000.

[4] V. Dagdilelis, "Introducing Secondary Education Students to Algorithms and Programming," *Education and Information Technologies*, vol. 9, 2004, pp. 159–173.

[5] D. Hagan, "The value of discussion classes for teaching introductory programming," Dublin City Univ., Ireland: ACM Press New York, NY, USA, 1998, pp. 108 - 111.

[6] T. Jenkins, "A participative approach to teaching programming," Dublin City Univ., Ireland: ACM Press New York, NY, USA, 1998, pp. 125 - 129.

[7] C. Bruce, "Ways of experiencing the act of learning to program: A phenomenographic study of introductory programming students at university," *Journal of Information Technology Education*, vol. 3, 2004, pág. 144.

[8] S. Booth, "Learning to program as entering the datalogical culture: a phenomenographic exploration," Fribourg Switzerland, August 2001: 2001.

[9] L. McIver, "GRAIL: A Zeroth Programming Language," Chiba, Japan: 1999, pp. 43-50.

[10] E. Kujansuu, "Using program visualisation learning objects with non-major students with different study background," Tampere, Finland: 2006, pp. 21-26.

[11] M.J. Conway, "Alice:Easy-to-Learn 3D Scripting for Novices," PhD. Thesis, University of Virginia, 1997.

[12] J. McKeown, "The use of a multimedia lesson to increase novice programmers' understanding of programming array concepts," *Journal of Computing Sciences in Colleges*, vol. Volume 19, Abr. 2004, pp. 39 - 50.

[13] M. de Raadt, R. Watson, y M. Toleman, "Language tug-of-war: industry demand and academic choice," Adelaide, Australia: 2003, pp. 137 - 142.

[14] K.S. Koong, L.C. Liu, y X. Liu, "A Study of the Demand for Information TechnologyProfessionals in Selected Internet Job Portals," *Journal of Information Systems Education*, vol. 13, 2002, pp. 21-28.

[15] C. Litecky, B. Prabhakatar, y K. Arnett, "The IT/IS job market: a longitudinal perspective," Claremont, California, USA: 2006, pp. 50-52.

[16] T. Boyle, "Improved success rates for students studying Programming," *Investigations in university teaching and learning*, vol. 1, 2003, pp. 52-54.

[17] M. Guzdial, "Teaching the Nintendo generation how to program," *Communications of the ACM*, vol. 45, Abr. 2002, pp. 17-21.

[18] V. Aleven y K. Koedinger, "An effective metacognitive strategy: learning by doing and explaining with a computer-based Cognitive Tutor," *Cognitive Science*, vol. 26, 2002, pp. 147–179.

[19] D.D.E.I. UAA, *Porcentajes de reprobación en la materia de Programación I, en las carreras de LI-LTI/ISC durante los años 2004 y 2007*, Aguascalientes: Universidad Autónoma de Aguascalientes, 2007.

[20] S. Shaffer, "Ludwig: An online programming tutoring and assesment system," *Inroads - The SIGCSE Bulletin*, vol. 37, 2005, pp. 56-60.

[21] A. Robins, J. Rountree, y N. Rountree, "Learning and Teaching Programming: A Review," *Computer Science Education*, vol. 13, 2003, pp. 137–172.

[22] M. Linn y M. Clancy, "The Case for Case Studies of Programming Problems," *Communications of the ACM*, vol. 35, 1992.

[23] M. Linn, "The Cognitive Consequences of Programming Instruction in Classrooms," *Educational Researcher*, vol. 14, 1985.

[24] T. McGill y S. Volet, "A Conceptual Framework for Analyzing Students' Knowledge of Programming," *Journal of Research on Computing in Education*, vol. 29, 1997.

[25] J. Mead, S. Gray, J. Hamer, R. James, J. Sorva, C. St.Clair, y L. Thomas, "A cognitive approach to identifying measurable milestones for programming skill acquisition," Bologna, Italy: ACM New York, NY, USA, 2006, pp. 182 - 194.

[26] S. Volet y C. Lund, "Metacognitive instruction in introductory computer programming: A better explanatory construct for performance than traditional factors," *Journal of educational computing research*, vol. 10, 1994.

[27] S. Papert, *Mindstorms, children, computers and powerful ideas*, Basic Books, New York., 1980.

[28] C. Kelleher y R. Pausch, "Lowering the Barriers to Programming: a survey of programming environments and languages for novice programmers," *ACM Computing surveys (CSUR)*, vol. 37, 2005, pp. 83 - 137.

[29] S. Pollack, "Selecting a Visualization System," Warwick, UK: 2004.

[30] A. Maries y A. Kumar, "Concept maps in intelligent tutors for programming," *Journal of Computing Sciences in Colleges*, vol. 22, 2007, pág. 54.

[31] A.N. Kumar, "Using Enhanced Concept Map for Student Modeling in Programming Tutors," Melbourne Beach, Florida: 2006, pp. 527-532.

[32] J.R. Anderson, A. Corbet, y K.R. Koedinger, "Cognitive Tutors, Lessons Learned," *The Journal of Learning Sciences*, vol. 4, 1995, pp. 167-207.

[33] V. Aleven, B. McLaren, J. Sewall, y K. Koedinger, "The Cognitive Tutor Authoring Tools (CTAT): Preliminary evaluation of efficiency gains," Maceio, Brazil: 2006.

[34] J.R. Anderson, F. Conrad, y A. Corbett, "Skill Aquisition and the Lisp Tutor," *Cognitive Science*, vol. 13, 1989, pp. 467-505.

[35] P. Bayman, "A diagnosis of beginning programmers' misconceptions of BASIC programming statements," *Communications of the ACM*, vol. 26, Sep. 1983, pp. 677 - 679.

[36] J. Larkin y H. Simon, "Why a diagram is (sometimes) worth ten thousand words," *Cognitive Science*, vol. 11, 1987, pp. 65-99.

[37] V. Ramalingam, "Self-Efficacy and mental models in learning to program," Leeds, United Kingdom: ACM Press New York, NY, USA, 2004, pp. 171 - 175.

[38] S. Wiedenbeck, D. LaBelle, y V. Kain, "Factors affecting course outcomes in introductory programming," Institute of Technology, Carlow, Ireland: 2004.

[39] D. Norman, *Some observations on mental models.*, Erlbaum, Hillsdale, NJ., 1983.

[40] P. Johnson-Laird, *Mental Models*, Cambridge University Press, 1983.

[41] L. Winslow, "Programming Pedagogy - A psychological Overview," *SIGCSE Bulletin*, vol. 28, 1996, pp. 17-22.

[42] R. Rist, "Learning to program: schema creation, application and evaluation," *Computer Science Education and Research*, Netherlands: Taylor & Francis Group, 2004, pp. 175-197.

[43] R. Brooks, "Categories of programming knowledge and their application," *International Journal of Man-Machine Studies*, vol. 18, 1990, pp. 543-554.

[44] R. Rist, "Program Structure and Design," *Cognitive Science*, vol. 19, 1995, pp. 507-562.

[45] S. Davies, "Models and Theories of programming strategy," *International Journal of Man-Machine Studies*, vol. 39, 1993, pp. 237-267.

[46] J. Flavell, "Metacognition and cognitive monitoring: A new area of cognitive developmental inquiry," *American Psychologist*, vol. 34, 1979, pp. 906–911.

[47] A.F. Gourgey, "Metacognition in basic skills instruction," *Instructional Science*, vol. 26, 1998, pp. 81-96.

[48] L. Williams y R.L. Upchurch, "In support of student pair-programming," *ACM SIGCSE Bulletin*, vol. 33, 2001, pp. 327 - 331.

[49] N. Arshad, "Teaching Programming and Problem Solving to CS2 Students using Think-Alouds," *ACM SIGCSE Bulletin*, vol. 41, 2009, pp. 372-376.

[50] K. Ericsson y H. Simon, *Protocol Analysis. Verbal reports as data*, Cambridge Massachusets.: MIT Press, 1993.

[51] A. Newell y H. Simon, *Human Problem Solving*, Englewood Cliffs, NJ: Prentice Hall, 1972.

[52] J. Russo, E. Jonson, y D. Stephens, "The validity of verbal methods," *Memory Cognition*, vol. 17, 1989, pp. 759-769.

[53] S. Knox, W. Bailey, y E. Lynch, "Directed dialogue protocols: verbal data for user interface design," ACM New York, NY, USA, 1989, pp. 283 - 287.

[54] J. Nielsen, T. Clemmensen, y Y. Carsten, "Getting access to what goes on in people's heads?: reflections on the think-aloud technique," Aarhus, Denmark: ACM New York, NY, USA, 2002, pp. 101 - 110.

[55] I. Vessey y S. Conger, "Requirements Specification: Learning Object, Process, and Data Methodologies," *Communications of the ACM*, vol. 37, 1994.

[56] M.A. Kuo y S. Hooper, "The Effects of Visual and Verbal Coding Mnemonics on Learning Chinese Characters in Computer-Based Instruction," *Educational Technology Research and Development*, vol. 52, 2004, pp. 23-34.

[57] R.E. Meyer y V.K. Sims, "For Whom Is a Picture Worth a Thousand Words? Extensions of a Dual-Coding Theory of Multimedia Learning," *Journal of Educational Psychology*, vol. 86, 1994, pp. 389-401.

[58] M. Sadoski y A. Paivio, "A Dual Coding Theoretical Model of Reading," *Theoretical models and processes of reading*, Newark, DE: International Reading Association, 2004.

[59] J. Kounios y P. Holcomb, "Concreteness effects in semantic processing: ERP evidence supporting dual-coding theory.," *Journal of Experimental Psychology: Learning, Memory and Cognition*, vol. 20, 1994, pp. 804-823.

[60] D. Hagan y S. Markham, "¿Does it help to have some programming experience before beginning of a computer degree program?," Helsinki, Finland: 2000, pp. 25 - 28.

[61] A.F. Blackwell, "First Steps in Programming: A Rationale for AttentionInvestment Models," Arlington, VA: 2002, pp. 2-10.

[62] J. Sweller, J. van Merrienboer, y F. Paas, "Cognitive Architecture and Instructional Design," *Educational Psychology Review*, vol. 10, 1998, pp. 251-295.

[63] M. Ward y J. Sweller, "Structuring effective worked examples," *Cognition and Instruction*, vol. 7, 1990, pp. 1-39.

[64] J. Sweller y G. Cooper, "The use of worked examples as a substitute for problemsolving in learning algebra," *Cognition and Instruction*, vol. 2, 1985, pp. 59-89.

[65] M.E. Caspersen y J. Bennedsen, "Instructional design of a programming course: a learning theoretic approach," Atlanta, Georgia, USA: ACM New York, NY, USA, 2007, pp. 111 - 122.

[66] J.R. Anderson, J. Fincham, y S. Douglass, "The role of examples and rules in the acquisition of a cognitive skill," *Journal of Experimental Psychology: Learning, Memory, and Cognition*, vol. 23, 1997, pp. 932-945.

[67] J. van Merrienboer y H. Krammer, "The "completion strategy" in programming instruction: Theoretical and empirical support," *Research on Instruction*, Englewood Cliffs, NJ: Educational Technology Publications, 1990, pp. 45-61.

[68] J. van Merrienboer, "Strategies for programming instruction in high school: Program completion vs. program generation," *Educ. Comput. Res.*, vol. 6, 1990, pp. 265-287.

AUTHORS

Carlos A. Arévalo, Ph.D., is a full time research professor at the Department of Information Systems within the Autonomous University of Aguascalientes, México. He earned his Doctoral degree in Software Engineering in the same University, and specializes in e-learning technologies (e-mail: carevalo@correo.uaa.mx).

Estela L. Muñoz, Ph.D., is a full time research professor at the Department of Electronic Systems within the Autonomous University of Aguascalientes, México. She is an experienced teacher of programming and data structures. (e-mail: elmunoz@correo.uaa.mx).

Juan M. Gómez, Ph.D., is a full time research professor, at the Department of Electronic Systems within the Autonomous University of Aguascalientes, México. He has made contributions in the field of design research (e-mail: jmgr@correo.uaa.mx).

31

Improving Educational Outcomes by Providing Educational Services through Mobile Technology

Hosam F. El-Sofany [1], Turky Al-Turki[1], Hasan El-Howimelm[1], Amjad Al-Sadoon[1], and Samir A. El-Seoud[2]
[1]Arab East Colleges for Graduate Studies, Riyadh, Kingdom of Saudi Arabia
[2]British University in Egypt-BUE, Cairo – Egypt

***Abstract*—The use of Computers, Networks, and Internet has successfully enabled educational institutions to provide their students and instructors with various online educational services. With the recent developments in M-learning and mobile technology, further possibilities are emerging to provide such services through mobile devices such as mobile phones and PDAs. By providing the educational services using wireless and mobile technologies, the educational institutions can potentially bring great convenience to those off-campus students who do not always have time to find Internet enabled computers to get the important educational information from their academic institutions. With the mobile or M-educational services, both the students and the instructors can access the services anytime and anywhere they want. This paper discusses those M-educational services that can be moved to the mobile platform and then presents the system prototype and architecture that integrate these services into the mobile technology platform. The paper will conclude with a description of the formative evaluation of the system prototype.**

***Index Terms*—Mobile Applications and Services, Computer Networks, Mobile Telephony, Educational Applications, E-learning, M-learning.**

I. Introduction

The number of mobile users is rapidly growing, so the demand for mobile services is becoming stronger and more diversified. Especially in the universities, the demands for mobile services that facilitate the educational services provided by the university are represent a strong challenge. Actually the widespread mobile technology is the digital cellular telephony and is emerging the need to integrate mobile telephony services and computer network services. The integration between information services (e.g. databases), notification services (e.g. E-mail, alarm systems) and telecommunications infrastructure (e.g. GSM, Fax, etc.) is one of the strategic issues to satisfy mobility needs [7]. Different kinds of protocols have been developed in recent years to permit the access to Internet Service from a mobile phone. In particular WAP (Wireless Application Protocol) became the "de facto" standard for this kind of application. Actually there are different reasons, for example different costs and diffusion of terminals that are WAP compliant, to prefer hybrid systems in which are embedded both WAP based service and SMS (Short Message Service) based services [8].

Use of WAP or SMS based tests through PDA, PALM, mobile phone or computer in higher education has been promoted due to the fact that they support the learning process of the students, offer exercise media and provide the opportunity to test the expected learning level achieved [4-6]. However, as stated by Homan and Wood [20], the cost of such services for the students should be decreased so that these new learning media can be widely used. Homan and Wood analyzed the achievement level and views of the students on the paper based and mobile based tests in internet based wireless and traditional classrooms. They found that there was no significant difference between students' achievement level and the students had positive views about the wireless test conducted on PDA, however the students stated that they wanted to use the infrastructural opportunities of the university, instead of their own devices [1].

Because most students today have some sort of internet-enabled mobile device (3G cell phone or PDA, smart phone), M-educational services system provides these users with fast, straightforward access to their academic lives. M-educational system has space for greater applications tailored to its strong points. For example, it has the possibility of sending notifications to the user whenever the course's instructor makes certain announcement, therefore notifying students of things such as course cancellations or room changes without risking the students not checking the system on their own.

The literature on M-learning points to a variety of benefits that mobile phones could have on the educational sector. For heuristic purposes, the impacts of mobile phones on educational outcomes that are identified in the M-learning literature can be classified into two broad categories. On the one hand, mobiles supposedly impact educational outcomes by improving access to education while maintaining the quality of education delivered. On the other hand, mobiles purportedly impact educational outcomes by facilitating alternative learning processes and instructional methods collectively known as new learning [2].

Our research study aims to introduce the system prototype and architecture that provide the educational services to the students through their mobile phone devices. Using short message service (SMS) services from the GSM network (Global System for Mobile communications). The student will be able to inquire all the information he need such as: the registration and admission date, open courses for the current semester, grades, registered course, short course preview, sessions period and class room for each course, course instructor, course prerequisites, and exams date, time and location. We choose SMS services because it is the most familiar service in the GSM network and it correspondent with all mobile phone brands. This system will increase the portability and accessibility of the registration system. The proposed architecture will used to integrate mobile telephony both with deferred applications, using an SMS Interface. For the immediate applications a WAP Proxy system can be used but also the SMS-Interface can give access to immediate applications to user with a no-WAP terminal.

The paper is organized as follows: in section two we present condensed survey and previous works related to the M-learning and M-educational services. In section three we present some advantages and disadvantages of M-learning. In section four we introduce an overview of mobile technology that consists of mobile devices and wireless communication technologies. In section five we introduce the architecture of the proposed M-educational system. The paper finally concluded in section six with conclusion and future works.

II. Previous Works

Samir Abou El-Seoud et al. presented a mobile learning platform (MLP) connected to a Moodle course management system build, using the Java2 Micro Edition (J2ME), the authors have developed the project in their Mobile Multimedia and Wireless Communication Lab (MMWCL). The MLP used wireless communications (Bluetooth, WI-FI, WIMAX) to connect with the Moodle server. The project applications provided the user with a bundle of services. These services include grades, course schedules, attendance records, course announcements, and university activities, as well as quality assessments [3].

Valk, J., Rashid et al. reviewed the evidence of the role of mobile phone-facilitated M-learning in contributing to improved educational outcomes in the developing countries of Asia by exploring the results of six M-learning pilot projects that took place in the Philippines, Mongolia, Thailand, India, and Bangladesh. In particular, this article examined the extent to which the use of mobile phones helped to improve educational outcomes in two specific ways: 1) in improving access to education, and 2) in promoting new learning. Analysis of the projects indicated that while there is important evidence of mobile phones facilitating increased access, much less evidence exists as to how mobiles promote new learning [2].

Marcelo and Michele in [9], illustrated a number of educational mobile services used by students in a university environment and discussed the patterns of use of these services on campus and other locations of their choice. Impact on learning itself was not measured, nor would it have been possible to measure meaningfully when the devices were used for such diverse purposes. Phone-optimized content was well used, and there was a clear request from students that more resources be made available in this format, including administrative information from the University. There is no conclusive evidence of the need for a specifically designed suite of tools in addition to those already included in the device, although students would like to send multimedia material of larger size than that allowed on an MMS (Multimedia Messaging Service).

They also illustrated that, greater institutional support is needed in order for the smart phones to be used more fully. Regular updates of timetables and content, as well as adequate training and hardware provision are needed. As more students bring the technology with them to the university, change will most likely be driven by their demands as consumers. It has been shown that the smart phones used in this study did not generate a radical change in students' styles or patterns of learning. As smart phones become a more intrinsic part of everyday life, it is important that these patterns and demands are known and understood.

III. Advantages and Disadvantages of M-Learning

M-learning has received a lot of attention in recent years as a growing segment of the educational and instructional technology field. Mobile learning through the use of wireless mobile technology allows anyone to access information and learning materials from anywhere and at any time. As a result, learners have control of when they want to learn and from which location they want to learn. Also, all humans have the right to access learning materials and information to improve their quality of life regardless of where they live, their status, and their culture. M-learning, through the use of mobile technology, will allow citizens of the world to access learning materials and information from anywhere and at any time. Also, learners do not have to learn what is prescribed to them. They can use the wireless mobile technology for formal and informal learning where they can access additional and personalized learning materials from the Internet or from the host organization. Graduate students who work can use the mobile technology to access course materials and educational information when they need it. Mobile learning having various benefits includes:

M-learning will not replace any form of learning (e.g. classroom learning or E-learning). M-learning is just providing another way of learning using new mobile technology. The strength of M-learning lies in a communication approaches rather than a content approaches (e.g. content creation). Each learning form (or service) has advantages and disadvantages, so the term of blended learning was introduced. Blended Learning refers to mix different learning environments in order to treat disadvantages of one form with advantages of another form. Therefore, it is better to blended M-learning with other forms of learning. Because of M-learning cannot be used as main learning form, but it can be an effective form that can be used in conjunction with other learning forms to improve and facilitate the learning process.

Many advantages and disadvantages related to use mobile technologies for learning have demonstrated in several studies. The key benefits of M-learning include [10-13]:

- Interaction: Learners can interact with others and/or instructors without restriction of time and place.
- Flexibility: M-learning can be just-in-time learning where learners can actually access learning and educational services at the place and time that they need it.
- Learner can take notes directly into the device in outdoor lessons, either by typed, handwritten or voice.
- Portability: PDAs or Smart phones holding notes and eBooks are lighter, less bulky and easier to carry than bags full of files, paper and textbooks, or even laptops.

- Enhances knowledge: contents and knowledge are adapting to meet individual needs and delivered on-demand.
- Mobile learning helps learners to improve their literacy and numeracy skills and to recognize their existing abilities.
- Mobile learning can be used to encourage both independent and collaborative learning experiences
- Mobile learning helps learners to identify areas where they need assistance and support.
- Mobile learning helps to combat resistance to the use of ICT and can help bridge the gap between mobile phone literacy and ICT literacy.
- Mobile learning helps to remove some of the formality from the learning experience and engages reluctant learners.
- Mobile learning helps learners to remain more focused for longer periods.
- Mobile learning helps to raise self-esteem.
- Mobile learning helps to raise self-confidence.

On the other hand, the defects (disadvantages) of M-learning include:

- Limit of devices capacities such as (small screens, memory size and computing power).
- Lack of common operating system and application programs like browsers.
- Wireless networks may down with a larger number of students when using it.

IV. Mobile Technology Overview

Mobile technology consists of mobile devices and wireless communication technologies. We firstly introduce an overview of general wireless communication technologies that can be used in M-learning systems.

A. *Wireless Communication Technologies*

1. *Global System for Mobile Communications (GSM)*

GSM is a worldwide mobile telephony system. GSM offers the best voice quality of any current digital wireless standard. GSM provides voice mail, high-speed data, fax, paging and short message services capabilities, as well as secure communication [14, 16].

2. *General Packet Radio Service (GPRS)*

GPRS is used for various data applications on phones, including Wireless Application Protocol (WAP), Multimedia Messaging Service (MMS), and software that connect to the Internet. We can say that, any network connection that is not voice or text messaging uses a data connection like GPRS. GPRS provides about four times greater speed than GSM systems [15].

3. *Third Generation (3G)*

3G is mobile phone standards and technologies targeting to multimedia communication through cell phone or smart phones. It supports increased bandwidth and high data transfer rates compared to 2 or 2.5G to allow internet browsing and exchange of audio and video files. In 3G, mobile phones became mobile devices combining a camera, video camera, stereo, MP3 player and radio into one device [17,18].

TABLE I.
Shows a comparison between M-learning and E-learning [13].

M-learning	E-learning
It can be used anywhere at any time.	Partially it can be used anywhere at any time.
Most mobile devices have lower prices than Desktop (PCs).	Desktop (PCs) are more expensive than mobile devices.
Mobility: mobile devices are smaller in size and lighter in weight than Desktops.	Desktops are not portable. They are not easily carried around due to their heavy weight.
Number of mobile users greater than number of PCs users	Number of PCs users less than number of mobile users
More easily lost or stolen.	Less easily lost or stolen
Poor learning contents	Rich learning contents (e.g. 3d animations)
Rich communication technologies (e.g. SMS, MMS, Bluetooth, etc)	Poor communication technologies

TABLE II.
Shows a comparison between existing wireless technologies [13].

	Transfer Rate (Mb/s)	Range (~meters)	Frequency Band (GHz)	Number of users
Bluetooth	1-2	10-100	2.4	Dozens
WiFi	10-54	100	2.4	Dozens
GPRS	0.17	30000	1.8	Thousands
3G	256 kbps	>1609344	Varies	Thousands
WiMax	10-54	49889	2-11	Thousands

4. *Wireless Fidelity (Wi-Fi)*

Wi-Fi is a wireless local area network (WLAN). Wi-Fi is technology based on the Institute of Electrical and Electronics Engineers' (IEEE) 802.11family of standards. Wi-Fi is a wireless network uses radio waves that communicate in two ways. A centralized network with access point, called infrastructure network and decentralized network without access point, called ad-hoc network. Wi-Fi is used for a variety of applications. One of the most common application is home and business WLAN where it can cover around 100 m with data transfer rate between 10 – 54 Mbps [19, 20].

5. *Worldwide Interoperability for Microwave Access (WiMax)*

WiMax provides high speed internet connection for a large number of users over a large area. WiMax can cover range around 31 miles and its data transfer rate between 10 – 54 Mbps. If there are too many users connected to WiMax at the same time, the range will be limited and speed lowered [21, 22].

6. *Short Message Service (SMS)*

SMS is the simplest of all the technologies available in the mobile environment. SMS allows single short messages of up to 160 characters to be passed between mobile phones, fax machines or email addresses. Messages are sent along a mobile phone network, and must pass through an SMS centre which handles and manages the messages. SMS is used in an M-learning scenario for dealing with variety of activities such as asking questions, providing answers, information delivery and providing feedback [23].

7. Multimedia Message Service (MMS)

MMS is a more advance technology than SMS which enables the sending and receiving multimedia messages that include multimedia objects (images, audio, video, rich text) between mobile phones. MMS content is longer than SMS and transport on WAP, GPRS or 3G. MMS can also be sent from a mobile phone to an email address [24].

B. Mobile Devices

The M-learning is impossible without the use of the mobile devices. They vary significantly in their abilities, sizes and prices. The common ability which united them is their mobility and possibility to make wireless connections. Mobile devices can be defined as any electronic devices that are small enough to fit in a shirt or jacket pocket. It would not include laptops, as while they are portable, they are not mobile. Mobile devices should fit in your pocket [13].

1) Cell Phone

A cell phone is an electronic telecommunication mobile device. Cell phones commonly named as mobile phones or cellular phones. A cell phone supports voice communication and SMS. The cellular phones from the higher class can access internet via WAP or GPRS technologies. They also can be used to send and receive MMS. A cell phone prices are decreasing continuously [25].

2) PDA

A Personal Digital Assistant (PDA) is pocket-sized computers that fits in hand and has a touch screen or a small keyboard. PDAs support voice communication, fax capability and email service via a wireless network connection. PDAs support software applications such as internet explorer (IE) and calendar. PDAs have evolved and incorporate cell phone and 3G functions. New PDAs can connect to internet with faster speed via 3G and WiFi or WiMax. Most PDAs have Bluetooth and the main operating systems used are Palm OS and Microsoft Pocket PC OS [26, 27].

3) Smart Phone

A smart phone is a mobile device which combines cell phone and PDA functions. They have smaller sizes than PDA and bigger than cellular phones. Smart phones include backlit color screen, Wi-Fi, Bluetooth, large RAM, large ROM and internet browsers. Smart phone use advanced operating systems (e.g. Symbian, Windows Mobile, etc) which can support advanced applications. It is also possible to insert memory card. Smart phones have potential to be successfully used in the M-learning field [28, 29].

4) Tablet PC

Tablet PC are one of the newest mobile devices. They also have full range of capabilities as personal computers. Some of them haven't keyboard but have software to recognize handwritten text and voice-recognition features that will continue to evolve and improve. Because of their large screens and high resolutions, Tablets PCs are good candidates for delivering online courses. They also offer rich data capture and image manipulation features, which can be useful for learning applications [30].

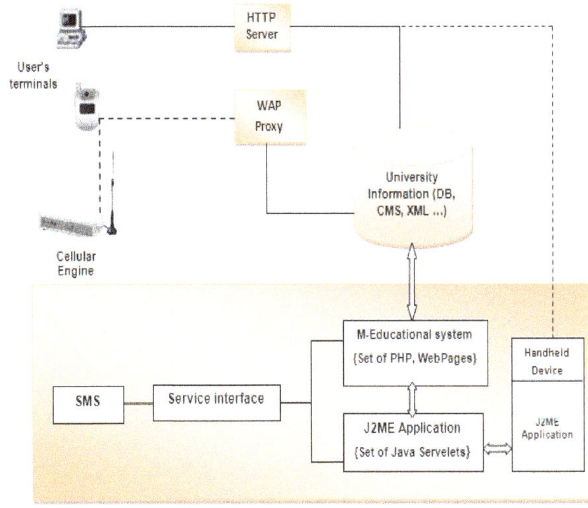

Figure 1. The general architecture of the M-education system

V. SYSTEM ARCHITECTURE

The system architecture is composed mainly of three components: the J2Me application, the interface between the J2ME application and the project server, and finally an instance of M-educational system server for back-end support and integration.

Figure 1, states clearly the system overview of the proposed M-educational services system. The mobile "handheld" device is composed of group of J2ME application that is designated to connect to J2ME application server to attain the provided services. The second component in our system is shortly an interface to match between the mobile applications and the M-educational server. This component is in charge of translating the mobile application requests into suitable format for M-educational system.

Various optimization techniques are applied on the mobile application in order to attain a good amount of mobile limited resource utilization. In addition, a novel technique is conducting to achieve high percentage of load balance on the M-educational system side. The load balance will guarantee a high degree of robustness and stable behavior for the whole platform give the great amount of users (students) who are using the system [31, 32].

The provided services include:

A. Security and confidentiality of personal information

- The possibility of changing the Username and Password.
- Displaying the user information and ability to update them through the system.

B. Registration

- Drop registered courses and/or add new courses.
- Displaying the weekly schedule (Time table).
- Displaying the whole schedule and the open courses/sections.
- Calculating and displaying the registration fee for academic term or for registered courses.
- Displaying the registration status for the student (continuous, withdraw ...).
- Displaying the historical record of student registration, and what has been registered and /or learned.

C. *Query for student information and displaying the following*

- The grade of midterm and final exams (if any).
- The details of the grades (i.e., homework, project, quizzes, …).
- The curriculum of the program, study plan, number of hours and courses description.
- The ability to obtain the academic transcript of the student from the registration office.

D. *Financial services for student*

- Displaying the summary of financial account.
- Displaying the deposit records of the student.

A pilot study is conducting to measure the efficiency of the proposed system. Expected results are to support our basic provided service usability and importance. In addition, a comparative study is taking place to cross examine the platform end-to-end efficiency.

VI. Conclusion and Future Works

By providing the educational services using wireless and mobile technologies, the educational institutions can potentially bring great convenience to those off-campus students who do not always have time to find Internet enabled computers to get the important educational information from their academic institutions. With the mobile or M-educational services, both the students and the instructors can access the services anytime and anywhere they want. The paper introduced the architecture of M-educational system that provides the educational services to student through the mobile devices. The study categorized the proposed educational services into four categories that include: 1) Security and confidentiality of personal information, 2) Registration, 3) Query for student information and displaying the following, and 4) Financial services for student.

To continue our work in the future, we will try to enhance the educational aspects of the mobile services by developing and implementing various solutions to specific problems we have identified based on our observations and the data we have collected from the students. Our future efforts will continue to refine both the technology and activities for providing learners with more meaningful experiences with regard to the use of smart phones in university settings.

Acknowledgment

The authors acknowledge the full support received from Arab East Colleges for Graduate Studies, especially they very thankful to Professor Dr. Abd Allah Alfaisal, Dean of the Colleges, for his encouragement and support.

References

[1] Hosam F. El-Sofany, and Samir A.El-Seoud, "Towards Development of Web-based Assessment System Based on Semantic Web Technology". International Journal of Interactive Mobile Technologies (iJIM), Vol. 5, Issue 1, Pages 22-30, , January 2011.

[2] Valk, J., Rashid, A. T., & Elder, L. Using Mobile Phones to Improve Educational Outcomes: An Analysis of Evidence from Asia. International Review of Research in Open and Distance Learning, 11(1), 117-140, March 2010.

[3] Samir Abou El-Seoud, Ashraf M.A. Ahmad and Hosam Farouk El-Sofany, "Mobile Learning Platform Connected to Moodle using J2ME", International Journal of Interactive Mobile Technologies (iJIM), Volume 3, Issue 2, P: 46-54, , April 2009.

[4] Dawabi P, Wessner M, Neuhold E. Using mobile devices for the classroom of the future. Proceedings of Mlearn 2003 Conference on Learning with Mobile Devices, London pp. 14-15, 2003.

[5] Scornavacca E, Marshall S. TXT-2-LRN: Improving students' learning experience in the classroom through interactive SMS. Proceedings of the 40th Hawaii International Conference on System Sciences, Hawaii, 2007.

[6] Wentling L, Park J, Peiper C. Learning gains associated with annotation and communication software designed for large undergraduate classes. J. Comp. Assist. Learn. 23(1), 36-46, 2007.

[7] Messerschmitt, D.G. "The convergence of telecommunications and computing: What are implications today?'', Proceeding of the IEEE, vol. 84, no. 8, pp. 1167-1186, Aug. 1996.

[8] Fasbender; A., Reichert; F., Geulen; J., Hjelm; J. And Wierlemann, T. "Any network, any terminal, anywhere'', IEEE Personal Communications, pp. 22-30, Apr. 1999.

[9] Marcelo Milrad, and Michele H. Jackson, "Designing and Implementing Educational Mobile Services in University Classrooms Using Smart Phones and Cellular Networks ". Int. J. Engng Ed. Vol. 24, No. 1, pp. 84-91, 2008.

[10] Balasundaram S. R. and Ramadoss B., SMS for Question-Answering in the m-Learning Scenario. Journal of Computer Science, 3(2): p. 119-121, 2007.

[11] Boyinbode O. K. and Akinyede R. O., MOBILE LEARNING: AN APPLICATION OF MOBILE AND WIRELESS TECHNOLOGIES IN NIGERIAN LEARNING SYSTEM. IJCSNS, 8(11): p. 386, 2008.

[12] Orr G., A Review of Literature in Mobile Learning: Affordances and Constraints, in The 6th IEEE International Conference on Wireless, Mobile, and Ubiquitous Technologies in Education. IEEE. p. 107-111, 2010.

[13] Amr A. M. Abozeid , "Toward the Development of an Adaptive Mobile Learning Environment", MS Thesis, Department of Mathematics, Faculty of Science, Ain Shams University, Cairo, Egypt, 2011.

[14] Egea-Lopez E., Martinez-Sala A., Vales-Alonso J., Garcia-Haro J. and Malgosa-Sanahuja J., Wireless communications deployment in industry: a review of issues, options and technologies. sciencedirect, 2005. 56(1): p. 29-53.

[15] Egea-Lopez E., Martinez-Sala A., Vales-Alonso J., Garcia-Haro J. and Malgosa-Sanahuja J., Wireless communications deployment in industry: a review of issues, options and technologies. Computers in industry, 2005. 56(1): p. 29-53.

[16] Georgiev T., Georgieva E. and Smrikarov A. M-learning-a New Stage of -Learning. in International Conference on Computer Systems and Technologies. 2006.

[17] J.L Gimenez Lópeza T. Magal Royoa, Jesus García Labordab, F. Garde Calvoa, Methods of adapting digital content for the learning process via mobile devices. Procedia Social and Behavioral Sciences,ScienceDirect, 2009. 1: p. 2673–2677.

[18] Keogh J., The Complete Reference J2ME. Published by McGrawHill OSBORNE Edition, 2003.

[19] 4G Americas. [cited 9-12-2010; Available from: http://www.4gamericas.org/index.cfm?fuseaction=page&sectionid=260

[20] Zahariadis T., Evolution of the Wireless PAN and LAN standards. Computer Standards & Interfaces, sciencedirect, 2004. 26(3): p. 175-185.

[21] Jacob S. M. and Issac B., Mobile technologies and its impact-an analysis in higher education context. International Journal of Interactive Mobile Technologies (iJIM), 2008. 2(1).

[22] Malek J., Laroussi M., Derycke A. and Ben Ghezala H., A Context-Aware Approach for Modeling Bijective Adaptations between Context and Activity in a Mobile and Collaborative learn-

ing. International Journal of Interactive Mobile Technologies (iJIM), 2007. 2(1).

[23] Balasundaram S. R. and Ramadoss B., SMS for Question-Answering in the m-Learning Scenario. Journal of Computer Science, 2007. 3(2): p. 119-121.

[24] Lee W. P.,Osman M. A.,Sabudin M.,Sulaiman S.,Zadeh L. A.,Kacprzyk J.,Mastorakis N.,Kuri-Morales A.,Borne P., and Kazovsky L. An Experience on Introducing a Mobile Community Platform: Flying 2 U. in WSEAS International Conference. Proceedings. Mathematics and Computers in Science and Engineering. 2010: World Scientific and Engineering Academy and Society.

[25] Economides A. A. and Nikolaou N., Evaluation of handheld devices for mobile learning. International Journal of Engineering Education, 2008. 24(1).

[26] Houser C., Thornton P. and Kluge D., Mobile learning: cell phones and PDAs for education, in Proceedings of the International Conference on Computers in Education (ICCE'02). 2003, IEEE. p. 1149-1150.

[27] Lindquist D., Denning T., Kelly M., Malani R., Griswold W. G. and Simon B., Exploring the potential of mobile phones for active learning in the classroom. ACM SIGCSE Bulletin, 2007. 39(1): p. 384-388.

[28] Holzinger A., Nischelwitzer A. and Meisenberger M., Mobile phones as a challenge for M-Learning: Examples for mobile interactive learning objects (MILOs), in Third IEEE International Conference on Pervasive Computing and Communications Workshops (PERCOMW'05). 2005. p. 307-311.

[29] Zheng P. and Ni L. M., The rise of the smart phone. IEEE Distributed Systems Online, 2006. 7(3): p. 1-14.

[30] Mahmoud S. S., A proposed model for distributing e-courses content through mobile technology architectures. International Journal of Human and Social Sciences, 2008. 27: p. 274-289.

[31] Aloysius K. Mok and Weijiang Yu, "TINMAN: A Resource Bound Security Checking System for Mobile Code," In Proceedings of 7th European Symposium on Research in Computer Security (ESORICS 2002), Zurich, Switzerland, pp 178-193, Springer LNCS Volume 2502, October, 2002.

[32] R. Kessl and P. Tvrdík, "Probabilistic Load Balancing Method for Parallel Mining of all Frequent Itemsets", Proceeding (513) Parallel and Distributed Computing and Systems - 2006.

AUTHORS

Hosam Farouk El-Sofany, *Assistant Professor of Computer Science*, Arab East Colleges for Graduate Studies, Department of Computer Science, Riyadh, Kingdom of Saudi Arabia. (hosam_elsofany@hotmail.com)

Samir A. El-Seoud, *Professor of Computer Science,* British University in Egypt-BUE, Faculty of Informatics and Computer Science, Cairo – Egypt. (Samir.elseoud@bue.edu.eg)

Turky, Al-Turki, *Student in MSc program*, Arab East Colleges for Graduate Studies, Department of Computer Science, Riyadh, Kingdom of Saudi Arabia.

Hasan, El-Howimel, *Student in MSc program*, Arab East Colleges for Graduate Studies, Department of Computer Science, Riyadh, Kingdom of Saudi Arabia.

Amjad, Al-Sadoon, *Student in MSc program*, Arab East Colleges for Graduate Studies, Department of Computer Science, Riyadh, Kingdom of Saudi Arabia.

A Proposed Pedagogical Mobile Application for Learning Sign Language

M. Samir Abou El-Seoud [1], Ann Nosseir[1], Islam Taj-Eddin[1], Hosam El-Sofany[2], Nadine Abu Rumman[3]
[1] British University in Egypt (BUE), Cairo, Egypt
[2] Arab East Colleges for Graduate Studies, Riyadh, Kingdom of Saudi Arabia
[3] Princes Sumaya University for Technology, Amman, Jordan

Abstract—**A handheld device system, such as cellular phone or a PDA, can be used in acquiring Sign Language (SL). The developed system uses graphic applications. The user uses the graphical system to view and to acquire knowledge about sign grammar and syntax based on the local vernacular particular to the country. This paper explores and exploits the possibility of the development of a mobile system to help the deaf and other people to communicate and learn using handheld devices. The pedagogical assessment of the prototype application that uses a recognition-based interface e.g., images and videos, gave evidence that the mobile application is memorable and learnable. Additionally, considering primary and recency effects in the interface design will improve memorability and learnability.**

Index Terms—**component; Sign Language; Wireless J2ME; Java Wireless Communication Technologies; JSF; JSP; JMS; SVG; J2SE; Client; Server; Handheld System; Mobile System; TCP/IP; Video Streaming; Adobe After Effects; HTTP Connectivity.**

I. Introduction

Cell phones became extremely popular devices considering their vast utility World Wide. However, making cell phones accessible to the deaf is still a challenge. Main available products on the market for this community offer no more than the possibility to boost/amplify volume. Many cellular provides individual cell phone models which are hearing aid compatible and possess speakerphone capabilities. However, if the user is completely deaf, these phones still tend to be somewhat complex or impossible to use [26].

Mobile technology consists of mobile devices and wireless communication technologies. An overview of general wireless communication technologies that can be used in M-learning systems will be introduced next, followed by Table I, which is a comparison between existing wireless technologies.

A. Wireless Communication Technologies

1) Global System for Mobile Communications (GSM)

GSM is a worldwide mobile telephony system. GSM offers the best voice quality of any current digital wireless standard. GSM provides voice mail, high-speed data, fax, paging and

short message services capabilities, as well as secure communication [10][39].

2) General Packet Radio Service (GPRS)

GPRS is used for various data applications on phones, including Wireless Application Protocol (WAP), Multimedia Messaging Service (MMS), and software that connect to the Internet. It could be said that, any network connection that is not voice or text messaging uses a data connection like GPRS. GPRS provides about four times greater speed than GSM systems [10].

3) Third Generation (3G)

3G is mobile phone standards and technologies targeting to multimedia communication through cell phone or smart phones. It supports increased bandwidth and high data transfer rates compared to 2 or 2.5G to allow internet browsing and exchange of audio and video files. In 3G, mobile phones became mobile devices combining a camera, video camera, stereo, MP3 player and radio into one device [18][17].

4) Wireless Fidelity (Wi-Fi)

Wi-Fi is a wireless local area network (WLAN). Wi-Fi is technology based on the Institute of Electrical and Electronics Engineers' (IEEE) 802.11family of standards. Wi-Fi is a wireless network uses radio waves that communicate in two ways. A centralized network with access point, called infrastructure network and decentralized network without access point, called ad-hoc network. Wi-Fi is used for a variety of applications. One of the most common application is home and business WLAN where it can cover around 100 m with data transfer rate between 10 – 54 Mbps [43][42].

5) Worldwide Interoperability for Microwave Access (WiMax)

WiMax provides high speed internet connection for a large number of users over a large area. WiMax can cover range around 31 miles and its data transfer rate between 10 – 54 Mbps. If there are too many users connected to WiMax at the same time, the range will be limited and speed lowered [36] [19].

6) Short Message Service (SMS)

SMS is the simplest of all the technologies available in the mobile environment. SMS allows single short messages of up to 160 characters to be passed between mobile phones, fax machines or email addresses. Messages are sent along a mobile phone network, and must pass through an SMS centre which handles and manages the messages. SMS is used in an M-learning scenario for dealing with variety of activities such as asking questions, providing answers, information delivery and providing feedback [37].

7) Multimedia Message Service (MMS)

MMS is a more advance technology than SMS which enables the sending and receiving multimedia messages that include multimedia objects (images, audio, video, rich text) between mobile phones. MMS content is longer than SMS and transport on WAP, GPRS or 3G. MMS can also be sent from a mobile phone to an email address [44].

B. Mobile Devices

The M-learning is impossible without the use of the mobile devices. They vary significantly in their abilities, sizes and prices. The common ability which united them is their mobility and possibility to make wireless connections. Mobile devices can be defined as any electronic devices that are small enough to fit in a shirt or jacket pocket. It would not include laptops, as while they are portable, they are not mobile. Mobile devices should fit in your pocket.

1) Cell Phone

A cell phone is an electronic telecommunication mobile device. Cell phones commonly named as mobile phones or cellular phones. A cell phone supports voice communication and SMS. The cellular phones from the higher class can access internet via WAP or GPRS technologies. They also can be used to send and receive MMS. A cell phone prices are decreasing continuously [1].

2) PDA

A Personal Digital Assistant (PDA) is pocket-sized computers that fits in hand and has a touch screen or a small keyboard. PDAs support voice communication, fax capability and email service via a wireless network connection. PDAs support software applications such as internet explorer (IE) and calendar. PDAs have evolved and incorporate cell phone and 3G functions. New PDAs can connect to internet with faster speed via 3G and WiFi or WiMax. Most PDAs have Bluetooth and the main operating systems used are Palm OS and Microsoft Pocket PC OS [6] [8].

3) Smart Phone

A smart phone is a mobile device which combines cell phone and PDA functions. They have smaller sizes than PDA and bigger than cellular phones. Smart phones include backlit color screen, Wi-Fi, Bluetooth, large RAM, large ROM and internet browsers. Smart phone use advanced operating systems (e.g. Symbian, Windows Mobile, etc) which can support advanced applications. It is also possible to insert memory card. Smart phones have potential to be successfully used in the M-learning field [4][32].

4) Tablet PC

Tablet PC are one of the newest mobile devices. They also have full range of capabilities as personal computers. Some of them haven't keyboard but have software to recognize handwritten text and voice-recognition features that will continue to evolve and improve. Because of their large screens and high resolutions, Tablets PCs are good candidates for delivering online courses. They also offer rich data capture and image manipulation features, which can be useful for learning applications [38].

TABLE I.
A COMPARISON BETWEEN EXISTING WIRELESS TECHNOLOGIES [12].

	Transfer Rate (Mb/s)	Range (~meters)	Frequency Band (GHz)	Number of users
Bluetooth	1-2	10-100	2.4	Dozens
WiFi	10-54	100	2.4	Dozens
GPRS	0.17	30000	1.8	Thousands
3G	256 kbps	>1609344	Varies	Thousands
WiMax	10-54	49889	2-11	Thousands

Sign languages, like oral languages, organize elementary, meaningless units (phonemes; once called cheremes in the case of sign languages) into meaningful semantic units. The elements of a sign are **H**andshape (or Handform), **O**rientation (or Palm Orientation), **L**ocation (or Place of Articulation), **M**ovement, and Non-manual markers (or Facial **E**xpression), summarized in the acronym **HOLME**. Common linguistic features of deaf sign languages are extensive use of classifiers, a high degree of inflection, and a topic-comment syntax. Many unique linguistic features emerge from sign languages' ability to produce meaning in different parts of the visual field simultaneously. For example, the recipient of a signed message can read meanings carried by the hands, the facial expression and the body posture in the same moment. This is in contrast to oral languages, where the sounds that comprise words are mostly sequential (tone being an exception) [35].

In a preliminary experimental result, it has been shown the effectiveness of the novel 3D agent sign language learning system. A 3D agent sign language learning system on mobile devices has been developed which provides an adaptive navigation support and multiple angles of 3D animation view for learners [45]. In [21], authors introduce a Chinese sign language animation system running on mobile devices, which includes three models: 3D virtual human model, word segment based on Chinese sign, and rendering. New wireless technology, educational environments, institution networks and future working and learning methods are studied actively in form of several interesting research projects [12].

The contribution in this paper is to explore the development of a mobile (client-server) system in order to help the deaf and other people to communicate and learn using handheld devices. Although the proposed system was developed to serve Arabic Sign Language users, the system could be adapted to be used with grammar and syntax of other vernacular sign languages. This system also could be developed to administers the network interface/programming, and conducts assessment measurements wirelessly [23]. The front-end (client side) was designed and implemented using a cellular phone or a Personal Digital Assistant (PDA) that have an adequate computational power. The back-end (server side) of the system was designed to run on a server or on the internet. The system is aimed at eventually providing an environment where the deaf can use to manipulate a database/dictionary that contains all the 'words' that are needed to be used regardless of the type of the mobile device. Java wireless communication technologies were suggested in the design. It implements a multi-tier architecture in order to integrate different fields of technology. The data generation and acquisition controlled by JSE.

The ultimate objective of the paper is to improve the scalability of the system and to allow for the use of Scalable Vector Graphics (SVG) which makes it possible to create and develop interactive graphical content, with the ability to zoom and resize on displays with different resolutions and aspect ratios.

II. Background

In Egypt, "a disabled person" means a person who need rehabilitation service to meet the basic needs in society because impairment such as movement-related function, sensory function and mental function brings physical, social, economical and psychological disability[11][9].

Like many other services, special education faces many challenges and limitations. Much effort is being made to meet these challenges; however, these efforts are considered insufficient and must be reconsidered in any future plans. These challenges and limitations are:

A. Financial Limitations:

Shortages in funding represent a vital challenge for the current delivery of services. Many services need to be revised, expanded, and improved to meet the needs of students with disabilities. Furthermore, new innovations in the field of assistive technology, technical aids, and communication services, demand an increase in financial input which might be difficult [24][11][5][40].

B. Practices and Instructional Limitations:

Many of the present practices used in providing special education services have quantitative, but not qualitative values. These practical limitations can be described as follow [24]:

- Most institutions and centers for students with disabilities have limited resources to provide effective and suitable services. These institutions have problems not only in the basic fundamental structural components such as inadequate building construction, transportation, class space, staff, health insurance and salaries, but also in their professional practice issues such as teaching methods, disregard of the role of parents, ignoring the importance of in service training, conducting research, providing assistive aids, and improving quality of services .
- Many of the teaching methods used have to be updated in order to improve the quality and quantity of services provided and to follow-up with any new supplementary instruction that increases the efficiency of services provided for a student with disabilities. Applications such as using new editions of diagnostic tools, applying sufficient behavioral management strategies, preparing Individualized Education Plans (IEP) with a good understanding of their roles, using variety of instructional methods, and improving teacher's evaluating strategies, must be included in any future improvement plans.
- The quality of services delivered for a student with disabilities, especially new services including vocational rehabilitation, prevention services, early medical, social, technological support services, and family services, must be encouraged throughout all public, non-profit and private service organizations.
- The government of Egypt places a high priority on disability, with governmental and non-governmental organizations working together to solve disability issues. However, current services cover only about 10% of the total number of persons with disabilities. The Ministry of Education provides special education services for children with disabilities. It introduced education services for the visually, hearing and mentally impaired through 165 specialized schools and 204 schools, with at least one or more special classrooms for children with disabilities [11].

C. Information Limitations:

This problem appears in the absence of accurate and adequate data on the prevalence rate of disability. The effectiveness and efficiency of services need to be evaluated (e.g., placement and identification procedures, early intervention services, assessment and diagnosis practices, and related services applications) and to be used in the future as guidelines for future planning policies and improvements.

According to the Egyptian Central Authority for Public Mobilization and Statistics, there are approximately two million persons with disabilities in Egypt, which represents about 3.5% of the total population [11], Table II, Table III and Figure 1. Unfortunately, no comprehensive data on disabilities is currently available.

D. Public Awareness Limitations:

There is a noted lack of public awareness and insufficient involvement of local communities in supporting, introducing, and conducting services for student with disabilities. Families' needs must be canvassed and addressed by all levels of service providers.

In Egypt, most youth engaging in socially productive work act through religious institutions. A large 67% of youth look for volunteering opportunities in religious institutions against only 23% in nongovernment organizations. For example, 47.7 % of those working with the deaf are in religious institutions in comparison to only about 19% in civil society organizations (CSOs); 38% help in teaching educational and computer skills in comparison to 17.7% in CSOs[40].

TABLE II.
PERCENT DISTRIBUTION OF DISABLED PERSONS BY TYPE OF DISABILITY IN 1996 AT EGYPT[5]

Type	M	F	T
1-Blind	10.7	11.1	10.8
2-Lose of one eye	1.8	1.3	1.6
3-Deaf-dumb	3.1	3.5	3.2
4-Deaf	1.6	1.9	1.7
5-Dumb	3.9	4.6	4.2
6-Lose of one or both arms	2.1	0.6	1.6
7-Lose of one or both legs	4.1	1.4	3.2
8- Mental Retardation	17.5	15.3	16.7
9-Poliomyelitis	12.8	12.1	12.5
10-Paralysis total or partial	19.4	17.6	18.7
11-Other disabilities	23.1	30.6	25.8
Total	**100**	**100**	**100**

TABLE III.
NUMBER OF PERSONS WITH DISABILITIES BY TYPES OF DISABILITY IN 1996 AT EGYPT[11]

Total Number of Persons with Disabilities	Visual Impairment	Hearing Impairment	Mental Disability	Mobility Impairment
1,060,536	151,510	90,906	1,515,100	303,020

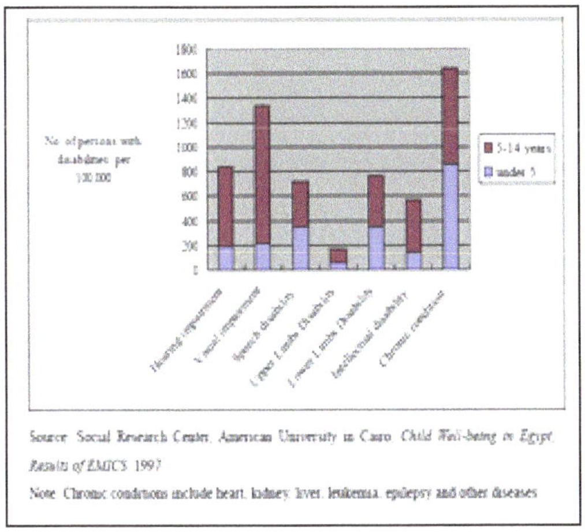

Figure 1. Number of Children with Disabilities per 100,000 by age in 1997 at Egypt [11]

E. Cooperation Limitations:

There is a lack of cooperation between service delivery systems. The absence of a higher authority to organize, unify, and integrate all efforts to provide comprehensive services, and to accelerate the creation of an effective system of accountability. That is very noticeable in Egypt [11][5][9][40].

III. THE MOBILE TUTORIAL SYSTEM

In this paper a digital deaf tutorial system to facilitate learning SL is proposed. The Deaf Tutorial System is defined as a computer-controlled external over a communication medium. It may be considered as a media demonstration, or process, running locally on a software platform but with the ability to be monitored and controlled over the Internet from within a web browser with the host computer connected to the Internet. The client can be any computer connected to the Internet running a browser or mobile device. Once connected, the client will see the same front panel (i.e. interface) as the local host and also have the same program functionality [13].

This system is designed to enable users to control a system via a handheld device, such as a cellular phone or a PDA, and to monitor its output anytime and anywhere, as shown in Figure 2. The hardware infrastructure includes a server and a client end. On the server side, the equipment may include any equipment that can be connected and controlled by a computer [41]. In addition, there is usually a camera on the server end which could be used to output videos activities.

Some of the functions done at the server end include: assessing user-added new words for acceptance or rejection, editing of graphics if needed before acceptance, and review of user reports or questions. Client end functions

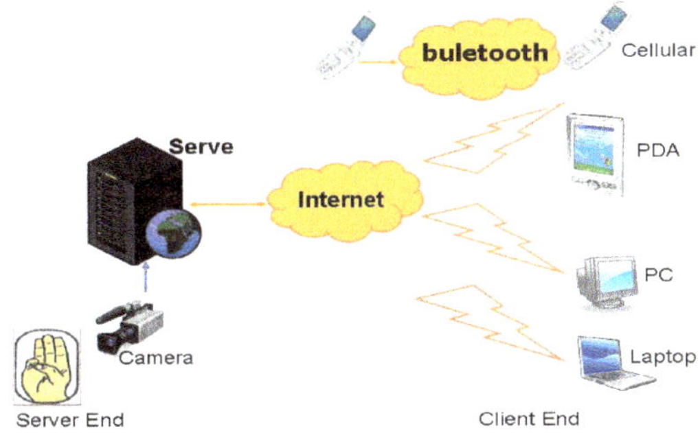

Figure 2. Mobile Deaf Tutorial System Architecture

include: viewing the word/phoneme lists, numbers, nouns/verb representations, pictures and videos, as well as playing games related to the SL for practice [2]. The two end users of the system (client end and server end) are integrated with each other to fulfill the function proposed. The system may be based on a set of tools that teachers in deaf school can combine together and personalize to suit appropriate [33], see Figure 2.

Java language was suggested for the development of the front-end model because it is well suited for interactive web applications and offers the feature of multi-threading to enable many people to work on the same area setup simultaneously. It is a very innovative language which allows one to write programs that can be embedded in Internet web pages. Java is a good candidate since the entire system of building remote tutorials relies on the Internet. The deaf child's parent in this environment could log-on from any system. As Java is a machine independent language, which creates programs that run on a wide variety of computers using a range of operating systems, the Java program does not execute directly on the computer and hence it will not interfere with the operating systems or user's data. Instead, as it runs on a standard hypothetical computer called 'Java virtual machine'. This solves the problems of security and unauthorized access. In addition, Java is an Object Oriented programming language unlike others like Common Gateway Interface (CGI) and HTML. Therefore, Java programs can be easily maintained, as reusability of code is possible [15].

The portability of a Java virtual machine is specifically designed from the start for small, resource-constrained devices. The high-level design goal for it was to create the smallest possible "complete" Java virtual machine that would maintain all the central aspects of the Java programming language, but would run in a resource-constrained device with only a few hundred kilobytes total memory budget [7].

The client design process could be based on J2ME (Java Micro Edition) through TCP/IP protocol to connect to the server. The PDA wireless functions were then utilized to connect to the Internet through which the information was transmitted to log-in to the server to monitor the system. This system not only has the PDA application monitor and control, but also has characteristic display, remote images display, and instruction materials. This system includes the server-end control media and client-end teaching media based on the PDA [15].

In order to find the most appropriate solution for the aforementioned problems, the paper proposes that the project could evolve through the following four phases:

A. The Analysis phase:

We analyzed the required information that had been gathered with the help of the deaf educators. This phase helped to discover what the project should include, what makes it useful, and how it can be created. After the analysis phase, building the database/dictionary was started.

B. The Design phase:

In this phase, the screen picture and the video tutorial had been designed. The video tutorial is suitable for both mobile and online sites. The screen picture and the video tutorial are the working base during the implementation phase.

In designing the Mobile Tutorial System (MTS), the View Control Data architecture has been employed to separate the authoring module from the learning module. The system architecture is based on the wireless J2ME communication technology for the client tier, J2SE for the server tier, and the Microsoft SQL server for the data tier. This SL dictionary/database includes a large number of word-concepts, images, and videos in SL. The site could allow the instructor to add undefined words, representations, and videos with certain conditions, (see the dataflow diagram in Figure 3).

HTTP connectivity and a data stream are used to implement the remote connectivity between client and server tier. The MTS has two use cases, the authoring case and learning case. In the authoring use case, the admin authority level uses an authoring tool on a PC to edit and to generate the required interactivity and consequently store the final media in the database.

In the end user learning use case, the user launches the mobile system client application on the target mobile device. A typical scenario of a mobile tutorial system media involves the following steps:

- Registration: a user inputs the URL to access the media. The system asks the user to first register for the web before trying to perform the tutorial. Once the user registers, a user login and password should be given for future use. It should be noted that self-registration is not allowed for any user.
- Logging-in: once the user has obtained a valid login and password, he/she can login to the server.
- Accessing the Media: when a user starts run, system host computer authenticates, and contacts the user's computer, and interacts with user.
- Admin Login: the instructor/teacher login is recognized, new picture and video can be added.

The flowchart of Figure 4 and the system screens in Figure 5 illustrate some of the necessary steps described above. In Figure 4, where (1) appears, it shows the list of options that are valid for this user. The tutorial system client tier application will send user credentials to the authentication servlet in the server tier. A servlet will authenticate user credential against the user profiles database and generate an XML file. The user will be asked for ID and Password as displayed on screen. The mobile client application will parse the XML file, and if it is a valid login, it will display a list of options dedicated for the specific user.

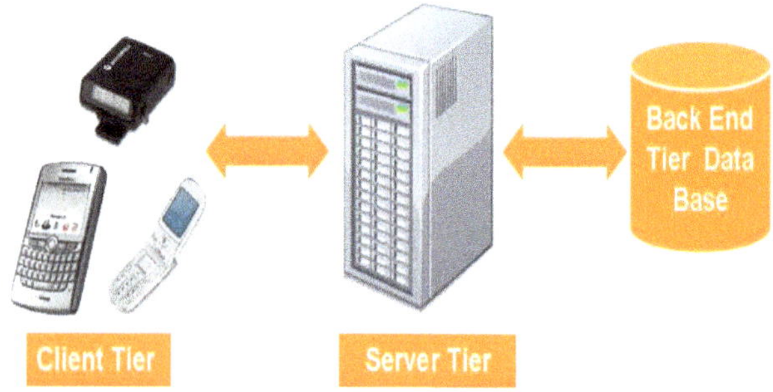

Figure 3. The Mobile Dataflow Diagram

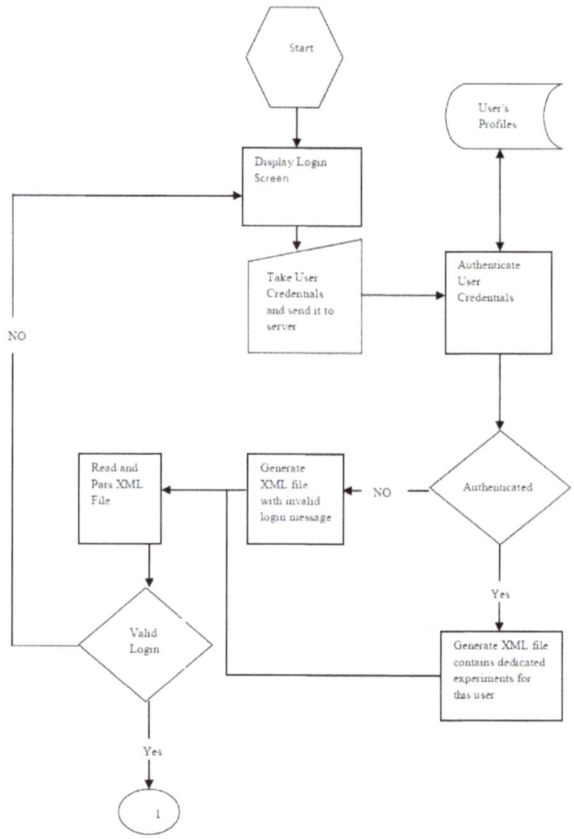

Figure 4. System Software Flowchart

The mobile system client will then extract the user selection and send it to the selection generation servlet in the server tier. The mobile client will eventually parse the XML file and generate the media interface for various types of mobile devices. The latter two figures in Figure 5 depict a video which the user views.

C. The Implementation phase

Based on the design phase, the screens could be implemented using Java Server Faces (JSF) or Java Server Page (JSP), and the middleware Java Message Service (JMS). For mobile telephone application J2ME could be used. In addition, Photoshop could be used to edit the pictures, and Adobe After Effects to convert the videos to gif animation [25].

Users on the client side can access and control the equipment using a PC, a laptop, a cellular phone, or PDA, or any similar device. The design could be based on transmit standards, including RS232, IrDA, and Wireless Local Area Network (WLAN); thus, students can use mobile devices from remote sites [22].

The client design process could be based on J2ME via TCP/IP protocol to connect to the server. The PDA wireless functions were then utilized to connect to the Internet through which the information was transmitted to log-in to the server. This system not only has the PDA application monitor and control, but also has characteristic display, images display, and instruction materials. This system includes the server-end control media and client-end teaching media, based on PDA remote monitoring and control, and on-line learning materials [22].

J2ME was suggested as the platform to implement the MTS because of its great connectivity and portability. What J2ME can do for mobile devices is just the same as what Java 2 Standard Edition (J2SE) and Java 2 Enterprise Edition (J2EE) did for desktop and server systems. In J2ME, the Connected Limited Device Configuration (CLDC) defines a generic "configuration" for a broad range of handheld devices through implementing a set of CLDC [46].

The main benefits, where CLDC devices are involved, include cross platform - where work is transferred between CLDC and other devices; dynamic content - where content is determined by user experience; and information transfer between CLDC and other devices, security, and the developer community - where the developer talent needed for these devices already exists and is readily available for CLDC devices [16]. A Java virtual machine implementation and a configuration specification CLDC are very closely aligned.

Together they are designed to capture just the essential capabilities of each category of device. On top of the CLDC, the Mobile Information Device Profile (MIDP) is defined specifically for wireless devices such as cell phones and PDAs. Wireless device manufacturers need to implement MIDP in order to support Java applications on their devices. A J2ME profile MIDP is layered on top of (and thus extends) a configuration CLDC. The main goal of a profile is to guarantee interoperability within a certain vertical device family or domain by defining a standard Java platform for that market. Profiles typically include classes are far more domain-specific than the class provided in a configuration CLDC. One main advantage of the mobile system is scalability. It may be improved further using the 2D Scalable Vector Graphics (SVG) API specification. SVG is a new vector-based, open-standard file format developed by the World Wide Web Consortium, which represents a new generation of dynamic, data driven, and interactive graphics [14].This SVG defines anAPI for rendering 2D graphics in the World Wide Web Consortium (W3C) SVG Tiny format [30]. The SVG also makes it possible for developers to create interactive graphical content, with the ability to zoom and resize on displays with different resolutions and aspect ratios. Furthermore, J2ME also defines a subset of the Micro Document Object Model (μDOM) API to allow user interaction and dynamic manipulation of SVG content. Developers can take advantage of the large amount of available SVG content. Because SVG is based on XML, a developer accustomed to this scripting environment can also take advantage of the robust features offered by a full programmatic environment, such as the Java language.

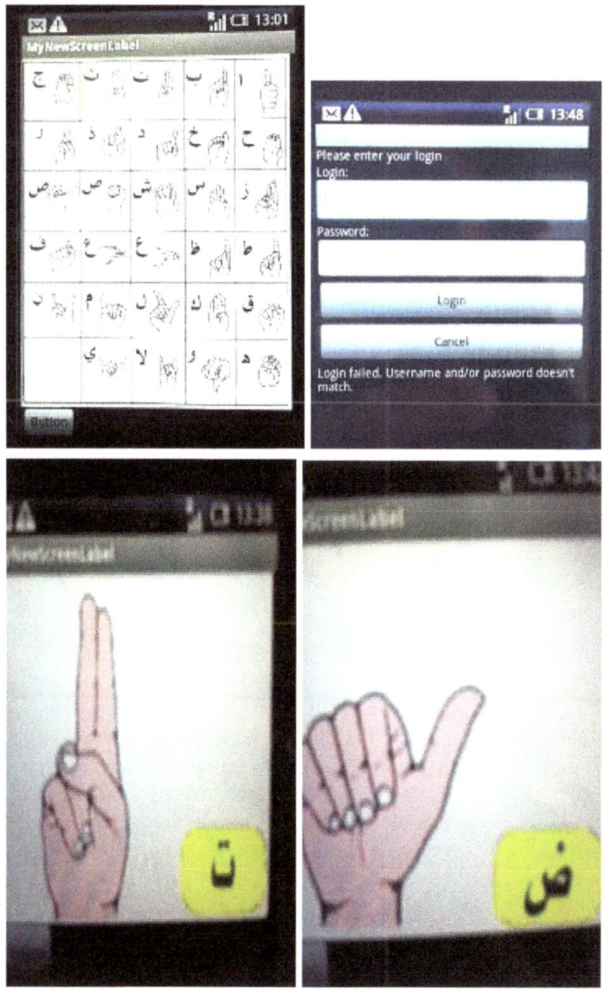

Figure 5. Some system screens

As a result, a user friendly client-side interface that allows the media configuration and parameters to be controlled interactively by the user could be developed. This interface could be realized by means of the SVG file format. An interactive web page can be created with SVG and J2ME, where the client may configure the measurement to be performed. Several setups may be prepared for direct selection, and the corresponding letter table pops up by clicking on the various elements in the sign language diagram.

D. *Testing*

After implementing the screens, testing was performed in order to discover errors without delaying the project process. The following section illustrates the results of some sample testing. The sample testing trying to answer the question of, would this application teach participants Arabic sign language?

E. *Pedagogical Evaluation*

In the area of Mobile learning, studies conducted to measure learnability, the evaluation was performed through repeating the task at least twice because the learnability of mobile phones can only be measured if tasks are performed twice[28]. Ziefle ([28] tested learnability in a study with 30 novice and 30 expert users who solved six tasks. In Bay's study [34], users were asked to solve four different tasks twice on a simulated Nokia 3210 mobile phone.

Learnability refers more to the beginning of the learning process whereas learnability is very often defined as being the time required learning to perform a specific set of tasks [3]. In our context, we defined learnability as the degree to which users feel they are able to start identifying and recalling the Arabic sign letters. The repetition is the re-learning, which is the situation where learning material a second time will typically and it will take less time and effort than initial learning. This process is represented by the learning curve. "*A learning curve is a graphical representation of the changing rate of learning (in the average person) for a given activity or tool. Typically, the increase in retention of information is sharpest after the initial attempts, and then gradually evens out, meaning that less and less new information is retained after each repetition. The learning curve can also represent at a glance the initial difficulty of learning something and, to an extent, how much there is to learn after initial familiarity*" [3].

Nielsen, Bevan, Macleod and Thagard [20][31][29] mention that memorability increases re-learnability to a great extent, but it does not replace it. Nielsen [20] states that recognition is better than recall. People tend to remember images better than text. The prototype application considers these issues and uses an image and video. The other issues that affect memorability are the primary and recency effect. People tend to remember facts or information at the beginning and the end of an experience or list. For example, in a list of 20 words, people usually remember the first and last two words from the list [27].

In that respect, it is important to test the learnability and memorability of the pedagogical application of the Arabic sign letter. The studies using mobile application prototype are answering the following two questions:

1. After how many trials will participants be able to learn all alphabetic?
2. Would the letter's location in the list affect memorability?

The mobile application prototype interface has an image that has of all Arabic letters and its mapping hand's signs and a video repeating the letter twice and showing the hand sign of each letter (see Figure 5). The prototype interface of the application had been installed on Sony Ericsson Xperia. At the beginning of the study, participants were informed with the procedures of the study. Participants see the image, watch the video, and answer five multiple-choice question's trial (see Figure 6). At the end of the study, comments about the application were collected.

In order to answer the 1st question, two studies have been conducted. The first study considered repeating the process of watching the image and the video and answering the questions five times. Based on the results of the first study, the decision was taken to make a second study that repeats the producers however seven times.

In order to answer the 2nd question, the design of the five multiple-choice questions per trial for the first study was as follow:

- The second question is randomly chosen from the first three letters presented in the video to assess the primary effect

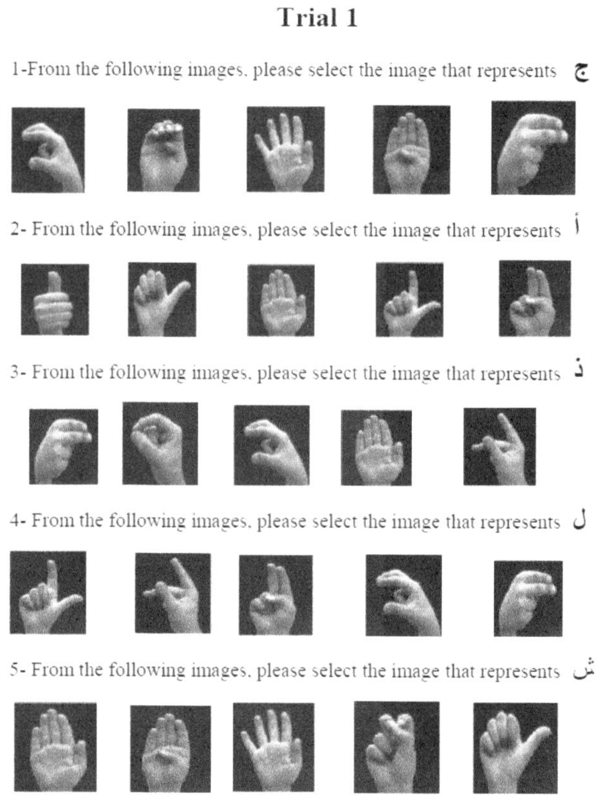

Figure 6. First Multiple-Choice Question Trial

- The fourth question is randomly chosen from the last three letters presented in the video to evaluate the recency effect,
- Questions one, three and five were randomly chosen from the rest of the alphabetic list, (see Figures 5 & 6).

Both studies were conducted in a quite office. On average, both studies took around 40-45 minutes to finish. For both studies, 31 participants took part. 20 participants were in the first study and 11 participants in the second one.

In the first study, 52% were male and 48% were female. 60% were from the age range 17-21 years old, 35% are from 21-30 years old and 5% from 31-40 years old. 85% of the participants are students and 15% are teaching assistants at the computer science department of the BUE University, see column (a) at Figure 7.

In the second study, 11 people participated. 60% are females and 40% are males. 60% are from 17-20 years old and 40%are from 21-30 years old. Majorities are students 90% and only 10% are teaching assistant, see column (b) at Figure 7.

The first study gives evidence that the second and fourth questions are memorable. The average correct answers are 96.2% and 96% for questions two and four respectively (see Figure 8). The learning curve in Figure 9 did not show that the first trials perform worse that the following ones. However, if the first trial is dismissed, there are some evidences that repeating the procedures improves learnability. Consequently, to check whether repeating the procedures increases the average of correct answers, the second study has been conducted and two trials had been added to become seven trials. The results

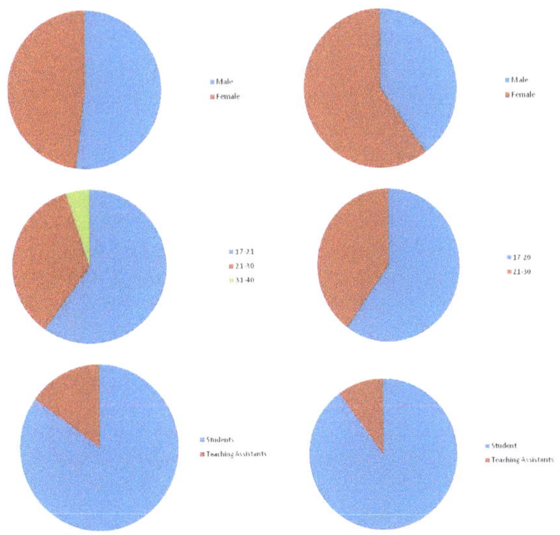

(a) (b)

Figure 7. a) First trial b) Second trial

proofed that repeating the steps of watching the image and the video improves learnability (see Figure 11). Figure 10 supports the argument of primary and recency effect. The second and fourth questions average scores are better that the questions one, three, and five. The average correct answers for questions two and four are 94% and 82% respectively.

The results gave evidence of the memorability and learnability of system. Furthermore, it was well received by the participants. There comments were:

- The pace of the video and in particular the repetition of each letter was perceived positively.
- A few number of participants commented on the angle of the hand sign picture (i.e., if the picture captured was a bit right or left to the hand, in some cases, the figures were not very clear). For example, letters such as (dal)-(د) and (thal)-(ذ) can look the same if the hand sign picture is captured from the right. Hence, for each letter, having different pictures of the hand sign that are captured from different angles can minimize confusion and speed up learning process.
- Very few participants commented on the sound however one participant said that a more friendly sound would improve the application.

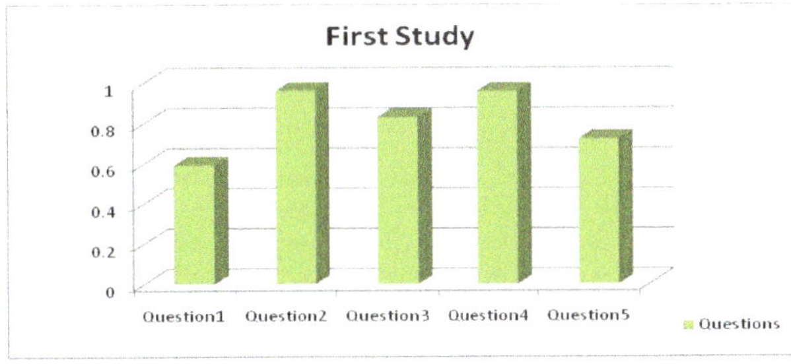

Figure 8. the avarage correct answer of each question for the first study

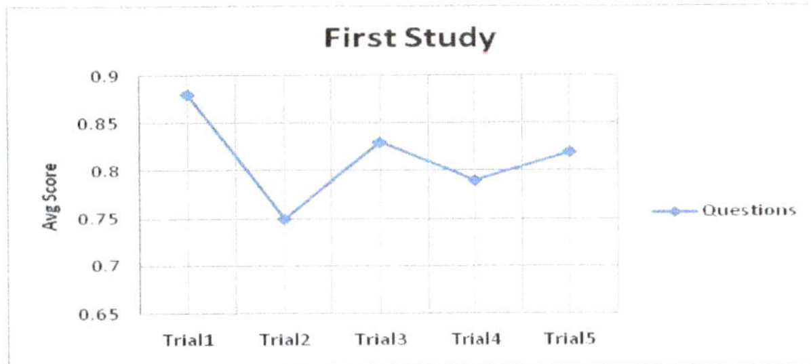

Figure 9. the avarage correct answer of each trial for the first study

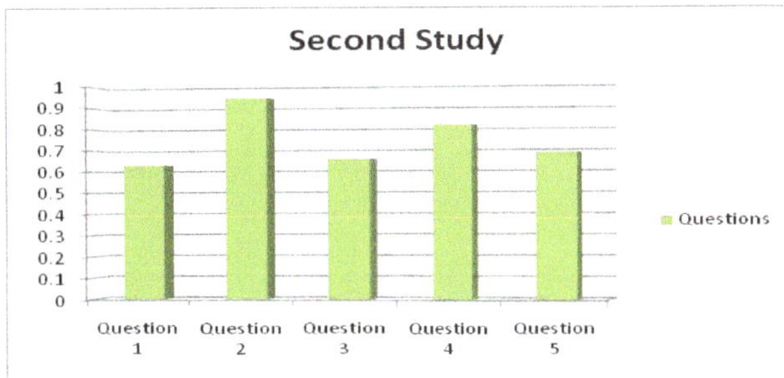

Figure 10. the avarage correct answer of each question for the second study

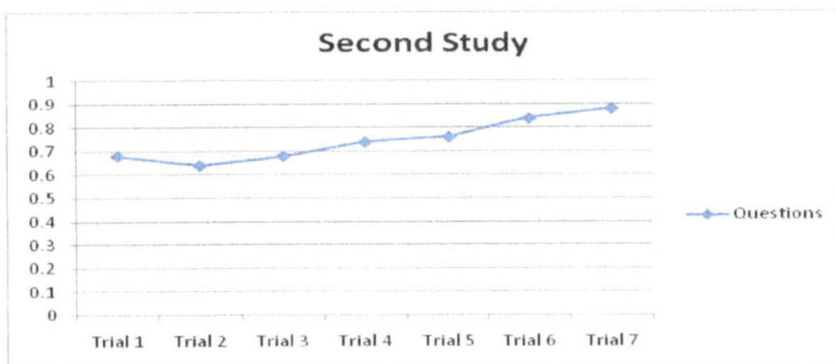

Figure 11. the avarage correct answer of each trial for the second study

IV. Conclusions

The MTS system was developed to enable deaf students to access media using cellular phones, PDAs, smart phones, Palms, Pocket PCs, Tablet PCs, and any other similar handheld devices. This allows for teaching to be conducted anytime, anywhere, and allows resources and equipment to be shared between various collaborating institutions. The system allows also for the use of Scalable Vector Graphics (SVG) which makes it possible to create and develop interactive graphical content. Moreover, MTS system could be developed to administer the network interface/programming, and conduct assessment measurements wirelessly [23].

To execute the aforementioned four phases of the MTS systems, we should proceed as follows:

A. MTS Implementation

MTS should be executed such that Adobe After Effects is used to convert the videos to gif animation, that method is used to put an animation of a video online without streaming the video itself. SVG user friendly client-side interface that allows the media configuration and parameters to be controlled interactively by the user will be added.

B. MTS upgrading

MTS should be upgraded such that the site will allow the instructor to add undefined words, representations, and videos with certain conditions, to the dictionary/database.

The interface of the application that is recognition-based has a good impact on learnability and memorability.

To improve the interface, the results of the primary and recency effects could be utilized by allowing users to choose any letter to watch its video or shuffling the letters before starting the application. Furthermore, after watching the image and the video, testing their knowledge by developing an exam similar to memory games could make the experience enjoyable and improve learnability.

The MTS ultimate goal is to develop the MTS system as an interactive graphical teaching tool for any subject, anytime, anywhere and to any person, hearing or deaf.

REFERENCES

[1] A. A. Economides and N. Nikolaou, "Evaluation of Handheld Devices for Mobile Learning", International Journal of Engineering Education, 24(1) , 2008.

[2] A. Alkouz, A. Y. Al-Zoubi, and M. Otair, "J2ME-Based Mobile Virtual Laboratory for Engineering Education", International Journal of Interactive Mobile Technologies (iJIM), Vol 2, No 2, 2008.

[3] A.G. Bills, *General experimental psychology*, Longmans Psychology Series, pp. 192-215, New York, NY: Longmans, Green and Co, 1934.

[4] A. Holzinger, A. Nischelwitzer, and M. Meisenberger, "Mobile Phones as a Challenge for M-Learning: Examples for Mobile Interactive Learning Objects (MILOs)", Third IEEE International Conference on Pervasive Computing and Communications Workshops (PERCOMW'05). p. 307-311, 2005.

[5] B. El-Deeb, "Disability Statistics in Egypt", Paper Presented at Washington Group meeting in Ottawa, Canada, Jan, 2002.

[6] C. Houser, P. Thornton, and D. Kluge, "Mobile Learning: Cell Phones and PDAs for Education" , Proceedings of the International Conference on Computers in Education (ICCE'02), IEEE. p. 1149-1150, 2003.

[7] C. W. Lund, and M. S. Norum, *The Peer2Me Framework - A Framework for Mobile Collaboration on Mobile Phones*, Master's Thesis, NTNU, 2005.

[8] D. Lindquist, T. Denning, M. Kelly, R. Malani, W.G. Griswold, and B. Simon, "Exploring the Potential of Mobile Phones for Active Learning in the Classroom", ACM SIGCSE Bulletin, 39(1): p. 384-388, 2007.

[9] Department of Social Rehabilitation, Ministry of Social Affairs, *Rehabilitation and Care of the Handicapped in A.R.E.*

[10] E. Egea-Lopez, A. Martinez-Sala, J. Vales-Alonso, J. Garcia-Haro, and J. Malgosa-Sanahuja, "Wireless communications deployment in industry: A Review of Issues, Options and Technologies", ScienceDirect, Computers in industry, 56(1): p. 29-53, 2005.

[11] Egyptian Planning and Evaluation Department, *Country Profile on Disability, Arab Republic of Egypt*, Japan International Cooperation Agency,March,2002.

[12] E. Koskinen, U. Veitonen, T. Lasksonen, and O. Vaananen, "New Generation Wireless Technologies Impact on Educational Environments Case: Utilizing Finnish Sign Language in Interpreter Training", Proceedings of the 2005 IEEE International Workshop on Wireless and Mobile Technologies in Education (WMTE'05), 2005.

[13] F. P. Bergenti, and A. Leap, "A FIPA Platform for Handheld and Mobile Devices", ATAL 2001 Conference, Seattle, USA, 2001.

[14] G. Alonso, F. Casati, H. Kuno, and V. Machiraju, *Web Services Concepts, Architectures and Applications*, Springer Verlag. page 8, 2004.

[15] J. Alja'm, S. A. El-Seoud, A. E. Block, M. Garcia-Ruiz, and A. M. Jaoua, "An Assistive Computerized Systm for Children with Intellectual and Learning Disabilities", Lecture Notes in Computer Science (LNCS), Springer, June, 2009.

[16] J. B. Freysen, "M-learning: an Educational Perspective", Mobile Learning Anytime Anywhere, A book of papers from MLEARN 2004, Edited by Jill Attewell and Carol Savill-Smith, Published by the Learning and Skillls Development Agency.

[17] J. Keogh, *The Complete Reference J2ME*, Published by McGrawHill OSBORNE Edition, 2003.

[18] J.L. Gimenez López, T. Magal-Royo, J. García Laborda, and F. Garde Calvo, "Methods of Adapting Digital Content for the Learning Process via Mobile Devices", Procedia – Social and Behavioral Sciences, Volume 1 (Issue 1), pp. 2673–2677, 2009.

[19] J. Malek, M. Laroussi, A. Derycke, and H. Ben Ghezala, "A Context-Aware Approach for Modeling Bijective Adaptations between Context and Activity in a Mobile and Collaborative learning", International Journal of Interactive Mobile Technologies (iJIM), 2(1) ,2007.

[20] J. Nielsen, *Coordinating user interfaces for consistency*. Boston, MA: Academic Press, 1989.

[21] J. Wang, Y. aun, and L. Wang, "Chinese Sign Language Animation System On Mobile Devices", Second International Conference on Information Technology and Computer Science, (DOI 10.1109/ITCS.2010.19), 2010.

[22] K. Smiley, *IT Trends : Mobile and Wireless Devices*, Giga Information Group, 2003.

[23] M. Abu Shaira, and I. A. Alzraigat, "The Effect of Sign Writing on the Achievement and Acquisition of Vocabulary by Deaf Students", Al-Amal School for Deaf, Amman-Jordan, May 2007.

[24] M. Al jabery, and M. Zumberg, "General and Special Education Systems in Jordan: Present and Future Perspectives", International Journal of Special Education, Vol. 23, No. 1, 2008.

[25] M. Brachtl, J. Slajs, and P. Slavik, "PDA Based Navigation System for a 3D Environment", In Third International Workshop on Intelligent Interactive Assistance and Mobile Multimedia Computing (IMC'2000), Rostock-Warnemu¨nde, Germany, 2000.

[26] M. Jemni, O. E. Ghoul, N. B. Yahia, and M. Boulares, "Sign Language MMS To make Cell Phones Accessible To DEAF AND Hard-of-Hearing Community", Conference & Workshop on Assistive Technologies for People with Vision & Hearing Impairments, Assistive Technology for All Ages, M.A. Hersh (ed.), CVHI 2007.

[27] M. J. Gelb and K. Howell, *Brain Power: Improve Your Mind as You Age*, New World Library Publisher, 2012.

[28] M. Ziefle, *The influence of user experience and phone complexity on performance, ease of use and learnability of different mobile phones*, Behaviour & Information technology, Vol. 21, No. 5, 303-311, 2002.

[29] N. Bevan, and M. Macleod, *Usability measurement in context*, Behaviour & Information Technology Vol. 13 No. 1 and 2, 132–145, 1994.

[30] P. Contreras, D. Zervas, and F. Murtagh, *Web Services in Natural Language: Towards an Integration of Web Service and Semantic Web Standards in the JXTA Peer to Peer Environment* , page 65, May 2005.

[31] P. Thagard, *Mind, introduction to cognitive science*. Cambridge, MA: The MIT Press, 1996.

[32] P. Zheng, L. M. Ni, "The Rise of the Smart Phone. IEEE Distributed Systems Online", 7(3): p. 1-14, 2006.

[33] R. J. Costa, and G. R. Alves.

9th IFAC Symposium on Automated Systems Based on Human Skill And Knowledge (ASBoHS'06), França-Nancy, 22nd to 24th of May, 2006.

[34] S. Bay, "Cellular Phone Manuals: Users' Benefit from Spatial Maps", Proceedings of ACM CHI 2003, Ft. Lauderdale, Florida, 662-663, 2003.

[35] Sign Language.

[36] S. M. Jacob, and B. Issac, "Mobile Technologies and its Impact-An Analysis in Higher Education Context", International Journal of Interactive Mobile Technologies (iJIM), 2(1), 2008.

[37] S. R. Balasundaram, and B. Ramadoss, "SMS for Question-Answering in the m-Learning Scenario", Journal of Computer Science, 3(2): p. 119-121, 2007.

[38] S.S. Mahmoud, "A proposed Model for Distributing e-Courses Content through Mobile Technology Architectures", Proceedings of World Academy of Science: Engineering & Technology, Vol. 39, p274, February 2008.

[39] T. Georgiev, E. Georgieva, and A. Smrikarov, "M-learning-a New Stage of –Learning", International Conference on Computer Systems and Technologies. 2006.

[40] The Institte of National Planning, *The Egypt Human Development Report*, the Human Development Project, Egypt, under the project document EGY/01/006 of technical cooperation with the United Nations Development Programme (UNDP), 2010.

[41] T. Menzies, *21st Century AI: Proud, Not Smug*, IEEE Intelligent Systems. 18–25, 2003.

[42] T. Zahariadis, "Evolution of the Wireless PAN and LAN standards", Computer Standards & Interfaces, Sciencedirect,. 26(3): p. 175-185, 2004.

[43] Wireless Fdelity (Wi-Fi): 4G Americas.

[44] W. P. Lee, M. A. Osman, M. Sabudin, S. Sulaiman, L. A. Zadeh ,J. Kacprzyk, N. Mastorakis, A. Kuri-Morales, P. Borne, and L. Kazovsky "An Experience on Introducing a Mobile Community Platform: Flying 2 U", in WSEAS International Conference. Proceedings. Mathematics and Computers in Science and Engineering, World Scientific and Engineering Academy and Society, 2010.

[45] Y. Lin, J.J. Leu, J. Huang, and Y. Huang, "Developing the Mobile 3D Agent Sign Language Learning System", The 6th IEEE International Conference on Wireless, Mobile, and Ubiquitous Technologies in Education, (DOI 10.1109/WMUTE.2010.25), 2010.

[46] Z. Wang, Z. Guo, and Y. Wang, "J2ME-Based Application of Vector Space in Query Optimization", ISECS International Colloquium on Computing, Communication, Control, and Management CCCM, Volume2, 2008.

AUTHORS

M. Samir Abou El-Seoud, **Ann Nosseir**, and **Islam Taj-Eddin** are with Faculty of Informatics and Computer Science, British University in Egypt-BUE, Cairo, Egypt. {samir.elseoud; ann.nosseir; islam.tajeddin}@bue.edu.eg

Hosam El-Sofany is with Arab East Colleges for Graduate Studies, Dept of Computer Science, Riyadh, Kingdom of Saudi Arabia. hosam_elsofany@hotmail.com

Nadine Abu Rumman is with The King Hussein School for Information Technology, Princess Sumaya University for Technology (PSUT), Jordan. Nadine@psut.edu.jo

33

Near Field Communication

Which Potentials Does NFC Bring for Teaching and Learning Materials?

M. Maierhuber and M. Ebner
Graz University of Technology, Graz, Austria

Abstract—**Near Field Communication (NFC) is deemed to be a future technology with a lot of potential in many areas. One of those areas, learning and teaching, will be covered in this article, showing possible usages of NFC with teaching and learning materials. With a lot of new NFC capable devices presented recently, this technology can be utilized in many areas, including the arbitrarily growing field of mobile learning.**

Mobile devices, especially smartphones, can help to close the gap between printed media and online media. Several methods, e.g. two-dimensional barcodes, have already been used to connect the digital world with printed media but almost all of them caused inconveniences or difficulties. NFC presents an easy to use way to share and communicate directly between capable devices or tags that can be applied almost anywhere.

In this publication a first insight to the potential of NFC for teaching and learning content is given. A prototype is programmed to allow data transformation between the print media and the smartphone. It can be summarized that NFC will be the logical successor of QR-Codes.

Index Terms—**Mobile Tagging, NFC, QR-Code, Technology Enhanced Learning**

I. INTRODUCTION

Information technology is one of the fastest growing branches; especially the mobile sector expands very quickly and is introduced in many new areas. According to a recent ITU report [11], very soon there will be as many mobile-cellular subscriptions as people inhabiting the planet, with the figure set to surpass the seven billion mark early in 2014. On the other side, printed materials are still needed in many cases and are enhanced by additional online content. Combining digital and printed materials is one of the biggest challenges of today as well as tomorrow. For example, mobile tagging is considered as one possibility to close the gap between digital and printed media [18]. One (publisher) can apply an optical code that encodes an URL to the online media on a printed media. The user (reader) scans the optical code with a mobile device and retrieves the decoded URL that can be opened on the mobile device without typing [12] [7].

However, doing so requires a scanning application on the mobile device that uses the integrated camera of the device to read the code. At first glance this sounds very easy and handy, but there are a couple of disadvantages. An optical code can be damaged easily or even get dirty enough to become unreadable even with improved error correction features as for example so called QR-Codes have, using Reed-Solomon Code with maximum 30% of code-words restorable [2]. Too, the process of scanning may take a considerable amount of time, depending on the code size, structure and error level of the code and the used camera hardware of the mobile device.

Our research idea is whether passive Near Field Communication (NFC) tags that can be applied to the printed media instead of an optical code can simplify that process or not (see Fig. 1). This paper gets granular on the potential of NFC as well as on a first prototype assisting daily teaching and learning processes.

NFC itself is a set of standards defined by the NFC Forum, which consists of many companies [14]. NFC tags are more robust than optical codes and are independent of the ambient light. The reading process is usually faster and can easily be done by simply getting the mobile device close to the tag. This method is technically similar to the well-known and established RFID (Radio-Frequency Identification), with mainly the data format and range specification being different. For further information read the tag specification published by the NFC Forum [15].

II. SIMILAR TECHNOLOGIES

However, nowadays two-dimensional codes are still the most common solution in mobile tagging applications mostly used for marketing [8]. Beside the widely used one-dimensional barcodes, frequently used types of two-dimensional codes are for example Data Matrix and QR-Code [21]. Both are nowadays used in printed media to provide URLs to websites or even handle shopping, as the Korean company Tesco introduced in a subway station in Seoul. Customers scan QR-Codes on products and order them this way [1]. Another example is using QR-Codes as Business Cards [3] or for additional information of tagged geo-locations [25]. Additionally, payments using QR-Codes have been implemented, for example at Starbucks [24].

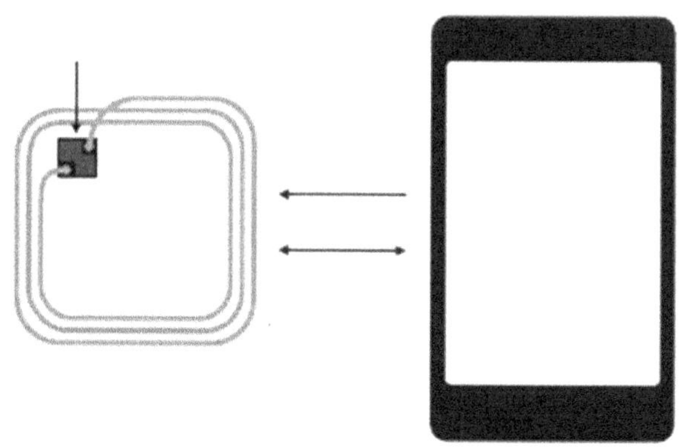

Figure 1. NFC tag (http://www.nxp.com)

Another technology that is very similar to NFC is RFID. Passive RFID tags are compatible to NFC as they work in the same way. An antenna is used to illuminate the tag with power from the radio field generated by the sender to send the response data [26]. RFID applications include any remote identification needs to be done, for example in logistics for stocktaking [17].

The following chapter points out how NFC can be used to connect digital and printed materials, using teaching and learning materials as an example.

III. Potentials of NFC for Teaching and Learning Materials

A. Mobile Learning

Mobile learning has become very important since mobile devices have recently been widely spread. Younger generations are usually very familiar with new technologies and quickly learn how to use them. Those new technologies have a lot of unused potentials in mobile learning and can induce additional values for all involved [10]. NFC potentials have not been closely examined but literature research yet; first practical tests have shown promising results.

According to [20] mobile learning adds the following values:

„The value of mobile learning

Tutors commented on the value of mobile learning as follows.

- It is important to bring new technology into the classroom.
- Mobile learning could be utilised as part of a learning approach which uses different types of activities (or a blended learning approach).
- Mobile learning supports the learning process rather than being integral to it.
- Mobile learning needs to be used appropriately, according to the groups of students involved.
- Mobile learning can be a useful add-on tool for students with special needs. However, for SMS and MMS this might be dependent on the students' specific disabilities or difficulties involved.
- Good IT support is needed.
- Mobile learning can be used as a 'hook' to re-engage disaffected youth.
- It is necessary to have enough devices for classroom use."

In the described survey of this paper, learning games have been developed. Those games use SMS for communication. This requires everyone being capable of writing SMS or MMS, which can cause problems especially with handicapped users. In addition to this, SMS and MMS are fairly "old" technologies and have some constraints that might not be suitable for the growing information flow. Nevertheless, it is a widely used possibility especially in developed countries because of missing infrastructure of Internet access [9].

One of the biggest challenges of mobile learning is to connect digital materials and printed materials. Thus, NFC is a technology that offers a way to close the gap between both media. It serves as a kind of interface for digital and printed materials to link each other.

However, several barriers currently (as of April 2013) block further expansion of the NFC technology, like incompatibilities between certain NFC tags and chips. That usually depends on the manufacturer of the chips and tags. For example, chips produced by NXP Semiconductors perfectly work with their own tags but other tags are not readable by chips produced by other manufacturers although their data format is the same.

The data format has been specified and published by the NFC Forum and is called NFC Data Exchange Format (NDEF) (NFC Forum, 2006). A current example of that case occurs in tags of the type of MifareClassic (sizes 1 and 4 KB). An NXP Semiconductors manufactured NFC chip is integrated in a Google Nexus S smartphone, while Google's current (as of April 2013) flagship Android smartphone Nexus 4 integrates a Broadcom NFC chip. Inspite of the fact that Google regards the Nexus series smartphones as the reference products for most of their published Android versions, those incompatibilities occur. However, those should not be hindering to define some potential for the future of learning and teaching contents.

B. Prototype implementation

A prototype for testing has been developed. As example, the NFC chip should enhance the well-known German lecture book on Technology Enhanced Learning, shortly L3T. This book was written by more than 130 authors in the German speaking area and all 50 chapters are available via open access at http://l3t.eu [5]. The whole project compromises not only the digital or printed book, but also apps for android, iPhone, and iPads as well as much additional material on the web like bookmarks, videos, or exercises. So there is also a big support for m-learning activities. [4].

The prototype for testing the NFC potential is written for the Android operating system that fully supports NFC and provides an additional feature called Android Beam, to transfer data via NFC directly between two Android devices. The prototype enables sharing of all L3T article links to NFC tags and devices and also reading from tags. It cannot display articles read from NFC tags or devices yet (the necessary reader is missing) but it shows the potential of the technology by simply listing different data types of received data. It is based on the boilerplate implementation published by the ndef-tools developer [23].

So from a given article, the user can use the options menu to choose a sharing method or even read from another tag. If a tag is read, data will be presented in a list (see Fig. 2), showing type information of the NDEF record.

The tag read from in Fig 2. was written by the tag writer of that app that currently writes a static text and a dynamic URI depending on which article is currently viewed. The written data could be extended by interface actions done by the user. This does not require the user having knowledge of the exact data that is written and thus simplifies any application involving writing data.

The following chapters point out how NFC can be additionally used and applied with some further use cases, according to the research work gained in the prototype test.

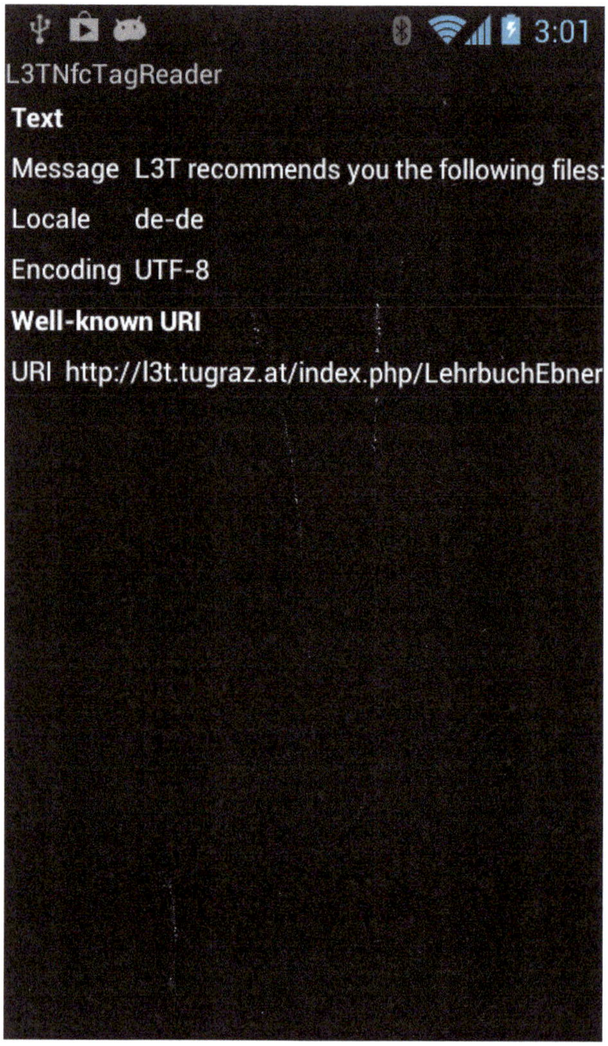

Figure 2. NFC tag content type information

C. Distributing materials

NFC can be used to distribute teaching and learning materials. Imagine the following scenario:

A lecturer wants to distribute current materials in his/her lecture hall where no IP-based network is available and he/she cannot upload them to a webserver. There are more than 300 students in the lecture hall who want to use digital materials in practical exercises. The most common way would be to deliver those materials as printed scripts to each single student. Although this can be accomplished in an appropriate time frame, it causes also extremely high costs to create 300 copies. Additionally, printed materials are not suited for some applications, for example technical studies that apply code fragments, scripts or other chiefly digitally used materials.

In that case, the lecturer can either place an URL to the materials, or the materials themselves on a NFC tag before the lesson starts. Students can retrieve the materials by simply touching the NFC tag with their mobile devices. Furthermore NFC tags will most likely increase in capacity to store large amounts of data, so the content can directly be put on the tag. Current tags are optionally rewritable, cheap and thus can significantly reduce cost of the distribution of materials.

Additionally, NFC tags featuring authentication methods can be protected against unauthorized writing or manipulation attempts.

D. Additional information to materials

Smartphones, tablets and other mobile devices will probably play an even more important role in the future. NFC simplifies several user interactions to a single touch action on a tag. For example, if a teacher presents a video in his/her class, there is certain equipment required. This equipment can also be expensive and thus it is shared in many cases and therefore it is not always available. To avoid that kind of problem, the students' mobile devices can be used to display multimedia content like a video that is linked to with a NFC tag. The students simply scan the tag which can be applied to printed materials as a sticker and their device loads the movie. It also enables the users to directly save the movies on their devices. This could be easily accomplished by a single NFC touch action using an application without requiring no other external equipment. This also applies to external materials of all kinds that cannot be delivered using the printed materials, shortly called learning objects (LO). So the teacher simply applies a NFC tag sticker to certain spots linking to the external data source that can also be updated, without replacing the contents of the tag.

E. Sharing of materials

With additional NFC related features like Android Beam users can easily share data directly between capable devices without using any other data transformation technology like wireless network, Bluetooth, or cell phone network. This may be useful for students who want to forward or share materials or URLs pointing to those materials. It is very simple to add Android Beam functionality to applications; so even people with less affinity to technology can easily use those apps.

In case of bigger data is needed to be shared, other connections should be used because NFC is limited to a bandwidth of 424kbit/s, as specified by the NFC Forum [16]. But NFC can play a major role in that case as well. NFC can be used to establish a connection like Bluetooth, which handles the transfer of the data afterwards. Bluetooth in its default mode is considerably slower at establishing connections than NFC and also might need user interaction like entering a key when pairing devices. That step can be skipped and simplified because the connection will be established using data exchanged via NFC while simply touching the devices. Nevertheless, to exchange larger amounts of data via NFC only, the specified bandwidth should be increased in the future because it may often not be feasible to keep devices very close to each other during the data transfer.

F. Delivery of practicals

As NFC tags can be configured to execute pre-defined actions, students may deliver their lesson results at the end of an exercise unit using their NFC capable device. The tag holds the necessary information about the delivery template and the device holds information about the student as well as the involved study and course he/she enrolled in. The way the information is shared and managed can be defined on the tag. The delivery itself than can be executed via email for example. In that email, all necessary information will be arranged and presented as defined on the tag and no other interaction is required as all actions can be conducted with the simple touch of the tag.

The lecturer can set up a filter that automatically assigns retrieved emails to a specified folder and also receives additional information about the student which he/she would need to look up elsewhere. This also reduces the time needed for administration related actions.

G. Integration of social networks

NFC tags could be used to integrate social networks like Facebook or Google+. There are numerous possibilities for using for example Twitter in academic environments [6].

Furthermore, a private social network for students could be created to support mobile learning and to offer an additional channel for sharing knowledge related to materials. This could be a linking system for bookmarks that simplifies sharing of single sentences or paragraphs.

For example, a student can bookmark a paragraph by means of a special marker tool on the user interface. That bookmark is added to the device local bookmark library and can be shared in the social network using NFC either to connect to another device directly with Android Beam or similar transfer methods or to a tag on his/her printed material which can be read by another student. Another possibility is to use a tag to identify the social network and upload the data using the connection information retrieved from that tag. In any case a shared learning network is created that allows students to simplify joint learning for exams. A group of students could easily split the chapters and share their thoughts when merging their elaborations using NFC. This would be very easy to use as students just need to tap an NFC tag and can use onscreen touch navigation on their mobile devices. The mobile device does not require any external tools. The user does neither need training to apply the described solution nor knowledge about the network connection as identification, authentication; establishing of the connection can be done in a single NFC touch event.

H. Access control to materials using NFC

As NFC tags support authentication methods a teacher/lecturer can control access to materials or parts of those.

NFC can be useful to create a standard access method that is independent of the used background system. That replaces the usual input form login mask that is not protected from phishing by secure NFC authentication that also provides identification.

Restricting access to materials could be executed by various methods. If a teacher/lecturer wants to give access to currently present students, he/she can require reading a NFC tag that holds access information to the linked materials, or even provides direct access to those protected materials. Additionally, he/she can use the access control combined to a payment system using NFC.

Also, a teacher can use persistent access control to check which student accessed which material at which time. The resulting data could influence the evaluation of training success that could help the teacher to adapt materials accordingly.

I. Examinations

NFC could replace or complement any previous used identity card. The teacher/lecturer could verify the identity using NFC and automatically connect to the administration system to check whether the student is allowed to participate in the examination or not and thereby verify his/her presence too. In that case, paper lists could be omitted and thus save costs. Even more, the verification process is accelerated and simplified.

NFC could also be used in examinations that allow the use of digital materials. In that case, NFC can be used to check who is accessing which part in the digital materials and measure how long a student needs that part. This additional knowledge could have impact on the evaluation of the given answer and the teacher can draw conclusions from the student's understanding of the related chapter. Especially from a Learning Analytics perspective this can be from high interest [22].

J. Links and Bookmarks

All hypertext links and bookmarks will be removed from learning materials. If you need to refer to an Internet email address or URL, currently you must type out the address or URL fully in regular font. With a NFC tag all URLs will be transmitted directly to the mobile devices and can be accessed easily.

IV. Other Uses of NFC

The current most known and most promising field of application of the NFC technology used to be payment systems. In the last years many companies ran experiments to test NFC to be used as wallet, cooperating with well-known payment providers like MasterCard and their PayPass system [13].

Transactions are rather small because, for example, the eurozone of the European Union limited payments via NFC to 25 EUR and the United Kingdom to 20 GBP, as reported by [19].

Additional uses are identification purposes and access control that can be done very simple and without any other action than approximating the device to a tag or other device. This removes possible unsafe input methods and allows secure identification of the device holder. However, protection against the loss of the device is still poor as many users do not block their device with a PIN code.

V. Conclusions

NFC offers several methods to connect digital media and printed media. It simplifies and reduces several interactions to a single action of narrow contact and can be easily used even by handicapped people. It allows integrating digital media into printed media using a mobile device with NFC as interface. As NFC is a standardized protocol, the used hardware following that standard is interchangeable.

Compared to other tagging methods using two-dimensional codes like QR-Code or DataMatrix, NFC offers more flexibility and is easier to use. Data on an NFC tag can easily be replaced without replacing the tag-hardware or the underlying structure it is applied to. This reusability can significantly reduce costs if the contained data is designed to be replaceable. Also, NFC tags offer more capacity than optical codes that are limited by the space they need and their specification.

Further, reading from a NFC tag is usually faster than scanning optical codes and is also independent of the environment. Reading optical codes requires a source of light and a not malformed or obstructed surface. Error correction of QR-Codes is still fairly good. According to the specification, it can restore about 30% of the codewords using „Level H" [2]. However, it also reduces the amount of data the QR-Code can store because of the consumed error correction symbols.

NFC tags instead can be applied to almost all objects. The only problem that has to be dealt with is that the tag cannot be applied to metal surfaces without modification because the antenna needs to act as inductor to create the electric power for the chip when using passive tags.

Table 1 summarizes the potential of NFC with a special focus on teaching and learning contents. It can be concluded that this contact-free technology will have a great impact on data exchange because of its simplicity and speed. The presented prototype gave a first insight to the forthcoming possibilities. Nevertheless, additional research work needs to be done to reduce the theoretical ideas to practice.

TABLE I.
POTENTIALS OF NFC AND THEIR IMPLEMENTATION

Potential	Implementation
Distribution of materials	NFC tag with URL to materials or materials themselves
Additional information to materials	NFC tag with URL to integrate external media
Sharing of materials	Peer-to-peer communication between devices
Release of lesson results	NFC as interface to an automatic delivery system
Integration of social networks	NFC as support for learning group with distribution of bookmarks in social networks
Access control to materials	Linking payment system and authentication methods using NFC to privide access to materials
Exams	NFC as interface to verify identity and access control to digital media that are permitted to be used during the exam

REFERENCES

[1] J. Bergen, *Korea's Tesco reinvents grocery shopping with QR-code "stores"*, 2013. http://www.geek.com/mobile/koreas-tesco-reinvents-grocery-shopping-with-qr-code-stores-1396025/ (last visited 3. May 2013).

[2] W. Denso, *Error Correction Feature*, 2013. http://www.qrcode.com/en/about/error_correction.html (last visited 20. April 2013).

[3] M. Ebner, "QR Code - The Business Card of Tomorrow," *Proceedings of FH Science Day, Linz, Shaker Verlag, Aachen*, pp. 431-453, 2008.

[4] M. Ebner and S. Schön, "L3T assits m-Learning," *Mobile learning: Crossing boundaries in convergent environments*, pp. 69–72, 2011.

[5] M. Ebner and S. Schön, *L3T – ein innovatives Lehrbuchprojekt im Detail: Gestaltung, Prozesse, Apps und Finanzierung*. Book on Demand, Norderstedt, 2012

[6] M. Ebner, "The Influence of Twitter on the Academic Environment," in Patrut, B., Patrut, M., Cmeciu, C. (ed.). *Social Media and the New Academic Environment: Pedagogical Challenges*. IGI Global, 2013, pp. 293-307.

[7] T. Falas and H. Kashani, "Two-Dimensional Bar-Code Decoding with Camera-Equipped Mobile Phones," In *Proceedings of the Fifth IEEE international Conference on Pervasive Computing and Communications Workshops* (March 19-23, 2007), PERCOMW, IEEE Computer Society, Washington, DC, 2007, pp. 597-600.

[8] J. Z. Gao, L. Prakash and R. Jagatesan, "Understanding 2D-BarCode Technology and Applications in M-Commerce - Design and Implementation of A 2D Barcode Processing Solution," In *Proceedings of the 31st Annual international Computer Software and Applications Conference*, vol. 2 - (COMPSAC 2007) - Volume 02 (July 24 - 27, 2007). COMPSAC. IEEE Computer Society, Washington, DC, 2007, pp. 49–56.

[9] M. Grimus, M. Ebner and A. Holzinger, "Mobile Learning as a chance to enhance education in developing countries – on the example of Ghana," *mLearn 2012 Conference Proceedings; Specht, M., Sharples, M., Multisilta, J. (Ed.), Helsinki, Finland*, Volume 955, ISSN 1613-073, pp. 340-345, 2013.

[10] S. Huber and M. Ebner, "iPad Human Interface Guidelines for M-Learning," In *Z.L. Berge and L.Y. Muilenburg (Eds.), Handbook of mobile learning*. New York: Routledge, 2013, pp. 318-328.

[11] ITU report, *The world in 2013: ICT facts and figures*, 2013 http://www.itu.int/ITU-D/ict/facts/material/ICTFactsFigures2013.pdf. (last visited May 2013)

[12] H. Kato and K. T. Tan, "2D barcodes for Mobile Phones" *Proceedings of 2nd International Conference on Mobile Technology, Applications and Systems*. pp.8, Nov. 2005.

[13] MasterCard, *What is PayPass NFC?* 2013. http://www.mastercard.com/us/paypass/phonetrial/whatispaypass.html (last visited 3. May 2013).

[14] NFC Forum, *About the NFC Forum*, 2013. http://www.nfc-forum.org/aboutus/ (last visited 27. February 2013).

[15] NFC Forum, *NFC Forum Type Tags*, 2009. http://www.nfc-forum.org/resources/white_papers/NXP_BV_Type_Tags_White_Paper-Apr_09.pdf (last visited 15. April 2013).

[16] C. E. Ortiz, *An Introduction to Near-Field Communication and the Contactless Communication API*, 2008. http://www.oracle.com/technetwork/articles/javame/nfc-140183.html (last visited 27. February 2013).

[17] H. Östman, *RFID – 5 most common applications on the shop floor*, 2012. http://www.rfidarena.com/2012/12/13/rfid-%E2%80%93-5-most-common-applications-on-the-shop-floor.aspx (last visited 3. May 2013).

[18] T. Pavlidis, A New Paper/Computer Interface: Two-Dimensional Symbologies. *icpr, 15th International Conference on Pattern Recognition (ICPR'00)*, vol. 2, 2000, p. 2145.

[19] qrcodetracking, *NFC payments up to £20 in the UK*, 2012. http://qrcodetracking.net/nfc-payments-up-to-20-in-the-uk/ (last visited 3. May 2013).

[20] C. Savill-Smith, J. Attewell and G. Stead, *Mobile learning in practice*. London: Learning and Skills Network, 2006.

[21] P. Schmidmayr, M. Ebner and F. Kappe, "What's the Power behind 2D Barcodes? Are they the Foundation of the Revival of Print Media?," Proceedings *I-Know08 and I-Media08, 6th International Conference on Knowledge Management and New Media Technology*, edited by H. Maurer and K. Tochtermann,Graz, 2008, pp. 234-24

[22] M. Schön, M. Ebner and G. Kothmeier, "It's Just About Learning the Multiplication Table," Proceedings *2nd International Conference on Learning Analytics and Knowledge (LAK '12)*, Simon Buckingham Shum, Dragan Gasevic, and Rebecca Ferguson (Eds.). ACM, New York, NY, USA, 2012, pp. 73-81

[23] Skjolberg and T. Rorvik, *Ndef Tools for Android*, 2013. http://code.google.com/p/ndef-tools-for-android/ (last visited 15. April 2013).

[24] Squareup, *Using Square Wallet at Starbucks*, 2013. https://squareup.com/help/en-us/article/5039-using-square-wallet-at-starbucks (last visited 18. February 2013).

[25] C. Strauß, J. Scholz, M. Ebner and P. Schmidmayer, „Einsatz von Quick Response Codes für ortsbezogene Dienstleistungen," *Konferenzband Geoinformatik 2009*, Universität Münster, 2009, pp. 57-63,

[26] B. Violino, *Genesis of the Versatile RFID Tag*, 2003. http://www.rfidjournal.com/articles/view?392 (last visited 26. February 2013).

AUTHORS

M. Maierhuber studies Informatics at Graz University of Technology (martin.maierhuber@student.tugraz.at).

M. Ebner is head of the Department Social Learning at TU Graz since 2006 as well as Senior researcher at the Institute of Information Systems and Computer Media (martin.ebner@tugraz.at).

Towards a Geographical Orchestration of Mobile Learning Activities

Nassim DENNOUNI [1], Yvan PETER [1], Luigi LANCIERI [1] and Zohra SLAMA [2]
[1] Lille1 University, Lille, France
[2] Djilali Liabes University, Sidi Bel Abbes, Algeria

Abstract—**We are in the transition to a new era where mobility extends to many aspects of our daily lives. Learning, for example, takes place throughout life and anywhere. One may wonder how the traditional orchestration of learning can be applied in a mobile context, for example, to better support students during field trips and museum visits. In this paper, we present a geographic orchestration of resources and activities associated to learning system. The objective is to consider collaboration in mobile learning scenario that depends heavily on the location of learners, their profiles and their activity logs. We also defined a software framework for the design and implementation of pedagogic scenarios of field trip. To validate our approach, we present the implementation of a mobile artifact dedicated to the support of new visitors in exploring the historic monuments of a city.**

Index Terms—**Web technologies, field trip, pervasive environment, mobile learning scenario, Collaborative process integration, geographic orchestration, Emerging technologies.**

I. Introduction

Today, mobile computing is already at the heart of many location-based software applications (GPS, RFID ...) that are integrated in specific terminals (Smartphone, tablet PC ...). Tomorrow, ambient computing, via the introduction of these new smart devices will be used in industries that require continuity of mobiles activities such as telemedicine, home automation and serious games. Field trips or visits to museums can improve our education system because they can better direct the learning process by enabling mobile learners to construct their own knowledge by collaborating with others.

In this scalable spatiotemporal context , the orchestration enable to manage real-time learning activities such as reading text, viewing video, testing, feedback, group work, etc. but it generates significant costs of tutoring.

On the other hand, we found that the centralized orchestration is not adapted to the context of a field trip because our learner must remain master of his learning choice and should not be subject to obligatory tutorials actions.

Therefore, our goal is to recommend automatically resources and activities based on learner's geographical context in order to minimize the cost of tutoring instructor (teacher) during the course of the field trip.

In this paper, we present an approach that allows implementing a decentralized orchestration to promote the collaboration of learners in a highly personalized and adaptive open environment. In section 2 of this article, we define our theoretical context, mobile learning and the concept of orchestration. Then we make a state of the art on the organization of field trips. Then, in Section 3, we introduce a design framework for mobile learning scenarios of field trip and a teaching system orchestration of activities and resources centered on the location of learners.

Finally, we describe a software framework for mobile learning scenarios of field trip through the implementation of an application that helps to discover the historical monuments of the city by new visitors.

II. Theoretical Context and State of the Art

A. Mobile Learning

Mobile learning has become a very popular topic of research because there are many areas of research that address the topic from different points of view. Y. Laouris defines mobile learning as a function which depends on several parameters such as time, space, the learning environment, content, technology, learner profile and the method used [15].

A.Derycke considers mobile learning as a e-Learning version that is suitable for learners mobile in the aim to deliver distance learning in other media as computer workstations. Thus, thanks to the many applications developed, the student can continue his training where it is with a mobile device (Smartphone, IPod, IPad, games console) [6].

According to M. Sharples, mobile learning can be defined as the process (at both personal and public) to arrive at knowledge through exploration and linking of different contexts involving other people and interactive technologies [21].

M. Laroussi think that mobile learning is only part of a whole which consists of learning tools, infrastructure support, contexts and people are distributed in time and space. It allows learning in context, but also through contexts [16].

In our opinion, the definition of mobile learning is always about to change because it must adapt each time to new physical, technological and educational contexts that are closely related to various fields of application.

B. The orchestration in the context of field trips

The scenario of field trips can be divided into three distinct phases: 1) The preparation phase which allows for planning the site visit, 2) the movement itself as a second phase, and finally 3) the finalization phase that integrate monitoring activities such as visit reports, presentations, tests and / or feedbacks. The first phase is dominated by the work of the teacher, while the other phase rely much more heavily on the activities of learners. The location of the visit can take place only outside the classroom, while the phases of preparation and finalization can occur inside or outside the classroom [8].

Goodyear et Yang define teaching screenwriting as a reflection on the tasks to be done by learners and human and material resources that can help them succeed in such tasks. According to them, learners transfer the responsibility for learning to the teacher or content creator [10]. In the case of field trips, it is neither reasonable nor possible, because students need to be active and creative in a disciplined manner so that the learning process is effective. The term 'orchestration' refers to the real-time management of educational graphs whose activities lie in different planes (individual, group, class, community ...) and link layers of digital and physical information. While instructional design models are concerned central constraints (who learns what and how), orchestration incorporates constraints less "noble" such as segmentation scheduling, discipline, producing traces for parents, the budget energy of the teacher, the physical organization of the classroom, etc. [7].

Specht and Glahn emphasize that the orchestration also covers personalization and adaptation in a learning environment [9]. In Virtual Learning Environments (VLEs), the orchestration concerns the organization of resources and learning tasks (forums, chat ...) to motivate the learner to use the media associated with a course. As part of a field trip, orchestration not only covers the preparation phase, but also the management of the implementation of learning tasks during the course of the visit and the analysis of the results after the field trip. This analysis helps to rearrange learning activities and examine orchestration strategies of learners in the aim to increase the success rate and reduce the dropout rate during the finalization phase. Figure 1 summarizes the concepts related to the orchestration of learning activities during a field trip.

C. *State of the art on Field trips*

In the "Ambient Wood" application, the context is mainly supported by the use of the GPS system to define the location of learners and provide relevant information and services. This is done for the activities of the exploration phase of the fauna and flora in the forest. PDA learners provided information in response to readings from sensors on humidity and light at a specific location or an indication of the physical presence of a student in a particular location (via GPS). This application integrates the notion of context management based on the geographical location but does not allow learners to interact with each other to build new knowledge through the exchange of comments or resources [20].

Chen, Kao et al. [2] [3] study the birds and butterflies using mobile devices. During the field trip, students will take pictures and try to compare them with those already stored in the database accessible via an ad hoc wireless network. This type of application focuses on outdoor activities but gives little interest to the enrichment of the learning scenario of the visit and evaluation activities to be performed by the learner and the analysis of feedback during the finalization phase.

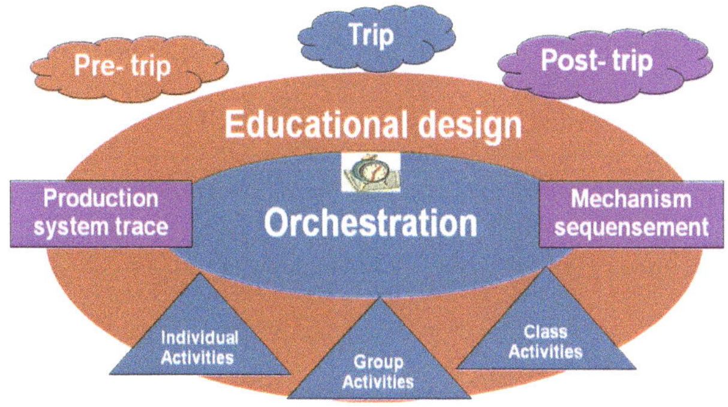

Figure 1. The main elements of orchestration of field trip.

The ENLACE project [22] focuses on enriching the mobile activities with learning resources because it allows a seamless integration of learning objects in the system. However, the architecture adopted for this project does not support real-time changes associated with different suggestions by learners.

In the RAFT project [14], a system was developed to support the creation of mobile applications that facilitate data collection during the field visit and the integration of support for collaborative monitoring activities after the field trip. This project can be considered as a new paradigm of learning, because it supports constructivist learning process in real-world contexts. One of its main objectives is to generate metadata for each new context of the visit. However, the difficulty lies in the combination of activities during the visit and activities to do in class.

In the MoULe project, sensitivity to context supports the location of learners who have a Smartphone with GPS. The system will automatically connect all learners' activities performed during their learning to specific locations within an area of interest. In this way, the system can find the current geographical position of learners and can acknowledge cultural sites in the aim to associate a set of predefined questions on these sites [1]. This project does not provide the opportunity for students to enrich their scenarios by proposing new evaluation activities and new resources for the system operates based on a predetermined scenario during the planning phase of the visit.

LEMONADE ("Learning Environment for Mobile Network-Able Devices") [9] is an environment that allows flexible creation of mobile learning scenarios. It offers flexibility to provide a generic approach for the description of scenarios associated with the full cycle of field trip. LEMONADE combines the activities on the ground with other classroom learning as brainstorming or discussion group activities. However, this tool fits just in the case of an open learning scenario that can not be planned in detail during the design phase.

Contrary to previous work on platforms such as CSCL (Computer Supported Collaborative Learning) that showed their incapacities to manage the activities of mobile learning, the CSCBL (Computer Supported Collaborative Blended Learning) can be used in field trips because they allow a focus on the use of an LMS collaboration and orchestration by the instructor. This kind of project can coordinate the activities taking place in different places thanks to new mobile technologies but it will meet many difficulties when accompanying groups and assigning contextual activities as well as for individualized learning [18].

After this state of the art, we found that applications of field trip are very specific to the context and linked to the objectives chosen by the mobile learning. The experiments presented previously do not include open mobile learning scenarios that support real-time collaboration in a spatio-temporal environment as shown in Table I.

To go further in this field, we propose a flexible solution that will support such collaborative activities.

III. Our Contribution

A. *Designing a field trip scenario*

The attitude of the mobile learners can not be predicted in detail during the design phase but it can be monitored and adjusted during the course of the field trip. For this we must develop a adaptable scenario, often in use, depending on the changing context and needs of visitors. When planning the field trip, we can imagine scenarios that adapt to the particularity of each learning situation with the possibility to integrate any new variants other than those already listed [19].

For this, we use the concept of point of interest (POI Point Of Interest) in the context of this scenario as a localizable physical location by GPS easily identifiable by the different actors of our device. We also classify POI into three categories: 1) POI of localization which is intended to help users while traveling like restaurants, subway stations... 2) the main POI used in the learning scenario such as historical monuments to visit during the field trip 3) additional POI introduced to complete the scenario predetermined by the instructor such as a historical monument that we visiting at the suggestion by one or more learners. Figure 2 shows the three phases associated to our design model.

Static layer of our scenario contains a set of main activities to do in the pedagogic scenario predetermined by the teacher (or orchestrator) and a number of POI to visit during the displacement of learners. This layer is to maintain the consistency of mobile scenario within a supervised learning mode. The variants layer by profile must be defined to be customized to meet the specific requirements of a particular spatio-temporal context which depends on the learner profile and historic of visit. In a given context, teachers can individualize the path of the mobile learner by eliminating variants that are no longer relevant to learning. The open part of our mobile scenario is executed in unexpected situations that are associated with unforeseen contexts during the design phase such as the addition of new centers of interest (POI) or the enrichment of scenario by new activities. The new variants of our scenario must be clearly identified during the finalization stage to better complement our training system.

These three parts form our mobile learning scenario that contains tasks to be done and the supports to be consulted in the context of field trip. To properly ensure respect for pedagogic objectives of our training system, we will rely on the orchestration for the choice of resources and relevant activities adapted to our open collaborative environment.

B. *Software framework of our scenario of field trip*

Field trip requires a collaborative scenario centered learning. This scenario can be based on a learning facilitator (instructor) or an expert that may be elected from among the active participants in mobile learning scenario.

The selection of experts is based on the number of comments by POI or the number of resources or activities proposed during the visit. It is a realistic learning situation based on active learning that encourages the learner to construct his own knowledge with the group help to achieve the formal objectives of pedagogic visit. The orchestration performed by the instructor or experts can manage real-time graphs relating to mobile learning activities at the level of the learner and within groups formed during the visit. For this, we propose a life cycle for the development of collaborative mobile applications capable of supporting scenarios of pedagogic visit. Figure 3 shows the four phases used in our approach.

TABLE I.
CHARACTERISTICS OF WORKS AROUND THE FIELD TRIP

Works / characteristics	Ambiant Wood Project [20]	Chen et.all. [2] [3]	ENLACE Project [22]	RAFT Project [14]	MoULe Project [1]	LEMO-NADE [9]	CSCBL [18]
Support the spatio-temporal context	Yes	Yes	Yes	Yes	Yes	Yes	Yes
Exchange comments	No	Yes	Yes	Yes	Yes	Yes	Yes
enrichment scenario	No	No	Yes	Yes	No	Yes	Yes
real-time collaboration	No	No	No	Yes	Yes	Yes	Yes
Integration of three phases of field trip	No	No	No	No	No	Yes	Yes
Using POI	No	No	No	No	Yes	No	No
support open or unexpected Context	No	No	No	No	No	No	No

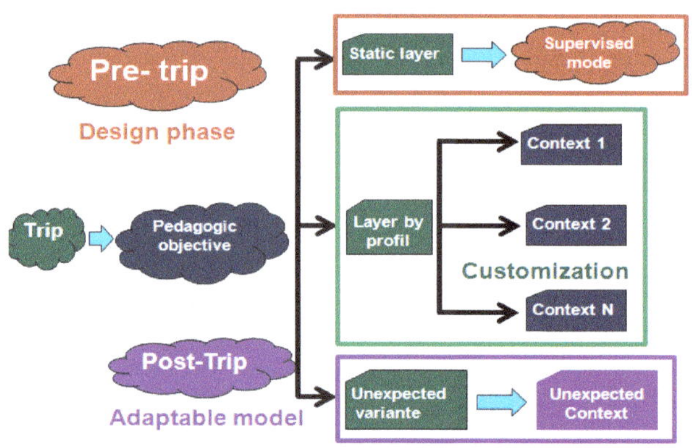

Figure 2. The layers of a fied trip mobile learning scenario.

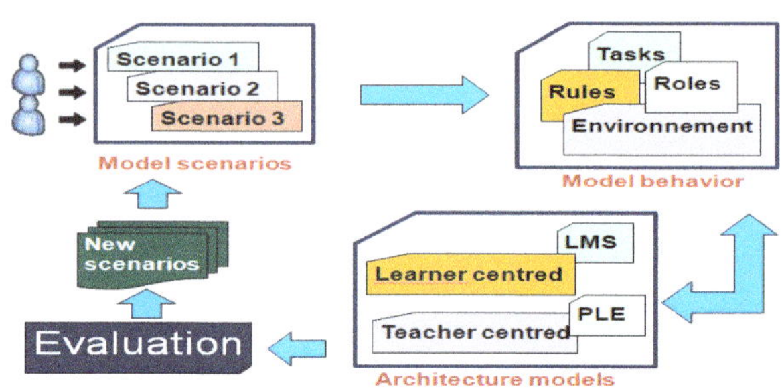

Figure 3. Lifecycle of our mobile collaborative scenario.

1) Collecting scenarios

This phase corresponds to the needs analysis. It consists of defining the scenario model as a graphical formalism that describes the roles of actors, artifacts, conditions in which the activities are performed and the relationships between these different elements [5].Our mobile scenario is divided into two parts: a fixed part developed by the instructor during the planning of fief trip and the second part enriched by learners and experts during the visit. In the first part, the instructor introduces significant POI (Main) to visit and see the resources and activities to do in the context of a pre-established supervised Mode scenario. In the second part, students and experts can introduce others POI different from those of the main timeline and join other media and activities. In the first part, the instructor introduces significant POI (Main) to visit and see the resources and activities to do in the context of a pre-established supervised Mode scenario. In the second part, students and experts can introduce other POI different from those of the main timeline and join other media and activities. Mobile users can interact with each other because they can post their comments for each resource or activity of the POI. During finalization phase, a test of knowledge acquired during the field trip is proposed to learners and experts. Finally, analysis of the results of this evaluation will allow us to have a good feedback on our system and identify the level of expertise of learners. The use case diagram of Figure 4 illustrates the role of each user of our system.

2) Development of a behavioral model

The behavioral model is an overview of the sequence of activities to be supported by the system and their characteristics (roles of actors, context of use, artifacts, etc.)[5]. This model must support spatio-temporal interactions and notifications of contextual events. It must also be sensitive to the context of communication (synchronous or asynchronous messages) and the availability of learning resources.

To ensure the different specificities of our mobile collaborative scenario, we proposed an orchestration system of resources and activity that uses a search algorithm by radius that is described as follows:

Algorithm of geographical Orchestration

Begin

1. Detecting the current position of the learner by GPS
2. Enter (R) the radius Research
3. Determine (Z) the area of interest of the learner.
4. Display POIs belonging to the area (Z)
5. *For* each POI of the area (**Z**) *Do*
   Show all tags related to POI
   End Do
6. *If* the learner selects a POI *Then*
   View activities to do by the learner and resources to consult for this POI
   End if

End.

This algorithm allows to define a pattern of behavior that can be adapted depending on the location of the learner. Moreover, profiles introduced in the database can be used to display the set of POIs that interests the mobile learner. Then, the learner selects manually the POI to visit according to its historical activities during the field trip. Finally, he can view the resources and activities associated with this POI. This information can be entered into the database by the experts, students or even by the instructor.

This database is updated in real time by all actors of our system. Thereby updating the same search with the same input parameters can give different results. However, our model behavior depends heavily on the willingness of learners to participate in the process through the quality of comments posted and groups created by the experts during the field trip.

3) The choice of an architecture model

The choice of collaborative mobile architecture associed to our behavioral model depends on the nature of the application and the specific needs of learners.

In general, the main learning platforms do not allow to teachers to anchor the activities or resources in the right places for the field trip.

On the other hand, our scenario is characterized by the absence of a global model for conduct learning activities and the freedom left to the learner. Therefore, our architecture model is based on a system that has as input elements: geographic location, learner profile, the history of the visit and the desired search radius.

Other hand, this system outputs the resources and activities associated with the selected POI by the learner as shown in Figure 5.

The choice of one or more POI is made by the learner according to his preferences and his historic during the visit. POI Selected by our learner can be a main point defined by the instructor in the scenario in supervised mode. Additional points added by an expert or another learner can be used in the aim to enrich our scenario. For this, the relevance of recommended activities and resources relative to the formal objectives of the visit are highly dependent of the learner path and his choice of POI. Other hand, the learner has a local view of all POI, resources and activities. It can also work with other students in order to construct their own knowledge, but this demarche can be biased by inappropriate comments.

Figure 4. The use case diagram associated at our scenario.

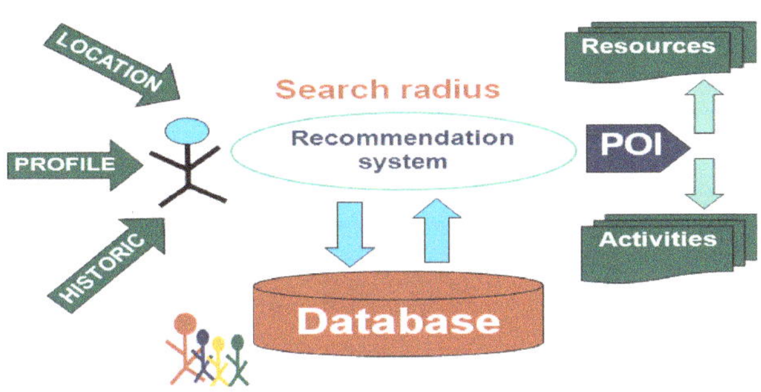

Figure 5. Architecture of our geographical orchestration system.

For this, the use of collaborative environments as DiGaE (Distributed Gathering Environment) [12] can support the exchange between learners and experts in the finalization phase of the visit.

4) The evaluation of our prototype

In this phase, we discuss the evaluation of our system based on the research and definition of indicators for the evaluation of our system. These indicators provide a basis for comparison between different learning systems and different configurations of the same system. There are several methods capable of assessing the behavior of collaborative work without worrying that it takes place in a pervasive environment or in a "classic" environment. However, we did not find any method for the specific evaluation of the pervasive aspect in collaboration [11]. Indeed, there are no indicators, no measure or method that should allow us to evaluate our mobile collaborative scenario in all these aspects. To try to remedy this, we calculate a set of measures that reflect the participation of the learner at training and collaboration with the others during the course of our pervasive scenario. These measures are classified into two categories:

1) Global indicators that focus on the evaluation of the contribution of learners or experts in relation to all the resources and activities of the scenario

2) Local indicators that evaluate the individual behavior of the learner.

IV. IMPLÉMENTATION

To validate our approach for developing field trip applications, we have implemented a project to help visitors of a city to better locate their points of interest (POI) on a Mapping support. This project also aims to provide information about the historical monuments and all places that can act as landmark for the city.

Our work also allows to assist new visitors of a city by providing them learning resources or activities that are associed to points of interest. This ensures a good learning with formal objectives. Our prototype uses mobile technologies found in a smartphone in order to exploit the context of locating a visitor (learner) and profile. to display multimedia tailored to their needs and can help during his visit. Our learners can visit the main POI added by the instructor, he also consults the associated resources and it can make the proposed activities. Our learner can visit the main POIs added by the instructor, he also consults the associated resources and he can make the proposed activities. The learner can become an expert in an area with one or more POI thanks to its potential contribution or its degree of interaction during the course of the visit. Experts can create groups based on their different areas of expertise and can power their workspaces by new resources or activities. The instructor or the orchestrator of our learning can create expert accounts based on the demands of learners.

Figure 6 explains the use cases to add to the diagram shown in Figure 4 to best answer the needs of collaboration between users of our application.

Other hand, users of our application can enrich database with new points of interest through the various updates made from their Smartphone as shown in Figure 7.

For this, they are classified according to different profiles: learner, expert orchestrator visitor... And according

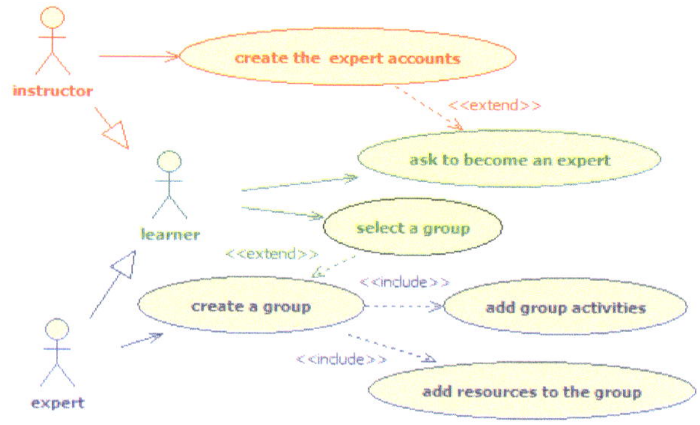

Figure 6. The specific use case in ours application.

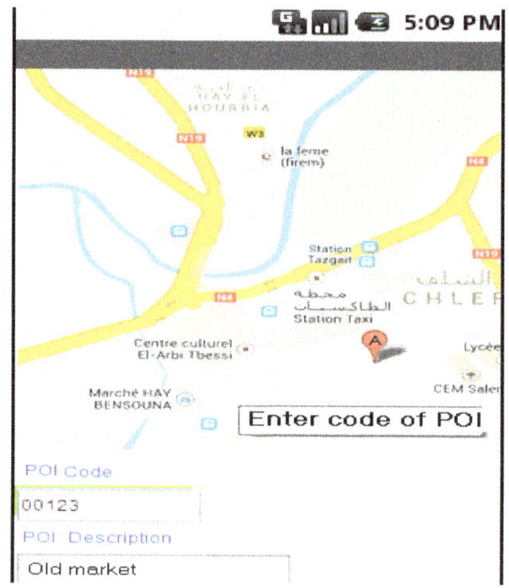

Figure 7. The technical architecture of our application.

to the rights granted by the administrator, they update the server from their mobile devices in order to feed the database with images / videos produced by their Smartphone during the visit as shown in Figure 8:

The architecture of our prototype uses REST Web services based on the HTTP methods (GET, POST, PUT and DELETE) and on the JSON format for the exchange between the server and the Smartphone. With the support of Google Maps, the use of tags is possible to display several information about points of interest (POI) on the smartphone. (See Figure 9)

To evaluate our system, we can calculate, for example, the rates of contributions by role as follows:

$$CRL = \sum RL \Big/ \left(\sum RL + \sum RO + \sum RE\right) \quad (1)$$

$$CRO = \sum RO \Big/ \left(\sum RL + \sum RO + \sum RE\right) \quad (2)$$

$$CRE = \sum RE \Big/ \left(\sum RL + \sum RO + \sum RE\right) \quad (3)$$

CRL: Contribution Rate of Learners.
CRO: Contribution Rate of Orchestrator.
CRE: Contribution Rate of Experts.
RL: Resources added by Learners.
RO: Resources added by the Orchestrator.
RE: Resources added by Experts.

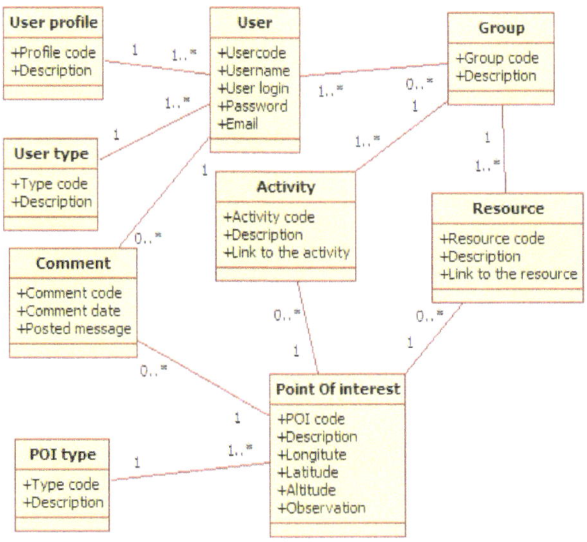

Figure 8. Class diagram associated to ours application.

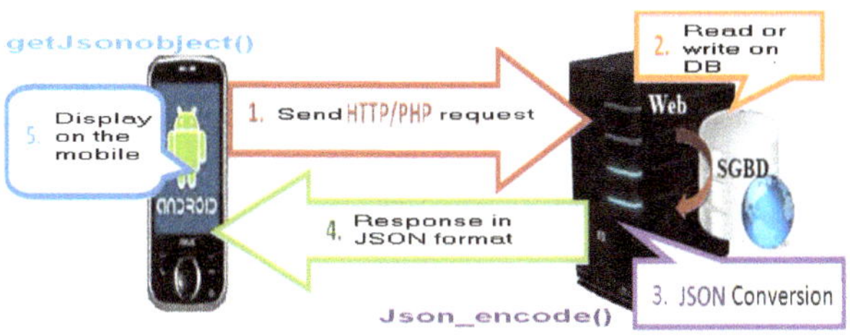

Figure 9. The technical architecture of our application.

These indicators are used to measure and compare the different active participation of different types of actors in our scenario.

Similarly, we can also calculate the contribution rate for Adding POI and for the exchange of different comments.

Most of these parameters that we want to calculate are simple enough to be extracted automatically by the system during our scenario. They allow us to facilitate the visualization of learners' activities in order to deduce new learning profiles. Then it will be possible to see the emergence of new scenarios for electing the experts or for evaluating learners.

V. Conclusion

Our objective was to define a system that allows an orchestration of mobile learning activities based on the geographical eposition of the learner while respecting the collaborative and open aspect of our mobile scenario.

In this article, we define a conceptual framework for a scenario of field trip and a software framework for the implementation of a development cycle of applications dedicated to mobile learning.

We also presented an implementation of a mobile artifact to validate our approach that is intended for a scenario of field trip.

Other hand, we have designed this artifact in order to implement a mobile learning located in near real time to make contact with a new city. However the limitations of our system may hamper the progress of our learning compared to formal objectives expected of our field trip.

For example, the use of geographic orchestration of resources and activities on the basis of a single location does not take into account the overall vision of pedagogical objectives of our scenario.

The use of ontologies can provide activities and resources based on their relevance with respect to the formal objectives of the mobile scenario [13].

To do this, we believe that the semantic orchestration of mobile learning activities can improve our solution in future work on field trips or visits museums.

Acknowledgment

This work is partially funded by the FUI Hermès project and Nord-Pas de Calais region

References

[1] M. Arrigo, O. Giuseppe, D. G. Fulantelli, M. Gentile, G. Novara, L. Seta. "A Collaborative MLearning Environment". Paper presented at the 6th Annual International Conference on Mobile Learning, mLearn 2007 pages 13-22., Melbourne Australia, 2007.

[2] Y. S. Chen, T. C. Kao and J. P. Sheu, "A Mobile Learning System for Scaffolding Bird Watching Learning." Journal of Computer Assisted Learning 19(3), 2003, pp. 347-359.

[3] Y. S. Chen, T. C. Kao and J. P. Sheu , "A Mobile Butterfly-Watching Learning System for Supporting Independent Learning", In Proceedings of the 2nd IEEE International Workshop on Wireless and Mobile Technologies in Education, JungLi, Taiwan, IEEE Computer Society, 2004

[4] C. Cobos, O. Rodriguez, J. Rivera , J. Betancourt , M. Mendoza ,E. León , E. Herrera-Viedma, "A hybrid system of pedagogical pattern recommendations based on singular value decomposition and variable data attributes" journal of Information Processing and Management,ScienceDirect,2013,pp.607-625

[5] O. Delotte, " CoCSys : une approche basée sur la construction d'un modèle comportemental pour la conception de systèmes collaboratifs mobiles'', thèse de doctorat, Ecole Centrale de Lyon, 2006.

[6] A. Derycke, " Tutoriel Ubimob 06, Du E-Learning au Pervasive-Learning: Concepts, Exemples et Questions de Recherche'' 3ème Journées Francophones Mobilité et Ubiquité, Conservatoire National des Arts et Métiers – Paris, 2006

[7] P. Dillenbourg, " Modéliser l'orchestration'', Conférence sur Éducation et sciences de l'apprendre Session 4, Ecole Normale Supérieure de Lyon, automne 2012

[8] A. Giemza, L. Bollen, P. Seydel, A. Overhagen, H. Ulrich Hoppe "LEMONADE: A Flexible Authoring Tool for Integrated Mobile Learning Scenarios'' The 6th IEEE International Conference on Wireless, Mobile, and Ubiquitous Technologies in Education, IEEE Computer Society, 2010, pp 73-80.

[9] C. Glahn and M. Specht. " Embedding moodle intoubiquitous computing environments". In Publications and Preprints. LMedia, 2010.

[10] P. Goodyear and D. F. Yang. "Patterns and pattern languages in educational design "In Handbook of Research on Learning Design and Learning Objects: Issues, Applications, and Technologies, volume 2, pages 167–187. IGI Global, 2009.

[11] K.Hamadache, L. Lancieri, "Indicateurs pour l'évaluation du travail collaboratif en environnement Pervasif ", ICT, 2010

[12] S. Kaddouci, Y. Peter, T. Vantroys, P. Laporte, Coupling pedagogical scenarios and location-based services for learning, IEEE Learning Technology Newsletter, October issue, 2010.

[13] S. Kaddouci, Y. Peter, T. Vantroys, Designing learning support for mobile workers, Proceedings 9th IEEE International Conference on Advanced Learning Technologies (ICALT'09).

[14] M. Kravcik, M. Specht, A. Kaibel, L. Terrenghi, "Collecting Data on Field Trips — RAFT Approach" Third IEEE International Conference on Advanced Learning Technologies (ICALT'03), 2003, pp. 478.

[15] Y. Laouris "We Need an Educationally Relevant Definition of Mobile Learning". 4th World conference on mLearning. Cape Town, South Africa; 2005.

[16] M. Laroussi, "Extension d'IMS/LIP pour supporter l'apprentissage pervasif". INSAT, université de Carthage, Tunis, Tunisie, décembre 2011

[17] N.Manouselis, H.Drachsler, R.Vuorikari, , H.Hummel, , & R.Koper, "Recommender systems in technology enhanced learning". In L. R. F. Ricci & B.Shapira (Eds.), Recommender systems handbook Berlin: Springer, 2011, pp. 387–415.

[18] M. Pérez-Sanagustın, G. Ramirez-Gonzalez, D. Hernandez-Leo, M. Munoz-Organero, P. Santos, P. Blata and C. Delgado Kloos. "Discovering the campus together: A mobile and computer-based learning experience", Journal of Network and Computer Applications pages 176–188, 2012.

[19] E. Ouraiba "Scénarisation pédagogique pour des EIAH ouverts Une approche dirigée par les modèles et spécifique au domaine métier", thèse de doctorat, Université du Maine, 2013.

[20] B. Russell & P. Lonsdale, "Mobile Context Aware Systems: TheIntelligence to Support Tasks and Effectively Utilise Resources ", Mobile Human-Computer Interaction, MobileHCI, 2004, pp. 240-251.

[21] M.Sharples, et al. (2007) 'Mobile Learning: Small devices, Big issues' (in Sharples, M., et al. (eds.) Technology-Enhanced Learning, 2009, Part IV)

[22] M. F. Verdejo, C. Celorrio, E. J. Lorenzo, M. Millan, S. Prades, & J. Velez, "Constructing Mobile Technology-Enabled Environments for an Integrated Learning Approach", En Innovative Mobile Learning: Techniques and Technologies , IGI Global, 2008, pp. 147-171.

AUTHORS

N. Dennouni is a PhD student in computer science and member in the NOCE team of the LIFL laboratory at Lille 1 university (dennouninas@gmail.com)

Y. Peter is an associate professor in computer science and member in the NOCE team of the LIFL laboratory at Lille 1 university. (Yvan.Peter @univ-lille1.fr)

L. Lancieri is a Professor of Computer Sciences and the Head of the NOCE team of the LIFL laboratory at Lille1 university (Luigi.lancieri @univ-lille1.fr)

Z. Slama is an associate professor in computer science and member in the ISIBA team of the EEDIS laboratory at Djilali Liabes university. (z.slama@yahoo.fr)

35

Designing an Intelligent Mobile Learning Tool for Grammar Learning (i-MoL)

Munir Shuib, Amelia Abdullah, Siti Norbaya Azizan, Thenmolli Gunasegaran
Universiti Sains Malaysia, Penang, Malaysia

Abstract—**English is the most important second language in most non-English speaking countries, including Malaysia. A good English proficiency comes from good grasp of grammar. To conquer the problems of low English proficiency among Malaysians, it is important to identify the key motivators that could facilitate the process of grammar learning. In this digital age, technology can play a very important role and mobile technology could be one of it. Thus, this study aims at designing a mobile learning tool, namely the Intelligent Mobile Learning Tool for Grammar Learning (i-MoL) to act as the "on-the-go" grammar learning support via mobile phones. i-MoL helps reinforce grammar learning through mobile phone with game-like applications, inquiry-based activities and flashcard-like information. The intelligent part of i-MoL lies in its ability to map the mobile-based grammar learning content to individual's preferred learning styles based on Felder-Silverman Learning Style Model (FSLSM). The instructional system design through the ADDIE model was used in this study as a systematic approach in designing a novel and comprehensive mobile learning tool for grammar learning. In terms of implications, this study provides insights on how mobile technologies can be utilized to meet the mobility demand among language learners today.**

Index Terms—**Mobile learning, English language, grammar, learning styles, FSLSM, instructional system design, ISD, ADDIE, higher learning institutions, graduates, Malaysia**

I. INTRODUCTION

In today's competitive and globalised world, there is a tremendous emphasis on the use of English in various working sectors. However, to date, there is still much room for improvement on English Language competency among Malaysian graduates, especially those from rural areas who are still struggling to grasp the rudiments of the language. While the government puts much effort to raise the level of English Language proficiency, the standard of English among Malaysian undergraduates is still declining. Amid this issue, effective efforts are urgently needed in the educational system to ensure that the future cohorts of Malaysians are able to master the English language in order to remain competitive in the job market. Since grammar is one of the areas that most students are still struggling with [1], it could be an area worth tackling in order to improve their English competency. The acquisition of English grammar by university students is important as accuracy in speech and writing is highly required in performing academic activities. Therefore, it is important for students not only to be able to communicate in English fluently but also be able to use and understand English grammar forms and functions accurately.

The fact is, today's students are vastly different from the way they were a few decades ago. According to Eaton [2], today's students are tech-savvy. They can access a world of resources and information just at their finger tips. They are hungry for motivation, inspiration, and guidance. They harness their creativity to express themselves and demonstrate what they know using technology. They are the creators, not simply consumers, of technology. In line with such tech-savvy attributes, the field of English language education has witnessed rapid changes in teaching methodology, material development, and assessment. Traditional beliefs and practices of language teaching are giving way to newer, more innovative ways of teaching. Old, authoritative, "teacher-centred" approaches to teaching are giving way to more collaborative and interactive approaches. Audio language labs, which were once considered a significant innovation, have begun to fall out of favour [3]. The reason for this most likely had its roots partly in the methodological move away from structural approaches to language learning, to a flurry of novel techniques for second learning acquisition [4]. Therefore, for grammar learning to be effective, the learning materials designed for students should be interesting and interactive. They also have to be in line with the learning needs of young generations in today's technological environment who would prefer more digital motivation, flexibility and mobility.

In a survey conducted by the Malaysian Communication and Multimedia Commission [5], it was reported that mobile phones penetration in Malaysia in the last four years kept growing tremendously in most states and the main mobile phone users were found to be those ranging from 20 to 49 years in age. Thus, it may not be impossible that mobile phones are used in learning as an alternative approach to complement the traditional and formal approaches to English grammar learning for the Malaysian graduates. There are many ways in which mobile learning may be utilized in HEIs. One of the ways concerns the teaching and learning of English grammar among undergraduates. Mobile-based approach to teaching and learning of English grammar is unique in that it allows the language learners to benefit the learning process in ubiquitous and more personalized manner. This approach can also enrich, enliven, or add variety to the conventional method of grammar learning. Teaching and learning of English grammar via mobile phone is seen as a viable alternative to enhance learners' proficiency in the language as it is digitally designed, flexible and mobile - i.e. anytime and anywhere.

For learning to be effective, is important for lesson delivery to match with students' style of learning. As Bertolami [6] has suggested, "one of the focal points of student frustration with the curriculum is the disparity between learning (content) and the delivery of instruction (form)". In line with this need, this study presents how a personalized intelligent mobile learning system for grammar learning, namely i-MoL, can be designed to support effective grammar learning via mobile phones which is tailored to students' preferred learning styles. The ADDIE instructional system design model was used in this study as a framework to guide the designing process of i-MoL. In this vein, this research attempts to marry the elements of learning styles with the concept of instructional system design in developing the intelligent mobile learning tool for grammar learning.

II. LITERATURE REVIEWS

A. M-Learning in General

M-learning is introduced globally as a flexible learning approach. It is a learning process that is delivered through wireless mobile handheld devices which let learning takes place at anytime and anywhere. Mobile learning takes learning to individuals, communities and countries that were previously too remote, socially or geographically, for other types of educational initiative [7]. In this way, it is more advantageous and convenient as compared to e-learning which always require computer and internet access. 'Mobile learning' is not simply the combination of 'mobile' and 'learning'. According to Traxler [7], it has always implicitly meant 'mobile e-learning' and its history and development have to be understood, thus many wider issues should be addressed in terms of explaining, understanding and conceptualizing it.

There has never been a specific definition of 'mobile learning. However, there are many evolving definitions that attempt at identifying and defining mobile learning. Quinn [8] defined m-learning as "e-learning through mobile computational devices: Palms, Windows CE machines, even your digital cell phone". Another view of mobile learning considers m-learning as an "exploitation of ubiquitous handheld hardware, wireless networking and mobile telephony to enhance and extend the reach of teaching and learning" [9]. A 2004 study conducted by Naismith et al. [10] suggested that mobile technologies can relate to six types of learning, or 'categories of activity', namely behaviourist, constructivist, situated, collaborative, informal/lifelong, and support/coordination. Jones et al. [11], by contrast, considers the motivational or affective aspects of mobile learning as defining characteristics, namely control (over goals), ownership, fun, communication, learning-in-context and continuity between contexts. The work of Winters [12] concluded that current perspectives on mobile learning generally fall into the following four broad categories:

- Techno-centric: mobile learning is primarily seen as learning that is supported by mobile devices
- An extension of e-learning: mobile learning is a subset of e-learning
- Augmenting formal education: mobile learning complements the formal education
- Learner-centred: mobile learning is about mobility and context

Mobile learning is often viewed not only as a silver bullet in education, but also as a tool for learners to access knowledge while they are on the move at anytime and anywhere. More and more educational institutions around the globe are now moving toward mobile learning as a way to benefit from the ubiquitous quality it may offer for educational purposes. An emerging body of literature that explores the potential of mobile learning for educational contexts has identified several significant benefits of mobile learning and these include cost effectiveness, convenience, motivation to learn, accessibility, flexibility, as well as immediacy of information and interaction [13][14][15][16].

A common type of m-learning which uses Short Messaging Service (SMS) is known as SMS-based learning. According to Gupta [17], SMS is a delivery mechanism of short messages; a store and forward way of transmitting messages over the mobile networks. The messages sent will be stored in a central short message center which then forwards it to the destination mobile. One may send up to 160 characters using a basic mobile phone. SMS-based learning is a new learning phenomenon and exists because of the drastic increase in the number of mobile devices used nowadays [18]. It is often defined as learning that takes place with the help of portable electronic tools [8]. SMS-based learning also offers a virtual learning environment. However, unlike e-learning which has been relatively established in many higher learning institutions, SMS-based learning is still at its infancy and has not been widely used yet.

B. M-Learning in Malaysia

M-learning in Malaysia is still in its infancy [19]. Despite this, many studies have already been carried out to determine its feasibility as a learning tool [20][20]. Generally, studies found that m-learning is accepted by students and feasible to be used by students and teachers. M-learning is also perceived well among scholars in terms of its potential in Malaysian educational institutions [24][25][26][27]. However, the majority of the studies conducted in Malaysia have focused on science and technical courses. Thus far, very few studies have been done involving students and teachers of non-science /non-technical courses including English language. Although the introduction of mobile phones for learning in Malaysia is considered a new idea, there was a forecast that smart phones would be used in secondary schools in Malaysia from the year 2016 to 2020 [22]. Siraj [23] estimated that Malaysia has a huge potential to implement mobile learning in its curriculum.

The mobile learning research team from Universiti Sains Malaysia (USM) has attempted to identify and fill the aforementioned gap by carrying out a few studies on mobile learning. The studies aimed to explore the potential of SMS-based learning in supporting the existing pedagogical practices in various fields, including information technology (IT), linguistics, education, education technology, management, distance education and informal education. As an outcome of this project, an SMS-based mobile learning system has been developed to serve as a flexible and affordable learning alternative for bottom billion students to learn at anytime and anywhere.

The team also conducted a study to determine if the SMS technology could be used as a supplementary tool to help students in developing their English language grammar competency. The research sought to explore students' and teachers' perception about the use of SMS in learning grammar which could offer some indication on the feasibility of the teaching of grammar using SMS. The research reported positive input from the participating students and teachers who are willing to try it out with their students. Teaching grammar via SMS is thus found to be a feasible pedagogical tool which will offer benefits to students. This research work has shed some light on the feasibility of SMS-based mobile learning in facilitating the teaching and learning of English grammar for undergraduate students, while tapping into the desire for more advanced and interactive features of mobile learning tools [28].

III. DESIGNING I-MOL

This paper seeks to introduce an intelligent mobile learning tool for grammar learning called i-MoL. The model is proposed as an alternative 'on-the-hand' approach to teaching and learning of English Grammar. The term 'i-MoL' is coined to express the mobile concept of grammar learning. The 'intelligent' part of the proposed model lies in its ability to create a personalized grammar learning environment for the learners. Specifically, personalization in this context is referring to the provision of learning contents that are tailored to suit each learner's learning style. In this regard, the proposed model is equipped with a mechanism to firstly model learners' learning styles and then adapt the grammar learning content to each type of learning styles. Through the i-MoL model, a mobile learning system is to be developed as a tool to facilitate a seamless ubiquitous learning environment for grammar learning. The system will include several mobile-based applications for grammar learning including notes, alerts, query, quizzes, polling, and forum.

i-MoL consists of two major parts: grammar content management and system management. Thus, it requires a joint initiative of linguistic and technological experts and educators to research the possibilities of using advanced mobile platforms for transferring English skills and competencies to the learners. This project aims to develop, test, and evaluate the usability of such educational approach for grammar learning. The i-MoL project can be seen as an extension of the previous work by the mobile learning team from USM on an SMS-based mobile learning project. The i-MoL model is proposed as an effective multimedia-based mobile learning tool for learning of English grammar.

The current i-MoL project intends to explore the next continuum; to design and develop a more diversified approach to pedagogical use of mobile devices for English grammar. This is in line with the advancement of mobile technologies, proliferation of mobile phone ownership in Malaysia [29] and increasing significant mobile learning projects among Malaysian scholars [25]. i-MoL is thus seen as an innovation in mobile learning that supports grammar learning through cross-platform user interface, i.e. from low-end to high-end mobile platform, including Android, iPhone, and basic mobile phones with SMS and MMS capabilities. By exploiting these various mobile platforms, i-MoL aims to provide a personalized and flexible learning environment to facilitate learners in learning grammar. In conjunction with this aim, i-MoL aims to enhance the existing teaching-learning modes by utilizing an adaptive mechanism for personalized contents that are made available to learners via mobile devices.

A. *i-MoL System Architecture*

i-MoL consists of two subsystems - the i-MoL server and the i-MoL tool. Figure 1 illustrates the system architecture of the i-MoL model. The i-MoL server is a learning server whereby lecturers and system administrators can connect to it via Internet connection through their personal computers. It consists of six modules, namely User Management, Content Management, Assessment Management, Feedback, Forum, and Push Modules. Each module is described as follow:-

- *User Management*: This is the heart of the i-MoL intelligent system that creates the personalization of learning content. The User Management module is composed by databases of learners' personal details, learning style, a map of the knowledge obtained from the domain module and a profile analyzer that helps to analyze learners' learning styles and map them to the appropriate content group
- *Content Management*: The Content Management module stores the learning materials in the mobile learning content database which are multimedia materials like texts, images, videos, and audios. The materials are fetched online from the web-based portal using a set of web services; packaged, structured, and then delivered to the learners according to their learning style group
- *Assessment Management*: The Assessment Management module includes interactive applications for lecturers to conduct quiz and polling in order to give assessment to learners to evaluate their learning results
- *Feedback*: This unit provides the mechanism for learners to give feedbacks or send course-related queries to their lecturers. Data is managed through the inbox database
- *Forum*: Through this unit, lecturers can organize group discussions for learners to participate and share their learning experiences
- *Push*: This unit allows lecturers to deliver and disseminate any alert, notifications, or learning tips to learners

The other subsystem of the i-MoL system architecture is the i-MoL tool. It is a set of software applications that provide learners access to the mobile-based grammar learning activities. The learners who are equipped with handheld devices will interact with the system by sending and responding to special instructions for each application. The system will deliver the required learning contents to the learners. The applications of i-MoL tool are as follow:-

- *Learning content*: access to various types of grammar learning contents (text-based and multimedia-based) and content managing tools like archive, download, bookmark, and notepage
- *Game-based application*: a selection of interactive game-like (e.g. quiz, poll, inquiry-based, ranking game) mobile modules.
- *Discussion room*: forum application to facilitate group-based learning
- *Alert and reminder*: a reinforcement tool to help students obtain instant information on grammar learning content

- *Learning style identifier*: SMS-based and mobile application that can automatically identify students' learning style through a series of questions
- *Query*: a student-centered mobile application that helps students to get instant feedback from their lecturer regarding the subjects

B. Tailoring i-MoL To Learning Styles

i-MoL is designed with a unique mechanism to adapt the grammar learning contents to learners' learning styles. The model promotes an adaptive approach for grammar learning whereby learning contents are tailored to adapt to an individual's learning style within a mobile learning environment. The i-MoL learning environment aims at providing the learners with a grammar learning experience that is not only flexible and mobile, but also convenient to them since it is personalized and adaptive to their learning styles.

Since personalization is a crucial aspect, the system will identify the characteristics of learners in terms of their preferred styles for grammar learning. The learning styles will be used as a basis for the adaptive mechanism which consequently aims at providing learners with personalized grammar learning contents that are mapped to their preferences and fit to their profile. Many learning style models exist in the literature [30][31][32][33]. However, for the purpose of this study, Felder and Silverman's Learning Styles Model (FSLSM) is to be used due to the following reasons:-

- Applicability and compatibility to the principles of interactive learning system designs [34]
- Suitability with a hypermedia learning system that incorporate varying types of multimedia-based learning materials [35]
- Practicality and convenience for establishing students' dominant preferences in a mobile learning environment [36]
- Appropriateness and feasibility for language learning [37] [38]

The information about learners' learning styles will be gathered through mobile-based questionnaires designed to accurately identify the preferred learning styles within mobile environment. The questionnaires used to provide the profile of learners' learning preferences are adapted from the Index of Learning Style (ILS) questionnaire developed by Felder and Soloman [39] based on the Felder-Silverman Learning Style Model (FSLSM).

C. Designing i-MoL via ADDIE Model

The ADDIE model is used as an Instructional System Design (ISD) approach to guide the development and production of the i-MoL mobile learning system. The term 'ADDIE' is actually an acronym for the five phases involved in this process, namely "Analysis", "Design", "Development", "Implementation", and "Evaluation". ISD refers to the systematic process which involves needs analysis and the development of an effective delivery system to meet the identified needs [40]. Many models of ISD exist in the literature, such as ADDIE, ASSURE, and Dick and Carrey's [41]. However, researchers agreed that ADDIE is a commonly used instructional model [42][43][40] [44]. In fact, most of the models are actually spin-offs or variations of the ADDIE model [45]. The five phases of ADDIE form a cyclical process; output from each phase will feed input to the subsequent phase [46]. Moreover, as Koneru [40] explains, ADDIE provides ample scope to integrate pedagogy, learning theories, and other instructional design principles. In this study, the researchers used the ADDIE model as an ISD approach to guide the development of the production design of i-MoL since it is the most generic, universal and simple ISD model that suits the purpose of a mobile learning system development [47]. Figure 2 illustrates the i-MoL system development framework utilizing the ADDIE model.

The first phase of ADDIE, i.e. Analysis, forms the foundation of the system design and development in this study. The Analysis phase is to define what is to be learned [48]. It generally involves a set of tasks, such as needs analysis and knowledge acquisition [46]. According to Zimnas et al. [43], the designer identifies the learning problem, goals and objectives, learners' needs, existing knowledge and any other relevant characteristics during this phase. In this study, the Analysis phase involves a needs analysis to identify system requirements for both system and content. It also includes students' learning style identification through questionnaires. By end of the phase, the i-MoL system requirement specification will be generated.

The second phase, i.e. the Design phase, receives outputs from the Analysis phase in order to plan for strategies in developing the instructional materials. Steps or activities in this phase may include tasks to write objectives, develop test items, plan instruction, and identify resources [49]. For the purpose of this study, the Design phase is divided into two stages, which are content design and system design. In the content design stage, the mobile-based grammar learning content is outlined to represent each type of learning style; while in the system design stage, the instructional materials comprising learning objectives, learning activities, and graphical user interface (GUI) of the system are planned according to the specific requirement of the mobile platforms.

The Development phase is where the design output from the previous phase is put to practice. According to Kulvietiene and Sileikiene [48], it is where the system designer or developer will author and produce the learning materials. The Development phase requires the designer or developer to use relevant software tools and applications for the production of the material, such as HTML editors, web-courseware, and web-authoring software [49]. It may also involve tasks of developing the GUI and user database [46], as well as preparing the multimedia components [51]. In this study, the Development phase comprises two tasks which are run concurrently. The first task is content development, whereby the designed grammar learning contents are produced in the forms of mobile-supported format. The mobile-based grammar learning contents are developed for each module of learning style. As for the second task, i.e. system development, the web-based portal and mobile applications are produced utilizing the relevant system development tool. The hardware components of the i-MoL server system will also be set up in this phase. The output from the Development phase will be a working prototype of i-MoL.

Implementation is the fourth phase of the ADDIE instructional process. In this phase, the designed and developed instructional materials are installed in the real-world setting. As Arkün and Akkoyunlu [51] noted, the Implementation phase is the process whereby the design is put into practice with the actual learners of the learning materials. In a 2011 study by Chen, Chen, and Tsai [52], the researchers distributed the training materials to the learners through the developed module platform and also conducted experimental teaching process during the Implementation phase. In this study, the Implementation phase involves the feasibility testing of the i-MoL prototype. The feasibility testing is for trial-running of the prototype in order to assess whether it is functional and feasible within the targeted audiences and specified learning objectives. In this phase, this study aims to identify the practicality of learning style elements and mobile-based grammar learning applications in the prototype.

The end of the Implementation phase leads to the final phase of ADDIE, which is the Evaluation phase. This phase determines the adequacy of the instructional materials [48] and verifies the extent that the materials are able to meet the specified learning objectives and learners' needs [51]. The Evaluation phase may be formative or summative and the tasks or activities may involve questionnaires, user focus groups, and interviews [50]. The Evaluation phase aims at evaluating the usability of the i-MoL prototype through questionnaires and focus group discussions. Then, the users' feedbacks are analyzed to determine the system's usability from various aspects. Technical strengths and weaknesses will also be identified for further enhancement of i-MoL. The output from this final phase would be the final and revised prototype of i-MoL.

IV. CONCLUSION

Based on existing literature, it is safe to say that this mobile learning project is the first of its kind: learning English grammar that is personalized to learning styles while 'on-the-go'. Mobile phones are now becoming people's most trusted personal devices and are receiving much attention by scholars, particularly educational researches. For this reason, it seems like a natural progression to introduce such ubiquitous technology to benefit the younger generations in learning English language. i-MoL as proposed in this paper differs in such a way that it provides learning contents that are adaptive to students' preferred learning styles. It intends to deliver a personalized grammar learning environment that fits individual's learning style and the mobility needs of the students. The project is still in its content development stages. Once the system has been fully developed, it is hoped that i-MoL will be able to contribute significantly to the pedagogical aspects of mobile learning, particularly for the purpose of English grammar learning. Nevertheless, it is believed that i-MoL will be able to address many needs of a wider and more diversified community of learners; not just in learning English grammar, but also in other skills and competencies.

ACKNOWLEDGMENT

The authors would like to thank the Malaysian Ministry of Higher Education (MOHE) for the provision of the Prototype Research Grant Scheme (203/PHUMANITI/ 6740011) in completing this project.

REFERENCES

[1] S.Y. Enxhi, B.H. Tan, M.F. Yong, "Speech disfluencies and mispronunciations in English oral communication among Malaysian undergraduates", International Journal of Applied Linguistics & English Literature, vol. 1, no. 7, pp. 19-32, 2012.

[2] S. E. Eaton, Global Trends in Language Learning in the 21st Century, Calgary: Onate Press, 2010.

[3] G. Davies, P. Bangs, R. Frisby, E. Walton, Setting up effective digital language laboratories and multimedia ICT suites for MFL, The National Centre for Languages as Part of the Languages ICT Initiative, 2005.

[4] C. Alexander, "Language Labs: An Overview of the Trends", Teaching English with Technology, vol.7, no. 3, 2007.

[5] Malaysian Communications and Multimedia Commission (MCMC), Report on handphone users survey 2008, 2008.

[6] C.N. Bertolami, "Rationalizing the Dental Curriculum in Light of Current Disease Prevalence and Patient Demand for Treatment: Form vs. Content", Journal of Dental Education, vol.65, pp. 725–735, 2001.

[7] J.Traxler, "Learning in a Mobile Age", International Journal of Mobile and Blended Learning, vol. 1, pp.1-12, 2009.

[8] Quinn, C., "Get Ready for M-learning", Training and Development, vol. 20,no. 2, pp. 20-21, 2001.

[9] MoLeNET, Mobile learning myths, 2010.

[10] L.Naismith, P. Lonsdale, G. Vavoula, M. Sharples, Literature Review in Mobile Technologies and Learning, Bristol: NESTA FutureLab, 2004.

[11] A. Jones, K. Issroff., E. Scanlon, G. Clough, G., P. Mcandrew, " Using mobile devices for learning in Informal Settings: Is it motivating?" In Proceedings of IADIS International conference Mobile Learning, July 14-16, 2006.

[12] N. Winters, "What is mobile learning?", In M. Sharples (Ed), Big Issues in Mobile Learning (pp. 1-34). Kaleidoscope Network of Excellence, Mobile Learning Initiative, 2006.

[13] D. Laurillard, "Pedagogical forms of mobile learning: Framing research questions", In N. Pachler(Ed.), Mobile learning: Towards a research agenda (Vol. 1, pp. 33-54). London: WLE Centre, Institute of Education, 2007.

[14] J.-H. Valk, A. T. Rashid, L. Elder, "Using Mobile Phones to Improve Educational Outcomes: An Analysis of Evidence from Asia", The International Review of Research in Open and Distance Learning, vol.11, no.1, pp.117-140, 2010.

[15] A.Valenta, D. Therriault, M. Dieter, R. Mrtek, "Identifying Student Attitudes and Learning Styles in Distance Education", Journal of asynchronous learning networks, vol. 5, no. 2, pp. 111-127, 2001.

[16] M. J. W. Lee, A. Chan, "Reducing the Effects of Isolation and Promoting Inclusivity for Distance Learners through Podcasting", Turkish Online Journal of Distance Education, TOJDE, vol. 8, no.1, pp. 85 – 105, 2007.

[17] P. Gupta, Short message service: What, how and where?, 2011.

[18] A. Hashim, W. Wan Ahmad, R. Ahmad, "Mobile learning implementation: Students' perception in UTP", In ICCESSE 2010 : International Conference on Computer, Electrical, and Systems Science, and Engineering, Feb 24-26, Pulau Pinang, 2010.

[19] I. Ismail, R.M.Idrus, "Development of SMS Mobile Technology for M-Learning for Distance Learners", International Journal of Interactive Mobile Technologies, iJIM, vol. 3,pp. 55-57, 2009.

[20] W.R. Penuel, J. Roschelle, L. Abrahamson, "Research on Classroom Networks for Whole - Class Activities", In IEEE International Workshop on Wireless and Mobile Technologies in Education (WMTE'05), pp.222-229, 2005.

[21] M. Freeman, P. Blayney, "Promoting interactive in-class learning environments: A comparison of an electronic response system

with a traditional alternative", In Proceedings of the 11th Australasian Teaching Economics Conference, July 11-12, Sydney, Australia, 2005.

[22] S.Siraj, M.P. Saleh, "Aplikasi Teknologi dalam Pengajaran dan Pembelajaran Peringkat Sekolah Menengah:Jangkaan Masa Depan (Technology Application in Teaching and Learning at Secondary School: A Future Projection)", Journal of Educational Research, vol.23, pp.123-139, 2003.

[23] S. Siraj, "Pembelajaran Mobile dalam Kurikulum Masa Depan (Mobile Learning in Future Curriculum", Masalah pendidikan, vol.27, pp. 128-142, 2004.

[24] S. Mahamad, M. N. Ibrahim, S. M. Taib, "M-learning: A New Paradigm of Learning Mathematics in Malaysia", International Journal of Computer Science and Information Technology, IJCSIT, vol. 2, no. 4,pp.76-86, 2010.

[25] M. Mohamad, J. Woollard, "Bringing change in secondary schools: can mobile learning via mobile phones be implemented in Malaysia?" Paper presented at the 4th International Malaysian Educational Technology Convention, Kuala Lumpur, Malaysia, 2010.

[26] T.Lim, M. Fadzil, N.Mansor, "Mobile Learning via SMS at Open University Malaysia: Equitable, Effective, and Sustainable", The International Review Of Research In Open And Distance Learning, vol. 12, no. 2, pp. 122-137, 2011.

[27] R. A. Rahman, M. H. M. Hashim, "Mobile learning in Malaysian Technical and Vocational Education (TVE): a qualitative case study", In: International Conference on Teaching and Learning in Higher Education 2011 (ICTLHE2011) , November 21-23, Melaka, 2011.

[28] M. Shuib, A. Abdullah, I. Ismail, S. N. A. Zahari, "The Feasibility of Teaching Grammar via SMS", SPECTRUM Studies in Language, Literature & Interpretation, vol. 9, 2012.

[29] Malaysian Communications and Multimedia Commission (MCMC), Report on communications and multimedia pocket book of statistics 2011 Q2, 2011.

[30] R.M. Felder, L.K. Silverman, "Learning and Teaching Styles in Engineering Education", Engineering education, vol. 78, no. 7, pp. 674-681, 1988.

[31] P. Honey, A. Mumford, The Learning Styles Helper's Guide, Maidenhead, UK: Peter Honey, 2000.

[32] D.A. Kolb, "Learning Styles and Disciplinary Differences", The Modern American College, pp. 232-255, 1981.

[33] N.D. Fleming, C. Mills, "Not Another Inventory, Rather a Catalyst for Reflection", To Improve The Academy, vol. 11, pp. 137-155, 1992.

[34] J. Kuljis, F. Liu, "A comparison of learning style theories on the suitability for elearning", In Proceedings of the IASTED Conference on Web Technologies, Applications, and Services, ACTA Press, pp. 191-197, 2005.

[35] N. B. Ahmad, S. M. Shamsuddin, A. Abraham, "Granular Mining of Student's Learning Behavior in Learning Management System Using Rough Set Technique", In F. Xhafa, S. Caballé, A. Abraham, T. Daradoumis & A. Juan Perez (Eds.), Computational Intelligence for Technology Enhanced Learning (Vol. 273, pp. 99-124): Springer Berlin Heidelberg, 2010.

[36] Kinshuk, & T. Lin, "Application of Learning Styles Adaptivity in Mobile Learning Environments", Third Pan Commonwealth Forum on Open Learning, pp. 4-8, 2004.

[37] N. Al-Jojo, Teacher assisting and subject adaptive material system: an Arabic adaptive learning environment (Doctoral dissertation, University of Portsmouth), 2012.

[38] L.Wang, "Variation in Learning Styles in a Group of Chinese English as a Foreign Language Learners", International Education Journal, vol. 8, no. 2, pp. 408-417, 2007.

[39] R. M. Felder, B.A. Soloman, Index of Learning Styles questionnaire,1997.

[40] I. Koneru, "Addie: Designing Web-Enabled Information Literacy Instructional Modules", *DESIDOC* Journal of Library & Information Technology, vol. 30, no. 3, pp. 23-34, 2010.

[41] A.L. Davis, "Using instructional design principles to develop effective information literacy instruction The ADDIE model", College & Research Libraries News , vol. 74, no. 4, pp. 205-207, 2013.

[42] R. M. Branch, Instructional design: The ADDIE approach. Springer, 2008.

[43] A. Zimnas, D. Kleftouris, N. Valkanos. "IDEL-A Simple Instructional Design Tool for E-Learning", World Academy of Science, Engineering and Technology, vol. 49, pp. 366-372, 2009.

[44] V. Soto, "Which Instructional Design Models are Educators Using to Design Virtual World Instruction", Journal of Online Learning and Teaching, vol. 9, no. 3, pp. 364-75,2013.

[45] C.-Y. Lai, W.-C. Liou. "Rapid ADDIE Curriculums Design Model Based on the Heterogeneous Multimedia Information Integration." Multimedia Workshops, 2007. ISMW'07. Ninth IEEE International Symposium on. IEEE, 2007.

[46] W. He, T. Means, G.Y. Lin, "Field Experience Tracking and Management in Teacher Development Programs", International Journal of Technology in Teaching and Learning, vol. 2, no. 2, pp. 134-147, 2006.

[47] P. Berking, T. Archibald, J. Haag, Mobile learning: Not just another delivery method, 2012, accessible at

[48] R. Kulvitiene, I. Sileikiene, " The blended learning delivery design model", In Proceedings of the 6th WSEAS International Conference on Distance Learning and Web Engineering, Lisbon, Portugal, September 22-24, 2006.

[49] S.J. McGriff,I nstructional system design (ISD): Using the ADDIE model, Penn State University, 2000.

[50] K. Passerini, M. J. Granger, "A Developmental Model for Distance Learning Using the Internet", Computers & Education, vol. 34, no. 1, pp. 1-15, 2001.

[51] S. Arkün, B. Akkoyunlu, "A Study on the development process of a multimedia learning environment according to the ADDIE model and students' opinions of the multimedia learning environment", Interactive Educational Multimedia, vol.17, pp. 1-19, 2008.

[52] Y.-T. Chen, T.-J. Chen, L.-Y. Tsai, "Development and Evaluation of Multimedia Reciprocal Representation Instructional Materials", International Journal of Physical Sciences, vol. 6, no. 6, pp. 1431-1439, 2011.

AUTHORS

Munir Shuib is with the National Higher Education Research Institute (NaHERI), sains@USM, 11900, Universiti Sains Malaysia, Penang, Malaysia (e-mail: munir@usm.my).

Amelia Abdullah is with the School of Educational Studies, 11800, Universiti Sains Malaysia, Penang, Malaysia (e-mail: amelia@usm.my).

Siti Norbaya Azizan is with the National Higher Education Research Institute (NaHERI), sains@USM, 11900, Universiti Sains Malaysia, Penang, Malaysia (e-mail: sitinorbaya49@gmail.com).

Thenmolli Gunasegaran is with the School of Educational Studies, 11800, Universiti Sains Malaysia, Penang, Malaysia (e-mail: gthenmolli@gmail.com).

Exploring the Potential of Mobile Learning Use Among Faculty Members

Mansour Alwraikat, Hiam Al Tokhaim
The University of Jordan, Amman, Jordan

Abstract—**The purpose of this study was to investigate faculty members' attitudes toward mobile learning in King Saud University. The sample of the study consisted of 362 faculty members from King Saud University in Saudi Arabia during the academic year 2012-2013. A questionnaire consisted of 37 items was developed to measure the attitudes. The results showed that the attitudes of faculty members towards mobile learning are positive and there are statistically significant differences attributed to gender in favor of (female faculty members), academic rank in favor of (instructor), and academic experience in favor of (21 years of experience and more).**

Index Terms—**Attitudes, Faculty, Mobile Learning, University.**

I. INTRODUCTION

In the light of developments subsequent to information age, educators seek to keep abreast of these developments through attempts and intensive efforts to transform the traditional learning environment directed by the teacher to a multiple sources learning environment, and due to the development of information and communication technology and its growing remarkable potentials, its employment in support of the educational process, and to develop long-term strategies for its integration in education, has become a fundamental goal of the contemporary educational reform.

The term m-learning refers to learning using mobile technologies, including but not limited to computers, mobile players, Personal Digital Pads, MP3, laptops and mobile phones. Mobile learning focuses on the learner to be mobile, and interacts with mobile technologies, and reflects a focus on how society and its institutions absorb and support mobile devices, which are growing.

With the growing of communication revolution, mobile devices such as mobile phones have become accessible for students and professors in schools, and universities. Common feature is that all these devices are easy to use and contains the ability to communicate through the Internet or a network of mobile operator; also the storage and software capacities for these devices allow reception of different programs. And we may exploit these advantages by developing files containing educational programs can be read, monitored, and traced by mobile phone, it can reach a large number of students at anytime and anywhere and then being able to manage part of the educational process with ease to thousands of students and trainees and the cost is very affordable for students.

One of the leading projects for m-learning in the Arab region was the Arab Open University in Bahrain learning project by the mobile phone (Mobile learning) during the academic semester (October 2008). Another project was The King Saud University project conducted by Dr. Mohammad Alharthi, for using SMS in university teaching. A project conducted by Manarat Educational Consultancy, MEC in the provision of m-learning service by which sending and downloading files and applied materials lessons and exams on a mobile phone for these service clients.

A. *Mobile Learning in King Saud University*

Recently King Saud University in Saudi Arabia began plans to employ mobile devices such as mobile phone and SMS service to take advantage of its applications to serve its educational programs, within the framework of the distance learning program, to ensure the provision of programs for distance education via cell phone. The employment of this service represents a new phase of the stages of distance education, which in turn will open up broad prospects for education for large segments of society, specially the educational system, and this has been done through e-learning, as well as the use of education via mobile phone, which would be a major shift that will raise the level of e-learning.

One of the most common applications used in the learning process was messaging. Therefore a Message Center was developed to provide a mechanism that allows all users the ability to send SMS for all university employees, or email, it supports interactive services to groups with an option that allows jointly or canceled at the request of the user. The Center has a system alerts advanced users, post messages to raise awareness and education, send administrative and regulatory decisions of the university, in addition to send congratulations in national, religious and social events. The user is enabled see the status of each transmission sent by a supervisor. A window showing the reason for the rejection in case a supervisor refused to send the message [1].

The system allows m-learning to provide educational services to the community through smart phones, tablet devices such as a database of students' names, dates of the attendees, and content of the lectures. In addition, the learning management system helps students' access scientific material, and communicates with the University anywhere, anytime [2]. The university has released the so-called system Tawasol, which is still in the experimental stage, which allows the subscriber to send SMS messages or e-mail from freelancers. Statistics indicate that the number of online reach (442), the total number of messages (23497618), the number of messages sent (23383237), the number of users (11122), the number of groups, (14 654), the number of recipients (656.460) [1].

B. *Characteristics of Mobile Learning*

The devices used in m-learning, such as mobile phones (Cell Phones), and smart devices (Smart Phones), and personal digital assistants (PDAs), and iPod, has the advantage of being portable and allows truly anywhere anytime personalized learning [3], from one place to another,these devices are used to transfer the educational content of the educational process, by the use of tools,such as audio and video players, etc. Which is characterized by its proximity to wireless application protocols and Wi-Fi that enable learners to link with the Internet anytime, anywhere [4]?

C. *Mobility of Learner*

The e-learning, which uses fixed computers requires the learners learn in a specific place and time to connect the computer on the Internet, but in the case of m-learning is different, learning may happens in anywhere and anytime, where the use of mobile technology devices in education as opposed to fixed heavy weight computers. This is one of the advantages of m-learning, as it is flexible, encourages individual student learning, and provide him with constant access to educational materials and activities. These merits keep the learner immersed in continuing educational activities, and increase the productivity and effectiveness rate [5]. The pros of m-learning is manifested in its ability to promote collaborative learning among students, since it help students enjoy a sense of ownership and monopoly in learning and innovation in educational content by active participation in cooperative interactive social activities available on mobile technology devices [6].

II. LITERATURE REVIEW

For the sake of integrating any educational innovations in the teaching-learning process successfully, we must conduct extensive studies for all related issues, and try to answer all its questions. Therefore, the introduction of m-learning within higher education should not be limited to equipment or educational resources, but must be accompanied by a focus on the human aspects of the mission, and on top of them the attitudes of learners and teachers about m-learning. While reviewing the literature some previous studies related to the subject were found. These studies were viewed as follows: The results of a project conducted in King Saud University by Dr. Mohammad Alharthi, about using SMS in university teaching, showed that learners' attitudes were positive ones. The project recommended that more support and upgrade of programs embodied in the e-learning management systems to support mobile learning, and more care using mobile technologies in education due to the large spread between the educated and especially the new generation of them [7].

Medical teachers and learners have been investigated how they are using m-learning devices such as the iPhone. Results showed that portability, flexibility, access to multimedia and look up for information were the main benefits they attained. Both medical students and physicians believed that the use of m-learning in medical education and practice will augment in the future [8].

The reasons behind the use of m-learning are a combination of factors: Mobile phone is the most commonly used technological devices in our hands, a growing of studies and reports indicated that keys, wallets, and mobile phones are among the more stuff we carry [9].

College instructors' and students' attitudes were examined toward the use of mobile devices (laptops and cell phone), for non-academic purposes, during lectures. Findings implied that some participants valued mobile devices as an acceptable and useful way; others found it unacceptable way and disturbing for both [10].

Teachers' perceptions toward m-learning in Cyprus were investigated. The findings entailed that teachers expressed above medium levels of perception toward m-learning. In addition, male teachers' perception of m-learning technologies was fairly higher than female teachers [11].

M-learning perceptions and m-learning levels of the prospective teachers were explored with regard to their departments and gender at a university in Turkish Republic of Northern Cyprus. The results showed that there was no significant difference according to teachers' department and gender. In addition, teachers' perceptions levels were low [12].

It is clear from the foregoing that the majority of previous studies indicated a positive change in faculty attitudes toward m-learning. In addition, there are a limited number of published studies in Arab countries especially with regard to faculty attitudes toward m-learning in university level.

A. *Problem of the Study*

Within this framework, which represents the orientation and transformation of King Saud University toward the use and integration of m-learning applications in learning, teaching and educational uses? This study came to reveal what kind of attitudes faculty members have toward m-learning. Since faculty members represent the driving force on most technological innovations within their universities. They play a major role in minds shift and shaping public opinions within their universities and societies. Therefore, this study seeks to answer the following questions:

- What are the attitudes of faculty members at King Saud University toward m-learning?
- Are there any statistical significant differences at the level of significance ($0.05 \geq \propto$) in faculty members attitudes at King Saud University towards m-learning with regard to their: Gender, academic rank, and academic experience?

B. Procedural Definitions of the Study

M-learning: A learning system that uses cell phones and students portable website used by King Saud University in teaching various courses.

Attitudes: Represent the state of mental preparedness and nervous system through experience, a direct response of individuals towards all things and situations that relate to them [13]. This is measured by the degree obtained by a faculty member on the scale to measure faculty attitudes toward m-learning within the questionnaire of the study.

III. Methodology of the Study

A descriptive survey methodology was employed for conducting the study. The interpretation of research results largely depend on the time of the study and the sample size, this type of research is more than just data, it involves that the researcher must discuss data to reach a convenient access to interpret, and discover the meanings and relationships within the data.

A. *Population and Sample of the Study*

Population of the study consisted of all the faculty members at King Saud University in Saudi Arabia during the academic year 2012/2013, totaling (6000). All faculty members who were teaching were chosen as a sample for the study. A questionnaire was distributed to faculty manually and electronically via their e-mail. After collecting questionnaires, only (362) were ready for analysis.

TABLE I.
DISTRIBUTION OF FACULTY ACCORDING TO GENDER, ACADEMIC RANK, AND EXPERIENCE

Variables	Interval of Variables	Number	Percent
Gender	Male	52	14.4%
	Female	310	85.6%
	Total	362	100%
Academic rank	Professor	42	11.6%
	Associate Professor	99	27.3%
	Assistant Professor	97	26.8%
	Instructor	124	34.3%
	Total	362	100%
Experience	Less than 5 years	105	29.0%
	5 to 10 years	97	26.8%
	11 to 15 years	69	19.1%
	16 to 20 years	53	14.6%
	More than 21 years	38	10.5%
	Total	362	100%

B. *Instrument of the Study*

A questionnaire was developed to measure faculty members attitudes towards the concept of m-learning through a review of the theoretical literature and previous studies that addressed the issue of the concept of m-learning. A four-point Likert scale ranging from strongly agrees to disagree was used to measure the degree of awareness of faculty.

C. *Validity*

To verify the content validity of the questionnaire it was presented to (9) experts in the field of curriculum and instruction, and educational technology in the Hashemite Kingdom of Jordan. Also, to determine the suitability and coverage of items for measuring attitudes towards the concept of m-learning, and the extent of affiliation of items to the dimensions used in the study. In addition, to achieve the clarity and integrity of items language, as well as mention of any proposed amendments, proposed items deem necessary, and delete unnecessary items. Proposed amendments made by the experts in their recommendations were added, such as re-wording and delete of some items because of redundancy. In light of the amendments, the instrument consisted in its final draft of (37) items to measure attitudes, distributed over one dimension.

D. *Reliability*

Cronbach's alpha coefficient was calculated to measure the internal consistency of the dimension, which was (0.93). It is obvious that this high value suitable for study purposes. Additionally, an item analysis was conducted to double check if items were highly correlated. The dimension consisted of negative items :(1 12 21 29 30 37) and the rest were positive items.

The negative wording of items was taken into account in the questionnaire when debugging. Positive items direction takes the mark as follows: strongly agree (4), agree (3), disagree (2), strongly disagree (1). While negative items direction as follows: strongly agree (1), agree (2), disagree (3), strongly disagree (4). For the purposes of the current study, the researchers adopted the views of experts in this field to measure faculty members' attitudes towards the concept of m-learning as follows:

The upper limit of alternatives for the scale in the instrument is (4), and a minimum of alternatives is (1). By subtracting the minimum upper limit equal to (3), and then dividing the difference between the two extremes on three levels, as shown in the following equation: 3 ÷ 3 levels (high, medium, low) = 1 and it will be: Minimum limit = 1 +1 = 2, average limit = 2 +1 = 3, and the upper limit = 3 or more. Thus, the weights for items as follows:

- Items that its means averaging between (3.01- 4.00) means that the attitudes of faculty members to the concept of m-learning is high.
- Items that its means averaging (2.01-3.00) means that the attitudes of faculty members to the concept of m-learning is medium.
- Items that its means averaging (1.00-2.00) means that the attitudes of faculty members to the concept of m-learning is low.

E. *Variables of the Study*

First: Independent variables: Gender: With two levels: Male, Female. Academic rank with four Levels: Professor, Associate Professor, Assistant Professor, Instructor. Experience: With five levels: Less of 5 years, 5 to 10 years, 11 to 15 years, 16 to 20 years, more than 21 years. Second: the dependent variables: the attitudes of faculty members towards the concept of m-learning measured through participants responses on the specified scale in the questionnaire.

IV. Results

The results for the first research question: To answer the first research question, means, standard deviations, and ranking for responses for the attitudes of faculty in general were calculated, and for the dimension as a whole. Table (2) shows the attitudes of faculty toward the concept of m-learning.

Table (2) shows that the attitudes of the faculty members towards m-learning for items of on the dimension of attitudes were all within the high degree, but some items were within the medium degree. The item "I think that m-learning provides diverse educational means and methods for delivering information to the student" was ranked first in terms of attitudes with a mean (3.46), a standard devia-

TABLE II.
MEANS, STANDARD DEVIATIONS, AND RANKING FOR FACULTY ATTITUDES AND THE TOTAL SCORE

No.	Items	Means	St. D	Ranking	Degree
1	I think that m-learning increases my educational duties	2.23	0.96	35	medium
2	I think that m-learning can assess students individually.	3.17	0.88	17	high
3	I think that m-learning provides me with ways and means of multiple evaluation.	3.18	0.90	16	high
4	I think that m-learning helps me to achieve my educational goals more effectively	3.13	0.86	22	high
5	I think that m-learning makes my educational role more flexible.	3.37	0.77	7	high
6	I think that m-learning requires me to develop my teaching skills.	3.39	0.82	6	high
7	I think that m-learning strengthens my capacity as a teacher.	3.14	0.90	21	high
8	I think that m-learning enables me to deliver feedback to the student in multiple formats.	3.43	0.78	2	high
9	I think that m-learning provides educational opportunities to meet the needs of different students.	3.25	0.83	12	high
10	I think that m-learning provides diverse educational means and methods for delivering information to the student.	3.46	0.71	1	high
11	I think that m-learning can follow each student's performance first hand.	3.25	0.80	13	high
12	I see that evaluation in the regular education is more effective and accurate than m-learning.	2.07	1.01	36	medium
13	I always make sure to evaluate e-learning methods to determine the suitability of the educational objectives.	3.27	0.75	11	high
14	I think that m-learning helps the student to improve his knowledge of digital skills.	3.43	0.77	3	high
15	I think that m-learning helps the student to get to know his skills and his own abilities.	3.35	0.71	9	high
16	I think that m-learning helps to raise the morale of the learner and appreciation for himself and his self-confidence.	3.35	0.75	8	high
17	I think that m-learning helps students to focus for longer periods.	3.09	0.90	24	high
18	I think that m-learning helps students to identify weaknesses that need to be developed.	3.07	0.88	25	high
19	I think that m-learning helps students to self-conclusion based on their own experiences.	3.17	0.79	18	high
20	I think that m-learning makes students responsible for his own learning.	3.40	0.74	5	high
21	I think that m-learning reduces the student's ability to dialogue and discussion.	2.06	1.03	37	medim
22	I think that students are more active in m-learning.	3.04	0.98	26	high
23	I think that m-learning offers students freedom to search for information.	3.41	0.64	4	high
24	I think that m-learning develops the student's ability to follow-up and well-absorption.	3.19	0.81	15	high
25	I think that m-learning helps students for good preparation for educational materials.	3.10	0.85	23	high
26	I think that m-learning enhances linking the process of learning activities of daily life more than the normal learning.	3.02	0.82	27	high
27	I think that m-learning facilitate managing the learning process between students and their teachers and educational resources more than normal education.	3.16	0.84	19	high
28	I think that the learning environment in the m-learning is more interesting than the normal learning.	3.31	0.73	10	high
29	I think that m-learning is inconsistent with the formal learning.	2.71	1.04	31	medium
30	I think that m-learning does not take into account ethical and privacy issues for students compared to the normal learning.	2.43	1.10	34	medium
31	I think that m-learning is a student-centered learning more than normal learning.	3.14	0.77	20	high
32	I think that m-learning is centered on educational content more than normal learning.	2.96	0.87	28	medium
33	I think that m-learning is centered on assessment methods more than normal learning.	2.80	0.87	30	medium
34	I think that m-learning is centered on issues related to community more than normal learning.	2.53	1.03	33	medium
35	I think that m-learning makes the student more actively engaged in the learning process, compared to the normal learning.	3.19	0.82	14	high
36	I think that m-learning fosters creativity of the student more than normal learning.	2.94	0.90	29	medium
37	I think that the role of teacher in m-learning and normal learning is the same.	2.59	1.12	32	medium
Total		3.05	0.44		high

tion (0.71), and a high degree. The item "I think that m-learning enables me to deliver feedback to the student in multiple formats" was ranked second with a mean (3.43), a standard deviation (0.78), and a high degree. While the item "I think that m-learning reduces the student's ability to dialogue and discussion" came in the last place with a mean (2.06), standard deviation (1.03), and a medium degree. The item "I see that evaluation in the regular education is more effective and accurate than m-learning" ranked before the last one with a mean (2.07), a standard deviation (1.01), and a medium degree. Reviewing the scores within table (2) showed high degree of attitudes by the faculty members at King Saud University.

The results for the second research question: Are there any statistical significant differences at the level of significance (0.05 ≥∝) in faculty members attitudes at King Saud towards m-learning with regard to their: Gender, academic rank, and academic experience?

To answer this question and an independent sample t-test was performed to examine the significance of any differences between the means with regard to faculty gender. In addition, a one-way analysis of variance (ANOVA) was conducted to find any statistically significant differences in mean scores between the faculty with regard to their academic rank and experience, and Tukey test for post comparisons. The following is a presentation of the results by each variable separately.

A. *Gender of faculty*

To determine any significant differences between the mean scores of faculty estimates to their attitudes toward m-learning, means, standard deviations of the sample estimates with regard to gender (male, female), were calculated. In addition, an independent samples t-test was performed to test the significance of any difference between the means. The results were as shown in the Table (3).

Table (3) implies that means scores indicate that there are differences in faculty estimates in their attitudes with regard to gender (male, female). The value of calculated t was (2.389) and the level of significance was (0.017), and its statistically significant at the level of (at $p \leq 0.05$), the differences were in favor of female faculty as shown in table (3).

B. *Academic rank of faculty*

To determine any significant differences between the mean estimates of faculty of their attitudes toward m-learning, means and standard deviations for faculty estimates with regard to their academic rank (Professor, Associate Professor, Assistant Professor, instructor) were calculated, and the results were as shown in (Table 4).

Table (4) indicates that there are differences in means scores of the faculty estimates in their attitudes toward m-learning with regard to their academic rank. To determine any significant differences between the mean estimates of faculty of their attitudes toward m-learning with regard to their academic rank (Professor, Associate Professor, Assistant Professor, instructor) a One way analysis of variance (ANOVA), was performed. The results were as shown in (Table 5).

Results in table (5) indicated that there were statistical significant differences for faculty estimates in their attitudes with regard to their academic rank, the value of calculated F was (4.078), and level of significant was (0.007), and its statistically significant at the level of (at $p \leq 0.05$). The Tukey test for post-hoc comparisons was used to determine where the differences in means lie in terms of faculty academic rank. The results showed that there were statistical differences between professors and instructors in their attitudes toward m-learning, the differences were in favor of instructors.

C. *Experience of faculty*

To determine any statistical significant differences between the means of faculty estimates of their attitudes to ward m-learning, means and standard deviation were calculated for faculty estimates with regard to their experience (less than 5 years, 5-10 years, 11-15 years, 16-20 years, more than 21 years), and the results were as shown in table (6).

Means scores in table (6) showed that indicated that there were statistical significant differences for faculty estimates in their attitudes with regard to their experience. A One way analysis of variance (ANOVA) was performed to know the statistical significant level in means differences. The results are shown in table (7).

TABLE III.
RESULTS OF T-TEST FOR FACULTY'S ATTITUDES TOWARD M-LEARNING WITH REGARD TO THEIR GENDER

Gender	Number	Means	St. D	t	Df	Sig.
Male	52	2.96	0.44	2.389	360	0.017
Female	310	3.12	0.46			

TABLE IV.
MEANS, STANDARD DEVIATION FOR FACULTY ATTITUDES WITH REGARD TO THEIR ACADEMIC RANK

Academic rank	Number	Means	St. D
Professor	42	2.93	0.32
Associate professor	99	3.13	0.41
Assistant professor	97	3.03	0.56
Instructor	124	3.18	0.43

TABLE V.
RESULTS OF ONE WAY ANALYSIS OF VARIANCE (ANOVA) FOR FACULTY ATTITUDES WITH REGARD TO THEIR ACADEMIC RANK

Difference of means	Sum of squares	Df	Mean square	F	P
Between groups	2.503	3	0.834	4.078	0.007
Within groups	73.260	358	0.205		
Total	75.763	361			

TABLE VI.
MEANS, STANDARD DEVIATION FOR FACULTY ATTITUDES WITH REGARD TO THEIR EXPERIENCE

Experience	Number	Means	St. D
Less than 5 years	105	3.23	0.43
5-10 years	97	3.08	0.47
11-15 years	69	3.08	0.36
16-20 years	53	3.11	0.39
More than 21 years	38	2.77	0.58

TABLE VII.
RESULTS OF ONE WAY ANALYSIS OF VARIANCE (ANOVA) FOR FACULTY ATTITUDES WITH REGARD TO THEIR EXPERIENCE

Difference of means	Sum of squares	Df	Mean square	F	P
Between groups	6.122	4	1.530	7.845	0.000
Within groups	69.641	357	0.195		
Total	75.763	361			

Results in table (7) indicated that there were statistical significant differences for faculty estimates in their attitudes with regard to their experience, the value of calculated F was (7.845), and level of significant was (0.000), and its statistically significant at the level of (at $p \leq 0.05$). The Tukey test for post-hoc comparisons was used to determine where the differences in means lie in terms of faculty experience. The results showed that there were statistical differences in attitudes of faculty with more 21 years of experience, and faculty with less than 21 years of experience (less than 5 years, 5-10 years, 11-15 years, and 16-20 years), the differences were in favor of faculty with more than 21 years of experience.

V. Discussion of Results

Discussion for the results of the first research question which stated: What are the attitudes of faculty members at King Saud University toward m-learning?

The results showed that the attitudes of faculty members toward m-learning were high. These findings are corresponding with the findings by [8], in their study, which indicated that the attitudes of the study sample toward m-learning were high, especially when it comes to future use of m-learning in medical education. These finding on our current study can be attributed to the fact that since the degree of awareness of faculty members therefore, this was manifested in their positive attitudes toward m-learning since their apparent view of the importance of this concept and its advantages and characteristics of the educational process. The researchers may attribute that to the revenue benefits that accrue to the faculty member through m-learning. Since these large returns benefit the educational environment with all its components, the faculty member and the student, and the buildings and their contents, which may reduce the use of, so came the attitudes of faculty members on mobile learning is high. In addition, the findings of the current study contradicted with the findings reached by [12], that teachers' perceptions levels were low toward m-learning.

Discussion for the results of the second research question which stated: Are there any statistical significant differences at the level of significance ($0.05 \geq \propto$) in faculty members attitudes at King Saud University towards m-learning with regard to their: Gender, academic rank, and academic experience?

The results indicated that there were statistically significant differences in faculty attitudes toward m-learning with regard to gender and in favor of female. Females seem to use the technology more competently than the males, thus that affects their perceptions. The researchers may attributed that to the fact that female faculty were represented more in the sample of the study. Theses findings contradicted with what [11], found in their study in Cyprus that male teachers' perception of m-learning technologies was comparatively higher than female teachers. In addition, the findings of the current study contradicted with the findings by [12], which indicated that there were no statistical significant differences according to teachers' gender. And contradicted with several researchers like [14], who believed that males use mobile technologies such as SMS, MMS, electronic mail, and the GPRS more than females.

The results of the study showed that there were differences in attitudes between faculty members who hold the rank of Professor and faculty members who hold the rank of instructor only, these differences were in favor of instructors. In addition, there differences in attitudes between faculty members, who have an experience of 21 years and more, and faculty members who are less than 21 years of experience (less than 5, 5-10, 11-15, and 16-20 years), these differences were in favor of faculty members who have a 21 years of experience and more. The researchers may attribute that to the fact the experience in teaching for a faculty may affect in his attitudes. While m-learning is one of the modern methods, still faculty members inclination toward the use of this technique is distinct, as a faculty member is aware of the benefits gained as a result of using m-learning in the learning process, especially since experience is an important factor in the process of shaping the attitudes and perceptions towards the use of m-learning.

VI. Conclusion

The integration of new technology programs in higher education institutions has become inevitable. But the success or failure of m-learning is determined largely on the degree of enthusiasm for optimal use of the new technology on the part of faculty members. Faculty members play an important role in using and teaching of modern technologies in m-learning programs, and for the success of the programs, emphasis on quality, and efficiency and effectiveness, we must overcome all the obstacles that stand in front of those who teach in these programs.

Therefore, there should be frameworks available that make higher education institutions able to meet the needs of m-learning, through satisfying the needs of prospective learners through balanced techniques to develop technology and the human element.

References

[1] Deanship of electronic transactions and communications. (2013). (Tawasol: Messages Center). King Saud University. Retrieved in March 11, 2013 from http://etc.ksu.edu.sa/message-center.

[2] Department of Public Relations and Media. (2012). Brief introduction, Deanship of e-learning and distance learning – toward amore some distinct e-learning environment, King Saud University.

[3] Attewell, J. (2011). From research and development to mobile learning: tools for education and training providers and their learners. Retrieved in January 07, 2013 from http://www.mlearn.org .za/CD/papers/Attewell.pdf.

[4] Trinder, J. (2005). Mobile Technologies and Systems. In A. & Kuklska- Hulme (Ed.) Mobile Leaning: A handbook for educators and trainers. USA: Taylor & Francis.

[5] Ting, Y .R. (2005). Mobile Learning: Current Trend and Future Challenges. Proceedings of the Fifth International Conference on Advanced Learning Technologies Los Alamitos, CA: IEEE Computer Society Press.

[6] Guralnich, D. (2008). The Importance of the Learner's Environmental Context in the Design of M- Learning Product. International Journal of Interactive Mobile Technologies, 2 (1), 36-39.

[7] Alharthi, M. (2008). The application of mobile learning using mobile phone in university, conference and exhibition of the Seventh International e-Learning-towards the knowledge society. The Association of technological and human development in the Arab Republic of Egypt.

[8] Wallace, S., Clark, M., & White, J. (2012). 'It's on my iPhone': attitudes to the use of mobile computing devices in medical education a mixed-methods study. bmjopen.bmj.com in February 10, 2013 - Published by group.bmj.com.

[9] Fathallah, M, A. (2012). Cellular educational technology, knowledge magazine site (09/02/2012).Retrieved in June 10, 2013 fromhttp://www.almarefh.net/show_content_sub.php?CUV=399&Model=M&SubModel=138&ID=1648&ShowAll=On.

[10] Hammer, R., Ronen, M., Sharon, A., Lankry, T., Huberman, Y., & Zamtsov, V. (2010).Mobile culture in college lectures: Instructors' and students' Perspectives. Interdisciplinary Journal of E-learning and Learning Objects, 6, 293-304. Retrieved on March 06,fromhttp://www.ijello.org/VOlume6/IJELLOv6p293-304Hammer709.pdf.

[11] Uzunboylu, H. & Ozdamli, F. (2011). Teacher Perception for M-Learning. Journal of Computer Assisted Learning, 27 (6), 544-556.

[12] Oğuz, S. (2012). Mobile Learning Perceptions of the Prospective Teachers. The Turkish Online Journal of Educational Technology, 11 (3), 222-233.

[13] Hamadat, M. (2006). The values of science and career commitment among managers and teachers in schools,(1), Amman: Dar Al-Hamed for publication and distribution.

[14] Anastasios, A. E., & Grousopoulou, A. (2009). Students' thoughts about the importance and costs of their mobile devices' features and services. Telematics and Informatics, 26(1), 57–84.

AUTHORS

Dr. Mansour Alwraikat is an Associate Professor in the Educational Technology Program, Curriculum and Instruction Department, College of Educational Sciences at the University of Jordan in Amman-Jordan. E-mail: mansouralwraikat@gmail.com, m.wraikat@ju.edu.jo

Hiam Al Tokhaim is a researcher specialized in Educational Technology. She holds an MA in Educational Technology from the University of Jordan. E-mail: h.altokhaim@ju.edu.jo.

Permissions

All chapters in this book were first published in iJIM, by International Association of Online Engineering; hereby published with permission under the Creative Commons Attribution License or equivalent. Every chapter published in this book has been scrutinized by our experts. Their significance has been extensively debated. The topics covered herein carry significant findings which will fuel the growth of the discipline. They may even be implemented as practical applications or may be referred to as a beginning point for another development.

The contributors of this book come from diverse backgrounds, making this book a truly international effort. This book will bring forth new frontiers with its revolutionizing research information and detailed analysis of the nascent developments around the world.

We would like to thank all the contributing authors for lending their expertise to make the book truly unique. They have played a crucial role in the development of this book. Without their invaluable contributions this book wouldn't have been possible. They have made vital efforts to compile up to date information on the varied aspects of this subject to make this book a valuable addition to the collection of many professionals and students.

This book was conceptualized with the vision of imparting up-to-date information and advanced data in this field. To ensure the same, a matchless editorial board was set up. Every individual on the board went through rigorous rounds of assessment to prove their worth. After which they invested a large part of their time researching and compiling the most relevant data for our readers.

The editorial board has been involved in producing this book since its inception. They have spent rigorous hours researching and exploring the diverse topics which have resulted in the successful publishing of this book. They have passed on their knowledge of decades through this book. To expedite this challenging task, the publisher supported the team at every step. A small team of assistant editors was also appointed to further simplify the editing procedure and attain best results for the readers.

Apart from the editorial board, the designing team has also invested a significant amount of their time in understanding the subject and creating the most relevant covers. They scrutinized every image to scout for the most suitable representation of the subject and create an appropriate cover for the book.

The publishing team has been an ardent support to the editorial, designing and production team. Their endless efforts to recruit the best for this project, has resulted in the accomplishment of this book. They are a veteran in the field of academics and their pool of knowledge is as vast as their experience in printing. Their expertise and guidance has proved useful at every step. Their uncompromising quality standards have made this book an exceptional effort. Their encouragement from time to time has been an inspiration for everyone.

The publisher and the editorial board hope that this book will prove to be a valuable piece of knowledge for researchers, students, practitioners and scholars across the globe.

List of Contributors

Hosam F. El-Sofany
Cairo Higher Institute for Engineering, Computer Science, and Management, Cairo, Egypt

Samir A. El-Seoud
Technische Universität Darmstadt, Darmstadt, Germany

Danish Nadeem
Center for Learning Sciences and Technologies (CELSTEC), Heerlen, The Netherlands

Slavi Stoyanov
Center for Learning Sciences and Technologies (CELSTEC), Heerlen, The Netherlands

Rob Koper
Center for Learning Sciences and Technologies (CELSTEC), Heerlen, The Netherlands

Ghassan F. Issa
Petra University, Amman, Jordan

Shakir M. Hussain
Petra University, Amman, Jordan

Hussein Al-Bahadili
Petra University, Amman, Jordan

S.L. Abrantes
Polytechnic Institute of Viseu, Viseu, Portugal

L.B. Gouveia
University Fernando Pessoa, Porto, Portugal

Arttu Perttula
Tampere University of Technology, Pori, Finland

Jari Multisilta
University of Helsinki, Helsinki, Finland

Pauliina Tuomi
Tampere University of Technology, Pori, Finland

H. Al Tair
Khalifa University, Sharjah, UAE

M. J. Zemerly
Khalifa University, Sharjah, UAE

M. A. Al Qutayri
Khalifa University, Sharjah, UAE

M. Leida
Etisalat-BT Innovation Centre (EBTIC), Abu Dhabi, UAE

Viet Anh NGUYEN
University of Engineering and Technology, VNU, Hanoi, Vietnam

Van Cong PHAM
University of Engineering and Technology, VNU, Hanoi, Vietnam

Roland Schwald
Albstadt-Sigmaringen University, Sigmaringen, Germany

N. Uday Bhaskar
Govt. Degree College, Pattikonda, A.P., India

U.S.N Raju
NIT Warangal, A.P., India

P.Govindarajulu
S.V. University, India

B.V.Ramana Reddy
MERITS, Udayagiri, A.P., India

Dimitrios Tektonidis
Research Programmes Division, ALTEC Software S.A., Thessaloniki, Greece

Adamantios Koumpis
Research Programmes Division, ALTEC Software S.A., Thessaloniki, Greece

Hend S. Al-Khalifa
King Saud University, Riyadh, Saudi Arabia

Ashraf Tahat
Princess Sumaya University for Technology, Amman, Jordan

Ahmad Jamal
Umniah Mobile Telecommunications, Amman, Jordan

Mohammad Kalbouneh
Princess Sumaya University for Technology, Amman, Jordan

Dr. György Molnár
Budapest University of Technology and Economics, Budapest, Hungary

Sami Vihavainen
Aalto University, Espoo, Finland

Kaisa Väänänen-Vainio-Mattila
Tampere University of Technology, Tampere, Finland

Noriko Uosaki
The University of Tokushima, Minami Josanjima, Tokushima, Japan

Hiroaki Ogata
The University of Tokushima, Minami Josanjima, Tokushima, Japan

Mengmeng Li
The University of Tokushima, Minami Josanjima, Tokushima, Japan

Bin Hou
The University of Tokushima, Minami Josanjima, Tokushima, Japan

Kousuke Mouri
The University of Tokushima, Minami Josanjima, Tokushima, Japan

Ferial Khaddage
Deakin University, Melbourne, Australia

Christoph Lattemann
Jacobs University, Bremen, Germany

P. Thinley
Queensland University of Technology, Brisbane, Australia

J. Reye
Queensland University of Technology, Brisbane, Australia

S. Geva
Queensland University of Technology, Brisbane, Australia

R. Meolic
University of Maribor, Maribor Slovenia

T. Dogša
University of Maribor, Maribor Slovenia

M. Seraj
Limkokwing University of Creative Technology, Cyberjaya, Selangor, Malaysia

C.Y. Wong
Multimedia University, Cyberjaya, Selangor, Malaysia

Jonas Vetterick
University of Rostock, Rostock, Germany

Martin Garbe
University of Rostock, Rostock, Germany

Andreas Dähn
University of Rostock, Rostock, Germany

Clemens H. Cap
University of Rostock, Rostock, Germany

Saida Latreche
Université Constantine 1, Constantine, Algeria

Zehira Ziari
Université Constantine 1, Constantine, Algeria

Smail Mouissat
Université Constantine 1, Constantine, Algeria

Hailay Kidu
University of Gondar, Gondar, Ethiopia

A.Serrano Tellería
Beira Interior University, Covilhã, Portugal

M. Oliveira
Beira Interior University, Covilhã, Portugal

L. Martinez
University of Deusto, Bilbao, Spain

P. Garaizar
University of Deusto, Bilbao, Spain

José Bidarra
Open University of Portugal, Lisbon, Portugal

Mauro Figueiredo
University of Algarve, Faro, Portugal

Carlos Natálio
Open University of Portugal, Lisbon, Portugal

R. Ennetta
Higher Institute of Industrial Systems of Gabes, University of Gabes, Salaheddine El Ayoubi Street, 6011, Gabes,Tunisia

I.Nasri
Higher Institute of Industrial Systems of Gabes, University of Gabes, Salaheddine El Ayoubi Street, 6011, Gabes,Tunisia

W. Daher
An-Najah National University, Nablus, Palestine
Al-Qasemi Academic College of Education, Baqa, Israel

N. Baya'a
Al-Qasemi Academic College of Education, Baqa, Israel

S.A. Shonola
University of Warwick, Coventry, United Kingdom

M.S. Joy
University of Warwick, Coventry, United Kingdom

Sulaiman Almutairy
University of Reading, United Kingdom

Trevor Davies
University of Reading, United Kingdom

Yota Dimitriadi
University of Reading, United Kingdom

Carlos Argelio Arévalo Mercado, Estela Lizbeth Muñoz Andrade and Juan Manuel Gómez Reynoso
Universidad Autónoma de Aguascalientes, Aguascalientes, México

Ghassan F. Issa, Shakir M. Hussain and Hussein Al-Bahadili
Petra University, Amman, Jordan

Hosam F. El-Sofany , Turky Al-Turki, Hasan El-Howimelm and Amjad Al-Sadoon
Arab East Colleges for Graduate Studies, Riyadh, Kingdom of Saudi Arabia

Samir A. El-Seoud
British University in Egypt-BUE, Cairo - Egypt

M. Samir Abou El-Seoud, Ann Nosseir and Islam Taj-Eddin
British University in Egypt (BUE), Cairo, Egypt

Hosam El-Sofany
Arab East Colleges for Graduate Studies, Riyadh, Kingdom of Saudi Arabia

Nadine Abu Rumman
Princes Sumaya University fo r Technology, Amman, Jordan

M. Maierhuber and M. Ebner
Graz University of Technology, Graz, Austria

Nassim Dennouni, Yvan Peter and Luigi Lancieri
Lille1 University, Lille, France

Zohra Slama
Djilali Liabes University, Sidi Bel Abbes, Algeria

Munir Shuib, Amelia Abdullah, Siti Norbaya Azizan and Thenmolli Gunasegaran
Universiti Sains Malaysia, Penang, Malaysia

Mansour Alwraikat and Hiam Al Tokhaim
The University of Jor dan, Amman, Jordan